Iterative Methods for Sparse Linear Systems

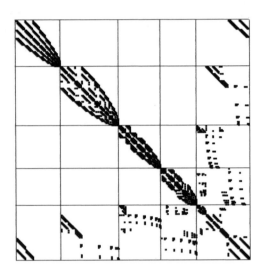

Iterative Methods for Sparse Linear Systems
SECOND EDITION

Yousef Saad
University of Minnesota
Minneapolis, Minnesota

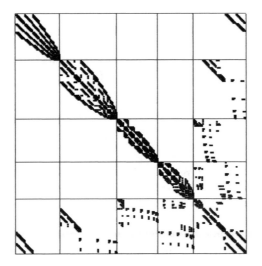

Society for Industrial and Applied Mathematics
Philadelphia

Copyright © 2003 by the Society for Industrial and Applied Mathematics.

10 9 8 7 6 5 4 3

All rights reserved. Printed in the United States of America. No part of this book may be reproduced, stored, or transmitted in any manner without the written permission of the publisher. For information, write to the Society for Industrial and Applied Mathematics, 3600 University City Science Center, Philadelphia, PA 19104-2688.

Library of Congress Cataloging-in-Publication Data

Saad, Y.
    Iterative methods for sparse linear systems / Yousef Saad. – 2nd ed.
      p. cm.
    Includes bibliographical references and index.
    ISBN 0-89871-534-2 (pbk.)
        1. Sparse matrices. 2. Iterative methods (Mathematics) 3. Differential equations, Partial–Numerical solutions. I. Title.

QA188.S17 2003
512.9'434–dc21

                                                                                                                    2002044644

PETSc is copyrighted by the University of Chicago. The University of Chicago makes no representations as to the suitability and operability of this software for any purpose. It is provided "as is" without express or implied warranty.

No warranties, express or implied, are made by the publisher, authors, and their employers that the programs contained in this volume are free of error. They should not be relied on as the sole basis to solve a problem whose incorrect solution could result in injury to person or property. If the programs are employed in such a manner, it is at the user's own risk and the publisher, authors, and their employers disclaim all liability for such misuse.

 is a registered trademark.

Contents

Preface to the Second Edition   xiii

Preface to the First Edition   xvii

1   Background in Linear Algebra   1
- 1.1 Matrices   1
- 1.2 Square Matrices and Eigenvalues   2
- 1.3 Types of Matrices   4
- 1.4 Vector Inner Products and Norms   5
- 1.5 Matrix Norms   7
- 1.6 Subspaces, Range, and Kernel   9
- 1.7 Orthogonal Vectors and Subspaces   10
- 1.8 Canonical Forms of Matrices   14
  - 1.8.1 Reduction to the Diagonal Form   15
  - 1.8.2 The Jordan Canonical Form   15
  - 1.8.3 The Schur Canonical Form   16
  - 1.8.4 Application to Powers of Matrices   18
- 1.9 Normal and Hermitian Matrices   20
  - 1.9.1 Normal Matrices   20
  - 1.9.2 Hermitian Matrices   23
- 1.10 Nonnegative Matrices, M-Matrices   25
- 1.11 Positive Definite Matrices   29
- 1.12 Projection Operators   32
  - 1.12.1 Range and Null Space of a Projector   32
  - 1.12.2 Matrix Representations   34
  - 1.12.3 Orthogonal and Oblique Projectors   34
  - 1.12.4 Properties of Orthogonal Projectors   36
- 1.13 Basic Concepts in Linear Systems   37
  - 1.13.1 Existence of a Solution   37
  - 1.13.2 Perturbation Analysis   38
- Exercises   39
- Notes and References   43

2 Discretization of Partial Differential Equations — 45
- 2.1 Partial Differential Equations — 45
  - 2.1.1 Elliptic Operators — 46
  - 2.1.2 The Convection Diffusion Equation — 48
- 2.2 Finite Difference Methods — 48
  - 2.2.1 Basic Approximations — 48
  - 2.2.2 Difference Schemes for the Laplacian Operator — 50
  - 2.2.3 Finite Differences for One-Dimensional Problems — 51
  - 2.2.4 Upwind Schemes — 52
  - 2.2.5 Finite Differences for Two-Dimensional Problems — 54
  - 2.2.6 Fast Poisson Solvers — 55
- 2.3 The Finite Element Method — 60
- 2.4 Mesh Generation and Refinement — 66
- 2.5 Finite Volume Method — 68
- Exercises — 71
- Notes and References — 72

3 Sparse Matrices — 73
- 3.1 Introduction — 73
- 3.2 Graph Representations — 75
  - 3.2.1 Graphs and Adjacency Graphs — 75
  - 3.2.2 Graphs of PDE Matrices — 76
- 3.3 Permutations and Reorderings — 77
  - 3.3.1 Basic Concepts — 77
  - 3.3.2 Relations with the Adjacency Graph — 79
  - 3.3.3 Common Reorderings — 81
  - 3.3.4 Irreducibility — 89
- 3.4 Storage Schemes — 89
- 3.5 Basic Sparse Matrix Operations — 92
- 3.6 Sparse Direct Solution Methods — 93
  - 3.6.1 MD Ordering — 93
  - 3.6.2 ND Ordering — 94
- 3.7 Test Problems — 94
- Exercises — 97
- Notes and References — 101

4 Basic Iterative Methods — 103
- 4.1 Jacobi, Gauss–Seidel, and Successive Overrelaxation — 103
  - 4.1.1 Block Relaxation Schemes — 106
  - 4.1.2 Iteration Matrices and Preconditioning — 110
- 4.2 Convergence — 111
  - 4.2.1 General Convergence Result — 112
  - 4.2.2 Regular Splittings — 115
  - 4.2.3 Diagonally Dominant Matrices — 116
  - 4.2.4 Symmetric Positive Definite Matrices — 119
  - 4.2.5 Property A and Consistent Orderings — 119

	4.3	Alternating Direction Methods 124
	Exercises 126
	Notes and References 128

5 Projection Methods — 129
- 5.1 Basic Definitions and Algorithms 129
  - 5.1.1 General Projection Methods 130
  - 5.1.2 Matrix Representation 131
- 5.2 General Theory 132
  - 5.2.1 Two Optimality Results 133
  - 5.2.2 Interpretation in Terms of Projectors 134
  - 5.2.3 General Error Bound 135
- 5.3 One-Dimensional Projection Processes 137
  - 5.3.1 Steepest Descent 138
  - 5.3.2 MR Iteration 140
  - 5.3.3 Residual Norm Steepest Descent 142
- 5.4 Additive and Multiplicative Processes 143
- Exercises 145
- Notes and References 149

6 Krylov Subspace Methods, Part I — 151
- 6.1 Introduction 151
- 6.2 Krylov Subspaces 152
- 6.3 Arnoldi's Method 153
  - 6.3.1 The Basic Algorithm 154
  - 6.3.2 Practical Implementations 156
- 6.4 Arnoldi's Method for Linear Systems 159
  - 6.4.1 Variation 1: Restarted FOM 160
  - 6.4.2 Variation 2: IOM and DIOM 161
- 6.5 Generalized Minimal Residual Method 164
  - 6.5.1 The Basic GMRES Algorithm 164
  - 6.5.2 The Householder Version 165
  - 6.5.3 Practical Implementation Issues 167
  - 6.5.4 Breakdown of GMRES 171
  - 6.5.5 Variation 1: Restarting 171
  - 6.5.6 Variation 2: Truncated GMRES Versions 172
  - 6.5.7 Relations Between FOM and GMRES 177
  - 6.5.8 Residual Smoothing 181
  - 6.5.9 GMRES for Complex Systems 184
- 6.6 The Symmetric Lanczos Algorithm 185
  - 6.6.1 The Algorithm 185
  - 6.6.2 Relation to Orthogonal Polynomials 186
- 6.7 The Conjugate Gradient Algorithm 187
  - 6.7.1 Derivation and Theory 187
  - 6.7.2 Alternative Formulations 191
  - 6.7.3 Eigenvalue Estimates from the CG Coefficients 192

	6.8	The Conjugate Residual Method	194
	6.9	Generalized Conjugate Residual, ORTHOMIN, and ORTHODIR	194
	6.10	The Faber–Manteuffel Theorem	196
	6.11	Convergence Analysis	198
		6.11.1 Real Chebyshev Polynomials	199
		6.11.2 Complex Chebyshev Polynomials	200
		6.11.3 Convergence of the CG Algorithm	203
		6.11.4 Convergence of GMRES	205
	6.12	Block Krylov Methods	208
	Exercises		212
	Notes and References		215

7 Krylov Subspace Methods, Part II — 217

	7.1	Lanczos Biorthogonalization	217
		7.1.1 The Algorithm	217
		7.1.2 Practical Implementations	220
	7.2	The Lanczos Algorithm for Linear Systems	221
	7.3	The Biconjugate Gradient and Quasi-Minimal Residual Algorithms	222
		7.3.1 The BCG Algorithm	222
		7.3.2 QMR Algorithm	224
	7.4	Transpose-Free Variants	228
		7.4.1 CGS	229
		7.4.2 BICGSTAB	231
		7.4.3 TFQMR	234
	Exercises		241
	Notes and References		243

8 Methods Related to the Normal Equations — 245

	8.1	The Normal Equations	245
	8.2	Row Projection Methods	247
		8.2.1 Gauss–Seidel on the Normal Equations	247
		8.2.2 Cimmino's Method	249
	8.3	Conjugate Gradient and Normal Equations	251
		8.3.1 CGNR	252
		8.3.2 CGNE	253
	8.4	Saddle-Point Problems	254
	Exercises		257
	Notes and References		259

9 Preconditioned Iterations — 261

	9.1	Introduction	261
	9.2	Preconditioned Conjugate Gradient	262
		9.2.1 Preserving Symmetry	262
		9.2.2 Efficient Implementations	265

9.3	Preconditioned Generalized Minimal Residual		267
	9.3.1	Left-Preconditioned GMRES	268
	9.3.2	Right-Preconditioned GMRES	269
	9.3.3	Split Preconditioning	270
	9.3.4	Comparison of Right and Left Preconditioning	271
9.4	Flexible Variants		272
	9.4.1	FGMRES	273
	9.4.2	DQGMRES	276
9.5	Preconditioned Conjugate Gradient for the Normal Equations		276
9.6	The Concus, Golub, and Widlund Algorithm		278
Exercises			279
Notes and References			280

10 Preconditioning Techniques — 283

10.1	Introduction		283
10.2	Jacobi, Successive Overrelaxation, and Symmetric Successive Overrelaxation Preconditioners		284
10.3	Incomplete LU Factorization Preconditioners		287
	10.3.1	ILU Factorizations	288
	10.3.2	Zero Fill-in ILU (ILU(0))	293
	10.3.3	Level of Fill and ILU(p)	296
	10.3.4	Matrices with Regular Structure	300
	10.3.5	MILU	305
10.4	Threshold Strategies and Incomplete LU with Threshold		306
	10.4.1	The ILUT Approach	307
	10.4.2	Analysis	308
	10.4.3	Implementation Details	310
	10.4.4	The ILUTP Approach	312
	10.4.5	The ILUS Approach	314
	10.4.6	The ILUC Approach	316
10.5	Approximate Inverse Preconditioners		320
	10.5.1	Approximating the Inverse of a Sparse Matrix	321
	10.5.2	Global Iteration	321
	10.5.3	Column-Oriented Algorithms	323
	10.5.4	Theoretical Considerations	324
	10.5.5	Convergence of Self-Preconditioned MR	326
	10.5.6	AINVs via Bordering	329
	10.5.7	Factored Inverses via Orthogonalization: AINV	331
	10.5.8	Improving a Preconditioner	333
10.6	Reordering for Incomplete LU		333
	10.6.1	Symmetric Permutations	333
	10.6.2	Nonsymmetric Reorderings	335
10.7	Block Preconditioners		337
	10.7.1	Block Tridiagonal Matrices	337
	10.7.2	General Matrices	339

	10.8	Preconditioners for the Normal Equations 339
		10.8.1 Jacobi, SOR, and Variants 340
		10.8.2 IC(0) for the Normal Equations 340
		10.8.3 Incomplete Gram–Schmidt and ILQ 342
	Exercises 345
	Notes and References 349

11 Parallel Implementations — 353

- 11.1 Introduction 353
- 11.2 Forms of Parallelism 354
  - 11.2.1 Multiple Functional Units 354
  - 11.2.2 Pipelining 354
  - 11.2.3 Vector Processors 355
  - 11.2.4 Multiprocessing and Distributed Computing 355
- 11.3 Types of Parallel Architectures 355
  - 11.3.1 Shared Memory Computers 356
  - 11.3.2 Distributed Memory Architectures 357
- 11.4 Types of Operations 359
- 11.5 Matrix-by-Vector Products 361
  - 11.5.1 The CSR and CSC Formats 362
  - 11.5.2 Matvecs in the Diagonal Format 364
  - 11.5.3 The Ellpack-Itpack Format 364
  - 11.5.4 The JAD Format 365
  - 11.5.5 The Case of Distributed Sparse Matrices 366
- 11.6 Standard Preconditioning Operations 369
  - 11.6.1 Parallelism in Forward Sweeps 369
  - 11.6.2 Level Scheduling: The Case of Five-Point Matrices . . 370
  - 11.6.3 Level Scheduling for Irregular Graphs 370
- Exercises 373
- Notes and References 375

12 Parallel Preconditioners — 377

- 12.1 Introduction 377
- 12.2 Block Jacobi Preconditioners 378
- 12.3 Polynomial Preconditioners 379
  - 12.3.1 Neumann Polynomials 380
  - 12.3.2 Chebyshev Polynomials 381
  - 12.3.3 Least-Squares Polynomials 383
  - 12.3.4 The Nonsymmetric Case 386
- 12.4 Multicoloring 389
  - 12.4.1 Red-Black Ordering 389
  - 12.4.2 Solution of Red-Black Systems 390
  - 12.4.3 Multicoloring for General Sparse Matrices 391
- 12.5 Multi-Elimination Incomplete LU 392
  - 12.5.1 Multi-Elimination 393
  - 12.5.2 ILUM 394

Contents

12.6	Distributed Incomplete LU and Symmetric Successive Overrelaxation	396
12.7	Other Techniques	399
	12.7.1 AINVs	399
	12.7.2 EBE Techniques	399
	12.7.3 Parallel Row Projection Preconditioners	401
Exercises		402
Notes and References		404

13 Multigrid Methods — 407

13.1	Introduction	407
13.2	Matrices and Spectra of Model Problems	408
	13.2.1 The Richardson Iteration	412
	13.2.2 Weighted Jacobi Iteration	414
	13.2.3 Gauss–Seidel Iteration	416
13.3	Intergrid Operations	419
	13.3.1 Prolongation	419
	13.3.2 Restriction	421
13.4	Standard Multigrid Techniques	422
	13.4.1 Coarse Problems and Smoothers	423
	13.4.2 Two-Grid Cycles	424
	13.4.3 V-Cycles and W-Cycles	426
	13.4.4 FMG	429
13.5	Analysis of the Two-Grid Cycle	433
	13.5.1 Two Important Subspaces	433
	13.5.2 Convergence Analysis	435
13.6	Algebraic Multigrid	437
	13.6.1 Smoothness in AMG	438
	13.6.2 Interpolation in AMG	439
	13.6.3 Defining Coarse Spaces in AMG	442
	13.6.4 AMG via Multilevel ILU	442
13.7	Multigrid versus Krylov Methods	445
Exercises		446
Notes and References		449

14 Domain Decomposition Methods — 451

14.1	Introduction	451
	14.1.1 Notation	452
	14.1.2 Types of Partitionings	453
	14.1.3 Types of Techniques	453
14.2	Direct Solution and the Schur Complement	456
	14.2.1 Block Gaussian Elimination	456
	14.2.2 Properties of the Schur Complement	457
	14.2.3 Schur Complement for Vertex-Based Partitionings	458
	14.2.4 Schur Complement for Finite Element Partitionings	460
	14.2.5 Schur Complement for the Model Problem	463

14.3	Schwarz Alternating Procedures		465
	14.3.1	Multiplicative Schwarz Procedure	465
	14.3.2	Multiplicative Schwarz Preconditioning	470
	14.3.3	Additive Schwarz Procedure	472
	14.3.4	Convergence	473
14.4	Schur Complement Approaches		477
	14.4.1	Induced Preconditioners	478
	14.4.2	Probing	480
	14.4.3	Preconditioning Vertex-Based Schur Complements	480
14.5	Full Matrix Methods		481
14.6	Graph Partitioning		483
	14.6.1	Basic Definitions	484
	14.6.2	Geometric Approach	484
	14.6.3	Spectral Techniques	486
	14.6.4	Graph Theory Techniques	487
Exercises			491
Notes and References			492

Bibliography   **495**

Index   **517**

Preface to the Second Edition

In the six years that have passed since the publication of the first edition of this book, iterative methods for linear systems have made good progress in scientific and engineering disciplines. This is due in great part to the increased complexity and size of the new generation of linear and nonlinear systems that arise from typical applications. At the same time, parallel computing has penetrated the same application areas, as inexpensive computer power has become broadly available and standard communication languages such as MPI have provided a much needed standardization. This has created an incentive to utilize iterative rather than direct solvers, because the problems solved are typically from three-dimensional models for which direct solvers often become ineffective. Another incentive is that iterative methods are far easier to implement on parallel computers.

Although iterative methods for linear systems have seen a significant maturation, there are still many open problems. In particular, it still cannot be stated that an arbitrary sparse linear system can be solved iteratively in an efficient way. If physical information about the problem can be exploited, more effective and robust methods can be tailored to the solutions. This strategy is exploited by multigrid methods. In addition, parallel computers necessitate different ways of approaching the problem and solution algorithms that are radically different from classical ones.

Several new texts on the subject of this book have appeared since the first edition. Among these are the books by Greenbaum [154] and Meurant [208]. The exhaustive five-volume treatise by G. W. Stewart [273] is likely to become the de facto reference in numerical linear algebra in years to come. The related multigrid literature has also benefited from a few notable additions, including a new edition of the excellent *Multigrid Tutorial* [65] and a new title by Trottenberg et al. [285].

Most notable among the changes from the first edition is the addition of a sorely needed chapter on multigrid techniques. The chapters that have seen the biggest changes are Chapters 3, 6, 10, and 12. In most cases, the modifications were made to update the material by adding topics that have been developed recently or gained importance in the last few years. In some instances some of the older topics were removed or shortened. For example, the discussion on parallel architecture has been shortened. In the mid-1990s hypercubes and "fat-trees" were important topics to teach. This is no longer the case, since manufacturers have taken steps to hide the topology from the user, in the sense that communication has become much less sensitive to the underlying architecture.

The bibliography has been updated to include work that has appeared in the last few years, as well as to reflect the change of emphasis when new topics have gained importance. Similarly, keeping in mind the educational side of this book, many new exercises have been added. The first edition suffered from many typographical errors, which have been corrected. Many thanks to those readers who took the time to point out errors.

I would like to reiterate my thanks to all my colleagues who helped make the first edition a success (see the preface to the first edition). I received support and encouragement from many students and colleagues to put together this revised volume. I also wish to thank those who proofread this book. I found that one of the best ways to improve clarity is to solicit comments and questions from students in a course that teaches the material. Thanks to all students in CSci 8314 who helped in this regard. Special thanks to Bernie Sheeham, who pointed out quite a few typographical errors and made numerous helpful suggestions.

My sincere thanks to Michele Benzi, Howard Elman, and Steve McCormick for their reviews of this edition. Michele proofread a few chapters thoroughly and caught a few misstatements. Steve's review of Chapter 13 helped ensure that my slight bias for Krylov methods (versus multigrid) was not too obvious. His comments were the origin of the addition of Section 13.7 (Multigrid versus Krylov Methods).

Finally, I would also like to express my appreciation to all SIAM staff members who handled this book, especially Linda Thiel and Lisa Briggeman.

Suggestions for Teaching

This book can be used as a text to teach a graduate-level course on iterative methods for linear systems. Selecting topics to teach depends on whether the course is taught in a mathematics department or a computer science (or engineering) department, and whether the course is over a semester or a quarter. Here are a few comments on the relevance of the topics in each chapter.

For a graduate course in a mathematics department, much of the material in Chapter 1 should be known already. For nonmathematics majors, most of the chapter must be covered or reviewed to acquire a good background for later chapters. The important topics for the rest of the book are in Sections 1.8.1, 1.8.3, 1.8.4, 1.9, and 1.11. Section 1.12 is best treated at the beginning of Chapter 5. Chapter 2 is essentially independent of the rest and could be skipped altogether in a quarter session, unless multigrid methods are to be included in the course. One lecture on finite differences and the resulting matrices would be enough for a nonmath course. Chapter 3 aims at familiarizing the student with some implementation issues associated with iterative solution procedures for general sparse matrices. In a computer science or engineering department, this can be very relevant. For mathematicians, a mention of the graph theory aspects of sparse matrices and a few storage schemes may be sufficient. Most students at this level should be familiar with a few of the elementary relaxation techniques covered in Chapter 4. The convergence theory can be skipped for nonmath majors. These methods are now often used as preconditioners, which may be the only motive for covering them.

Chapter 5 introduces key concepts and presents projection techniques in general terms. Nonmathematicians may wish to skip Section 5.2.3. Otherwise, it is recommended to start the theory section by going back to Section 1.12 on general definitions of projectors. Chapters 6 and 7 represent the heart of the matter. It is recommended to describe the first algorithms carefully and emphasize the fact that they generalize the one-dimensional methods covered in Chapter 5. It is also important to stress the optimality properties of those methods in Chapter 6 and the fact that these follow immediately from the properties

Preface to the Second Edition

of projectors seen in Section 1.12. Chapter 6 is rather long and the instructor will need to select what to cover among the nonessential topics as well as choose topics for reading.

When covering the algorithms in Chapter 7, it is crucial to point out the main differences between them and those seen in Chapter 6. The variants such as conjugate gradient squared (CGS), biconjugate gradient stabilized (BICGSTAB), and transpose-free quasi-minimal residual (TFQMR) can be covered in a short time, omitting details of the algebraic derivations or covering only one of the three. The class of methods based on the normal equations approach, i.e., Chapter 8, can be skipped in a math-oriented course, especially in the case of a quarter session. For a semester course, selected topics may be Sections 8.1, 8.2, and 8.4.

Preconditioning is known as the determining ingredient in the success of iterative methods in solving real-life problems. Therefore, at least some parts of Chapters 9 and 10 should be covered. Sections 9.2 and (very briefly) 9.3 are recommended. From Chapter 10, discuss the basic ideas in Sections 10.1 through 10.3. The rest could be skipped in a quarter course.

Chapter 11 may be useful to present to computer science majors, but may be skimmed through or skipped in a mathematics or an engineering course. Parts of Chapter 12 could be taught primarily to make the students aware of the importance of alternative preconditioners. Suggested selections are Sections 12.2, 12.4, and 12.7.2 (for engineers).

Chapters 13 and 14 present important research areas and are primarily geared toward mathematics majors. Computer scientists or engineers may cover this material in less detail.

To make these suggestions more specific, the following two tables are offered as sample course outlines. Numbers refer to sections in the text. A semester course represents approximately 30 lectures of 75 minutes each whereas a quarter course is approximately 20 lectures of 75 minutes each. Different topics are selected for a mathematics course and a nonmathematics course.

Weeks	Semester Course	
	Mathematics	Computer Science/Eng.
1–3	1.9–1.13 2.1–2.5 3.1–3.3	1.1–1.6 (Read); 1.7; 1.9; 1.11; 1.12; 2.1–2.2 3.1–3.6
4–6	4.1–4.2 5. 1–5.3; 6.1–6.4 6.5.1; 6.5.3–6.5.9	4.1–4.2.1; 4.2.3 5.1–5.2.1; 5.3 6.1–6.4; 6.5.1–6.5.5
7–9	6.6–6.8 6.9–6.11; 7.1–7.3 7.4.1; 7.4.2; 7.4.3 (Read)	6.7.1; 6.8–6.9 6.11.3; 7.1–7.3 7.4.1–7.4.2; 7.4.3 (Read)
10–12	8.1; 8.2; 9.1–9.4 10.1–10.3; 10.4.1 10.5.1–10.5.7	8.1–8.3; 9.1–9.3 10.1–10.3; 10.4.1–10.4.3 10.5.1–10.5.4; 10.5.7
13–15	12.2–12.4 13.1–13.5 14.1–14.6	11.1–11.4 (Read); 11.5–11.6 12.1–12.2; 12.4–12.7 14.1–14.3; 14.6

Quarter Course		
Weeks	Mathematics	Computer Science/Eng.
1–2	1.9–1.13 4.1–4.2; 5.1–5.4	1.1–1.6 (Read); 3.1–3.5 4.1; 1.12 (Read)
3–4	6.1–6.4 6.5.1; 6.5.3–6.5.5	5.1–5.2.1; 5.3 6.1–6.3
5–6	6.7.1; 6.11.3; 7.1–7.3 7.4.1–7.4.2; 7.4.3 (Read)	6.4; 6.5.1; 6.5.3–6.5.5 6.7.1; 6.11.3; 7.1–7.3
7–8	9.1–9.3 10.1–10.3; 10.5.1; 10.5.7	7.4.1–7.4.2 (Read); 9.1–9.3 10.1–10.3; 10.5.1; 10.5.7
9–10	13.1–13.5 14.1–14.4	11.1–11.4 (Read); 11.5; 11.6 12.1–12.2; 12.4–12.7

Preface to the First Edition

Iterative methods for solving general, large, sparse linear systems have been gaining popularity in many areas of scientific computing. Until recently, direct solution methods were often preferred over iterative methods in real applications because of their robustness and predictable behavior. However, a number of efficient iterative solvers were discovered and the increased need for solving very large linear systems triggered a noticeable and rapid shift toward iterative techniques in many applications.

This trend can be traced back to the 1960s and 1970s, when two important developments revolutionized solution methods for large linear systems. First was the realization that one can take advantage of *sparsity* to design special direct methods that can be quite economical. Initiated by electrical engineers, these *direct sparse solution methods* led to the development of reliable and efficient general-purpose direct solution software codes over the next three decades. Second was the emergence of preconditioned conjugate gradient–like methods for solving linear systems. It was found that the combination of preconditioning and Krylov subspace iterations could provide efficient and simple *general-purpose* procedures that could compete with direct solvers. Preconditioning involves exploiting ideas from sparse direct solvers. Gradually, iterative methods started to approach the quality of direct solvers. In earlier times, iterative methods were often special purpose in nature. They were developed with certain applications in mind and their efficiency relied on many problem-dependent parameters.

Now three-dimensional models are commonplace and iterative methods are almost mandatory. The memory and the computational requirements for solving three-dimensional partial differential equations, or two-dimensional ones involving many degrees of freedom per point, may seriously challenge the most efficient direct solvers available today. Also, iterative methods are gaining ground because they are easier to implement efficiently on high performance computers than direct methods.

My intention in writing this volume is to provide up-to-date coverage of iterative methods for solving large sparse linear systems. I focused the book on practical methods that work for general sparse matrices rather than for any specific class of problems. It is indeed becoming important to embrace applications not necessarily governed by partial differential equations, as these applications are on the rise. Apart from two recent volumes by Axelsson [14] and Hackbusch [163], few books on iterative methods have appeared since the excellent ones by Varga [292] and later Young [321]. Since then, researchers and practitioners have achieved remarkable progress in the development and use of effective iterative methods. Unfortunately, fewer elegant results have been discovered since the 1950s and 1960s. The field has moved in other directions. Methods have gained not only in efficiency but also in robustness and in generality. The traditional techniques, which required rather complicated procedures to determine optimal acceleration parameters, have yielded to the parameter-free conjugate gradient class of methods.

The primary aim of this book is to describe some of the best techniques available today, from both preconditioners and accelerators. One of the secondary aims of the book is to provide a good mix of theory and practice. It also addresses some of the current research issues, such as parallel implementations and robust preconditioners. The emphasis is on Krylov subspace methods, currently the most practical and common group of techniques used in applications. Although there is a tutorial chapter that covers the discretization of partial differential equations, the book is not biased toward any specific application area. Instead, the matrices are assumed to be general sparse and possibly irregularly structured.

The book has been structured in four distinct parts. The first part, Chapters 1 to 4, presents the basic tools. The second part, Chapters 5 to 8, presents projection methods and Krylov subspace techniques. The third part, Chapters 9 and 10, discusses preconditioning. The fourth part, Chapters 11 to 13, discusses parallel implementations and parallel algorithms.

Acknowledgments

I am grateful to a number of colleagues who proofread or reviewed different versions of the manuscript. Among them are Randy Bramley (University of Indiana at Bloomington), Xiao-Chuan Cai (University of Colorado at Boulder), Tony Chan (University of California at Los Angeles), Jane Cullum (IBM, Yorktown Heights), Alan Edelman (Massachusetts Institute of Technology), Paul Fischer (Brown University), David Keyes (Old Dominion University), Beresford Parlett (University of California at Berkeley), and Shang-Hua Teng (University of Minnesota). Their numerous comments and corrections and their encouragement were a highly appreciated contribution. In particular, they helped improve the presentation considerably and prompted the addition of a number of topics missing from earlier versions.

This book evolved from several successive improvements of a set of lecture notes for the course "Iterative Methods for Linear Systems," which I taught at the University of Minnesota in the last few years. I apologize to those students who used the earlier error-laden and incomplete manuscripts. Their input and criticism contributed significantly to improving the manuscript. I also wish to thank those students at MIT (with Alan Edelman) and UCLA (with Tony Chan) who used this book in manuscript form and provided helpful feedback. My colleagues at the University of Minnesota, staff and faculty members, have helped in different ways. I wish to thank in particular Ahmed Sameh for his encouragement and for fostering a productive environment in the department. Finally, I am grateful to the National Science Foundation for its continued financial support of my research, part of which is represented in this work.

Yousef Saad

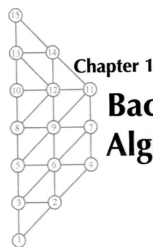

Chapter 1
Background in Linear Algebra

This chapter gives an overview of the relevant concepts in linear algebra that are useful in later chapters. It begins with a review of basic matrix theory and introduces the elementary notation used throughout the book. The convergence analysis of iterative methods requires a good level of knowledge in mathematical analysis and in linear algebra. Traditionally, many of the concepts presented specifically for these analyses have been geared toward matrices arising from the discretization of partial differential equations (PDEs) and basic relaxation-type methods. These concepts are now becoming less important because of the trend toward projection-type methods, which have more robust convergence properties and require different analysis tools. The material covered in this chapter will be helpful in establishing some theory for the algorithms and defining the notation used throughout the book.

1.1 Matrices

For the sake of generality, all vector spaces considered in this chapter are complex, unless otherwise stated. A complex $n \times m$ matrix A is an $n \times m$ array of complex numbers

$$a_{ij}, \quad i = 1, \ldots, n, \ j = 1, \ldots, m.$$

The set of all $n \times m$ matrices is a complex vector space denoted by $\mathbb{C}^{n \times m}$. The main operations with matrices are the following:

- Addition: $C = A + B$, where A, B, and C are matrices of size $n \times m$ and
$$c_{ij} = a_{ij} + b_{ij}, \quad i = 1, 2, \ldots, n, \quad j = 1, 2, \ldots, m.$$

- Multiplication by a scalar: $C = \alpha A$, where
$$c_{ij} = \alpha \, a_{ij}, \quad i = 1, 2, \ldots, n, \quad j = 1, 2, \ldots, m.$$

- Multiplication by another matrix:
$$C = AB,$$
where $A \in \mathbb{C}^{n \times m}$, $B \in \mathbb{C}^{m \times p}$, $C \in \mathbb{C}^{n \times p}$, and
$$c_{ij} = \sum_{k=1}^{m} a_{ik} b_{kj}.$$

Sometimes, a notation with column vectors and row vectors is used. The column vector a_{*j} is the vector consisting of the jth column of A:
$$a_{*j} = \begin{pmatrix} a_{1j} \\ a_{2j} \\ \vdots \\ a_{nj} \end{pmatrix}.$$

Similarly, the notation a_{i*} will denote the ith row of the matrix A:
$$a_{i*} = (a_{i1}, a_{i2}, \ldots, a_{im}).$$

For example, the following could be written:
$$A = (a_{*1}, a_{*2}, \ldots, a_{*m})$$
or
$$A = \begin{pmatrix} a_{1*} \\ a_{2*} \\ \vdots \\ a_{n*} \end{pmatrix}.$$

The *transpose* of a matrix A in $\mathbb{C}^{n \times m}$ is a matrix C in $\mathbb{C}^{m \times n}$ whose elements are defined by $c_{ij} = a_{ji}$, $i = 1, \ldots, m$, $j = 1, \ldots, n$. It is denoted by A^T. It is often more relevant to use the *transpose conjugate* matrix denoted by A^H and defined by
$$A^H = \bar{A}^T = \overline{A^T},$$
in which the bar denotes the (element-wise) complex conjugation.

Matrices are strongly related to linear mappings between vector spaces of finite dimension. This is because they represent these mappings with respect to two given bases: one for the initial vector space and the other for the image vector space, or *range*, of A.

1.2 Square Matrices and Eigenvalues

A matrix is *square* if it has the same number of columns and rows, i.e., if $m = n$. An important square matrix is the identity matrix
$$I = \{\delta_{ij}\}_{i,j=1,\ldots,n},$$
where δ_{ij} is the Kronecker symbol. The identity matrix satisfies the equality $AI = IA = A$ for every matrix A of size n. The inverse of a matrix, when it exists, is a matrix C such that
$$CA = AC = I.$$
The inverse of A is denoted by A^{-1}.

1.2. Square Matrices and Eigenvalues

The *determinant* of a matrix may be defined in several ways. For simplicity, the following recursive definition is used here. The determinant of a 1×1 matrix (a) is defined as the scalar a. Then the determinant of an $n \times n$ matrix is given by

$$\det(A) = \sum_{j=1}^{n} (-1)^{j+1} a_{1j} \det(A_{1j}),$$

where A_{1j} is an $(n-1) \times (n-1)$ matrix obtained by deleting the first row and the jth column of A. A matrix is said to be *singular* when $\det(A) = 0$ and *nonsingular* otherwise. We have the following simple properties:

- $\det(AB) = \det(A)\det(B)$.
- $\det(A^T) = \det(A)$.
- $\det(\alpha A) = \alpha^n \det(A)$.
- $\det(\bar{A}) = \overline{\det(A)}$.
- $\det(I) = 1$.

From the above definition of determinants it can be shown by induction that the function that maps a given complex value λ to the value $p_A(\lambda) = \det(A - \lambda I)$ is a polynomial of degree n; see Exercise 8. This is known as the *characteristic polynomial* of the matrix A.

Definition 1.1. *A complex scalar λ is called an eigenvalue of the square matrix A if a nonzero vector u of $\mathbb{C}^n$ exists such that $Au = \lambda u$. The vector u is called an eigenvector of A associated with λ. The set of all the eigenvalues of A is called the spectrum of A and is denoted by $\sigma(A)$.*

A scalar λ is an eigenvalue of A *if and only if* (iff hereafter) $\det(A - \lambda I) \equiv p_A(\lambda) = 0$. That is true iff λ is a root of the characteristic polynomial. In particular, there are at most n distinct eigenvalues.

It is clear that a matrix is singular iff it admits zero as an eigenvalue. A well-known result in linear algebra is stated in the following proposition.

Proposition 1.2. *A matrix A is nonsingular iff it admits an inverse.*

Thus, the determinant of a matrix determines whether or not the matrix admits an inverse.

The maximum modulus of the eigenvalues is called the *spectral radius* and is denoted by $\rho(A)$:

$$\rho(A) = \max_{\lambda \in \sigma(A)} |\lambda|.$$

The *trace* of a matrix is equal to the sum of all its diagonal elements:

$$\text{tr}(A) = \sum_{i=1}^{n} a_{ii}.$$

It can be easily shown that the trace of A is also equal to the sum of the eigenvalues of A counted with their multiplicities as roots of the characteristic polynomial.

Proposition 1.3. *If λ is an eigenvalue of A, then $\bar{\lambda}$ is an eigenvalue of A^H. An eigenvector v of A^H associated with the eigenvalue $\bar{\lambda}$ is called a left eigenvector of A.*

When a distinction is necessary, an eigenvector of A is often called a right eigenvector. Therefore, the eigenvalue λ as well as the right and left eigenvectors u and v satisfy the relations
$$Au = \lambda u, \quad v^H A = \lambda v^H$$
or, equivalently,
$$u^H A^H = \bar{\lambda} u^H, \quad A^H v = \bar{\lambda} v.$$

1.3 Types of Matrices

The choice of a method for solving linear systems will often depend on the structure of the matrix A. One of the most important properties of matrices is symmetry, because of its impact on the eigenstructure of A. A number of other classes of matrices also have particular eigenstructures. The most important ones are listed below:

- *Symmetric matrices:* $A^T = A$.
- *Hermitian matrices:* $A^H = A$.
- *Skew-symmetric matrices:* $A^T = -A$.
- *Skew-Hermitian matrices:* $A^H = -A$.
- *Normal matrices:* $A^H A = A A^H$.
- *Nonnegative matrices:* $a_{ij} \geq 0$, $i, j = 1, \ldots, n$ (similar definition for nonpositive, positive, and negative matrices).
- *Unitary matrices:* $Q^H Q = I$.

It is worth noting that a unitary matrix Q is a matrix whose inverse is its transpose conjugate Q^H, since
$$Q^H Q = I \quad \rightarrow \quad Q^{-1} = Q^H. \tag{1.1}$$
A matrix Q such that $Q^H Q$ is diagonal is often called orthogonal.

Some matrices have particular structures that are often convenient for computational purposes. The following list, though incomplete, gives an idea of these special matrices, which play an important role in numerical analysis and scientific computing applications.

- *Diagonal matrices:* $a_{ij} = 0$ for $j \neq i$. Notation:
$$A = \text{diag}(a_{11}, a_{22}, \ldots, a_{nn}).$$
- *Upper triangular matrices:* $a_{ij} = 0$ for $i > j$.
- *Lower triangular matrices:* $a_{ij} = 0$ for $i < j$.

- *Upper bidiagonal matrices:* $a_{ij} = 0$ for $j \neq i$ or $j \neq i+1$.
- *Lower bidiagonal matrices:* $a_{ij} = 0$ for $j \neq i$ or $j \neq i-1$.
- *Tridiagonal matrices:* $a_{ij} = 0$ for any pair i, j such that $|j - i| > 1$. Notation:
$$A = \text{tridiag}(a_{i,i-1}, a_{ii}, a_{i,i+1}).$$
- *Banded matrices:* $a_{ij} \neq 0$ only if $i - m_l \leq j \leq i + m_u$, where m_l and m_u are two nonnegative integers. The number $m_l + m_u + 1$ is called the bandwidth of A.
- *Upper Hessenberg matrices:* $a_{ij} = 0$ for any pair i, j such that $i > j+1$. Lower Hessenberg matrices can be defined similarly.
- *Outer product matrices:* $A = uv^H$, where both u and v are vectors.
- *Permutation matrices:* the columns of A are a permutation of the columns of the identity matrix.
- *Block diagonal matrices:* generalizes the diagonal matrix by replacing each diagonal entry with a matrix. Notation:
$$A = \text{diag}(A_{11}, A_{22}, \ldots, A_{nn}).$$
- *Block tridiagonal matrices:* generalizes the tridiagonal matrix by replacing each nonzero entry with a square matrix. Notation:
$$A = \text{tridiag}(A_{i,i-1}, A_{ii}, A_{i,i+1}).$$

The above properties emphasize structure, i.e., the positions of the nonzero elements with respect to the zeros. Also, they assume that there are many zero elements or that the matrix is of low rank. This is in contrast with the classifications listed earlier, such as symmetry and normality.

1.4 Vector Inner Products and Norms

An inner product on a (complex) vector space $\mathbb{X}$ is any mapping s from $\mathbb{X} \times \mathbb{X}$ into $\mathbb{C}$,
$$x \in \mathbb{X}, y \in \mathbb{X} \quad \rightarrow \quad s(x, y) \in \mathbb{C},$$
that satisfies the following conditions:

1. $s(x, y)$ is linear with respect to x; i.e.,
$$s(\lambda_1 x_1 + \lambda_2 x_2, y) = \lambda_1 s(x_1, y) + \lambda_2 s(x_2, y) \quad \forall x_1, x_2 \in \mathbb{X}, \forall \lambda_1, \lambda_2 \in \mathbb{C}.$$

2. $s(x, y)$ is *Hermitian*; i.e.,
$$s(y, x) = \overline{s(x, y)} \quad \forall x, y \in \mathbb{X}.$$

3. $s(x, y)$ is *positive definite*; i.e.,
$$s(x, x) > 0 \quad \forall x \neq 0.$$

Note that (2) implies that $s(x, x)$ is real and, therefore, (3) adds the constraint that $s(x, x)$ must also be positive for any nonzero x. For any x and y,

$$s(x, 0) = s(x, 0 \cdot y) = 0 \cdot s(x, y) = 0.$$

Similarly, $s(0, y) = 0$ for any y. Hence, $s(0, y) = s(x, 0) = 0$ for any x and y. In particular condition (3) can be rewritten as

$$s(x, x) \geq 0 \quad \text{and} \quad s(x, x) = 0 \quad \text{iff} \quad x = 0,$$

as can be readily shown. A useful relation satisfied by any inner product is the so-called Cauchy–Schwarz inequality

$$|s(x, y)|^2 \leq s(x, x)\, s(y, y). \tag{1.2}$$

The proof of this inequality begins by expanding $s(x - \lambda y, x - \lambda y)$ using the properties of s:

$$s(x - \lambda y, x - \lambda y) = s(x, x) - \bar{\lambda} s(x, y) - \lambda s(y, x) + |\lambda|^2 s(y, y).$$

If $y = 0$ then the inequality is trivially satisfied. Assume that $y \neq 0$ and take $\lambda = s(x, y)/s(y, y)$. Then, from the above equality, $s(x - \lambda y, x - \lambda y) \geq 0$ shows that

$$0 \leq s(x - \lambda y, x - \lambda y) = s(x, x) - 2\frac{|s(x, y)|^2}{s(y, y)} + \frac{|s(x, y)|^2}{s(y, y)}$$

$$= s(x, x) - \frac{|s(x, y)|^2}{s(y, y)},$$

which yields the result (1.2).

In the particular case of the vector space $\mathbb{X} = \mathbb{C}^n$, a *canonical* inner product is the *Euclidean inner product*. The Euclidean inner product of two vectors $x = (x_i)_{i=1,\ldots,n}$ and $y = (y_i)_{i=1,\ldots,n}$ of $\mathbb{C}^n$ is defined by

$$(x, y) = \sum_{i=1}^{n} x_i \bar{y}_i, \tag{1.3}$$

which is often rewritten in matrix notation as

$$(x, y) = y^H x. \tag{1.4}$$

It is easy to verify that this mapping does indeed satisfy the three conditions required for inner products listed above. A fundamental property of the Euclidean inner product in matrix computations is the simple relation

$$(Ax, y) = (x, A^H y) \quad \forall\, x, y \in \mathbb{C}^n. \tag{1.5}$$

The proof of this is straightforward. The *adjoint* of A *with respect to an arbitrary inner product* is a matrix B such that $(Ax, y) = (x, By)$ for all pairs of vectors x and y. A matrix is *self-adjoint*, or Hermitian, with respect to this inner product if it is equal to its adjoint. The following proposition is a consequence of equality (1.5).

1.5. Matrix Norms

Proposition 1.4. *Unitary matrices preserve the Euclidean inner product; i.e.,*

$$(Qx, Qy) = (x, y)$$

for any unitary matrix Q and any vectors x and y.

Proof. Indeed, $(Qx, Qy) = (x, Q^H Qy) = (x, y)$. □

A vector norm on a vector space $\mathbb{X}$ is a real-valued function $x \to \|x\|$ on $\mathbb{X}$ that satisfies the following three conditions:

1. $\|x\| \geq 0 \quad \forall\, x \in \mathbb{X}$ and $\|x\| = 0$ iff $x = 0$.
2. $\|\alpha x\| = |\alpha| \|x\| \quad \forall\, x \in \mathbb{X} \quad \forall\, \alpha \in \mathbb{C}$.
3. $\|x + y\| \leq \|x\| + \|y\| \quad \forall\, x, y \in \mathbb{X}$.

For the particular case when $\mathbb{X} = \mathbb{C}^n$, we can associate with the inner product (1.3) the *Euclidean norm* of a complex vector defined by

$$\|x\|_2 = (x, x)^{1/2}.$$

It follows from Proposition 1.4 that a unitary matrix preserves the Euclidean norm metric; i.e.,

$$\|Qx\|_2 = \|x\|_2 \; \forall\, x.$$

The linear transformation associated with a unitary matrix Q is therefore an *isometry*.

The most commonly used vector norms in numerical linear algebra are special cases of the Hölder norms

$$\|x\|_p = \left(\sum_{i=1}^n |x_i|^p \right)^{1/p}. \tag{1.6}$$

Note that the limit of $\|x\|_p$ when p tends to infinity exists and is equal to the maximum modulus of the x_i's. This defines a norm denoted by $\|\cdot\|_\infty$. The cases $p = 1$, $p = 2$, and $p = \infty$ lead to the most important norms in practice:

$$\|x\|_1 = |x_1| + |x_2| + \cdots + |x_n|,$$
$$\|x\|_2 = \left[|x_1|^2 + |x_2|^2 + \cdots + |x_n|^2 \right]^{1/2},$$
$$\|x\|_\infty = \max_{i=1,\ldots,n} |x_i|.$$

The Cauchy–Schwarz inequality of (1.2) becomes

$$|(x, y)| \leq \|x\|_2 \|y\|_2.$$

1.5 Matrix Norms

For a general matrix A in $\mathbb{C}^{n \times m}$, we define the following special set of norms:

$$\|A\|_{pq} = \max_{x \in \mathbb{C}^m,\, x \neq 0} \frac{\|Ax\|_p}{\|x\|_q}. \tag{1.7}$$

The norm $\|\cdot\|_{pq}$ is *induced* by the two norms $\|\cdot\|_p$ and $\|\cdot\|_q$. These norms satisfy the usual properties of norms; i.e.,

$$\|A\| \geq 0 \quad \forall\, A \in \mathbb{C}^{n \times m} \quad \text{and} \quad \|A\| = 0 \quad \text{iff} \quad A = 0, \tag{1.8}$$

$$\|\alpha A\| = |\alpha|\|A\| \quad \forall\, A \in \mathbb{C}^{n \times m}, \quad \forall\, \alpha \in \mathbb{C}, \tag{1.9}$$

$$\|A + B\| \leq \|A\| + \|B\| \quad \forall\, A, B \in \mathbb{C}^{n \times m}. \tag{1.10}$$

A norm that satisfies the above three properties is nothing but a *vector norm* applied to the matrix considered as a vector consisting of the m columns stacked into a vector of size nm.

The most important cases are again those associated with $p, q = 1, 2, \infty$. The case $q = p$ is of particular interest and the associated norm $\|\cdot\|_{pq}$ is simply denoted by $\|\cdot\|_p$ and called a *p-norm*. A fundamental property of a p-norm is that

$$\|AB\|_p \leq \|A\|_p \|B\|_p,$$

an immediate consequence of the definition (1.7). Matrix norms that satisfy the above property are sometimes called *consistent*. Often a norm satisfying the properties (1.8)–(1.10) that is consistent is called a *matrix norm*. A result of consistency is that, for any square matrix A,

$$\|A^k\|_p \leq \|A\|_p^k.$$

In particular the matrix A^k converges to zero if *any* of its p-norms is less than 1.

The Frobenius norm of a matrix is defined by

$$\|A\|_F = \left(\sum_{j=1}^{m}\sum_{i=1}^{n}|a_{ij}|^2\right)^{1/2}. \tag{1.11}$$

This can be viewed as the 2-norm of the column (or row) vector in $\mathbb{C}^{n^2}$ consisting of all the columns (resp., rows) of A listed from 1 to m (resp., 1 to n). It can be shown that this norm is also consistent, in spite of the fact that it is not induced by a pair of vector norms; i.e., it is not derived from a formula of the form (1.7); see Exercise 5. However, it does not satisfy some of the other properties of the p-norms. For example, the Frobenius norm of the identity matrix is not equal to one. To avoid these difficulties, *we will only use the term matrix norm for a norm that is induced by two norms, as in the definition* (1.7). Thus, we will not consider the Frobenius norm to be a proper matrix norm, according to our conventions, even though it is consistent.

The following equalities satisfied by the matrix norms defined above lead to alternative definitions that are often easier to work with:

$$\|A\|_1 = \max_{j=1,\ldots,m} \sum_{i=1}^{n} |a_{ij}|, \tag{1.12}$$

$$\|A\|_\infty = \max_{i=1,\ldots,n} \sum_{j=1}^{m} |a_{ij}|, \tag{1.13}$$

$$\|A\|_2 = \left[\rho(A^H A)\right]^{1/2} = \left[\rho(A A^H)\right]^{1/2}, \tag{1.14}$$

$$\|A\|_F = \left[\text{tr}(A^H A)\right]^{1/2} = \left[\text{tr}(A A^H)\right]^{1/2}. \tag{1.15}$$

1.6. Subspaces, Range, and Kernel

As will be shown later, the eigenvalues of $A^H A$ are nonnegative. Their square roots are called *singular values* of A and are denoted by σ_i, $i = 1, \ldots, m$. Thus, the relation (1.14) states that $\|A\|_2$ is equal to σ_1, the largest singular value of A.

Example 1.1. From the relation (1.14), it is clear that the spectral radius $\rho(A)$ is equal to the 2-norm of a matrix when the matrix is Hermitian. However, it is not a matrix norm in general. For example, the first property of norms is not satisfied, since, for

$$A = \begin{pmatrix} 0 & 1 \\ 0 & 0 \end{pmatrix},$$

we have $\rho(A) = 0$ while $A \neq 0$. Also, the triangle inequality is not satisfied for the pair A and $B = A^T$, where A is defined above. Indeed,

$$\rho(A+B) = 1 \quad \text{while} \quad \rho(A) + \rho(B) = 0.$$

1.6 Subspaces, Range, and Kernel

A subspace of $\mathbb{C}^n$ is a subset of $\mathbb{C}^n$ that is also a complex vector space. The set of all linear combinations of a set of vectors $G = \{a_1, a_2, \ldots, a_q\}$ of $\mathbb{C}^n$ is a vector subspace called the linear span of G:

$$\begin{aligned} \text{span}\{G\} &= \text{span}\{a_1, a_2, \ldots, a_q\} \\ &= \left\{ z \in \mathbb{C}^n \;\middle|\; z = \sum_{i=1}^{q} \alpha_i a_i, \; \{\alpha_i\}_{i=1,\ldots,q} \in \mathbb{C}^q \right\}. \end{aligned}$$

If the a_i's are linearly independent, then each vector of span$\{G\}$ admits a unique expression as a linear combination of the a_i's. The set G is then called a *basis* of the subspace span$\{G\}$.

Given two vector subspaces S_1 and S_2, their *sum* S is a subspace defined as the set of all vectors that are equal to the sum of a vector of S_1 and a vector of S_2. The intersection of two subspaces is also a subspace. If the intersection of S_1 and S_2 is reduced to $\{0\}$, then the sum of S_1 and S_2 is called their direct sum and is denoted by $S = S_1 \oplus S_2$. When S is equal to $\mathbb{C}^n$, then every vector x of $\mathbb{C}^n$ can be written in a unique way as the sum of an element x_1 of S_1 and an element x_2 of S_2. The transformation P that maps x into x_1 is a linear transformation that is *idempotent*, i.e., such that $P^2 = P$. It is called a *projector* onto S_1 along S_2.

Two important subspaces that are associated with a matrix A of $\mathbb{C}^{n \times m}$ are its *range*, defined by

$$\text{Ran}(A) = \{Ax \mid x \in \mathbb{C}^m\}, \qquad (1.16)$$

and its *kernel* or *null space*

$$\text{Ker}(A) = \{x \in \mathbb{C}^m \mid Ax = 0\}.$$

The range of A is clearly equal to the linear *span* of its columns. The *rank* of a matrix is equal to the dimension of the range of A, i.e., to the number of linearly independent columns. This *column rank* is equal to the *row rank*, the number of linearly independent rows of A. A matrix in $\mathbb{C}^{n \times m}$ is of *full rank* when its rank is equal to the smallest of m and n. A fundamental result of linear algebra is stated by the following relation:

$$\mathbb{C}^n = \text{Ran}(A) \oplus \text{Ker}(A^T). \tag{1.17}$$

The same result applied to the transpose of A yields $\mathbb{C}^m = \text{Ran}(A^T) \oplus \text{Ker}(A)$.

A subspace S is said to be *invariant* under a (square) matrix A whenever $AS \subset S$. In particular, for any eigenvalue λ of A the subspace $\text{Ker}(A - \lambda I)$ is invariant under A. The subspace $\text{Ker}(A - \lambda I)$ is called the eigenspace associated with λ and consists of all the eigenvectors of A associated with λ, in addition to the zero vector.

1.7 Orthogonal Vectors and Subspaces

A set of vectors $G = \{a_1, a_2, \ldots, a_r\}$ is said to be *orthogonal* if

$$(a_i, a_j) = 0 \quad \text{when} \quad i \neq j.$$

It is *orthonormal* if, in addition, every vector of G has a 2-norm equal to unity. A vector that is orthogonal to all the vectors of a subspace S is said to be orthogonal to this subspace. The set of all the vectors that are orthogonal to S is a vector subspace called the *orthogonal complement* of S and denoted by $S^\perp$. The space $\mathbb{C}^n$ is the direct sum of S and its orthogonal complement. Thus, any vector x can be written in a unique fashion as the sum of a vector in S and a vector in $S^\perp$. The operator that maps x into its component in the subspace S is the *orthogonal projector* onto S.

Every subspace admits an orthonormal basis that is obtained by taking any basis and *orthonormalizing* it. The orthonormalization can be achieved by an algorithm known as the Gram–Schmidt process, which we now describe.

Given a set of linearly independent vectors $\{x_1, x_2, \ldots, x_r\}$, first normalize the vector x_1, which means divide it by its 2-norm, to obtain the scaled vector q_1 of norm unity. Then x_2 is orthogonalized against the vector q_1 by subtracting from x_2 a multiple of q_1 to make the resulting vector orthogonal to q_1; i.e.,

$$x_2 \leftarrow x_2 - (x_2, q_1)q_1.$$

The resulting vector is again normalized to yield the second vector q_2. The ith step of the Gram–Schmidt process consists of orthogonalizing the vector x_i against all previous vectors q_j.

ALGORITHM 1.1. Gram–Schmidt

1. Compute $r_{11} := \|x_1\|_2$. If $r_{11} = 0$ Stop, else compute $q_1 := x_1/r_{11}$
2. For $j = 2, \ldots, r$, Do
3.   Compute $r_{ij} := (x_j, q_i)$ for $i = 1, 2, \ldots, j-1$
4. $\hat{q} := x_j - \sum_{i=1}^{j-1} r_{ij} q_i$
5. $r_{jj} := \|\hat{q}\|_2$
6.   If $r_{jj} = 0$ then Stop, else $q_j := \hat{q}/r_{jj}$
7. EndDo

1.7. Orthogonal Vectors and Subspaces

It is easy to prove that the above algorithm will not break down; i.e., all r steps will be completed iff the set of vectors $x_1, x_2, \ldots, x_r$ is linearly independent. From lines 4 and 5, it is clear that at every step of the algorithm the following relation holds:

$$x_j = \sum_{i=1}^{j} r_{ij} q_i.$$

If $X = [x_1, x_2, \ldots, x_r]$, $Q = [q_1, q_2, \ldots, q_r]$, and R denotes the $r \times r$ upper triangular matrix whose nonzero elements are the r_{ij}'s defined in the algorithm, then the above relation can be written as

$$X = QR. \tag{1.18}$$

This is called the QR decomposition of the $n \times r$ matrix X. From what was said above, the QR decomposition of a matrix exists whenever the column vectors of X form a linearly independent set of vectors.

The above algorithm is the standard Gram–Schmidt process. There are alternative formulations of the algorithm that have better numerical properties. The best known of these is the modified Gram–Schmidt (MGS) algorithm.

ALGORITHM 1.2. MGS

1. Define $r_{11} := \|x_1\|_2$. If $r_{11} = 0$ Stop, else $q_1 := x_1 / r_{11}$
2. For $j = 2, \ldots, r$, Do
3.   Define $\hat{q} := x_j$
4.   For $i = 1, \ldots, j-1$, Do
5. $r_{ij} := (\hat{q}, q_i)$
6. $\hat{q} := \hat{q} - r_{ij} q_i$
7.   EndDo
8.   Compute $r_{jj} := \|\hat{q}\|_2$
9.   If $r_{jj} = 0$ then Stop, else $q_j := \hat{q} / r_{jj}$
10. EndDo

Yet another alternative for orthogonalizing a sequence of vectors is the Householder algorithm. This technique uses Householder *reflectors*, i.e., matrices of the form

$$P = I - 2ww^T, \tag{1.19}$$

in which w is a vector of 2-norm unity. Geometrically, the vector Px represents a mirror image of x with respect to the hyperplane span$\{w\}^\perp$.

To describe the Householder orthogonalization process, the problem can be formulated as that of finding a QR factorization of a given $n \times m$ matrix X. For any vector x, the vector w for the Householder transformation (1.19) is selected in such a way that

$$Px = \alpha e_1,$$

where α is a scalar. Writing $(I - 2ww^T)x = \alpha e_1$ yields

$$2w^T x \, w = x - \alpha e_1. \tag{1.20}$$

This shows that the desired w is a multiple of the vector $x - \alpha e_1$:

$$w = \pm \frac{x - \alpha e_1}{\|x - \alpha e_1\|_2}.$$

For (1.20) to be satisfied, we must impose the condition

$$2(x - \alpha e_1)^T x = \|x - \alpha e_1\|_2^2,$$

which gives $2(\|x\|_1^2 - \alpha \xi_1) = \|x\|_2^2 - 2\alpha \xi_1 + \alpha^2$, where $\xi_1 \equiv e_1^T x$ is the first component of the vector x. Therefore, it is necessary that

$$\alpha = \pm \|x\|_2.$$

In order to avoid the resulting vector w being small, it is customary to take

$$\alpha = -\operatorname{sign}(\xi_1)\|x\|_2,$$

which yields

$$w = \frac{x + \operatorname{sign}(\xi_1)\|x\|_2 e_1}{\|x + \operatorname{sign}(\xi_1)\|x\|_2 e_1\|_2}. \tag{1.21}$$

Given an $n \times m$ matrix, its first column can be transformed to a multiple of the column e_1 by premultiplying it by a Householder matrix P_1:

$$X_1 \equiv P_1 X, \qquad X_1 e_1 = \alpha e_1.$$

Assume, inductively, that the matrix X has been transformed in $k - 1$ successive steps into the partially upper triangular form

$$X_k \equiv P_{k-1} \cdots P_1 X_1 = \begin{pmatrix} x_{11} & x_{12} & x_{13} & \cdots & \cdots & \cdots & x_{1m} \\ & x_{22} & x_{23} & \cdots & \cdots & \cdots & x_{2m} \\ & & x_{33} & \cdots & \cdots & \cdots & x_{3m} \\ & & & \ddots & \cdots & \cdots & \vdots \\ & & & & x_{kk} & \cdots & \vdots \\ & & & & x_{k+1,k} & \cdots & x_{k+1,m} \\ & & & & \vdots & \vdots & \vdots \\ & & & & x_{n,k} & \cdots & x_{n,m} \end{pmatrix}.$$

This matrix is upper triangular up to column number $k - 1$. To advance by one step, it must be transformed into one that is upper triangular up to the kth column, leaving the previous columns in the same form. To leave the first $k - 1$ columns unchanged, select a w-vector that has zeros in positions 1 through $k - 1$. So the next Householder reflector matrix is defined as

$$P_k = I - 2 w_k w_k^T, \tag{1.22}$$

in which the vector w_k is defined as

$$w_k = \frac{z}{\|z\|_2}, \tag{1.23}$$

1.7. Orthogonal Vectors and Subspaces

where the components of the vector z are given by

$$z_i = \begin{cases} 0 & \text{if } i < k, \\ \beta + x_{ii} & \text{if } i = k, \\ x_{ik} & \text{if } i > k, \end{cases} \quad (1.24)$$

with

$$\beta = \text{sign}(x_{kk}) \times \left(\sum_{i=k}^{n} x_{ik}^2 \right)^{1/2}. \quad (1.25)$$

We note in passing that the premultiplication of a matrix X by a Householder transform requires only a rank-one update since

$$(I - 2ww^T)X = X - wv^T, \quad \text{where} \quad v = 2X^T w.$$

Therefore, the Householder matrices need not, and should not, be explicitly formed. In addition, the vectors w need not be explicitly scaled.

Assume now that $m - 1$ Householder transforms have been applied to a certain matrix X of dimension $n \times m$ to reduce it into the upper triangular form

$$X_m \equiv P_{m-1} P_{m-2} \cdots P_1 X = \begin{pmatrix} x_{11} & x_{12} & x_{13} & \cdots & x_{1m} \\ & x_{22} & x_{23} & \cdots & x_{2m} \\ & & x_{33} & \cdots & x_{3m} \\ & & & \ddots & \vdots \\ & & & & x_{m,m} \\ & & & & 0 \\ & & & & \vdots \end{pmatrix}. \quad (1.26)$$

Recall that our initial goal was to obtain a QR factorization of X. We now wish to recover the Q- and R-matrices from the P_k's and the above matrix. If we denote by P the product of the P_i on the left side of (1.26), then (1.26) becomes

$$PX = \begin{pmatrix} R \\ O \end{pmatrix}, \quad (1.27)$$

in which R is an $m \times m$ upper triangular matrix and O is an $(n-m) \times m$ zero block. Since P is unitary, its inverse is equal to its transpose and, as a result,

$$X = P^T \begin{pmatrix} R \\ O \end{pmatrix} = P_1 P_2 \cdots P_{m-1} \begin{pmatrix} R \\ O \end{pmatrix}.$$

If E_m is the matrix of size $n \times m$ that consists of the first m columns of the identity matrix, then the above equality translates into

$$X = P^T E_m R.$$

The matrix $Q = P^T E_m$ represents the first m columns of P^T. Since

$$Q^T Q = E_m^T P P^T E_m = I,$$

Q and R are the matrices sought. In summary,

$$X = QR,$$

in which R is the triangular matrix obtained from the Householder reduction of X (see (1.26) and (1.27)) and

$$Qe_j = P_1 P_2 \cdots P_{m-1} e_j.$$

ALGORITHM 1.3. Householder Orthogonalization

---
1. Define $X = [x_1, \ldots, x_m]$
2. For $k = 1, \ldots, m$, Do
3.   If $k > 1$ compute $r_k := P_{k-1} P_{k-2} \cdots P_1 x_k$
4.   Compute w_k using (1.23), (1.24), (1.25)
5.   Compute $r_k := P_k r_k$ with $P_k = I - 2 w_k w_k^T$
6.   Compute $q_k = P_1 P_2 \cdots P_k e_k$
7. EndDo

Note that line 6 can be omitted since the q_i's are not needed in the execution of the next steps. It must be executed only when the matrix Q is needed at the completion of the algorithm. Also, the operation in line 5 consists only of zeroing the components $k+1, \ldots, n$ and updating the kth component of r_k. In practice, a work vector can be used for r_k, and its nonzero components after this step can be saved into an upper triangular matrix. Since the components 1 through k of the vector w_k are zero, the upper triangular matrix R can be saved in those zero locations that would otherwise be unused.

1.8 Canonical Forms of Matrices

This section discusses the reduction of square matrices into matrices that have simpler forms, such as diagonal, bidiagonal, or triangular. Reduction means a transformation that preserves the eigenvalues of a matrix.

Definition 1.5. *Two matrices A and B are said to be similar if there is a nonsingular matrix X such that*

$$A = XBX^{-1}.$$

The mapping $B \to A$ is called a similarity transformation.

It is clear that *similarity* is an equivalence relation. Similarity transformations preserve the eigenvalues of matrices. An eigenvector u_B of B is transformed into the eigenvector $u_A = X u_B$ of A. In effect, a similarity transformation amounts to representing the matrix B in a different basis.

We now introduce some terminology.

1. An eigenvalue λ of A has *algebraic multiplicity* μ if it is a root of multiplicity μ of the characteristic polynomial.

2. If an eigenvalue is of algebraic multiplicity one, it is said to be *simple*. A nonsimple eigenvalue is *multiple*.

1.8. Canonical Forms of Matrices

3. The *geometric multiplicity* γ of an eigenvalue λ of A is the maximum number of independent eigenvectors associated with it. In other words, the geometric multiplicity γ is the dimension of the eigenspace $\text{Ker}(A - \lambda I)$.

4. A matrix is *derogatory* if the geometric multiplicity of at least one of its eigenvalues is larger than one.

5. An eigenvalue is *semisimple* if its algebraic multiplicity is equal to its geometric multiplicity. An eigenvalue that is not semisimple is called *defective*.

Often, $\lambda_1, \lambda_2, \ldots, \lambda_p$ ($p \leq n$) are used to denote the *distinct* eigenvalues of A. It is easy to show that the characteristic polynomials of two similar matrices are identical; see Exercise 9. Therefore, the eigenvalues of two similar matrices are equal and so are their algebraic multiplicities. Moreover, if v is an eigenvector of B, then Xv is an eigenvector of A and, conversely, if y is an eigenvector of A, then $X^{-1}y$ is an eigenvector of B. As a result, the number of independent eigenvectors associated with a given eigenvalue is the same for two similar matrices; i.e., their geometric multiplicity is also the same.

1.8.1 Reduction to the Diagonal Form

The simplest form into which a matrix can be reduced is undoubtedly the diagonal form. Unfortunately, this reduction is not always possible. A matrix that can be reduced to the diagonal form is called *diagonalizable*. The following theorem characterizes such matrices.

Theorem 1.6. *A matrix of dimension n is diagonalizable iff it has n linearly independent eigenvectors.*

Proof. A matrix A is diagonalizable iff there exists a nonsingular matrix X and a diagonal matrix D such that $A = XDX^{-1}$ or, equivalently, $AX = XD$, where D is a diagonal matrix. This is equivalent to saying that n linearly independent vectors exist—the n column vectors of X—such that $Ax_i = d_i x_i$. Each of these column vectors is an eigenvector of A. □

A matrix that is diagonalizable has only semisimple eigenvalues. Conversely, if all the eigenvalues of a matrix A are semisimple, then A has n eigenvectors. It can be easily shown that these eigenvectors are linearly independent; see Exercise 2. As a result, we have the following proposition.

Proposition 1.7. *A matrix is diagonalizable iff all its eigenvalues are semisimple.*

Since every simple eigenvalue is semisimple, an immediate corollary of the above result is as follows: When A has n distinct eigenvalues, then it is diagonalizable.

1.8.2 The Jordan Canonical Form

From the theoretical viewpoint, one of the most important canonical forms of matrices is the well-known Jordan form. A full development of the steps leading to the Jordan form is beyond the scope of this book. Only the main theorem is stated. Details, including the proof, can be found in standard books of linear algebra such as [164]. In the following, m_i refers to the algebraic multiplicity of the individual eigenvalue λ_i and l_i is the *index* of the eigenvalue, i.e., the smallest integer for which $\text{Ker}(A - \lambda_i I)^{l_i+1} = \text{Ker}(A - \lambda_i I)^{l_i}$.

Theorem 1.8. *Any matrix A can be reduced to a block diagonal matrix consisting of p diagonal blocks, each associated with a distinct eigenvalue λ_i. Each of these diagonal blocks has itself a block diagonal structure consisting of γ_i sub-blocks, where γ_i is the geometric multiplicity of the eigenvalue λ_i. Each of the sub-blocks, referred to as a Jordan block, is an upper bidiagonal matrix of size not exceeding $l_i \leq m_i$, with the constant λ_i on the diagonal and the constant one on the superdiagonal.*

The ith diagonal block, $i = 1, \ldots, p$, is known as the ith Jordan submatrix (sometimes "Jordan box"). The Jordan submatrix number i starts in column $j_i \equiv m_1 + m_2 + \cdots + m_{i-1} + 1$. Thus,

$$X^{-1}AX = J = \begin{pmatrix} J_1 & & & & & \\ & J_2 & & & & \\ & & \ddots & & & \\ & & & J_i & & \\ & & & & \ddots & \\ & & & & & J_p \end{pmatrix},$$

where each J_i is associated with λ_i and is of size m_i, the algebraic multiplicity of λ_i. It has itself the following structure:

$$J_i = \begin{pmatrix} J_{i1} & & & \\ & J_{i2} & & \\ & & \ddots & \\ & & & J_{i\gamma_i} \end{pmatrix}, \text{ with } J_{ik} = \begin{pmatrix} \lambda_i & 1 & & \\ & \ddots & \ddots & \\ & & \lambda_i & 1 \\ & & & \lambda_i \end{pmatrix}.$$

Each of the blocks J_{ik} corresponds to a different eigenvector associated with the eigenvalue λ_i. Its size l_i is the index of λ_i.

1.8.3 The Schur Canonical Form

Here, it will be shown that any matrix is unitarily similar to an upper triangular matrix. The only result needed to prove the following theorem is that any vector having a 2-norm can be completed by $n - 1$ additional vectors to form an orthonormal basis of $\mathbb{C}^n$.

Theorem 1.9. *For any square matrix A, there exists a unitary matrix Q such that*

$$Q^H A Q = R$$

is upper triangular.

Proof. The proof is by induction over the dimension n. The result is trivial for $n = 1$. Assume that it is true for $n - 1$ and consider any matrix A of size n. The matrix admits

1.8. Canonical Forms of Matrices

at least one eigenvector u that is associated with an eigenvalue λ. Also assume without loss of generality that $\|u\|_2 = 1$. First, complete the set consisting of the vector u into an orthonormal set; i.e., find an $n \times (n-1)$ matrix V such that the $n \times n$ matrix $U = [u, V]$ is unitary. Then $AU = [\lambda u, AV]$ and, hence,

$$U^H A U = \begin{bmatrix} u^H \\ V^H \end{bmatrix} [\lambda u, AV] = \begin{pmatrix} \lambda & u^H A V \\ 0 & V^H A V \end{pmatrix}. \tag{1.28}$$

Now use the induction hypothesis for the $(n-1) \times (n-1)$ matrix $B = V^H A V$: There exists an $(n-1) \times (n-1)$ unitary matrix Q_1 such that $Q_1^H B Q_1 = R_1$ is upper triangular. Define the $n \times n$ matrix

$$\hat{Q}_1 = \begin{pmatrix} 1 & 0 \\ 0 & Q_1 \end{pmatrix}$$

and multiply both members of (1.28) by $\hat{Q}_1^H$ from the left and $\hat{Q}_1$ from the right. The resulting matrix is clearly upper triangular, which shows that the result is true for A, with $Q = \hat{Q}_1 U$, which is a unitary $n \times n$ matrix. □

A simpler proof that uses the Jordan canonical form and the QR decomposition is the subject of Exercise 7. Since the matrix R is triangular and similar to A, its diagonal elements are equal to the eigenvalues of A ordered in a certain manner. In fact, it is easy to extend the proof of the theorem to show that this factorization can be obtained with *any order* of the eigenvalues. Despite its simplicity, the above theorem has far-reaching consequences, some of which will be examined in the next section.

It is important to note that, for any $k \leq n$, the subspace spanned by the first k columns of Q is invariant under A. Indeed, the relation $AQ = QR$ implies that, for $1 \leq j \leq k$, we have

$$A q_j = \sum_{i=1}^{i=j} r_{ij} q_i.$$

If we let $Q_k = [q_1, q_2, \ldots, q_k]$ and if R_k is the principal leading submatrix of dimension k of R, the above relation can be rewritten as

$$A Q_k = Q_k R_k,$$

which is known as the partial Schur decomposition of A. The simplest case of this decomposition is when $k = 1$, in which case q_1 is an eigenvector. The vectors q_i are usually called Schur vectors. Schur vectors are not unique and depend, in particular, on the order chosen for the eigenvalues.

A slight variation of the Schur canonical form is the quasi-Schur form, also called the real Schur form. Here diagonal blocks of size 2×2 are allowed in the upper triangular matrix R. The reason for this is to avoid complex arithmetic when the original matrix is real. A 2×2 block is associated with each complex conjugate pair of eigenvalues of the matrix.

Example 1.2. Consider the 3×3 matrix

$$A = \begin{pmatrix} 1 & 10 & 0 \\ -1 & 3 & 1 \\ -1 & 0 & 1 \end{pmatrix}.$$

The matrix A has the pair of complex conjugate eigenvalues

$$2.4069\ldots \pm i \times 3.2110\ldots$$

and the real eigenvalue $0.1863\ldots$. The standard (complex) Schur form is given by the pair of matrices

$$V = \begin{pmatrix} 0.3381 - 0.8462i & 0.3572 - 0.1071i & 0.1749 \\ 0.3193 - 0.0105i & -0.2263 - 0.6786i & -0.6214 \\ 0.1824 + 0.1852i & -0.2659 - 0.5277i & 0.7637 \end{pmatrix}$$

and

$$S = \begin{pmatrix} 2.4069 + 3.2110i & 4.6073 - 4.7030i & -2.3418 - 5.2330i \\ 0 & 2.4069 - 3.2110i & -2.0251 - 1.2016i \\ 0 & 0 & 0.1863 \end{pmatrix}.$$

It is possible to avoid complex arithmetic by using the quasi-Schur form, which consists of the pair of matrices

$$U = \begin{pmatrix} -0.9768 & 0.1236 & 0.1749 \\ -0.0121 & 0.7834 & -0.6214 \\ 0.2138 & 0.6091 & 0.7637 \end{pmatrix}$$

and

$$R = \begin{pmatrix} 1.3129 & -7.7033 & 6.0407 \\ 1.4938 & 3.5008 & -1.3870 \\ 0 & 0 & 0.1863 \end{pmatrix}.$$

We conclude this section by pointing out that the Schur and the quasi-Schur forms of a given matrix are in no way unique. In addition to the dependence on the ordering of the eigenvalues, any column of Q can be multiplied by a complex sign $e^{i\theta}$ and a new corresponding R can be found. For the quasi-Schur form, there are infinitely many ways to select the 2×2 blocks, corresponding to applying arbitrary rotations to the columns of Q associated with these blocks.

1.8.4 Application to Powers of Matrices

The analysis of many numerical techniques is based on understanding the behavior of the successive powers A^k of a given matrix A. In this regard, the following theorem plays a fundamental role in numerical linear algebra, more particularly in the analysis of iterative methods.

Theorem 1.10. *The sequence* A^k, $k = 0, 1, \ldots$, *converges to zero iff* $\rho(A) < 1$.

Proof. To prove the necessary condition, assume that $A^k \to 0$ and consider u_1 a unit eigenvector associated with an eigenvalue λ_1 of maximum modulus. We have

$$A^k u_1 = \lambda_1^k u_1,$$

which implies, by taking the 2-norms of both sides,

$$|\lambda_1^k| = \|A^k u_1\|_2 \to 0.$$

This shows that $\rho(A) = |\lambda_1| < 1$.

1.8. Canonical Forms of Matrices

The Jordan canonical form must be used to show the sufficient condition. Assume that $\rho(A) < 1$. Start with the equality
$$A^k = X J^k X^{-1}.$$
To prove that A^k converges to zero, it is sufficient to show that J^k converges to zero. An important observation is that J^k preserves its block form. Therefore, it is sufficient to prove that each of the Jordan blocks converges to zero. Each block is of the form
$$J_i = \lambda_i I + E_i,$$
where E_i is a nilpotent matrix of index l_i; i.e., $E_i^{l_i} = 0$. Therefore, for $k \geq l_i$,
$$J_i^k = \sum_{j=0}^{l_1 - 1} \frac{k!}{j!(k-j)!} \lambda_i^{k-j} E_i^j.$$
Using the triangle inequality for any norm and taking $k \geq l_i$ yields
$$\|J_i^k\| \leq \sum_{j=0}^{l_1 - 1} \frac{k!}{j!(k-j)!} |\lambda_i|^{k-j} \|E_i^j\|.$$
Since $|\lambda_i| < 1$, each of the terms in this *finite* sum converges to zero as $k \to \infty$. Therefore, the matrix J_i^k converges to zero. □

An equally important result is stated in the following theorem.

Theorem 1.11. *The series*
$$\sum_{k=0}^{\infty} A^k$$
converges iff $\rho(A) < 1$. Under this condition, $I - A$ is nonsingular and the limit of the series is equal to $(I - A)^{-1}$.

Proof. The first part of the theorem is an immediate consequence of Theorem 1.10. Indeed, if the series converges, then $\|A^k\| \to 0$. By the previous theorem, this implies that $\rho(A) < 1$. To show that the converse is also true, use the equality
$$I - A^{k+1} = (I - A)(I + A + A^2 + \cdots + A^k)$$
and exploit the fact that, since $\rho(A) < 1$, then $I - A$ is nonsingular and, therefore,
$$(I - A)^{-1}(I - A^{k+1}) = I + A + A^2 + \cdots + A^k.$$
This shows that the series converges since the left-hand side will converge to $(I - A)^{-1}$. In addition, it shows the second part of the theorem. □

Another important consequence of the Jordan canonical form is a result that relates the spectral radius of a matrix to its matrix norm.

Theorem 1.12. *For any matrix norm $\|\cdot\|$, we have*
$$\lim_{k \to \infty} \|A^k\|^{1/k} = \rho(A).$$

Proof. The proof is a direct application of the Jordan canonical form and is the subject of Exercise 10. □

1.9 Normal and Hermitian Matrices

This section examines specific properties of normal and Hermitian matrices, including some optimality properties related to their spectra. The most common normal matrices that arise in practice are Hermitian and skew Hermitian.

1.9.1 Normal Matrices

By definition, a matrix is said to be normal if it commutes with its transpose conjugate, i.e., if it satisfies the relation

$$A^H A = A A^H. \tag{1.29}$$

An immediate property of normal matrices is stated in the following lemma.

Lemma 1.13. *If a normal matrix is triangular, then it is a diagonal matrix.*

Proof. Assume, for example, that A is upper triangular and normal. Compare the first diagonal element of the left-hand side matrix of (1.29) with the corresponding element of the matrix on the right-hand side. We obtain that

$$|a_{11}|^2 = \sum_{j=1}^{n} |a_{1j}|^2,$$

which shows that the elements of the first row are zeros except for the diagonal one. The same argument can now be used for the second row, the third row, and so on to the last row, to show that $a_{ij} = 0$ for $i \neq j$. $\square$

A consequence of this lemma is the following important result.

Theorem 1.14. *A matrix is normal iff it is unitarily similar to a diagonal matrix.*

Proof. It is straightforward to verify that a matrix that is unitarily similar to a diagonal matrix is normal. We now prove that any normal matrix A is unitarily similar to a diagonal matrix. Let $A = QRQ^H$ be the Schur canonical form of A, where Q is unitary and R is upper triangular. By the normality of A,

$$QR^H Q^H QRQ^H = QRQ^H QR^H Q^H$$

or

$$QR^H R Q^H = QRR^H Q^H.$$

Upon multiplication by Q^H on the left and Q on the right, this leads to the equality $R^H R = RR^H$, which means that R is normal and, according to the previous lemma, this is only possible if R is diagonal. $\square$

Thus, any normal matrix is diagonalizable and admits an orthonormal basis of eigenvectors, namely, the column vectors of Q.

The following result will be used in a later chapter. The question that is asked is, Assuming that any eigenvector of a matrix A is also an eigenvector of A^H, is A normal? If

1.9. Normal and Hermitian Matrices

A has a full set of eigenvectors, then the result is true and easy to prove. Indeed, if V is the $n \times n$ matrix of common eigenvectors, then $AV = VD_1$ and $A^H V = VD_2$, with D_1 and D_2 diagonal. Then $AA^H V = VD_1 D_2$ and $A^H AV = VD_2 D_1$ and, therefore, $AA^H = A^H A$. It turns out that the result is true in general, i.e., independent of the number of eigenvectors that A admits.

Lemma 1.15. *A matrix A is normal iff each of its eigenvectors is also an eigenvector of A^H.*

Proof. If A is normal, then its left and right eigenvectors are identical, so the sufficient condition is trivial. Assume now that a matrix A is such that each of its eigenvectors v_i, $i = 1, \ldots, k$ with $k \leq n$, is an eigenvector of A^H. For each eigenvector v_i of A, $Av_i = \lambda_i v_i$, and since v_i is also an eigenvector of A^H, then $A^H v_i = \mu v_i$. Observe that $(A^H v_i, v_i) = \mu(v_i, v_i)$ and, because $(A^H v_i, v_i) = (v_i, Av_i) = \bar{\lambda}_i (v_i, v_i)$, it follows that $\mu = \bar{\lambda}_i$. Next, it is proved by contradiction that there are no elementary divisors. Assume that the contrary is true for λ_i. Then the first principal vector u_i associated with λ_i is defined by

$$(A - \lambda_i I) u_i = v_i.$$

Taking the inner product of the above relation with v_i, we obtain

$$(Au_i, v_i) = \lambda_i (u_i, v_i) + (v_i, v_i). \tag{1.30}$$

On the other hand, it is also true that

$$(Au_i, v_i) = (u_i, A^H v_i) = (u_i, \bar{\lambda}_i v_i) = \lambda_i (u_i, v_i). \tag{1.31}$$

A result of (1.30) and (1.31) is that $(v_i, v_i) = 0$, which is a contradiction. Therefore, A has a full set of eigenvectors. This leads to the situation discussed just before the lemma, from which it is concluded that A must be normal. □

Clearly, Hermitian matrices are a particular case of normal matrices. Since a normal matrix satisfies the relation $A = QDQ^H$, with D diagonal and Q unitary, the eigenvalues of A are the diagonal entries of D. Therefore, if these entries are real it is clear that $A^H = A$. This is restated in the following corollary.

Corollary 1.16. *A normal matrix whose eigenvalues are real is Hermitian.*

As will be seen shortly, the converse is also true; i.e., a Hermitian matrix has real eigenvalues.
An eigenvalue λ of any matrix satisfies the relation

$$\lambda = \frac{(Au, u)}{(u, u)},$$

where u is an associated eigenvector. Generally, one might consider the complex scalars

$$\mu(x) = \frac{(Ax, x)}{(x, x)} \tag{1.32}$$

defined for any nonzero vector in $\mathbb{C}^n$. These ratios are known as *Rayleigh quotients* and are important for both theoretical and practical purposes. The set of all possible Rayleigh quotients as x runs over $\mathbb{C}^n$ is called the *field of values* of A. This set is clearly bounded since each $|\mu(x)|$ is bounded by the 2-norm of A; i.e., $|\mu(x)| \leq \|A\|_2$ for all x.

If a matrix is normal, then any vector x in $\mathbb{C}^n$ can be expressed as

$$\sum_{i=1}^{n} \xi_i q_i,$$

where the vectors q_i form an orthogonal basis of eigenvectors, and the expression for $\mu(x)$ becomes

$$\mu(x) = \frac{(Ax, x)}{(x, x)} = \frac{\sum_{k=1}^{n} \lambda_k |\xi_k|^2}{\sum_{k=1}^{n} |\xi_k|^2} \equiv \sum_{k=1}^{n} \beta_k \lambda_k, \tag{1.33}$$

where

$$0 \leq \beta_i = \frac{|\xi_i|^2}{\sum_{k=1}^{n} |\xi_k|^2} \leq 1 \quad \text{and} \quad \sum_{i=1}^{n} \beta_i = 1.$$

From a well-known characterization of convex hulls established by Hausdorff (known as *Hausdorff's convex hull theorem*), this means that the set of all possible Rayleigh quotients as x runs over all of $\mathbb{C}^n$ is equal to the convex hull of the λ_i's. This leads to the following theorem, which is stated without proof.

Theorem 1.17. *The field of values of a normal matrix is equal to the convex hull of its spectrum.*

The next question is whether or not this is also true for nonnormal matrices—the answer is no: the convex hull of the eigenvalues and the field of values of a nonnormal matrix are different in general. As a generic example, one can take any nonsymmetric real matrix that has real eigenvalues only. In this case, the convex hull of the spectrum is a real interval but its field of values will contain imaginary values. See Exercise 12 for another example. It has been shown (by a theorem shown by Hausdorff) that the field of values of a matrix is a convex set. Since the eigenvalues are members of the field of values, their convex hull is contained in the field of values. This is summarized in the following proposition.

Proposition 1.18. *The field of values of an arbitrary matrix is a convex set that contains the convex hull of its spectrum. It is equal to the convex hull of the spectrum when the matrix is normal.*

A useful definition based on the field of values is that of the *numerical radius*. The numerical radius $\nu(A)$ of an arbitrary matrix A is the radius of the smallest disk containing the field of values; i.e.,

$$\nu(A) = \max_{x \in \mathbb{C}^n} |\mu(x)|.$$

It is easy to see that

$$\rho(A) \leq \nu(A) \leq \|A\|_2.$$

1.9. Normal and Hermitian Matrices

The spectral radius and numerical radius are identical for normal matrices. It can also be easily shown (see Exercise 21) that $\nu(A) \geq \|A\|_2/2$, which means that

$$\frac{\|A\|_2}{2} \leq \nu(A) \leq \|A\|_2. \tag{1.34}$$

The numerical radius is a vector norm; i.e., it satisfies (1.8)–(1.10), but it is not consistent (see Exercise 22). However, it satisfies the power inequality (see [171, p. 333]):

$$\nu(A^k) \leq \nu(A)^k. \tag{1.35}$$

1.9.2 Hermitian Matrices

A first result on Hermitian matrices is the following.

Theorem 1.19. *The eigenvalues of a Hermitian matrix are real; i.e., $\sigma(A) \subset \mathbb{R}$.*

Proof. Let λ be an eigenvalue of A and u an associated eigenvector or 2-norm unity. Then

$$\lambda = (Au, u) = (u, Au) = \overline{(Au, u)} = \bar{\lambda},$$

which is the stated result. $\square$

It is not difficult to see that if, in addition, the matrix is real, then the eigenvectors can be chosen to be real; see Exercise 24. Since a Hermitian matrix is normal, the following is a consequence of Theorem 1.14.

Theorem 1.20. *Any Hermitian matrix is unitarily similar to a real diagonal matrix.*

In particular, a Hermitian matrix admits a set of orthonormal eigenvectors that form a basis of $\mathbb{C}^n$.

In the proof of Theorem 1.17 we used the fact that the inner products (Au, u) are real. Generally, it is clear that any Hermitian matrix is such that (Ax, x) is real for any vector $x \in \mathbb{C}^n$. It turns out that the converse is also true; i.e., it can be shown that if (Az, z) is real for all vectors z in $\mathbb{C}^n$, then the matrix A is Hermitian (see Exercise 15).

Eigenvalues of Hermitian matrices can be characterized by optimality properties of the Rayleigh quotients (1.32). The best known of these is the min-max principle. We now label all the eigenvalues of A in descending order:

$$\lambda_1 \geq \lambda_2 \geq \cdots \geq \lambda_n.$$

Here, the eigenvalues are not necessarily distinct and they are repeated, each according to its multiplicity. In the following theorem, known as the *min-max theorem*, S represents a generic subspace of $\mathbb{C}^n$.

Theorem 1.21. *The eigenvalues of a Hermitian matrix A are characterized by the relation*

$$\lambda_k = \min_{S,\ \dim(S)=n-k+1} \max_{x \in S, x \neq 0} \frac{(Ax, x)}{(x, x)}. \tag{1.36}$$

Proof. Let $\{q_i\}_{i=1,...,n}$ be an orthonormal basis of $\mathbb{C}^n$ consisting of eigenvectors of A associated with $\lambda_1, \ldots, \lambda_n$, respectively. Let S_k be the subspace spanned by the first k of these vectors and denote by $\mu(S)$ the maximum of $(Ax, x)/(x, x)$ over all nonzero vectors of a subspace S. Since the dimension of S_k is k, a well-known theorem of linear algebra shows that its intersection with any subspace S of dimension $n - k + 1$ is not reduced to $\{0\}$; i.e., there is a vector x in $S \cap S_k$. For this $x = \sum_{i=1}^{k} \xi_i q_i$, we have

$$\frac{(Ax, x)}{(x, x)} = \frac{\sum_{i=1}^{k} \lambda_i |\xi_i|^2}{\sum_{i=1}^{k} |\xi_i|^2} \geq \lambda_k,$$

so that $\mu(S) \geq \lambda_k$.

Consider, on the other hand, the particular subspace S_0 of dimension $n - k + 1$ that is spanned by $q_k, \ldots, q_n$. For each vector x in this subspace, we have

$$\frac{(Ax, x)}{(x, x)} = \frac{\sum_{i=k}^{n} \lambda_i |\xi_i|^2}{\sum_{i=k}^{n} |\xi_i|^2} \leq \lambda_k,$$

so that $\mu(S_0) \leq \lambda_k$. In other words, as S runs over all the $(n - k + 1)$-dimensional subspaces, $\mu(S)$ is never less than λ_k and there is at least one subspace S_0 for which $\mu(S_0) \leq \lambda_k$. This shows the desired result. □

The above result is often called the Courant–Fisher min-max principle or theorem. As a particular case, the largest eigenvalue of A satisfies

$$\lambda_1 = \max_{x \neq 0} \frac{(Ax, x)}{(x, x)}. \tag{1.37}$$

Actually, there are four different ways of rewriting the above characterization. The second formulation is

$$\lambda_k = \max_{S,\ \dim(S)=k} \min_{x \in S, x \neq 0} \frac{(Ax, x)}{(x, x)} \tag{1.38}$$

and the two other ones can be obtained from (1.36) and (1.38) by simply relabeling the eigenvalues increasingly instead of decreasingly. Thus, with our labeling of the eigenvalues in descending order, (1.38) tells us that the smallest eigenvalue satisfies

$$\lambda_n = \min_{x \neq 0} \frac{(Ax, x)}{(x, x)}, \tag{1.39}$$

with λ_n replaced by λ_1 if the eigenvalues are relabeled increasingly.

In order for all the eigenvalues of a Hermitian matrix to be positive, it is necessary and sufficient that

$$(Ax, x) > 0 \quad \forall\, x \in \mathbb{C}^n, \quad x \neq 0.$$

Such a matrix is called *positive definite*. A matrix that satisfies $(Ax, x) \geq 0$ for any x is said to be *positive semidefinite*. In particular, the matrix $A^H A$ is semipositive definite for any rectangular matrix, since

$$(A^H A x, x) = (Ax, Ax) \geq 0 \quad \forall\, x.$$

Similarly, AA^H is also a Hermitian semipositive definite matrix. The square roots of the eigenvalues of $A^H A$ for a general rectangular matrix A are called the *singular values* of A and are denoted by σ_i. In Section 1.5, we stated without proof that the 2-norm of any matrix A is equal to the largest singular value σ_1 of A. This is now an obvious fact, because

$$\|A\|_2^2 = \max_{x \neq 0} \frac{\|Ax\|_2^2}{\|x\|_2^2} = \max_{x \neq 0} \frac{(Ax, Ax)}{(x, x)} = \max_{x \neq 0} \frac{(A^H Ax, x)}{(x, x)} = \sigma_1^2,$$

which results from (1.37).

Another characterization of eigenvalues, known as the Courant characterization, is stated in the next theorem. In contrast with the min-max theorem, this property is recursive in nature.

Theorem 1.22. *The eigenvalue λ_i and the corresponding eigenvector q_i of a Hermitian matrix are such that*

$$\lambda_1 = \frac{(Aq_1, q_1)}{(q_1, q_1)} = \max_{x \in \mathbb{C}^n, x \neq 0} \frac{(Ax, x)}{(x, x)}$$

and, for $k > 1$,

$$\lambda_k = \frac{(Aq_k, q_k)}{(q_k, q_k)} = \max_{x \neq 0, q_1^H x = \cdots = q_{k-1}^H x = 0} \frac{(Ax, x)}{(x, x)}. \quad (1.40)$$

In other words, the maximum of the Rayleigh quotient over a subspace that is orthogonal to the first $k - 1$ eigenvectors is equal to λ_k and is achieved for the eigenvector q_k associated with λ_k. The proof follows easily from the expansion (1.33) of the Rayleigh quotient.

1.10 Nonnegative Matrices, *M*-Matrices

Nonnegative matrices play a crucial role in the theory of matrices. They are important in the study of convergence of iterative methods and arise in many applications, including economics, queuing theory, and chemical engineering.

A *nonnegative matrix* is simply a matrix whose entries are nonnegative. More generally, a partial order relation can be defined on the set of matrices.

Definition 1.23. *Let A and B be two $n \times m$ matrices. Then*

$$A \leq B$$

if, by definition, $a_{ij} \leq b_{ij}$ for $1 \leq i \leq n$, $1 \leq j \leq m$. If O denotes the $n \times m$ zero matrix, then A is nonnegative if $A \geq O$ and positive if $A > O$. Similar definitions hold in which "positive" is replaced by "negative."

The binary relation $\leq$ imposes only a *partial* order on $\mathbb{R}^{n \times m}$, since two arbitrary matrices in $\mathbb{R}^{n \times m}$ are not necessarily comparable by this relation. For the remainder of this section, we assume that only square matrices are involved. The next proposition lists a number of rather trivial properties regarding the partial order relation just defined.

Proposition 1.24. *The following properties hold:*

1. *The relation $\leq$ for matrices is reflexive ($A \leq A$), antisymmetric (if $A \leq B$ and $B \leq A$, then $A = B$), and transitive (if $A \leq B$ and $B \leq C$, then $A \leq C$).*

2. *If A and B are nonnegative, then so is their product AB and their sum $A + B$.*

3. *If A is nonnegative, then so is A^k.*

4. *If $A \leq B$, then $A^T \leq B^T$.*

5. *If $O \leq A \leq B$, then $\|A\|_1 \leq \|B\|_1$ and similarly $\|A\|_\infty \leq \|B\|_\infty$.*

The proof of these properties is left to Exercise 26.

A matrix is said to be *reducible* if there is a permutation matrix P such that PAP^T is block upper triangular. Otherwise, it is *irreducible*. An important result concerning nonnegative matrices is the following theorem known as the Perron–Frobenius theorem.

Theorem 1.25. *Let A be a real $n \times n$ nonnegative irreducible matrix. Then $\lambda \equiv \rho(A)$, the spectral radius of A, is a simple eigenvalue of A. Moreover, there exists an eigenvector u with positive elements associated with this eigenvalue.*

A relaxed version of this theorem allows the matrix to be reducible but the conclusion is somewhat weakened in the sense that the elements of the eigenvectors are only guaranteed to be *nonnegative*.

Next, a useful property is established.

Proposition 1.26. *Let A, B, C be nonnegative matrices, with $A \leq B$. Then*

$$AC \leq BC \quad \text{and} \quad CA \leq CB.$$

Proof. Consider the first inequality only, since the proof for the second is identical. The result that is claimed translates into

$$\sum_{k=1}^n a_{ik} c_{kj} \leq \sum_{k=1}^n b_{ik} c_{kj}, \quad 1 \leq i, j \leq n,$$

which is clearly true by the assumptions. □

A consequence of the proposition is the following corollary.

Corollary 1.27. *Let A and B be two nonnegative matrices, with $A \leq B$. Then*

$$A^k \leq B^k \quad \forall\, k \geq 0. \tag{1.41}$$

Proof. The proof is by induction. The inequality is clearly true for $k = 0$. Assume that (1.41) is true for k. According to the previous proposition, multiplying (1.41) from the left by A results in

$$A^{k+1} \leq AB^k. \tag{1.42}$$

1.10. Nonnegative Matrices, M-Matrices

Now, it is clear that if $B \geq 0$, then also $B^k \geq 0$, by Proposition 1.24. We now multiply both sides of the inequality $A \leq B$ by B^k to the right and obtain

$$AB^k \leq B^{k+1}. \tag{1.43}$$

The inequalities (1.42) and (1.43) show that $A^{k+1} \leq B^{k+1}$, which completes the induction proof. □

A theorem with important consequences on the analysis of iterative methods will now be stated.

Theorem 1.28. *Let A and B be two square matrices that satisfy the inequalities*

$$O \leq A \leq B. \tag{1.44}$$

Then

$$\rho(A) \leq \rho(B). \tag{1.45}$$

Proof. The proof is based on the following equality stated in Theorem 1.12:

$$\rho(X) = \lim_{k \to \infty} \|X^k\|^{1/k}$$

for any matrix norm. Choosing the 1-norm, for example, we have, from the last property in Proposition 1.24,

$$\rho(A) = \lim_{k \to \infty} \|A^k\|_1^{1/k} \leq \lim_{k \to \infty} \|B^k\|_1^{1/k} = \rho(B),$$

which completes the proof. □

Theorem 1.29. *Let B be a nonnegative matrix. Then $\rho(B) < 1$ iff $I - B$ is nonsingular and $(I - B)^{-1}$ is nonnegative.*

Proof. Define $C = I - B$. If it is assumed that $\rho(B) < 1$, then, by Theorem 1.11, $C = I - B$ is nonsingular and

$$C^{-1} = (I - B)^{-1} = \sum_{i=0}^{\infty} B^i. \tag{1.46}$$

In addition, since $B \geq 0$, all the powers of B as well as their sum in (1.46) are also nonnegative.

To prove the sufficient condition, assume that C is nonsingular and that its inverse is nonnegative. By the Perron–Frobenius theorem, there is a nonnegative eigenvector u associated with $\rho(B)$, which is an eigenvalue; i.e.,

$$Bu = \rho(B)u$$

or, equivalently,

$$C^{-1}u = \frac{1}{1 - \rho(B)}u.$$

Since u and C^{-1} are nonnegative and $I - B$ is nonsingular, this shows that $1 - \rho(B) > 0$, which is the desired result. □

Definition 1.30. *A matrix is said to be an M-matrix if it satisfies the following four properties:*

1. $a_{i,i} > 0$ for $i = 1, \ldots, n$.
2. $a_{i,j} \leq 0$ for $i \neq j$, $i, j = 1, \ldots, n$.
3. *A is nonsingular.*
4. $A^{-1} \geq 0$.

In reality, the four conditions in the above definition are somewhat redundant and equivalent conditions that are more rigorous will be given later. Let A be any matrix that satisfies properties (1) and (2) in the above definition and let D be the diagonal of A. Since $D > 0$,

$$A = D - (D - A) = D\left(I - (I - D^{-1}A)\right).$$

Now define

$$B \equiv I - D^{-1}A.$$

Using the previous theorem, $I - B = D^{-1}A$ is nonsingular and $(I - B)^{-1} = A^{-1}D \geq 0$ iff $\rho(B) < 1$. It is now easy to see that conditions (3) and (4) of Definition 1.30 can be replaced with the condition $\rho(B) < 1$.

Theorem 1.31. *Let a matrix A be given such that*

1. $a_{i,i} > 0$ for $i = 1, \ldots, n$;
2. $a_{i,j} \leq 0$ for $i \neq j$, $i, j = 1, \ldots, n$.

Then A is an M-matrix iff

3. $\rho(B) < 1$, where $B = I - D^{-1}A$.

Proof. From the above argument, an immediate application of Theorem 1.29 shows that properties (3) and (4) of Definition 1.30 are equivalent to $\rho(B) < 1$, where $B = I - C$ and $C = D^{-1}A$. In addition, C is nonsingular iff A is and C^{-1} is nonnegative iff A is. □

The next theorem shows that condition (1) of Definition 1.30 is implied by its other three conditions.

Theorem 1.32. *Let a matrix A be given such that*

1. $a_{i,j} \leq 0$ for $i \neq j$, $i, j = 1, \ldots, n$;
2. *A is nonsingular;*
3. $A^{-1} \geq 0$.

Then

4. $a_{i,i} > 0$ for $i = 1, \ldots, n$; i.e., *A is an M-matrix;*
5. $\rho(B) < 1$, where $B = I - D^{-1}A$.

Proof. Define $C \equiv A^{-1}$. Writing that $(AC)_{ii} = 1$ yields

$$\sum_{k=1}^{n} a_{ik}c_{ki} = 1,$$

which gives

$$a_{ii}c_{ii} = 1 - \sum_{\substack{k=1 \\ k \neq i}}^{n} a_{ik}c_{ki}.$$

Since $a_{ik}c_{ki} \leq 0$ for all k, the right-hand side is not less than 1 and, since $c_{ii} \geq 0$, then $a_{ii} > 0$. The second part of the result now follows immediately from an application of Theorem 1.31. □

Finally, this useful result follows.

Theorem 1.33. *Let A, B be two matrices that satisfy*

1. $A \leq B$,
2. $b_{ij} \leq 0$ *for all* $i \neq j$.

Then, if A is an M-matrix, so is the matrix B.

Proof. Assume that A is an M-matrix and let D_X denote the diagonal of a matrix X. The matrix D_B is positive because

$$D_B \geq D_A > 0.$$

Consider now the matrix $I - D_B^{-1}B$. Since $A \leq B$, then

$$D_A - A \geq D_B - B \geq O,$$

which, upon multiplying through by D_A^{-1}, yields

$$I - D_A^{-1}A \geq D_A^{-1}(D_B - B) \geq D_B^{-1}(D_B - B) = I - D_B^{-1}B \geq O.$$

Since the matrices $I - D_B^{-1}B$ and $I - D_A^{-1}A$ are nonnegative, Theorems 1.28 and 1.31 imply that

$$\rho(I - D_B^{-1}B) \leq \rho(I - D_A^{-1}A) < 1.$$

This establishes the result by using Theorem 1.31 once again. □

1.11 Positive Definite Matrices

A *real* matrix is said to be *positive definite* or *positive real* if

$$(Au, u) > 0 \quad \forall u \in \mathbb{R}^n, \ u \neq 0. \tag{1.47}$$

It must be emphasized that this definition is only useful when formulated entirely for real variables. Indeed, if u were not restricted to be real, then assuming that (Au, u) is real for all u complex would imply that A is Hermitian; see Exercise 15. If, in addition to the definition stated by (1.48), A is symmetric (real), then A is said to be *symmetric positive definite* (SPD). Similarly, if A is Hermitian, then A is said to be *Hermitian positive definite* (HPD). Some properties of HPD matrices were seen in Section 1.9, in particular with regard to their eigenvalues. Now the more general case where A is non-Hermitian and positive definite is considered.

We begin with the observation that any square matrix (real or complex) can be decomposed as

$$A = H + iS, \qquad (1.48)$$

in which

$$H = \frac{1}{2}(A + A^H), \qquad (1.49)$$

$$S = \frac{1}{2i}(A - A^H). \qquad (1.50)$$

Note that both H and S are Hermitian while the matrix iS in the decomposition (1.48) is skew Hermitian. The matrix H in the decomposition is called the *Hermitian part* of A, while the matrix iS is the *skew-Hermitian part* of A. The above decomposition is the analogue of the decomposition of a complex number z into $z = x + iy$:

$$x = \Re e(z) = \frac{1}{2}(z + \bar{z}), \quad y = \Im m(z) = \frac{1}{2i}(z - \bar{z}).$$

When A is real and u is a real vector, then (Au, u) is real and, as a result, the decomposition (1.48) immediately gives the equality

$$(Au, u) = (Hu, u). \qquad (1.51)$$

This results in the following theorem.

Theorem 1.34. *Let A be a real positive definite matrix. Then A is nonsingular. In addition, there exists a scalar $\alpha > 0$ such that*

$$(Au, u) \geq \alpha \|u\|_2^2 \qquad (1.52)$$

for any real vector u.

Proof. The first statement is an immediate consequence of the definition of positive definiteness. Indeed, if A were singular, then there would be a nonzero vector such that $Au = 0$ and, as a result, $(Au, u) = 0$ for this vector, which would contradict (1.47). We now prove the second part of the theorem. From (1.51) and the fact that A is positive definite, we conclude that H is HPD. Hence, from (1.39), based on the min-max theorem, we get

$$\min_{u \neq 0} \frac{(Au, u)}{(u, u)} = \min_{u \neq 0} \frac{(Hu, u)}{(u, u)} \geq \lambda_{min}(H) > 0.$$

Taking $\alpha \equiv \lambda_{min}(H)$ yields the desired inequality (1.52). $\square$

A simple yet important result that locates the eigenvalues of A in terms of the spectra of H and S can now be proved.

1.11. Positive Definite Matrices

Theorem 1.35. *Let A be any square (possibly complex) matrix and let $H = \frac{1}{2}(A + A^H)$ and $S = \frac{1}{2i}(A - A^H)$. Then any eigenvalue λ_j of A is such that*

$$\lambda_{min}(H) \leq \Re e(\lambda_j) \leq \lambda_{max}(H), \quad (1.53)$$
$$\lambda_{min}(S) \leq \Im m(\lambda_j) \leq \lambda_{max}(S). \quad (1.54)$$

Proof. When the decomposition (1.48) is applied to the Rayleigh quotient of the eigenvector u_j associated with λ_j, we obtain

$$\lambda_j = (Au_j, u_j) = (Hu_j, u_j) + i(Su_j, u_j), \quad (1.55)$$

assuming that $\|u_j\|_2 = 1$. This leads to

$$\Re e(\lambda_j) = (Hu_j, u_j),$$
$$\Im m(\lambda_j) = (Su_j, u_j).$$

The result follows using properties established in Section 1.9. □

Thus, the eigenvalues of a matrix are contained in a rectangle defined by the eigenvalues of its Hermitian and non-Hermitian parts. In the particular case where A is real, then iS is skew Hermitian and its eigenvalues form a set that is symmetric with respect to the real axis in the complex plane. Indeed, in this case, iS is real and its eigenvalues come in conjugate pairs.

Note that all the arguments herein are based on the field of values and, therefore, they provide ways to localize the eigenvalues of A from knowledge of the field of values. However, this approximation can be inaccurate in some cases.

Example 1.3. Consider the matrix

$$A = \begin{pmatrix} 1 & 1 \\ 10^4 & 1 \end{pmatrix}.$$

The eigenvalues of A are -99 and 101. Those of H are $1 \pm (10^4 + 1)/2$ and those of iS are $\pm i(10^4 - 1)/2$.

When a matrix B is SPD, the mapping

$$x, y \quad \rightarrow \quad (x, y)_B \equiv (Bx, y) \quad (1.56)$$

from $\mathbb{C}^n \times \mathbb{C}^n$ to $\mathbb{C}$ is a proper inner product on $\mathbb{C}^n$ in the sense defined in Section 1.4. The associated norm is often referred to as the *energy norm* or *A-norm*. Sometimes, it is possible to find an appropriate HPD matrix B that makes a given matrix A Hermitian, i.e., such that

$$(Ax, y)_B = (x, Ay)_B \quad \forall x, y,$$

although A is a non-Hermitian matrix with respect to the Euclidean inner product. The simplest examples are $A = B^{-1}C$ and $A = CB$, where C is Hermitian and B is HPD.

1.12 Projection Operators

Projection operators or *projectors* play an important role in numerical linear algebra, particularly in iterative methods for solving various matrix problems. This section introduces these operators from a purely algebraic point of view and gives a few of their important properties.

1.12.1 Range and Null Space of a Projector

A projector P is any linear mapping from $\mathbb{C}^n$ to itself that is idempotent, i.e., such that

$$P^2 = P.$$

A few simple properties follow from this definition. First, if P is a projector, then so is $(I - P)$, and the following relation holds:

$$\text{Ker}(P) = \text{Ran}(I - P). \quad (1.57)$$

In addition, the two subspaces $\text{Ker}(P)$ and $\text{Ran}(P)$ intersect only at the element zero. Indeed, if a vector x belongs to $\text{Ran}(P)$, then $Px = x$ by the idempotence property. If it is also in $\text{Ker}(P)$, then $Px = 0$. Hence, $x = Px = 0$, which proves the result. Moreover, every element of $\mathbb{C}^n$ can be written as $x = Px + (I - P)x$. Therefore, the space $\mathbb{C}^n$ can be decomposed as the direct sum

$$\mathbb{C}^n = \text{Ker}(P) \oplus \text{Ran}(P).$$

Conversely, every pair of subspaces M and S that forms a direct sum of $\mathbb{C}^n$ defines a unique projector such that $\text{Ran}(P) = M$ and $\text{Ker}(P) = S$. This associated projector P maps an element x of $\mathbb{C}^n$ into the component x_1, where x_1 is the M component in the unique decomposition $x = x_1 + x_2$ associated with the direct sum.

In fact, this association is unique; that is, an arbitrary projector P can be entirely determined by two subspaces: (1) the range M of P and (2) its null space S, which is also the range of $I - P$. For any x, the vector Px satisfies the conditions

$$Px \in M,$$
$$x - Px \in S.$$

The linear mapping P is said to project x *onto* M and *along* or *parallel to* the subspace S. If P is of rank m, then the range of $I - P$ is of dimension $n - m$. Therefore, it is natural to define S through its orthogonal complement $L = S^\perp$, which has dimension m. The above conditions, which define $u = Px$ for any x, become

$$u \in M, \quad (1.58)$$
$$x - u \perp L. \quad (1.59)$$

These equations define a projector P onto M and *orthogonal* to the subspace L. The first statement, (1.58), establishes the m degrees of freedom, while the second, (1.59), gives the m constraints that define Px from these degrees of freedom. The general definition of projectors is illustrated in Figure 1.1.

1.12. Projection Operators

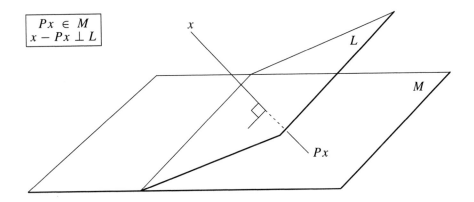

Figure 1.1. *Projection of x onto M and orthogonal to L.*

The question now is, Given two arbitrary subspaces M and L, both of dimension m, is it always possible to define a projector onto M orthogonal to L through the conditions (1.58) and (1.59)? The following lemma answers this question.

Lemma 1.36. *Given two subspaces M and L of the same dimension m, the following two conditions are mathematically equivalent:*

i. *No nonzero vector of M is orthogonal to L.*

ii. *For any x in $\mathbb{C}^n$ there is a unique vector u that satisfies the conditions (1.58) and (1.59).*

Proof. The first condition states that any vector that is in M and also orthogonal to L must be the zero vector. It is equivalent to the condition

$$M \cap L^\perp = \{0\}.$$

Since L is of dimension m, $L^\perp$ is of dimension $n - m$ and the above condition is equivalent to the condition that

$$\mathbb{C}^n = M \oplus L^\perp. \tag{1.60}$$

This in turn is equivalent to the statement that, for any x, there exists a unique pair of vectors u, w such that

$$x = u + w,$$

where u belongs to M and $w = x - u$ belongs to $L^\perp$, a statement that is identical to (ii). □

In summary, given two subspaces M and L satisfying the condition $M \cap L^\perp = \{0\}$, there is a projector P onto M orthogonal to L that defines the projected vector u of any vector x from (1.58) and (1.59). This projector is such that

$$\text{Ran}(P) = M, \qquad \text{Ker}(P) = L^\perp.$$

In particular, the condition $Px = 0$ translates into $x \in \text{Ker}(P)$, which means that $x \in L^\perp$. The converse is also true. Hence we have the following useful property:

$$Px = 0 \quad \text{iff} \quad x \perp L. \tag{1.61}$$

1.12.2 Matrix Representations

Two bases are required to obtain a matrix representation of a general projector: a basis $V = [v_1, \ldots, v_m]$ for the subspace $M = \text{Ran}(P)$ and a second one $W = [w_1, \ldots, w_m]$ for the subspace L. These two bases are *biorthogonal* when

$$(v_i, w_j) = \delta_{ij}. \tag{1.62}$$

In matrix form this means $W^H V = I$. Since Px belongs to M, let Vy be its representation in the V basis. The constraint $x - Px \perp L$ is equivalent to the condition

$$((x - Vy), w_j) = 0 \quad \text{for } j = 1, \ldots, m.$$

In matrix form, this can be rewritten as

$$W^H(x - Vy) = 0. \tag{1.63}$$

If the two bases are biorthogonal, then it follows that $y = W^H x$. Therefore, in this case, $Px = VW^H x$, which yields the matrix representation of P:

$$P = VW^H. \tag{1.64}$$

In case the bases V and W are not biorthogonal, then it is easily seen from the condition (1.63) that

$$P = V(W^H V)^{-1} W^H. \tag{1.65}$$

If we assume that no vector of M is orthogonal to L, then it can be shown that the $m \times m$ matrix $W^H V$ is nonsingular.

1.12.3 Orthogonal and Oblique Projectors

An important class of projectors is obtained in the case when the subspace L is equal to M, i.e., when

$$\text{Ker}(P) = \text{Ran}(P)^\perp.$$

Then the projector P is said to be the *orthogonal projector* onto M. A projector that is not orthogonal is *oblique*. Thus, an orthogonal projector is defined through the following requirements satisfied for any vector x:

$$Px \in M \quad \text{and} \quad (I - P)x \perp M \tag{1.66}$$

or, equivalently,

$$Px \in M \quad \text{and} \quad ((I - P)x, y) = 0 \quad \forall y \in M.$$

It is interesting to consider the mapping P^H defined as the adjoint of P:

$$(P^H x, y) = (x, Py) \quad \forall x, \forall y. \tag{1.67}$$

The above condition is illustrated in Figure 1.2.

1.12. Projection Operators

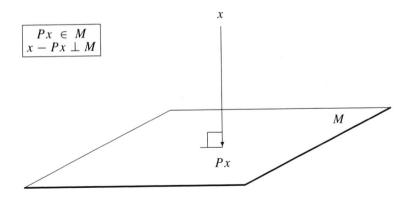

Figure 1.2. *Orthogonal projection of x onto a subspace M.*

First, note that P^H is also a projector because, for all x and y,

$$((P^H)^2 x, y) = (P^H x, Py) = (x, P^2 y) = (x, Py) = (P^H x, y).$$

A consequence of the relation (1.67) is

$$\text{Ker}(P^H) = \text{Ran}(P)^\perp, \tag{1.68}$$
$$\text{Ker}(P) = \text{Ran}(P^H)^\perp. \tag{1.69}$$

The above relations lead to the following proposition.

Proposition 1.37. *A projector is orthogonal iff it is Hermitian.*

Proof. By definition, an orthogonal projector is one for which $\text{Ker}(P) = \text{Ran}(P)^\perp$. Therefore, by (1.68), if P is Hermitian, then it is orthogonal. Conversely, if P is orthogonal, then (1.68) implies $\text{Ker}(P) = \text{Ker}(P^H)$ while (1.69) implies $\text{Ran}(P) = \text{Ran}(P^H)$. Since P^H is a projector and since projectors are uniquely determined by their range and null spaces, this implies that $P = P^H$. □

Given *any* unitary $n \times m$ matrix V whose columns form an orthonormal basis of $M = \text{Ran}(P)$, we can represent P by the matrix $P = VV^H$. This is a particular case of the matrix representation of projectors (1.64). In addition to being idempotent, the linear mapping associated with this matrix satisfies the characterization given above; i.e.,

$$VV^H x \in M \quad \text{and} \quad (I - VV^H)x \in M^\perp.$$

It is important to note that this representation of the orthogonal projector P is not unique. In fact, any orthonormal basis V will give a different representation of P in the above form. As a consequence, for any two orthogonal bases V_1, V_2 of M, we must have $V_1 V_1^H = V_2 V_2^H$, an equality that can also be verified independently; see Exercise 30.

1.12.4 Properties of Orthogonal Projectors

When P is an orthogonal projector, then the two vectors Px and $(I - P)x$ in the decomposition $x = Px + (I - P)x$ are orthogonal. The following relation results:

$$\|x\|_2^2 = \|Px\|_2^2 + \|(I - P)x\|_2^2.$$

A consequence of this is that, for any x,

$$\|Px\|_2 \leq \|x\|_2.$$

Thus, the maximum of $\|Px\|_2/\|x\|_2$ for all x in $\mathbb{C}^n$ does not exceed one. In addition, the value one is reached for any element in Ran(P). Therefore,

$$\|P\|_2 = 1$$

for any orthogonal projector P.

An orthogonal projector has only two eigenvalues: zero or one. Any vector of the range of P is an eigenvector associated with the eigenvalue one. Any vector of the null space is obviously an eigenvector associated with the eigenvalue zero.

Next, an important optimality property of orthogonal projectors is established.

Theorem 1.38. *Let P be the orthogonal projector onto a subspace M. Then, for any given vector x in $\mathbb{C}^n$, the following is true:*

$$\min_{y \in M} \|x - y\|_2 = \|x - Px\|_2. \tag{1.70}$$

Proof. Let y be any vector of M and consider the square of its distance from x. Since $x - Px$ is orthogonal to M, to which $Px - y$ belongs, then

$$\|x - y\|_2^2 = \|x - Px + (Px - y)\|_2^2 = \|x - Px\|_2^2 + \|(Px - y)\|_2^2.$$

Therefore, $\|x - y\|_2 \geq \|x - Px\|_2$ for all y in M. Thus, we establish the result by noticing that the minimum is reached for $y = Px$. □

By expressing the conditions that define $y^* \equiv Px$ for an orthogonal projector P onto a subspace M, it is possible to reformulate the above result in the form of necessary and sufficient conditions that enable us to determine the best approximation to a given vector x in the least-squares sense.

Corollary 1.39. *Let a subspace M and a vector x in $\mathbb{C}^n$ be given. Then*

$$\min_{y \in M} \|x - y\|_2 = \|x - y^*\|_2 \tag{1.71}$$

iff the following two conditions are satisfied:

$$\begin{cases} y^* & \in \quad M, \\ x - y^* & \perp \quad M. \end{cases}$$

1.13 Basic Concepts in Linear Systems

Linear systems are among the most important and common problems encountered in scientific computing. From the theoretical point of view, it is well understood when a solution exists, when it does not, and when there are infinitely many solutions. In addition, explicit expressions of the solution using determinants exist. However, the numerical viewpoint is far more complex. Approximations may be available but it may be difficult to estimate how accurate they are. This clearly will depend on the data at hand, i.e., primarily the coefficient matrix. This section gives a very brief overview of the existence theory as well as the sensitivity of the solutions.

1.13.1 Existence of a Solution

Consider the *linear system*

$$Ax = b. \tag{1.72}$$

Here, x is termed the *unknown* and b the *right-hand side*. When solving the linear system (1.72), we distinguish three situations.

Case 1 The matrix A is nonsingular. There is a unique solution given by $x = A^{-1}b$.

Case 2 The matrix A is singular and $b \in \text{Ran}(A)$. Since $b \in \text{Ran}(A)$, there is an x_0 such that $Ax_0 = b$. Then $x_0 + v$ is also a solution for any v in $\text{Ker}(A)$. Since $\text{Ker}(A)$ is at least one-dimensional, there are infinitely many solutions.

Case 3 The matrix A is singular and $b \notin \text{Ran}(A)$. There are no solutions.

Example 1.4. The simplest illustration of the above three cases is with small diagonal matrices. Let

$$A = \begin{pmatrix} 2 & 0 \\ 0 & 4 \end{pmatrix}, \quad b = \begin{pmatrix} 1 \\ 8 \end{pmatrix}.$$

Then A is nonsingular and there is a unique x given by

$$x = \begin{pmatrix} 0.5 \\ 2 \end{pmatrix}.$$

Now let

$$A = \begin{pmatrix} 2 & 0 \\ 0 & 0 \end{pmatrix}, \quad b = \begin{pmatrix} 1 \\ 0 \end{pmatrix}.$$

Then A is singular and, as is easily seen, $b \in \text{Ran}(A)$. For example, a particular element x_0 such that $Ax_0 = b$ is $x_0 = \begin{pmatrix} 0.5 \\ 0 \end{pmatrix}$. The null space of A consists of all vectors whose first component is zero, i.e., all vectors of the form $\begin{pmatrix} 0 \\ \alpha \end{pmatrix}$. Therefore, there are infinitely many solutions given by

$$x(\alpha) = \begin{pmatrix} 0.5 \\ \alpha \end{pmatrix} \quad \forall \alpha.$$

Finally, let A be the same as in the previous case, but define the right-hand side as

$$b = \begin{pmatrix} 1 \\ 1 \end{pmatrix}.$$

In this case there are no solutions because the second equation cannot be satisfied.

1.13.2 Perturbation Analysis

Consider the linear system (1.72), where A is an $n \times n$ nonsingular matrix. Given any matrix E, the matrix $A(\epsilon) = A + \epsilon E$ is nonsingular for ϵ small enough, i.e., for $\epsilon \leq \alpha$, where α is some small number; see Exercise 37. Assume that we perturb the data in the above system, i.e., that we perturb the matrix A by ϵE and the right-hand side b by ϵe. The solution $x(\epsilon)$ of the perturbed system satisfies the equation

$$(A + \epsilon E)x(\epsilon) = b + \epsilon e. \quad (1.73)$$

Let $\delta(\epsilon) = x(\epsilon) - x$. Then

$$(A + \epsilon E)\delta(\epsilon) = (b + \epsilon e) - (A + \epsilon E)x$$
$$= \epsilon(e - Ex),$$
$$\delta(\epsilon) = \epsilon(A + \epsilon E)^{-1}(e - Ex).$$

As an immediate result, the function $x(\epsilon)$ is differentiable at $\epsilon = 0$ and its derivative is given by

$$x'(0) = \lim_{\epsilon \to 0} \frac{\delta(\epsilon)}{\epsilon} = A^{-1}(e - Ex). \quad (1.74)$$

The size of the derivative of $x(\epsilon)$ is an indication of the size of the variation that the solution $x(\epsilon)$ undergoes when the data, i.e., the pair $[A, b]$, are perturbed in the direction $[E, e]$. In absolute terms, a small variation $[\epsilon E, \epsilon e]$ will cause the solution to vary by roughly $\epsilon x'(0) = \epsilon A^{-1}(e - Ex)$. The relative variation is such that

$$\frac{\|x(\epsilon) - x\|}{\|x\|} \leq \epsilon \|A^{-1}\| \left(\frac{\|e\|}{\|x\|} + \|E\| \right) + o(\epsilon).$$

Using the fact that $\|b\| \leq \|A\|\|x\|$ in the above equation yields

$$\frac{\|x(\epsilon) - x\|}{\|x\|} \leq \epsilon \|A\|\|A^{-1}\| \left(\frac{\|e\|}{\|b\|} + \frac{\|E\|}{\|A\|} \right) + o(\epsilon), \quad (1.75)$$

which relates the relative variation in the solution to the relative sizes of the perturbations. The quantity

$$\kappa(A) = \|A\| \|A^{-1}\|$$

is called the *condition number* of the linear system (1.72) with respect to the norm $\|\cdot\|$. The condition number is relative to a norm. When using the standard norms $\|\cdot\|_p$, $p = 1, \ldots, \infty$, it is customary to label $\kappa(A)$ with the same label as the associated norm. Thus,

$$\kappa_p(A) = \|A\|_p \|A^{-1}\|_p.$$

For large matrices, the determinant of a matrix is almost never a good indication of "near" singularity or degree of sensitivity of the linear system. The reason is that $\det(A)$ is the product of the eigenvalues, which depends very much on a scaling of a matrix, whereas the condition number of a matrix is scaling invariant. For example, for $A = \alpha I$, the determinant is $\det(A) = \alpha^n$, which can be very small if $|\alpha| < 1$, whereas $\kappa(A) = 1$ for any of the standard norms.

In addition, small eigenvalues do not always give a good indication of poor conditioning. Indeed, a matrix can have all its eigenvalues equal to one yet be poorly conditioned.

Example 1.5. The simplest example is provided by matrices of the form
$$A_n = I + \alpha e_1 e_n^T$$
for large α. The inverse of A_n is
$$A_n^{-1} = I - \alpha e_1 e_n^T$$
and for the ∞-norm we have
$$\|A_n\|_\infty = \|A_n^{-1}\|_\infty = 1 + |\alpha|,$$
so that
$$\kappa_\infty(A_n) = (1 + |\alpha|)^2.$$
For a large α, this can give a very large condition number, whereas all the eigenvalues of A_n are equal to unity.

When an iterative procedure is used to solve a linear system, we typically face the problem of choosing a good stopping procedure for the algorithm. Often a residual norm,
$$\|r\| = \|b - A\tilde{x}\|,$$
is available for some current approximation $\tilde{x}$ and an estimate of the absolute error $\|x - \tilde{x}\|$ or the relative error $\|x - \tilde{x}\|/\|x\|$ is desired. The following simple relation is helpful in this regard:
$$\frac{\|x - \tilde{x}\|}{\|x\|} \leq \kappa(A) \frac{\|r\|}{\|b\|}.$$
It is necessary to have an estimate of the condition number $\kappa(A)$ in order to exploit the above relation.

Exercises

1. Verify that the Euclidean inner product defined by (1.4) does indeed satisfy the general definition of inner products on vector spaces.

2. Show that two eigenvectors associated with two distinct eigenvalues are linearly independent. In a more general sense, show that a family of eigenvectors associated with distinct eigenvalues forms a linearly independent family.

3. Show that, if λ is any nonzero eigenvalue of the matrix AB, then it is also an eigenvalue of the matrix BA. Start with the particular case where A and B are square and B is nonsingular, then consider the more general case where A, B may be singular or even rectangular (but such that AB and BA are square).

4. Let A be an $n \times n$ orthogonal matrix, i.e., such that $A^H A = D$, where D is a diagonal matrix. Assuming that D is nonsingular, what is the inverse of A? Assuming that $D > 0$, how can A be transformed into a unitary matrix (by operations on its rows or columns)?

5. Show that the Frobenius norm is consistent. Can this norm be associated with two vector norms via (1.7)? What is the Frobenius norm of a diagonal matrix? What is the p-norm of a diagonal matrix (for any p)?

6. Find the Jordan canonical form of the matrix
$$A = \begin{pmatrix} 1 & 2 & -4 \\ 0 & 1 & 2 \\ 0 & 0 & 2 \end{pmatrix}.$$
Repeat the question for the matrix obtained by replacing the element a_{33} with 1.

7. Give an alternative proof of Theorem 1.9 on the Schur form by starting from the Jordan canonical form. [Hint: Write $A = XJX^{-1}$ and use the QR decomposition of X.]

8. Show from the definition of determinants used in Section 1.2 that the characteristic polynomial is a polynomial of degree n for an $n \times n$ matrix.

9. Show that the characteristic polynomials of two similar matrices are equal.

10. Show that
$$\lim_{k \to \infty} \|A^k\|^{1/k} = \rho(A)$$
for any matrix norm. [Hint: Use the Jordan canonical form.]

11. Let X be a nonsingular matrix and, for any matrix norm $\|\cdot\|$, define $\|A\|_X = \|AX\|$. Show that this is indeed a matrix norm. Is this matrix norm consistent? Show the same for $\|XA\|$ and $\|YAX\|$, where Y is also a nonsingular matrix. These norms are not, in general, associated with any vector norms; i.e., they can't be defined by a formula of the form (1.7). Why? What can you say when $Y = X^{-1}$? Is $\|X^{-1}AX\|$ associated with a vector norm in this particular case?

12. Find the field of values of the matrix
$$A = \begin{pmatrix} 0 & 1 \\ 0 & 0 \end{pmatrix}$$
and verify that it is not equal to the convex hull of its eigenvalues.

13. Show that, for a skew-Hermitian matrix S,
$$\Re e(Sx, x) = 0 \quad \text{for any } x \in \mathbb{C}^n.$$

14. Given an arbitrary matrix S, show that, if $(Sx, x) = 0$ for all x in $\mathbb{C}^n$, then it is true that
$$(Sy, z) + (Sz, y) = 0 \quad \forall y, z \in \mathbb{C}^n. \tag{1.76}$$
[Hint: Expand $(S(y+z), y+z)$.]

15. Using the results of the previous two exercises, show that, if (Ax, x) is real for all x in $\mathbb{C}^n$, then A must be Hermitian. Would this result be true if the assumption were to be replaced with (Ax, x) *is real for all real x*? Explain.

Exercises

16. Show that, if $(Sx, x) = 0$ for all complex vectors x, then S is zero. [Hint: Start by doing Exercise 14. Then, selecting $y = e_k$, $z = e^\theta e_j$ in (1.76) for an arbitrary θ, establish that $s_{kj} e^{2\theta} = -s_{jk}$ and conclude that $s_{jk} = s_{jk} = 0$.] Is the result true if $(Sx, x) = 0$ for all *real* vectors x?

17. The definition of a positive definite matrix is that (Ax, x) is real and positive for all real vectors x. Show that this is equivalent to requiring that the Hermitian part of A, namely, $\frac{1}{2}(A + A^H)$, be (Hermitian) positive definite.

18. Let $A_1 = B^{-1}C$ and $A_2 = CB$, where C is a Hermitian matrix and B is an HPD matrix. Are A_1 and A_2 Hermitian *in general*? Show that A_1 and A_2 are Hermitian (self-adjoint) with respect to the B inner product.

19. Let a matrix A be such that $A^H = p(A)$, where p is a polynomial. Show that A is normal. Given a diagonal complex matrix D, show that there exists a polynomial of degree less than n such that $\bar{D} = p(D)$. Use this to show that a normal matrix satisfies $A^H = p(A)$ for a certain polynomial of p of degree less than n. As an application, use this result to provide an alternative proof of Lemma 1.13.

20. Show that A is normal iff its Hermitian and skew-Hermitian parts, as defined in Section 1.11, commute.

21. The goal of this exercise is to establish the relation (1.34). Consider the numerical radius $\nu(A)$ of an arbitrary matrix A. Show that $\nu(A) \leq \|A\|_2$. Show that, for a normal matrix, $\nu(A) = \|A\|_2$. Consider the decomposition of a matrix into its Hermitian and skew-Hermitian parts, as shown in (1.48), (1.49), and (1.50). Show that $\|A\|_2 \leq \nu(H) + \nu(S)$. Now, using this inequality and the definition of the numerical radius, show that $\|A\|_2 \leq 2\nu(A)$.

22. Show that the numerical radius is a vector norm in the sense that it satisfies the three properties (1.8)–(1.10) of norms. [Hint: For (1.8) solve Exercise 16 first.] Find a counter-example to show that the numerical radius is not a (consistent) matrix norm, i.e., that $\nu(AB)$ can be larger than $\nu(A)\nu(B)$.

23. Let A be a Hermitian matrix and B an HPD matrix defining a B inner product. Show that A is Hermitian (self-adjoint) with respect to the B inner product iff A and B commute. What condition must B satisfy for the same condition to hold in the more general case where A is not Hermitian?

24. Let A be a real symmetric matrix and λ an eigenvalue of A. Show that, if u is an eigenvector associated with λ, then so is $\bar{u}$. As a result, prove that, for any eigenvalue of a real symmetric matrix, there is an associated eigenvector that is real.

25. Show that a Hessenberg matrix H such that $h_{j+1,j} \neq 0$, $j = 1, 2, \ldots, n-1$, cannot be derogatory.

26. Prove all the properties listed in Proposition 1.24.

27. Let A be an M-matrix and u, v two nonnegative vectors such that $v^T A^{-1} u < 1$. Show that $A - uv^T$ is an M-matrix.

28. Show that if $0 \leq A \leq B$, then $0 \leq A^T A \leq B^T B$. Conclude that, under the same assumption, we have $\|A\|_2 \leq \|B\|_2$.

29. Consider the subspace M of $\mathbb{R}^4$ spanned by the vectors
$$v_1 = \begin{pmatrix} 1 \\ 0 \\ 1 \\ 1 \end{pmatrix}, \quad v_2 = \begin{pmatrix} 1 \\ -1 \\ 0 \\ -1 \end{pmatrix}.$$

   a. Write down the matrix representing the orthogonal projector onto M.
   b. What is the null space of P?
   c. What is its range?
   d. Find the vector x in S that is the closest in the 2-norm sense to the vector $c = [1, 1, 1, 1]^T$.

30. Show that, for two orthonormal bases V_1, V_2 of the same subspace M of $\mathbb{C}^n$, we have $V_1 V_1^H x = V_2 V_2^H x \ \forall\, x$.

31. What are the eigenvalues of a projector? What about its eigenvectors?

32. Show that, if two projectors P_1 and P_2 commute, then their product $P = P_1 P_2$ is a projector. What are the range and kernel of P?

33. Theorem 1.32 shows that condition (2) in Definition 1.30 is not needed; i.e., it is implied by (4) (and the other conditions). One is tempted to say that only one of (2) or (4) is required. Is this true? In other words, does (2) also imply (4)? [Prove or show a counter-example.]

34. Consider the matrix A of size $n \times n$ and the vector $x \in \mathbb{R}^n$:
$$A = \begin{pmatrix} 1 & -1 & -1 & -1 & \cdots & -1 \\ 0 & 1 & -1 & -1 & \cdots & -1 \\ 0 & 0 & 1 & -1 & \cdots & -1 \\ \vdots & \vdots & \vdots & \ddots & & \vdots \\ \vdots & \vdots & \vdots & & \ddots & \vdots \\ 0 & 0 & 0 & \cdots & 0 & 1 \end{pmatrix}, \quad x = \begin{pmatrix} 1 \\ 1/2 \\ 1/4 \\ 1/8 \\ \vdots \\ 1/2^{n-1} \end{pmatrix}.$$

   a. Compute Ax, $\|Ax\|_2$, and $\|x\|_2$.
   b. Show that $\|A\|_2 \geq \sqrt{n}$.
   c. Give a lower bound for $\kappa_2(A)$.

35. What is the inverse of the matrix A of the previous exercise? Give an expression for $\kappa_1(A)$ and $\kappa_\infty(A)$ based on this.

36. Find a small rank-one perturbation that makes the matrix A in Exercise 34 singular. Derive a lower bound for the singular values of A.

37. Consider a nonsingular matrix A. Given any matrix E, show that there exists α such that the matrix $A(\epsilon) = A + \epsilon E$ is nonsingular for all $\epsilon < \alpha$. What is the largest possible value for α satisfying the condition? [Hint: Consider the eigenvalues of the generalized eigenvalue problem $Au = \lambda E u$.]

Notes and References

For additional reading on the material presented in this chapter, see Golub and Van Loan [149], Meyer [209], Demmel [99], Datta [93], Stewart [272], and Varga [292]. Volume 2 ("Eigensystems") of the series [273] offers up-to-date coverage of algorithms for eigenvalue problems. The excellent treatise of nonnegative matrices in the book by Varga [292] remains a good reference on this topic and on iterative methods four decades after its first publication. State-of-the-art coverage of iterative methods up to the beginning of the 1970s can be found in the book by Young [321], which covers M-matrices and related topics in great detail. For a good overview of the linear algebra aspects of matrix theory and a complete proof of Jordan's canonical form, Halmos [164] is recommended.

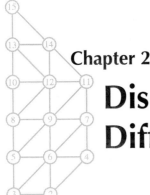

Chapter 2
Discretization of Partial Differential Equations

Partial differential equations (PDEs) constitute by far the biggest source of sparse matrix problems. The typical way to solve such equations is to *discretize* them, i.e., to approximate them by equations that involve a finite number of unknowns. The matrix problems that arise from these discretizations are generally large and sparse; i.e., they have very few nonzero entries. There are several different ways to discretize a PDE. The simplest method uses *finite difference* approximations for the partial differential operators. The *finite element method* replaces the original function with a function that has some degree of smoothness over the global domain but is piecewise polynomial on simple cells, such as small triangles or rectangles. This method is probably the most general and well understood discretization technique available. In between these two methods, there are a few conservative schemes called *finite volume methods*, which attempt to emulate continuous *conservation laws* of physics. This chapter introduces these three different discretization methods.

2.1 Partial Differential Equations

Physical phenomena are often modeled by equations that relate several partial derivatives of physical quantities, such as forces, momentums, velocities, energy, temperature, etc. These equations rarely have a *closed-form* (explicit) solution. In this chapter, a few types of PDEs are introduced, which will serve as models throughout the book. Only one- and two-dimensional problems are considered, and the space variables are denoted by x in the case of one-dimensional problems and x_1 and x_2 for two-dimensional problems. In two dimensions, x denotes the "vector" of components (x_1, x_2).

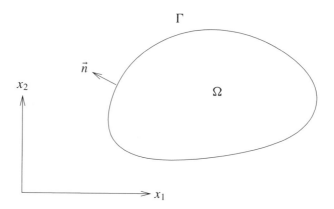

Figure 2.1. *Domain Ω for Poisson's equation.*

2.1.1 Elliptic Operators

One of the most common PDEs encountered in various areas of engineering is Poisson's equation:

$$\frac{\partial^2 u}{\partial x_1^2} + \frac{\partial^2 u}{\partial x_2^2} = f \quad \text{for} \quad x = \begin{pmatrix} x_1 \\ x_2 \end{pmatrix} \quad \text{in } \Omega, \tag{2.1}$$

where Ω is a bounded, open domain in $\mathbb{R}^2$. Here, x_1, x_2 are the two space variables.

The above equation is to be satisfied only for points that are located at the interior of the domain Ω. See Figure 2.1. Equally important are the conditions that must be satisfied on the *boundary* Γ of Ω. These are termed *boundary conditions*, which come in three common types:

$$\begin{aligned}
\text{Dirichlet condition:} \quad & u(x) = \phi(x). \\
\text{Neumann condition:} \quad & \tfrac{\partial u}{\partial \vec{n}}(x) = 0. \\
\text{Cauchy condition:} \quad & \tfrac{\partial u}{\partial \vec{n}}(x) + \alpha(x)u(x) = \gamma(x).
\end{aligned}$$

The vector $\vec{n}$ usually refers to a unit vector that is normal to Γ and directed outward. Note that the Neumann boundary conditions are a particular case of the Cauchy conditions with $\gamma = \alpha = 0$. For a given unit vector $\vec{v}$ with components v_1 and v_2, the directional derivative $\partial u / \partial \vec{v}$ is defined by

$$\begin{aligned}
\frac{\partial u}{\partial \vec{v}}(x) &= \lim_{h \to 0} \frac{u(x + h\vec{v}) - u(x)}{h} \\
&= \frac{\partial u}{\partial x_1}(x) v_1 + \frac{\partial u}{\partial x_2}(x) v_2 \tag{2.2} \\
&= \nabla u \cdot \vec{v}, \tag{2.3}
\end{aligned}$$

where ∇u is the gradient of u:

$$\nabla u = \begin{pmatrix} \frac{\partial u}{\partial x_1} \\ \frac{\partial u}{\partial x_2} \end{pmatrix}, \tag{2.4}$$

and the dot in (2.3) indicates a dot product of two vectors in $\mathbb{R}^2$.

2.1. Partial Differential Equations

In reality, Poisson's equation is often a limit case of a time-dependent problem. Its solution can, for example, represent the steady-state temperature distribution in a region Ω when there is a heat source f that is constant with respect to time. The boundary conditions should then model heat loss across the boundary Γ.

The particular case where $f(x) = 0$, i.e., the equation

$$\Delta u = 0,$$

to which boundary conditions must be added, is called the *Laplace equation* and its solutions are called *harmonic functions*.

Many problems in physics have boundary conditions of *mixed type*, e.g., of Dirichlet type in one part of the boundary and of Cauchy type in another. Another observation is that the Neumann conditions do not define the solution uniquely. Indeed, if u is a solution, then so is $u + c$ for any constant c.

The operator

$$\Delta = \frac{\partial^2}{\partial x_1^2} + \frac{\partial^2}{\partial x_2^2}$$

is called the *Laplacian operator* and appears in many models of physical and mechanical phenomena. These models often lead to more general elliptic operators of the form

$$\begin{aligned} L &= \frac{\partial}{\partial x_1}\left(a \frac{\partial}{\partial x_1}\right) + \frac{\partial}{\partial x_2}\left(a \frac{\partial}{\partial x_2}\right) \\ &= \nabla \cdot (a\nabla), \end{aligned} \tag{2.5}$$

where the scalar function a depends on the coordinate and may represent some specific parameter of the medium, such as density, porosity, etc. At this point it may be useful to recall some notation that is widely used in physics and mechanics. The ∇ operator can be considered as a vector consisting of the components $\frac{\partial}{\partial x_1}$ and $\frac{\partial}{\partial x_2}$. When applied to a scalar function u, this operator is nothing but the *gradient* operator, since it yields a vector with the components $\frac{\partial u}{\partial x_1}$ and $\frac{\partial u}{\partial x_2}$, as is shown in (2.4). The dot notation allows dot products of vectors in $\mathbb{R}^2$ to be defined. These vectors can include partial differential operators. For example, the dot product $\nabla \cdot u$ of ∇ with $u = \binom{u_1}{u_2}$ yields the scalar quantity

$$\frac{\partial u_1}{\partial x_1} + \frac{\partial u_2}{\partial x_2},$$

which is called the *divergence* of the vector function $\vec{u} = \binom{u_1}{u_2}$. Applying this *divergence operator* to $u = a\nabla$, where a is a scalar function, yields the L operator in (2.5). The divergence of the vector function $\vec{v}$ is often denoted by div $\vec{v}$ or $\nabla \cdot \vec{v}$. Thus,

$$\text{div } \vec{v} = \nabla \cdot \vec{v} = \frac{\partial v_1}{\partial x_1} + \frac{\partial v_2}{\partial x_2}.$$

The closely related operator

$$\begin{aligned} L &= \frac{\partial}{\partial x_1}\left(a_1 \frac{\partial}{\partial x_1}\right) + \frac{\partial}{\partial x_2}\left(a_2 \frac{\partial}{\partial x_2}\right) \\ &= \nabla \left(\vec{a} \cdot \nabla \right) \end{aligned} \tag{2.6}$$

is a further generalization of the Laplacian operator Δ in the case where the medium is *anisotropic* and *inhomogeneous*. The coefficients a_1, a_2 depend on the space variable x and reflect the position as well as the directional dependence of the material properties, such as porosity in the case of fluid flow or dielectric constants in electrostatics. In fact, the above operator can be viewed as a particular case of $L = \nabla \cdot (A \nabla)$, where A is a 2×2 matrix that acts on the two components of ∇.

2.1.2 The Convection Diffusion Equation

Many physical problems involve a combination of *diffusion* and *convection* phenomena. Such phenomena are modeled by the convection diffusion equation

$$\frac{\partial u}{\partial t} + b_1 \frac{\partial u}{\partial x_1} + b_2 \frac{\partial u}{\partial x_2} = \nabla \cdot (a \nabla) u + f$$

or

$$\frac{\partial u}{\partial t} + \vec{b} \cdot \nabla u = \nabla \cdot (a \nabla) u + f,$$

the steady-state version of which can be written as

$$-\nabla \cdot (a \nabla) u + \vec{b} \cdot \nabla u = f. \tag{2.7}$$

Problems of this type are often used as model problems because they represent the simplest form of conservation of mass in fluid mechanics. Note that the vector $\vec{b}$ is sometimes quite large, which may cause some difficulties either to the discretization schemes or to the iterative solution techniques.

2.2 Finite Difference Methods

The *finite difference* method is based on local approximations of the partial derivatives in a PDE, which are derived by low order Taylor series expansions. The method is quite simple to define and rather easy to implement. Also, it is particularly appealing for simple regions, such as rectangles, and when uniform meshes are used. The matrices that result from these discretizations are often well structured, which means that they typically consist of a few nonzero diagonals. Another advantage is that there are a number of *fast Poisson solvers* (FPS) for constant coefficient problems, which can deliver the solution in logarithmic time per grid point. This means the total number of operations is of the order of $n \log(n)$, where n is the total number of discretization points. This section gives an overview of finite difference discretization techniques.

2.2.1 Basic Approximations

The simplest way to approximate the first derivative of a function u at the point x is via the formula

$$\left(\frac{du}{dx} \right)(x) \approx \frac{u(x+h) - u(x)}{h}. \tag{2.8}$$

2.2. Finite Difference Methods

When u is differentiable at x, then the limit of the above ratio when h tends to zero is the derivative of u at x. For a function that is C^4 in the neighborhood of x, we have, by Taylor's formula,

$$u(x+h) = u(x) + h\frac{du}{dx} + \frac{h^2}{2}\frac{d^2u}{dx^2} + \frac{h^3}{6}\frac{d^3u}{dx^3} + \frac{h^4}{24}\frac{d^4u}{dx^4}(\xi_+) \qquad (2.9)$$

for some ξ_+ in the interval $(x, x+h)$. Therefore, the approximation (2.8) satisfies

$$\frac{du}{dx} = \frac{u(x+h) - u(x)}{h} - \frac{h}{2}\frac{d^2u(x)}{dx^2} + O(h^2). \qquad (2.10)$$

The formula (2.9) can be rewritten with h replaced by $-h$ to obtain

$$u(x-h) = u(x) - h\frac{du}{dx} + \frac{h^2}{2}\frac{d^2u}{dx^2} - \frac{h^3}{6}\frac{d^3u}{dx^3} + \frac{h^4}{24}\frac{d^4u(\xi_-)}{dx^4}, \qquad (2.11)$$

in which ξ_- belongs to the interval $(x-h, x)$. Adding (2.9) and (2.11), dividing through by h^2, and using the mean value theorem for the fourth order derivatives results in the following approximation of the second derivative:

$$\frac{d^2u(x)}{dx^2} = \frac{u(x+h) - 2u(x) + u(x-h)}{h^2} - \frac{h^2}{12}\frac{d^4u(\xi)}{dx^4}, \qquad (2.12)$$

where $\xi_- \leq \xi \leq \xi_+$. The above formula is called a *centered difference approximation* of the second derivative since the point at which the derivative is being approximated is the center of the points used for the approximation. The dependence of this derivative on the values of u at the points involved in the approximation is often represented by a "stencil" or "molecule," shown in Figure 2.2.

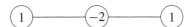

Figure 2.2. *The three-point stencil for the centered difference approximation to the second order derivative.*

The approximation (2.8) for the first derivative is *forward* rather than centered. Also, a *backward* formula can be used that consists of replacing h with $-h$ in (2.8). The two formulas can also be averaged to obtain the *centered difference* formula

$$\frac{du(x)}{dx} \approx \frac{u(x+h) - u(x-h)}{2h}. \qquad (2.13)$$

It is easy to show that the above centered difference formula is of the second order, while (2.8) is only first order accurate. Denoted by δ^+ and δ^-, the forward and backward difference operators are defined by

$$\delta^+ u(x) = u(x+h) - u(x), \qquad (2.14)$$
$$\delta^- u(x) = u(x) - u(x-h). \qquad (2.15)$$

All previous approximations can be rewritten using these operators.

In addition to standard first order and second order derivatives, it is sometimes necessary to approximate the second order operator

$$\frac{d}{dx}\left[a(x)\frac{d}{dx}\right].$$

A centered difference formula for this with second order accuracy is given by

$$\frac{d}{dx}\left[a(x)\frac{du}{dx}\right] = \frac{1}{h^2}\delta^+\left(a_{i-1/2}\,\delta^- u\right) + O(h^2) \quad (2.16)$$

$$\approx \frac{a_{i+1/2}(u_{i+1} - u_i) - a_{i-1/2}(u_i - u_{i-1})}{h^2}.$$

2.2.2 Difference Schemes for the Laplacian Operator

If the approximation (2.12) is used for both the $\frac{\partial^2}{\partial x_1^2}$ and $\frac{\partial^2}{\partial x_2^2}$ terms in the Laplacian operator, using a mesh size of h_1 for the x_1 variable and h_2 for the x_2 variable, the following second order accurate approximation results:

$$\Delta u(x) \approx \frac{u(x_1 + h_1, x_2) - 2u(x_1, x_2) + u(x - h_1, x_2)}{h_1^2}$$
$$+ \frac{u(x_1, x_2 + h_2) - 2u(x_1, x_2) + u(x_1, x_2 - h_2)}{h_2^2}.$$

In the particular case where the mesh sizes h_1 and h_2 are the same and equal to a mesh size h, the approximation becomes

$$\Delta u(x) \approx \frac{1}{h^2}[u(x_1 + h, x_2) + u(x_1 - h, x_2) + u(x_1, x_2 + h)$$
$$+ u(x_1, x_2 - h) - 4u(x_1, x_2)], \quad (2.17)$$

which is called the five-point centered approximation to the Laplacian. The stencil of this finite difference approximation is illustrated in (a) of Figure 2.3.

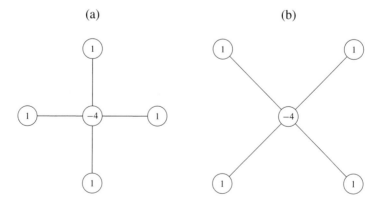

Figure 2.3. *Five-point stencils for the centered difference approximation to the Laplacian operator:* (a) *the standard stencil,* (b) *the skewed stencil.*

2.2. Finite Difference Methods

Another approximation may be obtained by exploiting the four points $u(x_1 \pm h, x_2 \pm h)$ located on the two diagonal lines from $u(x_1, x_2)$. These points can be used in the same manner as in the previous approximation except that the mesh size has changed. The corresponding stencil is illustrated in (b) of Figure 2.3.

The approximation (2.17) is second order accurate and the error takes the form

$$\frac{h^2}{12}\left(\frac{\partial^4 u}{\partial^4 x_1} + \frac{\partial^4 u}{\partial^4 x_2}\right) + O(h^3).$$

There are other schemes that utilize nine-point formulas as opposed to five-point formulas. Two such schemes obtained by combining the standard and skewed stencils described above are shown in Figure 2.4. Both approximations (c) and (d) are second order accurate. However, (d) is sixth order for harmonic functions, i.e., functions whose Laplacian is zero.

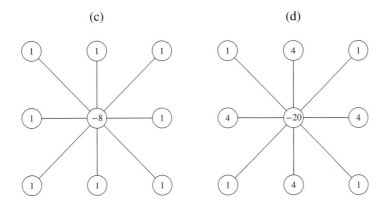

Figure 2.4. *Two nine-point centered difference stencils for the Laplacian operator.*

2.2.3 Finite Differences for One-Dimensional Problems

Consider the one-dimensional equation

$$-u''(x) = f(x) \text{ for } x \in (0, 1), \tag{2.18}$$
$$u(0) = u(1) = 0. \tag{2.19}$$

The interval [0, 1] can be discretized uniformly by taking the $n + 2$ points

$$x_i = i \times h, \quad i = 0, \ldots, n+1,$$

where $h = 1/(n + 1)$. Because of the Dirichlet boundary conditions, the values $u(x_0)$ and $u(x_{n+1})$ are known. At every other point, an approximation u_i is sought for the exact solution $u(x_i)$.

If the centered difference approximation (2.12) is used, then, by (2.18) expressed at the point x_i, the unknowns u_i, u_{i-1}, u_{i+1} satisfy the relation

$$-u_{i-1} + 2u_i - u_{i+1} = h^2 f_i,$$

in which $f_i \equiv f(x_i)$. Notice that, for $i = 1$ and $i = n$, the equation will involve u_0 and u_{n+1}, which are known quantities, both equal to zero in this case. Thus, for $n = 6$, the linear system obtained is of the form
$$Ax = f,$$
where
$$A = \frac{1}{h^2} \begin{pmatrix} 2 & -1 & & & & \\ -1 & 2 & -1 & & & \\ & -1 & 2 & -1 & & \\ & & -1 & 2 & -1 & \\ & & & -1 & 2 & -1 \\ & & & & -1 & 2 \end{pmatrix}.$$

2.2.4 Upwind Schemes

Consider now the one-dimensional version of the convection diffusion equation (2.7) in which the coefficients a and b are constant and $f = 0$, using Dirichlet boundary conditions:
$$\begin{cases} -a\, u'' + b\, u' = 0, & 0 < x < L = 1, \\ u(0) = 0, \ u(L) = 1. \end{cases} \quad (2.20)$$

In this particular case, it is easy to verify that the exact solution to the above equation is given by
$$u(x) = \frac{1 - e^{Rx}}{1 - e^R},$$
where R is the so-called Péclet number defined by $R = bL/a$. Now consider the approximate solution provided by using the centered difference schemes seen above for both the first and second order derivatives. The equation for the unknown number i becomes
$$b\frac{u_{i+1} - u_{i-1}}{2h} - a\frac{u_{i+1} - 2u_i + u_{i-1}}{h^2} = 0$$
or, defining $c = Rh/2$,
$$-(1 - c)u_{i+1} + 2u_i - (1 + c)u_{i-1} = 0. \quad (2.21)$$

This is a second order homogeneous linear difference equation and the usual way to solve it is to seek a general solution in the form $u_j = r^j$. Substituting in (2.21), r must satisfy
$$(1 - c)r^2 - 2r + (c + 1) = 0.$$

Therefore, $r_1 = 1$ is a root and the second root is $r_2 = (1 + c)/(1 - c)$. The general solution of the above difference equation is now sought as a linear combination of the two solutions corresponding to these two roots:
$$u_i = \alpha r_1^i + \beta r_2^i = \alpha + \beta \left(\frac{1 + c}{1 - c}\right)^i.$$

Because of the boundary condition $u_0 = 0$, it is necessary that $\beta = -\alpha$. Likewise, the boundary condition $u_{n+1} = 1$ yields
$$\alpha = \frac{1}{1 - \sigma^{n+1}}, \quad \text{with} \quad \sigma \equiv \frac{1 + c}{1 - c}.$$

2.2. Finite Difference Methods

Thus, the solution is
$$u_i = \frac{1 - \sigma^i}{1 - \sigma^{n+1}}.$$

When $h > 2/R$ the factor σ becomes negative and the above approximations will oscillate around zero. In contrast, the exact solution is positive and monotone in the range $[0, 1]$. In this situation the solution is very inaccurate regardless of the arithmetic. In other words, the scheme itself creates the oscillations. To avoid this, a small enough mesh h can be taken to ensure that $c < 1$. The resulting approximation is in much better agreement with the exact solution. Unfortunately, this condition can limit the mesh size too drastically for large values of b.

Note that, when $b < 0$, the oscillations disappear, since $\sigma < 1$. In fact, a linear algebra interpretation of the oscillations comes from comparing the tridiagonal matrices obtained from the discretization. Again, for the case $n = 6$, the tridiagonal matrix resulting from discretizing (2.7) takes the form

$$A = \frac{1}{h^2} \begin{pmatrix} 2 & -1+c & & & & \\ -1-c & 2 & -1+c & & & \\ & -1-c & 2 & -1+c & & \\ & & -1-c & 2 & -1+c & \\ & & & -1-c & 2 & -1+c \\ & & & & -1-c & 2 \end{pmatrix}.$$

The above matrix is no longer a diagonally dominant M-matrix. Observe that, if the backward difference formula for the first order derivative is used, we obtain

$$b \frac{u_i - u_{i-1}}{h} - a \frac{u_{i-1} - 2u_i + u_{i+1}}{h^2} = 0.$$

Then (weak) diagonal dominance is preserved if $b > 0$. This is because the new matrix obtained for the above backward scheme is

$$A = \frac{1}{h^2} \begin{pmatrix} 2+c & -1 & & & & \\ -1-c & 2+c & -1 & & & \\ & -1-c & 2+c & -1 & & \\ & & -1-c & 2+c & -1 & \\ & & & -1-c & 2+c & -1 \\ & & & & -1-c & 2+c \end{pmatrix},$$

where c is now defined by $c = Rh$. Each diagonal term a_{ii} is reinforced by the positive term c, while each subdiagonal term $a_{i,i-1}$ increases by the same amount in absolute value. In the case where $b < 0$, the forward difference formula

$$b \frac{u_{i+1} - u_i}{h} - a \frac{u_{i-1} - 2u_i + u_{i+1}}{h^2} = 0$$

can be used to achieve the same effect. Generally speaking, if b depends on the space variable x, the effect of weak diagonal dominance can be achieved by simply adopting the following discretization, known as an "upwind scheme":

$$b \frac{\delta_i^* u_i}{h} - a \frac{u_{i-1} - 2u_i + u_{i+1}}{h^2} = 0,$$

where

$$\delta_i^* = \begin{cases} \delta_i^- & \text{if } b > 0, \\ \delta_i^+ & \text{if } b < 0. \end{cases}$$

The above difference scheme can be rewritten by introducing the sign function $\text{sign}(b) = |b|/b$. The approximation to u' at x_i is then defined by

$$u'(x_i) \approx \frac{1}{2}(1 - \text{sign}(b))\frac{\delta^+ u_i}{h} + \frac{1}{2}(1 + \text{sign}(b))\frac{\delta^- u_i}{h}.$$

Making use of the notation

$$(x)^+ = \frac{1}{2}(x + |x|), \quad (x)^- = \frac{1}{2}(x - |x|), \quad (2.22)$$

a slightly more elegant formula can be obtained by expressing the approximation of the product $b(x_i)u'(x_i)$:

$$b(x_i)u'(x_i) \approx \frac{1}{2}(b_i - |b_i|)\frac{\delta^+ u_i}{h} + \frac{1}{2}(b_i + |b_i|)\frac{\delta^- u_i}{h}$$

$$\approx \frac{1}{h}\left[-b_i^+ u_{i-1} + |b_i| u_i + b_i^- u_{i+1}\right], \quad (2.23)$$

where b_i stands for $b(x_i)$. The diagonal term in the resulting tridiagonal matrix is nonnegative, the off-diagonal terms are nonpositive, and the diagonal term is the negative sum of the off-diagonal terms. This property characterizes upwind schemes.

A notable disadvantage of upwind schemes is the low order of approximation that they yield. An advantage is that upwind schemes yield linear systems that are easier to solve by iterative methods.

2.2.5 Finite Differences for Two-Dimensional Problems

Consider this simple problem, which is similar to the previous case:

$$-\left(\frac{\partial^2 u}{\partial x_1^2} + \frac{\partial^2 u}{\partial x_2^2}\right) = f \quad \text{in } \Omega, \quad (2.24)$$

$$u = 0 \quad \text{on } \Gamma, \quad (2.25)$$

where Ω is now the rectangle $(0, l_1) \times (0, l_2)$ and Γ its boundary. Both intervals can be discretized uniformly by taking $n_1 + 2$ points in the x_1 direction and $n_2 + 2$ points in the x_2 direction:

$$x_{1,i} = i \times h_1, i = 0, \ldots, n_1 + 1, \quad x_{2,j} = j \times h_2, j = 0, \ldots, n_2 + 1,$$

where

$$h_1 = \frac{l_1}{n_1 + 1}, \quad h_2 = \frac{l_2}{n_2 + 1}.$$

Since the values at the boundaries are known, we number only the interior points, i.e., the points $(x_{1,i}, x_{2,j})$ with $0 < i < n_1$ and $0 < j < n_2$. The points are labeled from the bottom up, one horizontal line at a time. This labeling is called *natural ordering* and is shown in Figure 2.5 for the very simple case when $n_1 = 7$ and $n_2 = 5$. The pattern of the matrix corresponding to the above equations appears in Figure 2.6.

2.2. Finite Difference Methods

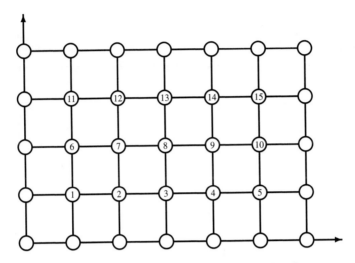

Figure 2.5. *Natural ordering of the unknowns for a 7×5 two-dimensional grid.*

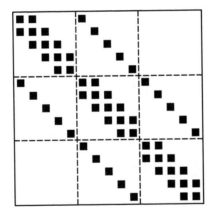

Figure 2.6. *Pattern of matrix associated with the 7×5 finite difference mesh of Figure 2.5.*

In the case when $h_1 = h_2 = h$, the matrix has the following block structure:

$$A = \frac{1}{h^2} \begin{pmatrix} B & -I & \\ -I & B & -I \\ & -I & B \end{pmatrix}, \quad \text{with} \quad B = \begin{pmatrix} 4 & -1 & & & \\ -1 & 4 & -1 & & \\ & -1 & 4 & -1 & \\ & & -1 & 4 & -1 \\ & & & -1 & 4 \end{pmatrix}.$$

2.2.6 Fast Poisson Solvers

A number of special techniques have been developed for solving linear systems arising from finite difference discretizations of the Poisson equation on rectangular grids. These are termed fast Poisson solvers (FPS) because of the relatively low number of arithmetic operations that they require, typically of the order of $O(N \log(N))$, where N is the size of the matrix.

Consider first the linear systems seen in the previous subsection, which have the form (after scaling by h^2)

$$\begin{pmatrix} B & -I & & & \\ -I & B & -I & & \\ & \ddots & \ddots & \ddots & \\ & & -I & B & -I \\ & & & -I & B \end{pmatrix} \begin{pmatrix} u_1 \\ u_2 \\ \vdots \\ u_{m-1} \\ u_m \end{pmatrix} = \begin{pmatrix} b_1 \\ b_2 \\ \vdots \\ b_{m-1} \\ b_m \end{pmatrix}, \quad (2.26)$$

in which

$$B = \begin{pmatrix} 4 & -1 & & & \\ -1 & 4 & -1 & & \\ & \ddots & \ddots & \ddots & \\ & & -1 & 4 & -1 \\ & & & -1 & 4 \end{pmatrix}. \quad (2.27)$$

The notation has changed slightly in that we call p and m the mesh sizes in the x_1 and x_2 directions, respectively. Therefore, each u_i is of dimension p and corresponds to a block of solution components along one horizontal line.

Fourier methods exploit the knowledge of the eigenvalues and eigenvectors of the matrix B. The eigenvalues are known to be

$$\lambda_j = 4 - \cos\left(\frac{j\pi}{m+1}\right), \quad j = 1, \ldots, m,$$

and, defining $\theta_j \equiv (j\pi)/(m+1)$, the corresponding eigenvectors are given by

$$q_j = \sqrt{\frac{2}{m+1}} \times \left[\sin \theta_j, \sin(2\theta_j), \ldots, \sin(m\theta_j)\right]^T.$$

Defining

$$Q = [q_1, \ldots, q_m],$$

it is clear that $Q^T B Q = \Lambda = \text{diag}(\lambda_j)$. The jth (block) row of the system (2.26), which can be written as

$$-u_{j-1} + B u_j - u_{j+1} = b_j,$$

will now be transformed by applying the similarity transformation Q to the above equation, leading to

$$-Q^T u_{j-1} + (Q^T B Q) Q^T u_j - Q^T u_{j+1} = Q^T b_j.$$

If we denote by a bar quantities expressed in the Q basis, then the above equation becomes

$$-\bar{u}_{j-1} + \Lambda \bar{u}_j - \bar{u}_{j+1} = \bar{b}_j.$$

Note that the transformation from u_j to $\bar{u}_j$ can be performed with a (real) fast Fourier transform (FFT), which will be exploited shortly. Together the above equations yield the

2.2. Finite Difference Methods

large system

$$\begin{pmatrix} \Lambda & -I & & & \\ -I & \Lambda & -I & & \\ & \ddots & \ddots & \ddots & \\ & & -I & \Lambda & -I \\ & & & -I & \Lambda \end{pmatrix} \begin{pmatrix} \bar{u}_1 \\ \bar{u}_2 \\ \vdots \\ \bar{u}_{m-1} \\ \bar{u}_p \end{pmatrix} = \begin{pmatrix} \bar{b}_1 \\ \bar{b}_2 \\ \vdots \\ \bar{b}_{m-1} \\ \bar{b}_p \end{pmatrix}. \tag{2.28}$$

As it turns out, the above system disguises a set of m independent tridiagonal systems. Indeed, taking the ith row of each block yields

$$\begin{pmatrix} \lambda_i & -1 & & & \\ -1 & \lambda_i & -1 & & \\ & \ddots & \ddots & \ddots & \\ & & -1 & \lambda_i & -1 \\ & & & -1 & \lambda_i \end{pmatrix} \begin{pmatrix} \bar{u}_{i1} \\ \bar{u}_{i2} \\ \vdots \\ \bar{u}_{im-1} \\ \bar{u}_{ip} \end{pmatrix} = \begin{pmatrix} \bar{b}_{i1} \\ \bar{b}_{i2} \\ \vdots \\ \bar{b}_{im-1} \\ \bar{b}_{ip} \end{pmatrix}, \tag{2.29}$$

where u_{ij} and b_{ij} represent the jth components of the vectors u_j and b_j, respectively.

The procedure becomes clear and is described in the next algorithm.

ALGORITHM 2.1. FFT-based FPS

1. Compute $\bar{b}_j = Q^T b_j$, $j = 1, \ldots, p$
2. Solve the tridiagonal systems (2.29) for $i = 1, \ldots, m$
3. Compute $u_j = Q \bar{u}_j$, $j = 1, \ldots, p$

The operations in lines 1 and 3 are performed by FFT transforms and require a total of $O(m \log_2 m)$ operations each, leading to a total of $O(p \times m \log_2 m)$ operations. Solving the m tridiagonal systems requires a total of $8 \times p \times m$ operations. As a result, the complexity of the algorithm is $O(N \log N)$, where $N = p \times m$.

A second class of FPS utilize block cyclic reduction (BCR). For simplicity, assume that $p = 2^\mu - 1$. Denoting 2^r by h, at the rth step of BCR, the system is of the following form:

$$\begin{pmatrix} B^{(r)} & -I & & & \\ -I & B^{(r)} & -I & & \\ & \ddots & \ddots & \ddots & \\ & & -I & B^{(r)} & -I \\ & & & -I & B^{(r)} \end{pmatrix} \begin{pmatrix} u_h \\ u_{2h} \\ \vdots \\ u_{(p_r-1)h} \\ u_{p_r h} \end{pmatrix} = \begin{pmatrix} b_h \\ b_{2h} \\ \vdots \\ b_{(p_r-1)h} \\ b_{p_r h} \end{pmatrix}. \tag{2.30}$$

Equations whose block index j is odd are now eliminated by multiplying each equation indexed $2jh$ by $B^{(r)}$ and adding it to equations $(2j-1)h$ and $(2j+1)h$. This would yield a system with a size half that of (2.30), which involves only the equations with indices that are even multiples of h:

$$-u_{(2j-2)h} + \left[(B^{(r)})^2 - 2I\right] u_{2jh} - u_{(2j+2)h} = B^{(r)} b_{2jh} + b_{(2j-1)h} + b_{(2j+1)h}.$$

The process can then be repeated until we have only one system of m equations. This could then be solved and the other unknowns recovered from it in a back substitution. The method based on this direct approach is not stable.

A stable modification due to Buneman [69] consists of writing the right-hand sides differently. Each b_{jh} is written as

$$b_{jh}^{(r)} = B^{(r)} p_{jh}^{(r)} + q_{jh}^{(r)}. \tag{2.31}$$

Initially, when $r = 0$, the vector $p_i^{(0)}$ is zero and $q_i^{(0)} \equiv b_j$. The elimination of block row jh proceeds in the same manner as was described above, leading to

$$-u_{(2j-2)h} + \left[(B^{(r)})^2 - 2I\right] u_{2jh} - u_{(2j+2)h} = (B^{(r)})^2 p_{2jh}^{(r)}$$
$$+ B^{(r)}(q_{2jh}^{(r)} + p_{(2j-1)h}^{(r)} + p_{(2j+1)h}^{(r)}) + q_{(2j-1)h}^{(r)} + q_{(2j+1)h}^{(r)}. \tag{2.32}$$

It is clear that the diagonal block matrix for the next step is

$$B^{(r+1)} = (B^{(r)})^2 - 2I. \tag{2.33}$$

It remains to recast (2.32) in such a way that the right-hand side blocks are again in the form (2.31). The new right-hand side is rewritten as

$$b_{2jh}^{(r+1)} = (B^{(r)})^2 \left[p_{2jh}^{(r)} + (B^{(r)})^{-1}(q_{2jh}^{(r)} + p_{(2j-1)h}^{(r)} + p_{(2j+1)h}^{(r)})\right] + q_{(2j-1)h}^{(r)} + q_{(2j+1)h}^{(r)}.$$

The term in the brackets is defined as $p_{2jh}^{(r+1)}$:

$$p_{2jh}^{(r+1)} = p_{2jh}^{(r)} + (B^{(r)})^{-1}(q_{2jh}^{(r)} + p_{(2j-1)h}^{(r)} + p_{(2j+1)h}^{(r)}), \tag{2.34}$$

so that

$$b_{2jh}^{(r+1)} = (B^{(r)})^2 p_{2jh}^{(r+1)} + q_{(2j-1)h}^{(r)} + q_{(2j+1)h}^{(r)}$$
$$= [(B^{(r)})^2 - 2I] p_{2jh}^{(r+1)} + 2 p_{2jh}^{(r+1)} + q_{(2j-1)h}^{(r)} + q_{(2j+1)h}^{(r)}.$$

Then it becomes clear that $q_{2jh}^{(r+1)}$ should be defined as

$$q_{2jh}^{(r+1)} = 2 p_{2jh}^{(r+1)} + q_{(2j-1)h}^{(r)} + q_{(2j+1)h}^{(r)}. \tag{2.35}$$

After $\mu - 1$ steps of the above transformation, the original system (2.26) is reduced to a system with a single block that can be solved directly. The other unknowns are then obtained by back substitution, computing the u_{jh}'s with odd values of j from the u_{jh}'s with even values of j:

$$u_{jh}^{(r+1)} = (B^{(r)})^{-1}[b_{jh}^r + u_{(j-1)h} + u_{(j+1)h}]$$
$$= (B^{(r)})^{-1}[B^{(r)} p_{jh}^r + q_{jh}^r + u_{(j-1)h} + u_{(j+1)h}]$$
$$= p_{jh}^r + (B^{(r)})^{-1}[q_{jh}^r + u_{(j-1)h} + u_{(j+1)h}].$$

These substitutions are done for $h = 2^r$ decreasing from $h = 2^\mu$ to $h = 2^0$. Buneman's algorithm is described below.

2.2. Finite Difference Methods

ALGORITHM 2.2. BCR (Buneman's Version)

1. *Initialize:* $p_i^{(0)} = 0, q_j^{(0)} = b_j, j = 1, \ldots, p$ and $h = 1, r = 0$
2. *Forward solution:* While ($h = 2^r < p$), Do
3.   *Form the matrix Y_r with columns*
$$q_{2jh}^{(r)} + p_{(2j-1)h}^{(r)} + p_{(2j+1)h}^{(r)}, \quad j = 1, \ldots, (p+1)/2h - 1$$
4.   *Solve the (multi)-linear system* $B^{(r)} X_r = Y_r$
5.   *Update the vectors p and q according to* (2.34) *and* (2.35)
6. $r := r + 1$
7. EndWhile
8.   *Solve for u:* $B^{(r)} u = q_1^{(r)}$ *and set* $u_h = p_h + u$
9. *Backward substitution:* While $h \geq 1$, Do
10. $h := h/2$
11.   *Form the matrix Y_r with column vectors*
$$q_{jh}^{(r)} + u_{(j-1)h} + u_{(j+1)h}, \ j = 1, 3, 5, \ldots, n/h$$
12.   *Solve the (multi)-linear system* $B^{(r)} W_r = Y_r$
13.   *Update the solution vectors* $u_{jh}, j = 1, 3, \ldots,$ *by*
$$U_r = P_r + W_r, \text{ where } U_r \text{ (resp., } P_r\text{) is the matrix with vector columns } u_{jh} \text{ (resp., } p_{jh}\text{)}$$
14. EndWhile

The bulk of the work in the above algorithm lies in lines 4 and 12, where systems of equations with multiple right-hand sides are solved with the same coefficient matrix $B^{(r)}$. For this purpose the matrix $B^{(r)}$ is not formed explicitly. Instead, it is observed that $B^{(r)}$ is a known polynomial in B, specifically

$$B^{(r)} \equiv p_h(A) = 2C_h(B/2) = \prod_{i=1}^{h} (B - \lambda_i^{(r)} I),$$

where C_k denotes the Chebyshev polynomial of degree k of the first kind. (See Section 6.11.1 of Chapter 6 for a brief discussion of Chebyshev polynomials.) The roots λ_i of the polynomials p_h are easily determined from those of C_h:

$$\lambda_i^{(r)} = 2 \cos\left(\frac{(2i-1)\pi}{2h}\right), \quad i = 1, \ldots, h.$$

Thus, if $p = 2^\mu - 1$, the systems in line 4 can be written as

$$\prod_{i=1}^{2^r} (A - \lambda_i^{(r)} I)[x_1| \cdots |x_{2^{\mu-r-1}-1}] = [y_1| \cdots |y_{2^{\mu-r-1}-1}]. \tag{2.36}$$

An interesting and more efficient technique consists of combining BCR with the FFT approach [279, 170]. In this technique a small number of cyclic reduction steps are taken and the resulting system is then solved using the Fourier-based approach described earlier. The cost of the algorithm is still of the form $O(mp \log p)$ but the constant in the cost is smaller.

BCR can also be applied for solving general *separable* equations using the algorithm described by Swartzrauber [278]. However, the roots of the polynomial must be computed since they are not known in advance.

2.3 The Finite Element Method

The finite element method is best illustrated with the solution of a simple elliptic PDE in a two-dimensional space. Consider again Poisson's equation (2.24) with the Dirichlet boundary condition (2.25), where Ω is a bounded open domain in $\mathbb{R}^2$ and Γ its boundary. The Laplacian operator

$$\Delta = \frac{\partial^2}{\partial x_1^2} + \frac{\partial^2}{\partial x_2^2}$$

appears in many models of physical and mechanical phenomena. Equations involving the more general elliptic operators (2.5) and (2.6) can be treated in the same way as Poisson's equation (2.24) and (2.25), at least from the viewpoint of numerical solution techniques.

An essential ingredient for understanding the finite element method is *Green's formula*. The setting for this formula is an open set Ω whose boundary consists of a closed and smooth curve Γ, as illustrated in Figure 2.1. A vector-valued function $\vec{v} = \binom{v_1}{v_2}$, which is continuously differentiable in Ω, is given. The *divergence theorem* in two-dimensional space states that

$$\int_\Omega \operatorname{div} \vec{v} \, dx = \int_\Gamma \vec{v} \cdot \vec{n} \, ds. \tag{2.37}$$

The dot on the right-hand side represents a dot product of two vectors in $\mathbb{R}^2$. In this case it is between the vector $\vec{v}$ and the unit vector $\vec{n}$ that is normal to Γ at the point of consideration and oriented outward. To derive Green's formula, consider a scalar function v and a vector function $\vec{w} = \binom{w_1}{w_2}$. By standard differentiation,

$$\nabla \cdot (v\vec{w}) = (\nabla v) \cdot \vec{w} + v \nabla \cdot \vec{w},$$

which expresses $\nabla v \cdot \vec{w}$ as

$$\nabla v \cdot \vec{w} = -v \nabla \cdot \vec{w} + \nabla \cdot (v\vec{w}). \tag{2.38}$$

Integrating the above equality over Ω and using the divergence theorem, we obtain

$$\int_\Omega \nabla v \cdot \vec{w} \, dx = -\int_\Omega v \nabla \cdot \vec{w} \, dx + \int_\Omega \nabla \cdot (v\vec{w}) \, dx$$

$$= -\int_\Omega v \nabla \cdot \vec{w} \, dx + \int_\Gamma v\vec{w} \cdot \vec{n} \, ds. \tag{2.39}$$

The above equality can be viewed as a generalization of the standard integration by parts formula in calculus. Green's formula results from (2.39) by simply taking a vector $\vec{w}$, which is itself a gradient of a scalar function u; namely, $\vec{w} = \nabla u$:

$$\int_\Omega \nabla v \cdot \nabla u \, dx = -\int_\Omega v \nabla \cdot \nabla u \, dx + \int_\Gamma v \nabla u \cdot \vec{n} \, ds.$$

Observe that $\nabla \cdot \nabla u = \Delta u$. Also, the function $\nabla u \cdot \vec{n}$ is called the *normal derivative* and is denoted by

$$\nabla u \cdot \vec{n} = \frac{\partial u}{\partial \vec{n}}.$$

2.3. The Finite Element Method

With this, we obtain Green's formula

$$\int_\Omega \nabla v \cdot \nabla u \; dx = -\int_\Omega v \Delta u \; dx + \int_\Gamma v \frac{\partial u}{\partial \vec{n}} \; ds. \quad (2.40)$$

We now return to the initial problem (2.24)–(2.25). To solve this problem approximately, it is necessary to (1) take approximations to the unknown function u and (2) translate the equations into a system that can be solved numerically. The options for approximating u are numerous. However, the primary requirement is that these approximations be in a (small) finite dimensional space. There are also some additional desirable numerical properties. For example, it is difficult to approximate high degree polynomials numerically. To extract systems of equations that yield the solution, it is common to use the *weak formulation* of the problem. Let us define

$$a(u, v) \equiv \int_\Omega \nabla u \cdot \nabla v \; dx = \int_\Omega \left(\frac{\partial u}{\partial x_1} \frac{\partial v}{\partial x_1} + \frac{\partial u}{\partial x_2} \frac{\partial v}{\partial x_2} \right) dx,$$

$$(f, v) \equiv \int_\Omega f v \; dx.$$

An immediate property of the functional a is that it is *bilinear*. That means that it is linear with respect to u and v; namely,

$$a(\mu_1 u_1 + \mu_2 u_2, v) = \mu_1 a(u_1, v) + \mu_2 a(u_2, v) \quad \forall \mu_1, \mu_2 \in \mathbb{R},$$
$$a(u, \lambda_1 v_1 + \lambda_2 v_2) = \lambda_1 a(u, v_1) + \lambda_2 a(u, v_2) \quad \forall \lambda_1, \lambda_2 \in \mathbb{R}.$$

Notice that (u, v) denotes the L_2 inner product of u and v in Ω; i.e.,

$$(u, v) = \int_\Omega u(x) v(x) dx.$$

Then, for functions satisfying the Dirichlet boundary conditions, which are at least twice differentiable, Green's formula (2.40) shows that

$$a(u, v) = -(\Delta u, v).$$

The weak formulation of the initial problem (2.24)–(2.25) consists of selecting a subspace of reference V of L^2 and then defining the following problem:

$$\text{Find} \quad u \in V \quad \text{such that} \quad a(u, v) = (f, v) \quad \forall v \in V. \quad (2.41)$$

In order to understand the usual choices for the space V, note that the definition of the weak problem only requires the dot products of the gradients of u and v and the functions f and v to be L_2-integrable. The most general V under these conditions is the space of all functions whose derivatives up to the first order are in L_2. This is known as $H^1(\Omega)$. However, this space does not take into account the boundary conditions. The functions in V must be restricted to have zero values on Γ. The resulting space is called $H_0^1(\Omega)$.

The finite element method consists of approximating the weak problem by a finite dimensional problem obtained by replacing V with a subspace of functions that are defined as low degree polynomials on small pieces (elements) of the original domain.

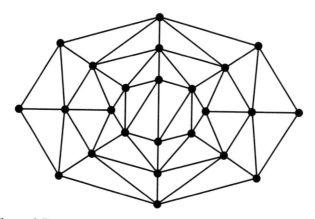

Figure 2.7. *Finite element triangulation of a domain.*

Consider a region Ω in the plane that is triangulated, as shown in Figure 2.7. In this example, the domain is simply an ellipse but the external enclosing curve is not shown. The original domain is thus approximated by the union Ω_h of m triangles K_i:

$$\Omega_h = \bigcup_{i=1}^{m} K_i.$$

For the triangulation to be valid, these triangles must have no vertex that lies on the edge of any other triangle. The *mesh size h* is defined by

$$h = \max_{i=1,\ldots,m} \operatorname{diam}(K_i),$$

where $\operatorname{diam}(K)$, the diameter of a triangle K, is the length of its longest side.

Then the finite dimensional space V_h is defined as the space of all functions that are piecewise linear and continuous on the polygonal region Ω_h and that vanish on the boundary Γ. More specifically,

$$V_h = \{\phi \mid \phi_{|\Omega_h} \text{ continuous}, \ \phi_{|\Gamma_h} = 0, \ \phi_{|K_j} \text{ linear } \forall \ j\}.$$

Here, $\phi_{|X}$ represents the restriction of the function ϕ to the subset X. If $x_j, j = 1, \ldots, n$, are the nodes of the triangulation, then a function ϕ_j in V_h can be associated with each node x_j so that the family of functions ϕ_j satisfies the following conditions:

$$\phi_j(x_i) = \delta_{ij} = \begin{cases} 1 & \text{if } x_i = x_j, \\ 0 & \text{if } x_i \neq x_j. \end{cases} \qquad (2.42)$$

These conditions define $\phi_i, i = 1, \ldots, n$, uniquely. In addition, the ϕ_i's form a basis of the space V_h.

Each function of V_h can be expressed as

$$\phi(x) = \sum_{i=1}^{n} \xi_i \phi_i(x).$$

2.3. The Finite Element Method

The finite element approximation consists of writing the Galerkin condition (2.41) for functions in V_h. This defines the following approximate problem:

$$\text{Find } u \in V_h \text{ such that } a(u, v) = (f, v) \ \forall \ v \in V_h. \tag{2.43}$$

Since u is in V_h, there are n degrees of freedom. By the linearity of a with respect to v, it is only necessary to impose the condition $a(u, \phi_i) = (f, \phi_i)$ for $i = 1, \ldots, n$. This results in n constraints.

Writing the desired solution u in the basis $\{\phi_i\}$ as

$$u = \sum_{i=1}^{n} \xi_i \phi_i(x)$$

and substituting in (2.43) give the linear problem

$$\sum_{j=1}^{n} \alpha_{ij} \xi_i = \beta_i, \tag{2.44}$$

where

$$\alpha_{ij} = a(\phi_i, \phi_j), \quad \beta_i = (f, \phi_i).$$

The above equations form a linear system of equations

$$Ax = b,$$

in which the coefficients of A are the α_{ij}'s; those of b are the β_j's. In addition, A is a *symmetric positive definite* (SPD) matrix. Indeed, it is clear that

$$\int_\Omega \nabla \phi_i \nabla \phi_j \, dx = \int_\Omega \nabla \phi_j \nabla \phi_i \, dx,$$

which means that $\alpha_{ij} = \alpha_{ji}$. To see that A is positive definite, first note that $a(u, u) \geq 0$ for any function u. If $a(\phi, \phi) = 0$ for a function in V_h, then it must be true that $\nabla \phi = 0$ *almost everywhere* in Ω_h. Since ϕ is linear in each triangle and continuous, then it is clear that it must be constant on all Ω. Since, in addition, it vanishes on the boundary, then it must be equal to zero on all of Ω. The result follows by exploiting the relation

$$(A\xi, \xi) = a(\phi, \phi), \quad \text{with} \quad \phi = \sum_{i=1}^{n} \xi_i \phi_i,$$

which is valid for any vector $\{\xi_i\}_{i=1,\ldots,n}$.

Another important observation is that the matrix A is also sparse. Indeed, α_{ij} is nonzero only when the two basis functions ϕ_i and ϕ_j have common support triangles or, equivalently, when the nodes i and j are the vertices of a common triangle. Specifically, for a given node i, the coefficient α_{ij} will be nonzero only when the node j is one of the nodes of a triangle that is adjacent to node i.

In practice, the matrix is built by summing up the contributions of all triangles by applying the formula

$$a(\phi_i, \phi_j) = \sum_{K} a_K(\phi_i, \phi_j),$$

in which the sum is over all the triangles K and

$$a_K(\phi_i, \phi_j) = \int_K \nabla \phi_i \, \nabla \phi_j \, dx.$$

Note that $a_K(\phi_i, \phi_j)$ is zero unless the nodes i and j are both vertices of K. Thus, a triangle contributes nonzero values to its three vertices from the above formula. The 3×3 matrix

$$A_K = \begin{pmatrix} a_K(\phi_i, \phi_i) & a_K(\phi_i, \phi_j) & a_K(\phi_i, \phi_k) \\ a_K(\phi_j, \phi_i) & a_K(\phi_j, \phi_j) & a_K(\phi_j, \phi_k) \\ a_K(\phi_k, \phi_i) & a_K(\phi_k, \phi_j) & a_K(\phi_k, \phi_k) \end{pmatrix}$$

associated with the triangle $K(i, j, k)$ with vertices i, j, k is called an *element stiffness matrix*. In order to form the matrix A, it is necessary to sum up all the contributions $a_K(\phi_k, \phi_m)$ to the position k, m of the matrix. This process is called an *assembly* process. In the assembly, the matrix is computed as

$$A = \sum_{e=1}^{nel} A^{[e]}, \tag{2.45}$$

in which *nel* is the number of elements. Each of the matrices $A^{[e]}$ is of the form

$$A^{[e]} = P_e A_{K_e} P_e^T,$$

where A_{K_e} is the element matrix for the element K_e, as defined above. Also, P_e is an $n \times 3$ Boolean connectivity matrix that maps the coordinates of the 3×3 matrix A_{K_e} into the coordinates of the full matrix A.

Example 2.1. The assembly process can be illustrated with a very simple example. Consider the finite element mesh shown in Figure 2.8. The four elements are numbered from bottom to top, as indicated by the labels located at their centers. There are six nodes in this

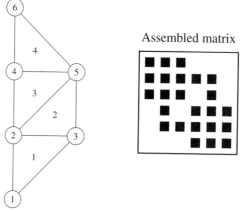

Figure 2.8. *A simple finite element mesh and the pattern of the corresponding assembled matrix.*

2.3. The Finite Element Method

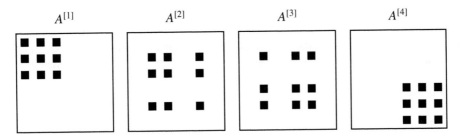

Figure 2.9. *The element matrices $A^{[e]}$, $e = 1, \ldots, 4$, for the finite element mesh shown in Figure* 2.8.

mesh and their labeling is indicated in the circled numbers. The four matrices $A^{[e]}$ associated with these elements are shown in Figure 2.9. Thus, the first element will contribute to the nodes 1, 2, 3; the second to nodes 2, 3, 5; the third to nodes 2, 4, 5; and the fourth to nodes 4, 5, 6.

In fact there are two different ways to represent and use the matrix A. We can form all the element matrices one by one and then we can store them, e.g., in an $nel \times 3 \times 3$ rectangular array. This representation is often called the *unassembled form* of A. Then the matrix A may be assembled if it is needed. However, element stiffness matrices can also be used in different ways without having to assemble the matrix. For example, *frontal techniques* are direct solution methods that take the linear system in unassembled form and compute the solution by a form of Gaussian elimination.

There are also iterative solution techniques that work directly with unassembled matrices. One of the main operations required in many iterative methods is to compute $y = Ax$, the product of the matrix A and an arbitrary vector x. In unassembled form, this can be achieved as follows:

$$y = Ax = \sum_{e=1}^{nel} A^{[e]} x = \sum_{e=1}^{nel} P_e A_{K_e} (P_e^T x). \tag{2.46}$$

Thus, the product $P_e^T x$ gathers the x data associated with the e element into a 3-vector consistent with the ordering of the matrix A_{K_e}. After this is done, this vector must be multiplied by A_{K_e}. Finally, the result is added to the current y-vector in appropriate locations determined by the P_e array. This sequence of operations must be done for each of the nel elements.

A more common, and somewhat more appealing, technique is to perform the assembly of the matrix. All the elements are scanned one by one and the nine associated contributions $a_K(\phi_k, \phi_m), k, m \in \{i, j, k\}$, added to the corresponding positions in the global stiffness matrix. The assembled matrix must now be stored but the element matrices may be discarded. The structure of the assembled matrix depends on the ordering of the nodes. To facilitate the computations, a widely used strategy transforms all triangles into a reference triangle with vertices $(0, 0)$, $(0, 1)$, $(1, 0)$. The area of the triangle is then simply the determinant of the Jacobian of the transformation that allows passage from one set of axes to the other.

Simple boundary conditions such as Neumann or Dirichlet do not cause any difficulty. The simplest way to handle Dirichlet conditions is to include boundary values as unknowns and modify the assembled system to incorporate the boundary values. Thus, each equation associated with the boundary point in the assembled system is replaced with the equation $u_i = f_i$. This yields a small identity block hidden within the linear system.

For Neumann conditions, Green's formula will give rise to the equations

$$\int_\Omega \nabla u \cdot \nabla \phi_j \, dx = \int_\Omega f \phi_j dx + \int_\Gamma \phi_j \frac{\partial u}{\partial n} \, ds, \qquad (2.47)$$

which will involve the Neumann data $\frac{\partial u}{\partial n}$ over the boundary. Since the Neumann data is typically given at some points only (the boundary nodes), linear interpolation (trapezoidal rule) or the mid-line value (midpoint rule) can be used to approximate the integral. Note that (2.47) can be viewed as the jth equation of the linear system. Another important point is that, if the boundary conditions are only of Neumann type, then the resulting system is singular. An equation must be removed or the linear system must be solved by taking this singularity into account.

2.4 Mesh Generation and Refinement

Generating a finite element triangulation can be done easily by exploiting some initial grid and then refining the mesh a few times either uniformly or in specific areas. The simplest refinement technique consists of taking the three midpoints of a triangle, thus creating four smaller triangles from a larger triangle and losing one triangle, namely, the original one. A systematic use of one level of this strategy is illustrated for the mesh in Figure 2.8 and is shown in Figure 2.10.

This approach has the advantage of preserving the angles of the original triangulation. This is an important property since the angles of a good-quality triangulation must satisfy certain bounds. On the other hand, the indiscriminate use of the uniform refinement strategy may lead to some inefficiencies. It is desirable to introduce more triangles in areas where the solution is likely to have large variations. In terms of vertices, midpoints should be introduced only where needed. To obtain standard finite element triangles, the points that have been created on the edges of a triangle must be linked to existing vertices in the triangle. This is because no vertex of a triangle is allowed to lie on the edge of another triangle.

Figure 2.11 shows three possible cases that can arise. The original triangle is (a). In (b), only one new vertex (numbered 4) has appeared on one edge of the triangle and it is joined to the vertex opposite to it. In (c), two new vertices appear inside the original triangle. There is no alternative but to join vertices 4 and 5. However, after this is done, either vertices 4 and 3 or vertices 1 and 5 must be joined. If angles are desired that will not become too small with further refinements, the second choice is clearly better in this case. In fact, various strategies for improving the quality of the triangles have been devised. The final case (d) corresponds to the *uniform refinement* case, where all edges have been split in two. There are three new vertices and four new elements and the larger initial element is removed.

2.4. Mesh Generation and Refinement

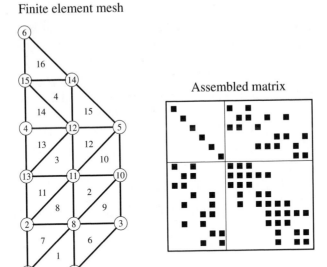

Figure 2.10. *The simple finite element mesh of Figure 2.8 after one level of refinement and the corresponding matrix.*

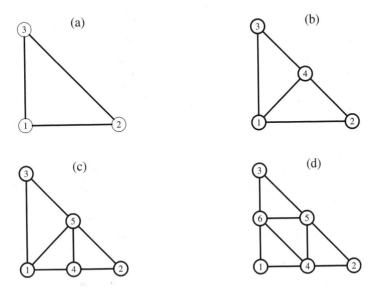

Figure 2.11. *Original triangle* (a) *and three possible refinement scenarios.*

2.5 Finite Volume Method

The finite volume method is geared toward the solution of conservation laws of the form

$$\frac{\partial u}{\partial t} + \nabla \cdot \vec{F} = Q. \tag{2.48}$$

In the above equation, $\vec{F}(u, t)$ is a certain vector function of u and time, possibly nonlinear. This is called the *flux vector*. The *source term* Q is a function of space and time. We now apply the principle used in the weak formulation, described before. Multiply both sides by a test function w and take the integral:

$$\int_\Omega w \frac{\partial u}{\partial t} dx + \int_\Omega w \nabla \cdot \vec{F} dx = \int_\Omega w Q dx.$$

Then integrate by parts using formula (2.39) for the second term on the left-hand side to obtain

$$\int_\Omega w \frac{\partial u}{\partial t} dx - \int_\Omega \nabla w \cdot \vec{F} dx + \int_\Gamma w \vec{F} \cdot \vec{n} ds = \int_\Omega w Q dx.$$

Consider now a *control volume* consisting, for example, of an elementary triangle K_i in the two-dimensional case, such as those used in the finite element method. Take for w a function w_i whose value is one on the triangle and zero elsewhere. The second term in the above equation vanishes and the following relation results:

$$\int_{K_i} \frac{\partial u}{\partial t} dx + \int_{\Gamma_i} \vec{F} \cdot \vec{n} ds = \int_{K_i} Q dx. \tag{2.49}$$

The above relation is at the basis of the finite volume approximation. To go a little further, the assumptions will be simplified slightly by taking a vector function $\vec{F}$ that is linear with respect to u. Specifically, assume

$$\vec{F} = \begin{pmatrix} \lambda_1 u \\ \lambda_2 u \end{pmatrix} \equiv \vec{\lambda} u.$$

Note that, in this case, the term $\nabla \cdot \vec{F}$ in (2.48) becomes $\vec{F}(u) = \vec{\lambda} \cdot \nabla u$. In addition, the right-hand side and the first term on the left-hand side of (2.49) can be approximated as follows:

$$\int_{K_i} \frac{\partial u}{\partial t} dx \approx \frac{\partial u_i}{\partial t} |K_i|, \quad \int_{K_i} Q dx \approx q_i |K_i|.$$

Here, $|K_i|$ represents the volume of K_i and q_i is some average value of Q in the cell K_i. (Note that, in two dimensions, volume is considered to mean area.) These are crude approximations but they serve the purpose of illustrating the scheme.

The finite volume equation (2.49) yields

$$\frac{\partial u_i}{\partial t} |K_i| + \vec{\lambda} \cdot \int_{\Gamma_i} u \vec{n} ds = q_i |K_i|. \tag{2.50}$$

The contour integral

$$\int_{\Gamma_i} u \vec{n} ds$$

2.5. Finite Volume Method

is the sum of the integrals over all edges of the control volume. Let the value of u on each edge j be approximated by some "average" $\bar{u}_j$. In addition, s_j denotes the length of each edge, and a common notation is

$$\vec{s}_j = s_j \vec{n}_j.$$

Then the contour integral is approximated by

$$\vec{\lambda} \cdot \int_{\Gamma_i} u\vec{n}\, ds \approx \sum_{edges} \bar{u}_j \vec{\lambda} \cdot \vec{n}_j s_j = \sum_{edges} \bar{u}_j \vec{\lambda} \cdot \vec{s}_j. \qquad (2.51)$$

The situation in the case where the control volume is a simple triangle is depicted in Figure 2.12. The unknowns are the approximations u_i of the function u associated with each cell. These can be viewed as approximations of u at the centers of gravity of each cell i. This type of model is called a *cell-centered* finite volume approximation. Other techniques based on using approximations of the vertices of the cells are known as *cell-vertex* finite volume techniques.

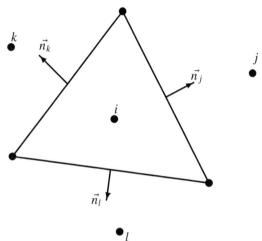

Figure 2.12. *Finite volume cell associated with node i and three neighboring cells.*

The value $\bar{u}_j$ required in (2.51) can be taken simply as the average of the approximation u_i of u in cell i and the approximation u_j in the cell j on the other side of the edge:

$$\bar{u}_j = \frac{1}{2}(u_j + u_i). \qquad (2.52)$$

This gives

$$\frac{\partial u_i}{\partial t}|K_i| + \frac{1}{2}\sum_j (u_i + u_j)\vec{\lambda} \cdot \vec{s}_j = q_i |K_i|.$$

One further simplification takes place by observing that

$$\sum_j \vec{s}_j = 0$$

and therefore
$$\sum_j u_i \vec{\lambda} \cdot \vec{s}_j = u_i \vec{\lambda} \cdot \sum_j \vec{s}_j = 0.$$

This yields
$$\frac{\partial u_i}{\partial t} |K_i| + \frac{1}{2} \sum_j u_j \vec{\lambda} \cdot \vec{s}_j = q_i |K_i|.$$

In the above equation, the summation is over all the neighboring cells j. One problem with such simple approximations is that they do not account for large gradients of u in the components. In finite volume approximations, it is typical to exploit upwind schemes, which are more suitable in such cases. By comparing with one-dimensional upwind schemes, it can be easily seen that the suitable modification to (2.52) is as follows:

$$\bar{u}_j = \frac{1}{2}(u_j + u_i) - \frac{1}{2} \operatorname{sign}(\vec{\lambda} \cdot \vec{s}_j)(u_j - u_i). \tag{2.53}$$

This gives
$$\frac{\partial u_i}{\partial t} |K_i| + \sum_j \vec{\lambda} \cdot \vec{s}_j \left(\frac{1}{2}(u_j + u_i) - \frac{1}{2} \operatorname{sign}(\vec{\lambda} \cdot \vec{s}_j)(u_j - u_i) \right) = q_i |K_i|.$$

Now write
$$\frac{\partial u_i}{\partial t} |K_i| + \sum_j \left(\frac{1}{2}(u_j + u_i)\vec{\lambda} \cdot \vec{s}_j - \frac{1}{2}|\vec{\lambda} \cdot \vec{s}_j|(u_j - u_i) \right) = q_i |K_i|,$$

$$\frac{\partial u_i}{\partial t} |K_i| + \sum_j \left(u_i(\vec{\lambda} \cdot \vec{s}_j)^+ + u_j(\vec{\lambda} \cdot \vec{s}_j)^- \right) = q_i |K_i|,$$

where
$$(z)^{\pm} \equiv \frac{z \pm |z|}{2}.$$

The equation for cell i takes the form
$$\frac{\partial u_i}{\partial t} |K_i| + \beta_i u_i + \sum_j \alpha_{ij} u_j = q_i |K_i|,$$

where
$$\beta_i = \sum_j (\vec{\lambda} \cdot \vec{s}_j)^+ \geq 0, \tag{2.54}$$

$$\alpha_{ij} = (\vec{\lambda} \cdot \vec{s}_j)^- \leq 0. \tag{2.55}$$

Thus, the diagonal elements of the matrix are nonnegative, while its off-diagonal elements are nonpositive. In addition, the row sum of the elements, i.e., the sum of all elements in the same row, is equal to zero. This is because

$$\beta_i + \sum_j \alpha_{ij} = \sum_j (\vec{\lambda} \cdot \vec{s}_j)^+ + \sum_j (\vec{\lambda} \cdot \vec{s}_j)^- = \sum_j \vec{\lambda} \cdot \vec{s}_j = \vec{\lambda} \cdot \sum_j \vec{s}_j = 0.$$

Exercises

The matrices obtained have the same desirable property of weak diagonal dominance seen in the one-dimensional case. A disadvantage of upwind schemes, whether in the context of irregular grids or in one-dimensional equations, is the loss of accuracy due to the low order of the schemes.

Exercises

1. Derive second and third order forward difference formulas similar to (2.8), i.e., involving $u(x), u(x+h), u(x+2h), \ldots$. Write down the discretization errors explicitly.

2. Derive a centered difference formula of at least third order for the first derivative, similar to (2.13).

3. Show that the upwind difference scheme described in Section 2.2.4, when a and $\vec{b}$ are constant, is stable for the model problem (2.7).

4. Develop the two nine-point formulas illustrated in Figure 2.4. Find the corresponding discretization errors. [Hint: Combine $\frac{1}{3}$ of the five-point formula (2.17) plus $\frac{2}{3}$ of the same formula based on the diagonal stencil $\{(x, y), (x+h, y+h) + (x+h, y-h), (x-h, y+h), (x-h, y-h)\}$ to get one formula. Use the reverse combination $\frac{2}{3}$, $\frac{1}{3}$ to get the other formula.]

5. Consider a (two-dimensional) rectangular mesh that is discretized as in the finite difference approximation. Show that the finite volume approximation to $\vec{\lambda} \cdot \nabla u$ yields the same matrix as an upwind scheme applied to the same problem. What would be the mesh of the equivalent upwind finite difference approximation?

6. Show that the right-hand side of (2.16) can also be written as

$$\frac{1}{h^2} \delta^- \left(a_{i+\frac{1}{2}} \delta^+ u \right).$$

7. Show that the formula (2.16) is indeed second order accurate for functions that are in C^4.

8. Show that the functions ϕ_i defined by (2.42) form a basis of V_h.

9. Develop the equivalent of Green's formula for the elliptic operator L defined in (2.6).

10. Write a short FORTRAN or C program to perform a matrix-by-vector product when the matrix is stored in unassembled form.

11. Consider the finite element mesh of Example 2.1. Compare the number of operations required to perform a matrix-by-vector product when the matrix is in assembled and unassembled forms. Compare also the storage required in each case. For a general finite element matrix, what can the ratio be between the two in the worst case (consider only linear approximations on triangular elements) for arithmetic? Express the number of operations in terms of the number of nodes and edges of the mesh. You may make the assumption that the maximum number of elements that are adjacent to a given node is p (e.g., $p = 8$).

12. Let K be a polygon in $\mathbb{R}^2$ with m edges and let $\vec{s}_j = s_j \vec{n}_j$ for $j = 1, \ldots, m$, where s_j is the length of the jth edge and $\vec{n}_j$ is the unit outward normal at the jth edge. Use the divergence theorem to prove that $\sum_{j=1}^{m} \vec{s}_j = 0$.

Notes and References

The books by Johnson [178], Ciarlet [84], and Strang and Fix [276] are recommended for good coverage of the finite element method. Axelsson and Barker [15] discuss solution techniques for finite element problems emphasizing iterative methods. For finite difference and finite volume methods, see Hirsch's book [168], which also discusses equations and solution methods for fluid flow problems. A 1965 article by Hockney [169] describes a one-level BCR method that seems to be the first FPS. BCR was developed by Buneman [69] and Hockney [170] for Poisson's equations and extended by Swartzrauber [278] to separable elliptic equations. An efficient combination of BCR and Fourier analysis known as FACR(l) was developed by Hockney [170] and later extended in [279] and [170]. Parallel BCR algorithms were considered in [138, 280].

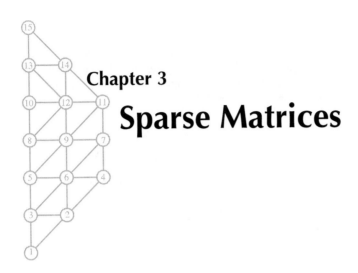

Chapter 3
Sparse Matrices

As described in the previous chapter, standard discretizations of partial differential equations (PDEs) typically lead to large and *sparse* matrices. A sparse matrix is defined, somewhat vaguely, as a matrix with very few nonzero elements. But, in fact, a matrix can be termed sparse whenever special techniques can be utilized to take advantage of the large number of zero elements and their locations. These sparse matrix techniques begin with the idea that the zero elements need not be stored. One of the key issues is to define data structures for these matrices that are well suited for efficient implementation of standard solution methods, whether direct or iterative. This chapter gives an overview of sparse matrices, their properties, their representations, and the data structures used to store them.

3.1 Introduction

The natural idea to take advantage of the zeros of a matrix and their location was initiated by engineers in various disciplines. In the simplest case involving banded matrices, special techniques are straightforward to develop. Electrical engineers dealing with electrical networks in the 1960s were the first to exploit sparsity to solve general sparse linear systems for matrices with irregular structure. The main issue, and the first addressed by sparse matrix technology, was to devise direct solution methods for linear systems. These had to be economical in terms of both storage and computational effort. Sparse direct solvers can handle very large problems that cannot be tackled by the usual "dense" solvers.

Essentially, there are two broad types of sparse matrices: *structured* and *unstructured*. A structured matrix is one whose nonzero entries form a regular pattern, often along a small number of diagonals. Alternatively, the nonzero elements may lie in blocks (dense submatrices) of the same size, which form a regular pattern, typically along a small number of (block) diagonals. A matrix with irregularly located entries is said to be irregularly structured. The best example of a regularly structured matrix is a matrix that consists of only a few diagonals. Finite difference matrices on rectangular grids, such as the ones seen in the previous chapter, are typical examples of matrices with regular structure. Most finite element or finite volume techniques applied to complex geometries lead to irregularly structured matrices. Figure 3.2 shows a small irregularly structured sparse matrix associated with the finite element grid problem shown in Figure 3.1.

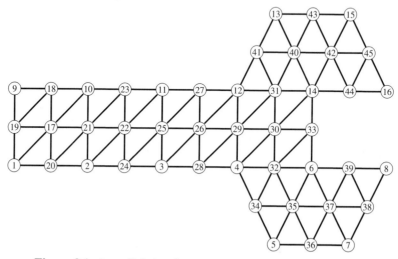

Figure 3.1. *A small finite element grid model.*

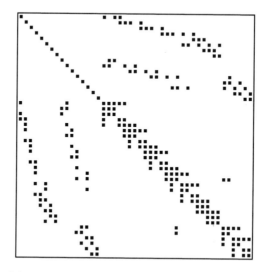

Figure 3.2. *Sparse matrix associated with the finite element grid of Figure 3.1.*

The distinction between the two types of matrices may not noticeably affect direct solution techniques and it has not received much attention in the past. However, this

3.2. Graph Representations

distinction can be important for iterative solution methods. In these methods, one of the essential operations is matrix-by-vector products. The performance of these operations can differ significantly on high performance computers, depending on whether or not they are regularly structured. For example, on vector computers, storing the matrix by diagonals is ideal, but the more general schemes may suffer because they require indirect addressing.

The next section discusses graph representations of sparse matrices. This is followed by an overview of some of the storage schemes used for sparse matrices and an explanation of how some of the simplest operations with sparse matrices can be performed. Then sparse linear system solution methods will be covered. Finally, Section 3.7 discusses test matrices.

3.2 Graph Representations

Graph theory is an ideal tool for representing the structure of sparse matrices and for this reason plays a major role in sparse matrix techniques. For example, graph theory is the key ingredient used in unraveling parallelism in sparse Gaussian elimination and in preconditioning techniques. In the following section, graphs are discussed in general terms and then their applications to finite element and finite difference matrices are discussed.

3.2.1 Graphs and Adjacency Graphs

Remember that a graph is defined by two sets, a set of vertices

$$V = \{v_1, v_2, \ldots, v_n\}$$

and a set of edges E that consists of pairs (v_i, v_j), where v_i, v_j are elements of V; i.e.,

$$E \subseteq V \times V.$$

This graph $G = (V, E)$ is often represented by a set of points in the plane linked by a directed line between the points that are connected by an edge. A graph is a way of representing a binary relation between objects of a set V. For example, V can represent the major cities of the world. A line is drawn between any two cities that are linked by a nonstop airline connection. Such a graph will represent the relation "there is a nonstop flight from city A to city B." In this particular example, the binary relation is likely to be symmetric; i.e., when there is a nonstop flight from A to B there is also a nonstop flight from B to A. In such situations, the graph is said to be undirected, as opposed to a general graph, which is directed.

Going back to sparse matrices, the *adjacency graph* of a sparse matrix is a graph $G = (V, E)$ whose n vertices in V represent the n unknowns. Its edges represent the binary relations established by the equations in the following manner: There is an edge from node i to node j when $a_{ij} \neq 0$. This edge will therefore represent the binary relation *equation i involves unknown j*. Note that the adjacency graph is an undirected graph when the matrix pattern is symmetric, i.e., when $a_{ij} \neq 0$ iff $a_{ji} \neq 0$ for all $1 \leq i, j \leq n$. See Figure 3.3 for an illustration.

When a matrix has a symmetric nonzero pattern, i.e., when a_{ij} and a_{ji} are always nonzero at the same time, then the graph is *undirected*. Thus, for undirected graphs, every edge points in both directions. As a result, undirected graphs can be represented with nonoriented edges.

As an example of the use of graph models, parallelism in Gaussian elimination can be extracted by finding unknowns that are independent at a given stage of the elimination. These are unknowns that do not depend on each other according to the above binary relation. The rows corresponding to such unknowns can then be used as pivots simultaneously. Thus, in one extreme, when the matrix is diagonal, then all unknowns are independent. Conversely, when a matrix is dense, each unknown will depend on all other unknowns. Sparse matrices lie somewhere between these two extremes.

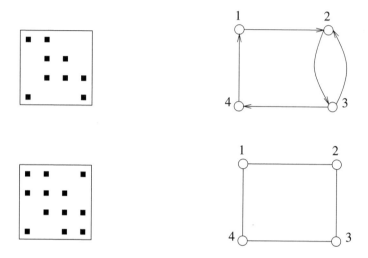

Figure 3.3. *Graphs of two* 4×4 *sparse matrices.*

There are a few interesting simple properties of adjacency graphs. The graph of A^2 can be interpreted as an n-vertex graph whose edges are the pairs (i, j) for which there exists at least one path of length exactly two from node i to node j in the original graph of A. Similarly, the graph of A^k consists of edges that represent the binary relation "there is at least one path of length k from node i to node j." For details, see Exercise 4.

3.2.2 Graphs of PDE Matrices

For PDEs involving only one physical unknown per mesh point, the adjacency graph of the matrix arising from the discretization is often the graph represented by the mesh itself. However, it is common to have several unknowns per mesh point. For example, the equations modeling fluid flow may involve the two velocity components of the fluid (in two dimensions) as well as energy and momentum at each mesh point.

In such situations, there are two choices when labeling the unknowns. They can be labeled contiguously at each mesh point. Thus, for the example just mentioned, we can label all four variables (two velocities followed by momentum and then pressure) at a given mesh point as $u(k), \ldots, u(k + 3)$. Alternatively, all unknowns associated with one type of variable can be labeled first (e.g., first velocity components), followed by those associated with the second type of variables (e.g., second velocity components), etc. In either case, it is clear that there is redundant information in the graph of the adjacency matrix.

The *quotient* graph corresponding to the *physical mesh* can be used instead. This results in substantial savings in storage and computation. In the fluid flow example mentioned above, the storage can be reduced by a factor of almost 16 for the integer arrays needed to represent the graph. This is because the number of edges has been reduced by this much, while the number of vertices, which is usually much smaller, remains the same.

3.3 Permutations and Reorderings

Permuting the rows or the columns, or both the rows and columns, of a sparse matrix is a common operation. In fact, *reordering* rows and columns is one of the most important ingredients used in *parallel* implementations of both direct and iterative solution techniques. This section introduces the ideas related to these reordering techniques and their relations to the adjacency graphs of the matrices. Recall the notation introduced in Chapter 1 that the jth column of a matrix is denoted by a_{*j} and the ith row by a_{i*}.

3.3.1 Basic Concepts

We begin with a definition and new notation.

Definition 3.1. *Let A be a matrix and* $\pi = \{i_1, i_2, \ldots, i_n\}$ *a permutation of the set* $\{1, 2, \ldots, n\}$. *Then the matrices*

$$A_{\pi,*} = \{a_{\pi(i),j}\}_{i=1,\ldots,n;\, j=1,\ldots,m},$$
$$A_{*,\pi} = \{a_{i,\pi(j)}\}_{i=1,\ldots,n;\, j=1,\ldots,m}$$

are called the row π-permutation and column π-permutation of A, respectively.

It is well known that any permutation of the set $\{1, 2, \ldots, n\}$ results from at most n interchanges, i.e., elementary permutations in which only two entries have been interchanged. An *interchange matrix* is the identity matrix with two of its rows interchanged. Denote by X_{ij} such matrices, with i and j being the numbers of the interchanged rows. Note that, in order to interchange rows i and j of a matrix A, we need only premultiply it by the matrix X_{ij}. Let $\pi = \{i_1, i_2, \ldots, i_n\}$ be an arbitrary permutation. This permutation is the product of a sequence of n consecutive interchanges $\sigma(i_k, j_k), k = 1, \ldots, n$. Then the rows of a matrix can be permuted by interchanging rows i_1, j_1, then rows i_2, j_2 of the resulting matrix, etc., and finally by interchanging i_n, j_n of the resulting matrix. Each of these operations can be achieved by a premultiplication by X_{i_k, j_k}. The same observation can be made regarding the columns of a matrix: In order to interchange columns i and j of a matrix, postmultiply it by X_{ij}. The following proposition follows from these observations.

Proposition 3.2. *Let π be a permutation resulting from the product of the interchanges* $\sigma(i_k, j_k), k = 1, \ldots, n$. *Then*

$$A_{\pi,*} = P_\pi A, \quad A_{*,\pi} = A Q_\pi,$$

where

$$P_\pi = X_{i_n,j_n} X_{i_{n-1},j_{n-1}} \cdots X_{i_1,j_1}, \tag{3.1}$$
$$Q_\pi = X_{i_1,j_1} X_{i_2,j_2} \cdots X_{i_n,j_n}. \tag{3.2}$$

Products of interchange matrices are called *permutation matrices*. Clearly, a permutation matrix is nothing but the identity matrix with its rows (or columns) permuted.

Observe that $X_{i,j}^2 = I$; i.e., the square of an interchange matrix is the identity or, equivalently, the inverse of an interchange matrix is equal to itself, a property that is intuitively clear. It is easy to see that the matrices (3.1) and (3.2) satisfy

$$P_\pi Q_\pi = X_{i_n,j_n} X_{i_{n-1},j_{n-1}} \cdots X_{i_1,j_1} \times X_{i_1,j_1} X_{i_2,j_2} \cdots X_{i_n,j_n} = I,$$

which shows that the two matrices Q_π and P_π are nonsingular and that they are the inverse of one another. In other words, permuting the rows and the columns of a matrix, *using the same permutation*, actually performs a similarity transformation. Another important consequence arises because the products involved in the definitions (3.1) and (3.2) of P_π and Q_π occur in reverse order. Since each of the elementary matrices X_{i_k,j_k} is symmetric, the matrix Q_π is the transpose of P_π. Therefore,

$$Q_\pi = P_\pi^T = P_\pi^{-1}.$$

Since the inverse of the matrix P_π is its own transpose, permutation matrices are unitary.

Another way of deriving the above relationships is to express the permutation matrices P_π and P_π^T in terms of the identity matrix whose columns or rows are permuted. It can easily be seen (see Exercise 3) that

$$P_\pi = I_{\pi,*}, \quad P_\pi^T = I_{*,\pi}.$$

It is then possible to verify directly that

$$A_{\pi,*} = I_{\pi,*} A = P_\pi A, \quad A_{*,\pi} = A I_{*,\pi} = A P_\pi^T.$$

It is important to interpret permutation operations for the linear systems to be solved. When the rows of a matrix are permuted, the order in which the equations are written is changed. On the other hand, when the columns are permuted, the unknowns are in effect *relabeled*, or *reordered*.

Example 3.1. Consider, for example, the linear system $Ax = b$, where

$$A = \begin{pmatrix} a_{11} & 0 & a_{13} & 0 \\ 0 & a_{22} & a_{23} & a_{24} \\ a_{31} & a_{32} & a_{33} & 0 \\ 0 & a_{42} & 0 & a_{44} \end{pmatrix}$$

and $\pi = \{1, 3, 2, 4\}$. Then the (column-) permuted linear system is

$$\begin{pmatrix} a_{11} & a_{13} & 0 & 0 \\ 0 & a_{23} & a_{22} & a_{24} \\ a_{31} & a_{33} & a_{32} & 0 \\ 0 & 0 & a_{42} & a_{44} \end{pmatrix} \begin{pmatrix} x_1 \\ x_3 \\ x_2 \\ x_4 \end{pmatrix} = \begin{pmatrix} b_1 \\ b_2 \\ b_3 \\ b_4 \end{pmatrix}.$$

3.3. Permutations and Reorderings

Note that only the unknowns have been permuted, not the equations and, in particular, the right-hand side has not changed.

In the above example, only the columns of A have been permuted. Such one-sided permutations are not as common as two-sided permutations in sparse matrix techniques. In reality, this is often related to the fact that the diagonal elements in linear systems play a distinct and important role. For instance, diagonal elements are typically large in PDE applications and it may be desirable to preserve this important property in the permuted matrix. In order to do so, it is typical to apply the same permutation to both the columns and the rows of A. Such operations are called *symmetric permutations* and, if denoted by $A_{\pi,\pi}$, then the result of such symmetric permutations satisfies the relation

$$A_{\pi,\pi} = P_\pi A P_\pi^T.$$

The interpretation of the symmetric permutation is quite simple. The resulting matrix corresponds to renaming, relabeling, or reordering the unknowns and then reordering the equations in the same manner.

Example 3.2. For the previous example, if the rows are permuted with the same permutation as the columns, the linear system obtained is

$$\begin{pmatrix} a_{11} & a_{13} & 0 & 0 \\ a_{31} & a_{33} & a_{32} & 0 \\ 0 & a_{23} & a_{22} & a_{24} \\ 0 & 0 & a_{42} & a_{44} \end{pmatrix} \begin{pmatrix} x_1 \\ x_3 \\ x_2 \\ x_4 \end{pmatrix} = \begin{pmatrix} b_1 \\ b_3 \\ b_2 \\ b_4 \end{pmatrix}.$$

Observe that the diagonal elements are now diagonal elements from the original matrix, placed in a different order on the main diagonal.

3.3.2 Relations with the Adjacency Graph

From the point of view of graph theory, another important interpretation of a symmetric permutation is that *it is equivalent to relabeling the vertices of the graph* without altering the edges. Indeed, let (i, j) be an edge in the adjacency graph of the original matrix A and let A' be the permuted matrix. Then $a'_{ij} = a_{\pi(i),\pi(j)}$ and, as a result, (i, j) is an edge in the adjacency graph of the permuted matrix A' iff $(\pi(i), \pi(j))$ is an edge in the graph of the original matrix A. In essence, it is as if we simply relabel each node having the "old" label $\pi(i)$ with the "new" label i. This is pictured in the following diagram:

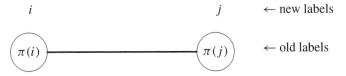

Thus, the graph of the permuted matrix has not changed; rather, the labeling of the vertices has. In contrast, nonsymmetric permutations do not preserve the graph. In fact, they can transform an undirected graph into a directed one. Symmetric permutations change the order in which the nodes are considered in a given algorithm (such as Gaussian elimination), which may have a tremendous impact on the performance of the algorithm.

Example 3.3. Consider the matrix illustrated in Figure 3.4 together with its adjacency graph. Such matrices are sometimes called "arrow" matrices because of their shape, but it would probably be more accurate to term them "star" matrices because of the structure of their graphs. If the equations are reordered using the permutation 9, 8, ..., 1, the matrix and graph shown in Figure 3.5 are obtained.

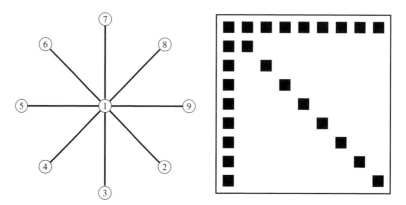

Figure 3.4. *Pattern of a 9×9 arrow matrix and its adjacency graph.*

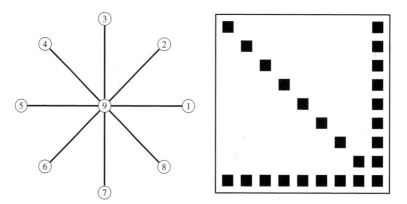

Figure 3.5. *Adjacency graph and matrix obtained from Figure 3.4 after permuting the nodes in reverse order.*

Although the difference between the two graphs may seem slight, the matrices have a completely different structure, which may have a significant impact on the algorithms. As an example, if Gaussian elimination is used on the reordered matrix, no fill-in will occur; i.e., the L and U parts of the LU factorization will have the same structure as the lower and upper parts of A, respectively.

On the other hand, Gaussian elimination on the original matrix results in disastrous fill-ins. Specifically, the L and U parts of the LU factorization are now dense matrices after the first step of Gaussian elimination. With direct sparse matrix techniques, it is important

3.3. Permutations and Reorderings

to find permutations of the matrix that will have the effect of reducing fill-ins during the Gaussian elimination process.

To conclude this section, it should be mentioned that two-sided nonsymmetric permutations may also arise in practice. However, they are more common in the context of direct methods.

3.3.3 Common Reorderings

The type of reordering, or permutations, used in applications depends on whether a direct or an iterative method is being considered. The following is a sample of such reorderings, which are more useful for iterative methods.

Level-Set Orderings. This class of orderings contains a number of techniques that are based on traversing the graph by *level sets*. A level set is defined recursively as the set of all unmarked neighbors of all the nodes of a previous level set. Initially, a level set consists of one node, although strategies with several starting nodes are also important and will be considered later. As soon as a level set is traversed, its nodes are marked and numbered. They can, for example, be numbered in the order in which they are traversed. In addition, the order in which each level itself is traversed gives rise to different orderings. For instance, the nodes of a certain level can be visited in the natural order in which they are listed. The neighbors of each of these nodes are then inspected. Each time a neighbor of a visited vertex that is not numbered is encountered, it is added to the list and labeled as the next element of the next level set. This simple strategy is called *breadth first search* (BFS) traversal in graph theory. The ordering will depend on the way in which the nodes are traversed in each level set. In BFS the elements of a level set are always traversed in the natural order in which they are listed. In the *Cuthill–McKee (CMK) ordering* the nodes adjacent to a visited node are always traversed from lowest to highest degree.

ALGORITHM 3.1. BFS (G, v)

1. Initialize $S = \{v\}$, $seen = 1$, $\pi(seen) = v$; Mark v
2. While $seen < n$, Do
3. $S_{new} = \emptyset$
4.   For each node v in S, Do
5.   For each unmarked w in $adj(v)$, Do
6.   Add w to S_{new}
7.   Mark w
8. $\pi(++seen) = w$
9.   EndDo
10. $S := S_{new}$
11.   EndDo
12. EndWhile

In the above algorithm, the notation $\pi(++seen) = w$ in line 8 uses a style borrowed from the C/C++ language. It states that *seen* should be first incremented by one, and then

$\pi(seen)$ assigned w. Two important modifications will be made to this algorithm to obtain the CMK ordering. The first concerns the selection of the first node to begin the traversal. The second, mentioned above, is the order in which the nearest neighbors of a given node are traversed.

ALGORITHM 3.2. CMK (G)

0. *Find an initial node v for the traversal*
1. *Initialize $S = \{v\}, seen = 1, \pi(seen) = v$; Mark v*
2. *While $seen < n$, Do*
3. $S_{new} = \emptyset$
4.   *For each node v, Do*
5. $\pi(++seen) = v$
6.   *For each unmarked w in adj(v), going from lowest to highest degree, Do*
7.   *Add w to S_{new}*
8.   *Mark w*
9.   *EndDo*
10. $S := S_{new}$
11.   *EndDo*
12. *EndWhile*

The π array obtained from the procedure lists the nodes in the order in which they are visited and can, in a practical implementation, be used to store the level sets in succession. A pointer is needed to indicate where each set starts.

The main property of level sets is that, with the exception of the first and the last levels, they are *graph separators*. A graph separator is a set of vertices whose removal separates the graph into two disjoint components. In fact, if there are l levels and $V_1 = S_1 \cup S_2 \cdots S_{i-1}$, $V_2 = S_{i+1} \cup \cdots S_l$, then the nodes of V_1 and V_2 are not coupled. This is easy to prove by contradiction. A major consequence of this property is that the matrix resulting from the CMK (or BFS) ordering is block tridiagonal, with the ith block being of size $|S_i|$.

In order to explain the concept of level sets, the previous two algorithms were described with the explicit use of level sets. A more common, and somewhat simpler, implementation relies on *queues*. The queue implementation is as follows.

ALGORITHM 3.3. CMK (G)—Queue Implementation

0. *Find an initial node v for the traversal*
1. *Initialize $Q = \{v\}$; Mark v*
2. *While $|Q| < n$, Do*
3. $head + +$
4.   *For each unmarked w in adj(h), going from lowest to highest degree, Do*
5.   *Append w to Q*
6.   *Mark w*
7.   *EndDo*
8. *EndWhile*

The final array Q will give the desired permutation π. Clearly, this implementation can also be applied to BFS. As an example, consider the finite element mesh problem illustrated

3.3. Permutations and Reorderings

in Figure 2.10 of Chapter 2 and assume that $v = 3$ is the initial node of the traversal. The state of the Q array after each step along with the head vertex *head* and its adjacency list are shown in the following table. Note that the adjacency lists in the third column are listed by increasing degree.

Q	head	$adj(head)$
3	3	7, 10, 8
3, 7, 10, 8	7	1, 9
3, 7, 10, 8, 1, 9	10	5, 11
3, 7, 10, 8, 1, 9, 5, 11	8	2
3, 7, 10, 8, 1, 9, 5, 11, 2	1	-
3, 7, 10, 8, 1, 9, 5, 11, 2	9	-
3, 7, 10, 8, 1, 9, 5, 11, 2	5	14, 12
3, 7, 10, 8, 1, 9, 5, 11, 2, 14, 12	11	13
3, 7, 10, 8, 1, 9, 5, 11, 2, 14, 12, 13	2	-
3, 7, 10, 8, 1, 9, 5, 11, 2, 14, 12, 13	14	6, 15
3, 7, 10, 8, 1, 9, 5, 11, 2, 14, 12, 13, 6, 15	12	4
3, 7, 10, 8, 1, 9, 5, 11, 2, 14, 12, 13, 6, 15, 4		

An implementation using explicit levels would find the sets $S_1 = \{3\}$, $S_2 = \{7, 8, 10\}$, $S_3 = \{1, 9, 5, 11, 2\}$, $S_4 = \{14, 12, 13\}$, and $S_5 = \{6, 15, 4\}$. The new labeling of the graph along with the corresponding matrix pattern are shown in Figure 3.6. The partitioning of the matrix pattern corresponds to the levels.

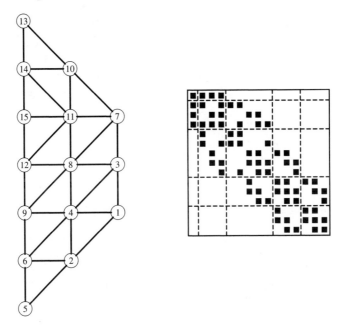

Figure 3.6. *Graph and matrix pattern for example of Figure 2.10 after CMK ordering.*

In 1971, George [142] observed that *reversing* the CMK ordering yields a better scheme for sparse Gaussian elimination. The simplest way to understand this is to look at the two graphs produced by these orderings. The results of the standard and reverse CMK (RCM) orderings on the sample finite element mesh problem seen earlier are shown in Figures 3.6 and 3.7, when the initial node is $i_1 = 3$ (relative to the labeling of the original ordering of Figure 2.10). The case of the figure corresponds to a variant of CMK in which the traversal in line 6 is done in a random order instead of according to the degree. A large part of the structure of the two matrices consists of little "arrow" submatrices, similar to the ones seen in Example 3.3. In the case of the regular CMK ordering, these arrows point upward, as in Figure 3.4, a consequence of the level-set labeling. These blocks are similar to the star matrices of Figure 3.4. As a result, Gaussian elimination will essentially fill in the square blocks that they span. As was indicated in Example 3.3, a remedy is to reorder the nodes backward, as is done globally in the RCM strategy. For the RCM ordering, the arrows are pointing downward, as in Figure 3.5, and Gaussian elimination yields much less fill-in.

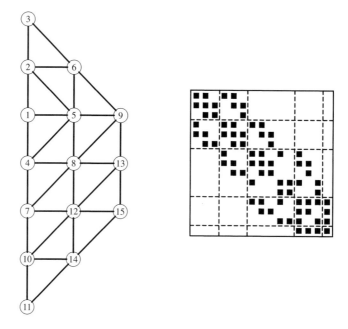

Figure 3.7. *RCM ordering.*

Example 3.4. The choice of the initial node in the CMK and RCM orderings may be important. Referring to the original ordering of Figure 2.10, the previous illustration used $i_1 = 3$. However, it is clearly a poor choice if matrices with small bandwidth or *profile* are desired. If $i_1 = 1$ is selected instead, then the RCM algorithm produces the matrix in Figure 3.8, which is more suitable for banded or *skyline* solvers.

Independent Set Orderings (ISOs). The matrices that arise in the model finite element problems seen in Figures 2.7, 2.10, and 3.2 are all characterized by an upper-left block that

3.3. Permutations and Reorderings

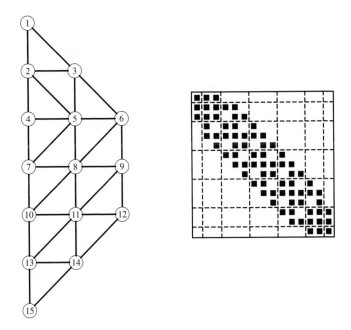

Figure 3.8. *RCM starting with* $i_1 = 1$.

is diagonal; i.e., they have the structure

$$A = \begin{pmatrix} D & E \\ F & C \end{pmatrix}, \qquad (3.3)$$

in which D is diagonal and C, E, and F are sparse matrices. The upper diagonal block corresponds to unknowns from the previous levels of refinement and its presence is due to the ordering of the equations in use. As new vertices are created in the refined grid, they are given new numbers and the initial numbering of the vertices is unchanged. Since the old connected vertices are "cut" by new ones, they are no longer related by equations. Sets such as these are called *independent sets*. Independent sets are especially useful in parallel computing for implementing both direct and iterative methods.

Referring to the adjacency graph $G = (V, E)$ of the matrix and denoting by (x, y) the edge from vertex x to vertex y, an *independent set* S is a subset of the vertex set V such that,

$$\text{if } x \in S, \quad \text{then} \quad \{(x, y) \in E \text{ or } (y, x) \in E\} \rightarrow y \notin S.$$

In other words, elements of S are not allowed to be connected to other elements of S either by incoming or outgoing edges. An independent set is *maximal* if it cannot be augmented by elements in its complement to form a larger independent set. Note that a maximal independent set is by no means the largest possible independent set that can be found. In fact, finding the independent set of maximum cardinality is *NP*-hard [182]. In the following, the term *independent set* always refers to *maximal independent set*.

There are a number of simple and inexpensive heuristics for finding large maximal independent sets. A greedy heuristic traverses the nodes in a given order and, if a node is not already marked, it selects the node as a new member of S. Then this node is marked along with its nearest neighbors. Here, a nearest neighbor of a node x means any node linked to x by an incoming or an outgoing edge.

ALGORITHM 3.4. Greedy Algorithm for Independent Set Ordering

1. Set $S = \emptyset$
2. For $j = 1, 2, \ldots, n$, Do
3.   If node j is not marked then
4. $S = S \cup \{j\}$
5.   Mark j and all its nearest neighbors
6.   EndIf
7. EndDo

In the above algorithm, the nodes are traversed in the natural order $1, 2, \ldots, n$, but they can also be traversed in any permutation $\{i_1, \ldots, i_n\}$ of $\{1, 2, \ldots, n\}$. Since the size of the reduced system is $n - |S|$, it is reasonable to try to maximize the size of S in order to obtain a small reduced system. It is possible to give a rough idea of the size of S. Assume that the maximum degree of each node does not exceed ν. Whenever the above algorithm accepts a node as a new member of S, it potentially puts all its nearest neighbors, i.e., at most ν nodes, in the complement of S. Therefore, if s is the size of S, the size of its complement, $n - s$, is such that $n - s \leq \nu s$ and, as a result,

$$s \geq \frac{n}{1 + \nu}.$$

This lower bound can be improved slightly by replacing ν with the maximum degree ν_S of *all the vertices that constitute S*. This results in the inequality

$$s \geq \frac{n}{1 + \nu_S},$$

which suggests that it may be a good idea to first visit the nodes with smaller degrees. In fact, this observation leads to a general heuristic regarding a good order of traversal. The algorithm can be viewed as follows: Each time a node is visited, remove it and its nearest neighbors from the graph, and then visit a node from the remaining graph. Continue in the same manner until all nodes are exhausted. Every node that is visited is a member of S and its nearest neighbors are members of $\bar{S}$. As a result, if ν_i is the degree of the node visited at step i, adjusted for all the edge deletions resulting from the previous visitation steps, then the number n_i of nodes that are left at step i satisfies the relation

$$n_i = n_{i-1} - \nu_i - 1.$$

The process adds a new element to the set S at each step and stops when $n_i = 0$. In order to maximize $|S|$, the number of steps in the procedure must be maximized. The difficulty in the analysis arises from the fact that the degrees are updated at each step i because of the removal of the edges associated with the removed nodes. If the process is to be lengthened, a rule of thumb would be to visit first the nodes that have the smallest degree.

3.3. Permutations and Reorderings

ALGORITHM 3.5. Increasing-Degree Traversal for ISO

1. Set $S = \emptyset$. Find an ordering $i_1, \ldots, i_n$ of the nodes by increasing degree
2. For $j = 1, 2, \ldots n$, Do
3.   If node i_j is not marked then
4. $S = S \cup \{i_j\}$
5.   Mark i_j and all its nearest neighbors
6.   EndIf
7. EndDo

A refinement to the above algorithm would be to update the degrees of all nodes involved in a removal and dynamically select the one with the smallest degree as the next node to be visited. This can be implemented efficiently using a min-heap data structure. A different heuristic is to attempt to maximize the number of elements in S by a form of local optimization that determines the order of traversal dynamically. In the following, removing a vertex from a graph means deleting the vertex and all edges incident to/from this vertex.

Example 3.5. The algorithms described in this section were tested on the same example used before, namely, the finite element mesh problem of Figure 2.10. Here, all strategies used yield the initial independent set in the matrix itself, which corresponds to the nodes of all the previous levels of refinement. This may well be optimal in this case; i.e., a larger independent set may not exist.

Multicolor Orderings. Graph coloring is a familiar problem in computer science that refers to the process of labeling (coloring) the nodes of a graph in such a way that no two adjacent nodes have the same label (color). The goal of graph coloring is to obtain a colored graph that uses the smallest possible number of colors. However, optimality in the context of numerical linear algebra is a secondary issue and simple heuristics do provide adequate colorings. Basic methods for obtaining a multicoloring of an arbitrary grid are quite simple. They rely on greedy techniques, a simple version of which is as follows.

ALGORITHM 3.6. Greedy Multicoloring Algorithm

1. For $i = 1, \ldots, n$, Do set $Color(i) = 0$
2. For $i = 1, 2, \ldots, n$, Do
3.   Set $Color(i) = \min\ \{k > 0 \mid k \neq Color(j)\ \forall\ j \in \text{Adj}(i)\}$
4. EndDo

Line 3 assigns the smallest *allowable* color number to node i. Allowable means a positive number that is different from the colors of the neighbors of node i. The procedure is illustrated in Figure 3.9. The node being colored in the figure is indicated by an arrow. It will be assigned color number 3, the smallest positive integer different from 1, 2, 4, 5.

In the above algorithm, the order $1, 2, \ldots, n$ has been arbitrarily selected for traversing the nodes and coloring them. Instead, the nodes can be traversed in any order $\{i_1, i_2, \ldots, i_n\}$. If a graph is *bipartite*, i.e., if it can be colored with two colors, then the algorithm will find the optimal two-color (red-black) ordering for *breadth first* traversals. In addition, if a graph is bipartite, it is easy to show that the algorithm will find two colors for any traversal

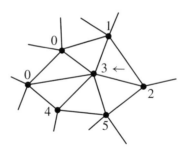

Figure 3.9. *The greedy multicoloring algorithm.*

that, at a given step, visits an unmarked node that is adjacent to at least one visited node. In general, the number of colors needed does not exceed the maximum degree of each node +1. These properties are the subject of Exercises 11 and 10.

Example 3.6. Figure 3.10 illustrates the algorithm for the same example used earlier, i.e., the finite element mesh problem of Figure 2.10. The dashed lines separate the different color sets found. Four colors are found in this example.

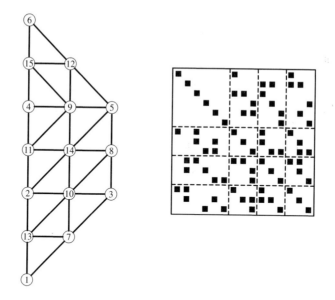

Figure 3.10. *Graph and matrix corresponding to mesh of Figure 2.10 after multicolor ordering.*

Once the colors have been found, the matrix can be permuted to have a block structure in which the diagonal blocks are diagonal. Alternatively, the color sets $S_j = [i_1^{(j)}, \ldots, i_{n_j}^{(j)}]$ and the permutation array in the algorithms can be used.

3.3.4 Irreducibility

Remember that a *path* in a graph is a sequence of vertices $v_1, v_2, \ldots, v_k$ such that (v_i, v_{i+1}) is an edge for $i = 1, \ldots, k-1$. Also, a graph is said to be *connected* if there is a path between any pair of vertices in V. A *connected component* in a graph is a *maximal subset* of vertices that can all be connected to one another by paths in the graph. Now consider matrices whose graphs may be *directed*. A matrix is *reducible* if its graph is not connected and *irreducible* otherwise. When a matrix is reducible, then it can be permuted by means of *symmetric* permutations into a block upper triangular matrix of the form

$$\begin{pmatrix} A_{11} & A_{12} & A_{13} & \cdots \\ & A_{22} & A_{23} & \cdots \\ & & \ddots & \vdots \\ & & & A_{pp} \end{pmatrix},$$

where each partition corresponds to a connected component. It is clear that linear systems with the above matrix can be solved through a sequence of subsystems with the matrices $A_{ii}, i = p, p-1, \ldots, 1$.

3.4 Storage Schemes

In order to take advantage of the large number of zero elements, special schemes are required to store sparse matrices. The main goal is to represent only the nonzero elements and to be able to perform the common matrix operations. In the following, Nz denotes the total number of nonzero elements.

The simplest storage scheme for sparse matrices is the so-called coordinate format. The data structure consists of three arrays: (1) a real array containing all the real (or complex) values of the nonzero elements of A in any order, (2) an integer array containing their row indices, and (3) a second integer array containing their column indices. All three arrays are of length Nz, the number of nonzero elements.

Example 3.7. The matrix

$$A = \begin{pmatrix} 1. & 0. & 0. & 2. & 0. \\ 3. & 4. & 0. & 5. & 0. \\ 6. & 0. & 7. & 8. & 9. \\ 0. & 0. & 10. & 11. & 0. \\ 0. & 0. & 0. & 0. & 12. \end{pmatrix}$$

will be represented (for example) by

AA	12.	9.	7.	5.	1.	2.	11.	3.	6.	4.	8.	10.
JR	5	3	3	2	1	1	4	2	3	2	3	4
JC	5	5	3	4	1	4	4	1	1	2	4	3

In the above example, the elements are listed in arbitrary order. In fact, they are usually listed by row or column. If the elements were listed by row, the array JC, which contains redundant information, might be replaced with an array that points to the beginning of each row instead. This would involve nonnegligible savings in storage. The new data structure has three arrays with the following functions:

- A real array AA contains the real values a_{ij} stored row by row, from row 1 to n. The length of AA is Nz.

- An integer array JA contains the column indices of the elements a_{ij} as stored in the array AA. The length of JA is Nz.

- An integer array IA contains the pointers to the beginning of each row in the arrays AA and JA. Thus, the content of $IA(i)$ is the position in arrays AA and JA where the ith row starts. The length of IA is $n+1$ with $IA(n+1)$ containing the number $IA(1) + Nz$, i.e., the address in A and JA of the beginning of a fictitious row number $n+1$.

Thus, the above matrix may be stored as follows:

AA	1.	2.	3.	4.	5.	6.	7.	8.	9.	10.	11.	12.
JA	1	4	1	2	4	1	3	4	5	3	4	5

IA	1	3	6	10	12	13

This format is probably the most popular for storing general sparse matrices. It is called the *compressed sparse row* (CSR) format. This scheme is preferred over the coordinate scheme because it is often more useful for performing typical computations. On the other hand, the coordinate scheme is advantageous for its simplicity and its flexibility. It is often used as an *entry* format in sparse matrix software packages.

There are a number of variations for the CSR format. The most obvious variation is storing the columns instead of the rows. The corresponding scheme is known as the *compressed sparse column* (CSC) scheme.

Another common variation exploits the fact that the diagonal elements of many matrices are all usually nonzero and/or that they are accessed more often than the rest of the elements. As a result, they can be stored separately. The *modified sparse row* (MSR) format has only two arrays: a real array AA and an integer array JA. The first n positions in AA contain the diagonal elements of the matrix in order. The unused position $n+1$ of the array AA may sometimes carry information concerning the matrix.

Starting at position $n+2$, the nonzero entries of AA, excluding its diagonal elements, are stored by row. For each element $AA(k)$, the integer $JA(k)$ represents its column index on the matrix. The $n+1$ first positions of JA contain the pointer to the beginning of each row in AA and JA. Thus, for the above example, the two arrays are as follows:

AA	1.	4.	7.	11.	12.	*	2.	3.	5.	6.	8.	9.	10.
JA	7	8	10	13	14	14	4	1	4	1	4	5	3

3.4. Storage Schemes

The star denotes an unused location. Notice that $JA(n) = JA(n+1) = 14$, indicating that the last row is a zero row, once the diagonal element has been removed.

Diagonally structured matrices are matrices whose nonzero elements are located along a small number of diagonals. These diagonals can be stored in a rectangular array `DIAG(1:n,1:Nd)`, where `Nd` is the number of diagonals. The offsets of each of the diagonals with respect to the main diagonal must be known. These will be stored in an array `IOFF(1:Nd)`. Thus, the element $a_{i,i+\text{ioff}(j)}$ of the original matrix is located in position (i, j) of the array `DIAG`; i.e.,

$$\text{DIAG}(i, j) \leftarrow a_{i,i+\text{ioff}(j)}.$$

The order in which the diagonals are stored in the columns of `DIAG` is generally unimportant, though if several more operations are performed with the main diagonal, storing it in the first column may be slightly advantageous. Note also that all the diagonals except the main diagonal have fewer than n elements, so there are positions in `DIAG` that will not be used.

Example 3.8. For example, the following matrix:

$$A = \begin{pmatrix} 1. & 0. & 2. & 0. & 0. \\ 3. & 4. & 0. & 5. & 0. \\ 0. & 6. & 7. & 0. & 8. \\ 0. & 0. & 9. & 10. & 0. \\ 0. & 0. & 0. & 11. & 12. \end{pmatrix},$$

which has three diagonals, will be represented by the two arrays

$$\text{DIAG} = \begin{bmatrix} * & 1. & 2. \\ 3. & 4. & 5. \\ 6. & 7. & 8. \\ 9. & 10. & * \\ 11 & 12. & * \end{bmatrix}, \quad \text{IOFF} = \begin{bmatrix} -1 & 0 & 2 \end{bmatrix}.$$

A more general scheme that is popular on vector machines is the so-called Ellpack-Itpack format. The assumption in this scheme is that there are at most `Nd` nonzero elements per row, where `Nd` is small. Then two rectangular arrays of dimension $n \times \text{Nd}$ each are required (one real and one integer). The first, `COEF`, is similar to `DIAG` and contains the nonzero elements of A. The nonzero elements of each row of the matrix can be stored in a row of the array `COEF(1:n,1:Nd)`, completing the row by zeros as necessary. Together with `COEF`, an integer array `JCOEF(1:n,1:Nd)` must be stored that contains the column positions of each entry in `COEF`.

Example 3.9. Thus, for the matrix of the previous example, the Ellpack-Itpack storage scheme is

$$\text{COEF} = \begin{bmatrix} 1. & 2. & 0. \\ 3. & 4. & 5. \\ 6. & 7. & 8. \\ 9. & 10. & 0. \\ 11 & 12. & 0. \end{bmatrix}, \quad \text{JCOEF} = \begin{bmatrix} 1 & 3 & 1 \\ 1 & 2 & 4 \\ 2 & 3 & 5 \\ 3 & 4 & 4 \\ 4 & 5 & 5 \end{bmatrix}.$$

A certain column number must be chosen for each of the zero elements that must be added to pad the shorter rows of A, i.e., rows 1, 4, and 5. In this example, those integers are selected to be equal to the row numbers, as can be seen in the JCOEF array. This is somewhat arbitrary and, in fact, any integer between 1 and n would be acceptable. However, there may be good reasons for not inserting the same integers too often, e.g., a constant number, for performance considerations.

3.5 Basic Sparse Matrix Operations

The matrix-by-vector product is an important operation that is required in most of the iterative solution algorithms for solving sparse linear systems. This section shows how these can be implemented for a small subset of the storage schemes considered earlier.

The following FORTRAN 90 segment shows the main loop of the matrix-by-vector operation for matrices stored in the CSR stored format:

```
DO I=1, N
   K1 = IA(I)
   K2 = IA(I+1)-1
   Y(I) = DOTPRODUCT(A(K1:K2),X(JA(K1:K2)))
ENDDO
```

Notice that each iteration of the loop computes a different component of the resulting vector. This is advantageous because each of these components can be computed independently. If the matrix is stored by columns, then the following code could be used instead:

```
DO J=1, N
   K1 = IA(J)
   K2 = IA(J+1)-1
   Y(JA(K1:K2)) = Y(JA(K1:K2))+X(J)*A(K1:K2)
ENDDO
```

In each iteration of the loop, a multiple of the jth column is added to the result, which is assumed to have been initially set to zero. Notice now that the outer loop is no longer parallelizable. An alternative to improve parallelization is to try to split the vector operation in each inner loop. The inner loop has few operations, in general, so this is unlikely to be a sound approach. This comparison demonstrates that data structures may have to change to improve performance when dealing with high performance computers.

Now consider the matrix-by-vector product in diagonal storage:

```
DO J=1, NDIAG
   JOFF = IOFF(J)
   DO I=1, N
      Y(I) = Y(I)+DIAG(I,J)*X(JOFF+I)
   ENDDO
ENDDO
```

Here, each of the diagonals is multiplied by the vector x and the result added to the vector y. It is again assumed that the vector y has been filled with zeros at the start of the loop. From the point of view of parallelization and/or vectorization, the above code is probably the better to use. On the other hand, it is not general enough.

Solving a lower or upper triangular system is another important kernel in sparse matrix computations. The following segment of code shows a simple routine for solving a unit lower triangular system $Lx = y$ for the CSR storage format:

```
X(1) = Y(1)
DO I = 2, N
    K1 = IAL(I)
    K2 = IAL(I+1)-1
    X(I) =Y(I) -DOTPRODUCT(AL(K1:K2),X(JAL(K1:K2)))
ENDDO
```

At each step, the inner product of the current solution x with the ith row is computed and subtracted from $y(i)$. This gives the value of $x(i)$. The DOTPRODUCT function computes the dot product of two arbitrary vectors u(k1:k2) and v(k1:k2). The vector AL(K1:K2) is the ith row of the matrix L in sparse format and X(JAL(K1:K2)) is the vector of the components of X *gathered* into a short vector that is consistent with the column indices of the elements in the row AL(K1:K2).

3.6 Sparse Direct Solution Methods

Most direct methods for sparse linear systems perform an LU factorization of the original matrix and try to reduce cost by minimizing fill-ins, i.e., nonzero elements introduced during the elimination process in positions that were initially zeros. The data structures employed are rather complicated. The early codes relied heavily on *linked lists*, which are convenient for inserting new nonzero elements. Linked-list data structures were dropped in favor of other more dynamic schemes that leave some initial elbow room in each row for the insertions and then adjust the structure as more fill-ins are introduced.

A typical sparse direct solution solver for positive definite matrices consists of four phases. First, preordering is applied to reduce fill-in. Two popular methods are used: minimum degree (MD) ordering and nested dissection (ND) ordering. Second, a symbolic factorization is performed. This means that the factorization is processed only symbolically, i.e., without numerical values. Third, the numerical factorization, in which the actual factors L and U are formed, is processed. Finally, the forward and backward triangular sweeps are executed for each different right-hand side. In a code where numerical pivoting is necessary, the symbolic phase cannot be separated from the numerical factorization.

3.6.1 MD Ordering

The MD algorithm is perhaps the most popular strategy for minimizing fill-in in sparse Gaussian elimination, specifically for symmetric positive definite (SPD) matrices. At a given step of Gaussian elimination, this strategy selects the node with the smallest degree as the next pivot row. This will tend to reduce fill-in. To be exact, it will minimize (locally) an upper bound for the number of fill-ins that will be introduced at the corresponding step of Gaussian elimination.

In contrast with the CMK ordering, MD ordering does not have, nor does it attempt to have, a banded structure. While the algorithm is excellent for sparse direct solvers, it has been observed that it does not perform as well as the RCM ordering when used in conjunction with preconditioning (Chapter 10).

The multiple MD algorithm is a variation due to Liu [203, 143] that exploits independent sets of pivots at each step. Degrees of nodes adjacent to any vertex in the independent set are updated only after all vertices in the set are processed.

3.6.2 ND Ordering

ND is used primarily to reduce fill-in in sparse direct solvers for SPD matrices. The technique is easily described with the help of recursivity and by exploiting the concept of *separators*. A set S of vertices in a graph is called a separator if the removal of S results in the graph being split into two disjoint subgraphs. For example, each of the intermediate levels in the BFS algorithm is in fact a separator. The ND algorithm can be succinctly described by the following algorithm.

ALGORITHM 3.7. ND(G, nmin)

1. If $|V| \leq nmin$
2.   Label nodes of V
3. Else
4.   Find a separator S for V
5.   Label the nodes of S
6.   Split V into G_L, G_R by removing S
7.   ND(G_L, nmin)
8.   ND(G_R, nmin)
9. End

The labeling of the nodes in lines 2 and 5 usually proceeds in sequence, so, for example, in line 5, the nodes of S are labeled in a certain order, starting from the last labeled node so far in the procedure. The main step of the ND procedure is to separate the graph into three parts, two of which have no coupling between each other. The third set has couplings with vertices from both of the first sets and is referred to as a separator. The key idea is to separate the graph in this way and then repeat the process recursively in each subgraph. The nodes of the separator are numbered last. An illustration is shown in Figure 3.11.

3.7 Test Problems

For comparison purposes it is important to use a common set of test matrices that represent a wide spectrum of applications. There are two distinct ways of providing such data sets. The first approach is to collect sparse matrices in a well-specified standard format from various applications. This approach is used in the Harwell–Boeing collection of test matrices. The second approach is to generate these matrices with a few sample programs, such as those provided in the SPARSKIT library [244]. The coming chapters will use examples from these

3.7. Test Problems

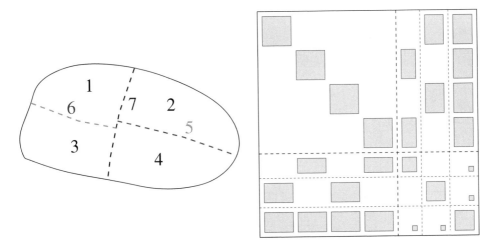

Figure 3.11. *ND ordering and corresponding reordered matrix.*

two sources. In particular, five test problems will be emphasized for their varying degrees of difficulty.

The SPARSKIT package can generate matrices arising from the discretization of the two- or three-dimensional PDEs

$$-\frac{\partial}{\partial x}\left(a\frac{\partial u}{\partial x}\right) - \frac{\partial}{\partial y}\left(b\frac{\partial u}{\partial y}\right) - \frac{\partial}{\partial z}\left(c\frac{\partial u}{\partial z}\right)$$
$$+\frac{\partial (du)}{\partial x} + \frac{\partial (eu)}{\partial y} + \frac{\partial (fu)}{\partial z} + gu = h$$

on rectangular regions with general mixed-type boundary conditions. In the test problems, the regions are the square $\Omega = (0, 1)^2$ or the cube $\Omega = (0, 1)^3$; the Dirichlet condition $u = 0$ is always used on the boundary. Only the discretized matrix is of importance, since the right-hand side will be created artificially. Therefore, the right-hand side, h, is not relevant.

Problem 1: F2DA. In the first test problem, which will be labeled F2DA, the domain is two-dimensional, with
$$a(x, y) = b(x, y) = 1.0$$
and
$$d(x, y) = \gamma(x + y), \quad e(x, y) = \gamma(x - y), \quad f(x, y) = g(x, y) = 0.0, \quad (3.4)$$

where the constant γ is equal to 10. The domain and coefficients for this problem are shown in Figure 3.12. If the number of points in each direction is 34, then there are $n_x = n_y = 32$ interior points in each direction and a matrix of size $n = n_x \times n_y = 32^2 = 1024$ is obtained. In this test example, as well as the other ones described below, the right-hand side is generated as
$$b = Ae,$$

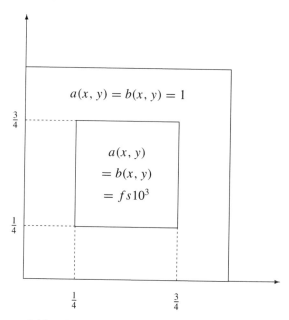

Figure 3.12. *Physical domain and coefficients for Problem 1.*

in which $e = (1, 1, \ldots, 1)^T$. The initial guess is always taken to be a vector of pseudo-random values.

Problem 2: F2DB. The second test problem is similar to the previous one but involves discontinuous coefficient functions a and b. Here, $n_x = n_y = 32$ and the functions d, e, f, g are also defined by (3.4). However, the functions a and b now both take the value 1000 inside the subsquare of width $\frac{1}{2}$ centered at $(\frac{1}{2}, \frac{1}{2})$, and 1 elsewhere in the domain; i.e.,

$$a(x, y) = b(x, y) = \begin{cases} 10^3 & \text{if } \frac{1}{4} < x, y < \frac{3}{4}, \\ 1 & \text{otherwise.} \end{cases}$$

Problem 3: F3D. The third test problem is three-dimensional with $n_x = n_y = n_z = 16$ internal mesh points in each direction, leading to a problem of size $n = 4096$. In this case, we take

$$a(x, y, z) = b(x, y, z) = c(x, y, z) = 1,$$
$$d(x, y, z) = \gamma e^{xy}, \quad e(x, y, z) = \gamma e^{-xy},$$

and

$$f(x, y, z) = g(x, y, z) = 0.0.$$

The constant γ is taken to be equal to 10.0 as before.

The Harwell–Boeing collection is a large data set consisting of test matrices that have been contributed by researchers and engineers from many different disciplines. These have often been used for test purposes in the literature [108]. The collection provides a data structure that constitutes an excellent medium for exchanging matrices. The matrices are stored as ASCII files with a very specific format consisting of a four- or five-line header. Then the data containing the matrix are stored in CSC format together with any right-hand sides, initial guesses, and exact solutions when available. The SPARSKIT library also provides routines for reading and generating matrices in this format.

Only one matrix from the collection was selected for testing the algorithms described in the coming chapters. The matrices in the last two test examples are both irregularly structured.

Problem 4: ORS. The matrix selected from the Harwell–Boeing collection is ORSIRR1. This matrix arises from a reservoir engineering problem. Its size is $n = 1030$ and it has a total of $Nz = 6858$ nonzero elements. The original problem is based on a $21 \times 21 \times 5$ irregular grid. In this case and the next one, the matrices are preprocessed by scaling their rows and columns.

Problem 5: FID. This test matrix is extracted from the well-known fluid flow simulation package FIDAP [120]. It is actually test example number 36 from this package and features a two-dimensional chemical vapor deposition in a horizontal reactor. The matrix has a size of $n = 3079$ and has $Nz = 53843$ nonzero elements. It has a symmetric pattern and few diagonally dominant rows or columns. The rows and columns are prescaled in the same way as in the previous example. Figure 3.13 shows the patterns of the matrices ORS and FID.

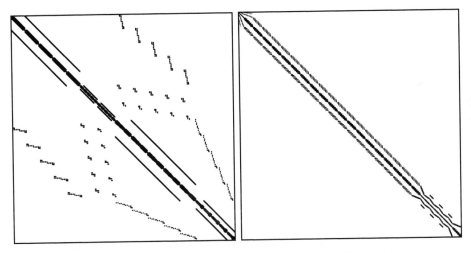

Figure 3.13. *Patterns of the matrices ORS (left) and FID (right).*

Exercises

1. Consider the mesh of a discretized PDE. In which situations is the graph representing this mesh the same as the adjacency graph of the matrix? Give examples from both finite difference and finite element discretizations.

2. Let A and B be two sparse (square) matrices of the same dimension. How can the graph of $C = A + B$ be characterized with respect to the graphs of A and B?

3. Consider the matrix defined as
$$P_\pi = I_{\pi,*}.$$
Show directly (without using Proposition 3.2 or interchange matrices) that the following three relations hold:
$$A_{\pi,*} = I_{\pi,*} A,$$
$$I_{*,\pi} = P_\pi^T,$$
$$A P_\pi^T = A_{*,\pi}.$$

4. Consider the two matrices
$$A = \begin{pmatrix} \star & \star & 0 & \star & 0 & 0 \\ 0 & \star & 0 & 0 & 0 & \star \\ 0 & \star & \star & 0 & 0 & 0 \\ 0 & \star & 0 & 0 & \star & 0 \\ 0 & 0 & 0 & 0 & \star & 0 \\ 0 & 0 & 0 & 0 & 0 & \star \end{pmatrix}, \quad B = \begin{pmatrix} \star & 0 & 0 & 0 & 0 & 0 \\ \star & 0 & \star & 0 & \star & 0 \\ 0 & \star & 0 & 0 & 0 & 0 \\ \star & \star & 0 & 0 & 0 & 0 \\ 0 & \star & 0 & \star & \star & 0 \\ 0 & 0 & \star & 0 & 0 & \star \end{pmatrix},$$
where a star represents an arbitrary nonzero element.

   a. Show the adjacency graphs of the matrices A, B, AB, and BA. (Assume that there are no numerical cancellations in computing the products AB and BA.) Since there are zero diagonal elements, represent explicitly the cycles corresponding to the (i,i) edges when they are present.

   b. Consider the matrix $C = AB$. Give an interpretation of an edge in the graph of C in terms of edges in the graphs of A and B. Verify this answer using the above matrices.

   c. Consider the particular case in which $B = A$. Give an interpretation of an edge in the graph of C in terms of paths of length two in the graph of A. The paths must take into account the cycles corresponding to nonzero diagonal elements of A.

   d. Now consider the case where $B = A^2$. Give an interpretation of an edge in the graph of $C = A^3$ in terms of paths of length three in the graph of A. Generalize the result to arbitrary powers of A.

5. Consider two matrices A and B of dimension $n \times n$ whose diagonal elements are all nonzero. Let E_X denote the set of edges in the adjacency graph of a matrix X (i.e., the set of pairs (i,j) such that $X_{ij} \neq 0$). Then show that
$$E_{AB} \supset E_A \cup E_B.$$
Give extreme examples when $|E_{AB}| = n^2$ while $E_A \cup E_B$ is of order n. What practical implications does this have on ways to store products of sparse matrices? (Is it better to store the product AB or the matrices A, B separately? Consider both the computational cost for performing matrix-by-vector products and the cost of memory.)

Exercises

6. Consider a 6×6 matrix with the pattern

$$A = \begin{pmatrix} \star & \star & & & \star & \\ \star & \star & \star & & \star & \\ & \star & \star & & & \\ & & & \star & \star & \\ \star & & & \star & \star & \star \\ & \star & & & \star & \star \end{pmatrix}.$$

   a. Show the adjacency graph of A.
   b. Consider the permutation $\pi = \{1, 3, 4, 2, 5, 6\}$. Show the adjacency graph and new pattern for the matrix obtained from a symmetric permutation of A based on the permutation array π.

7. You are given an 8×8 matrix with the following pattern:

$$\begin{pmatrix} x & x & & & & & x & \\ x & x & x & & & x & x & \\ & x & x & x & & & x & \\ & & x & x & x & & & \\ & & & x & x & x & & \\ & x & x & & x & x & x & \\ & x & & & & x & x & x \\ x & & & & & & x & x \end{pmatrix}.$$

   a. Show the adjacency graph of A.
   b. Find the CMK ordering for the matrix (break ties by giving priority to the node with lowest index). Show the graph of the matrix permuted according to the CMK ordering.
   c. What is the RCM ordering for this case? Show the matrix reordered according to the RCM ordering.
   d. Find a multicoloring of the graph using the greedy multicolor algorithm. What is the minimum number of colors required for multicoloring the graph?
   e. Consider the variation of the CMK ordering in which the first level L_0 consists of several vertices instead of only one vertex. Find the CMK ordering with this variant with the starting level $L_0 = \{1, 8\}$.

8. Consider a matrix with the pattern

$$A = \begin{pmatrix} \star & \star & & & \star & & & \star & \\ \star & \star & \star & & & \star & & & \\ & \star & \star & \star & & & \star & & \\ & & \star & \star & \star & & & \star & \\ \star & & & \star & \star & \star & & & \\ & \star & & & \star & \star & \star & & \\ & & \star & & & \star & \star & \star & \\ \star & & & \star & & & \star & \star & \\ & & & & & & & \star & \star \end{pmatrix}.$$

a. Show the adjacency graph of A. (Place the eight vertices on a circle.)

b. Consider the permutation $\pi = \{1, 3, 5, 7, 2, 4, 6, 8\}$. Show the adjacency graph and new pattern for the matrix obtained from a symmetric permutation of A based on the permutation array π.

c. Show the adjacency graph and new pattern for the matrix obtained from an RCM ordering of A starting with the node 1. (Assume that the vertices adjacent to a given vertex are always listed in increasing order in the data structure that describes the graph.)

d. Find a multicolor ordering for A (list the vertex labels for color 1, followed by those for color 2, etc.).

9. Given a five-point finite difference graph, show that the greedy algorithm will always find a coloring of the graph with two colors.

10. Prove that the total number of colors found by the greedy multicoloring algorithm does not exceed $\nu_{max} + 1$, where ν_{max} is the maximum degree of all the vertices of a graph (not counting the cycles (i, i) associated with diagonal elements).

11. Consider a graph that is bipartite, i.e., 2-colorable. Assume that the vertices of the graph are colored by a variant of Algorithm 3.6, in which the nodes are traversed in a certain order $i_1, i_2, \ldots, i_n$.

    a. Is it true that for any permutation $i_1, \ldots, i_n$ the number of colors found will be two?

    b. Consider now a permutation satisfying the following property: For each j at least one of the nodes $i_1, i_2, \ldots, i_{j-1}$ is adjacent to i_j. Show that the algorithm will find a 2-coloring of the graph.

    c. Among the following traversals indicate which ones satisfy the property of part (b): (1) BFS, (2) random traversal, (3) traversal defined by i_j = any node adjacent to i_{j-1}.

12. Given a matrix that is irreducible and with a symmetric pattern, show that its structural inverse is dense. Structural inverse means the pattern of the inverse, regardless of the values, or, otherwise stated, is the union of all patterns of the inverses for all possible values. [Hint: Use the Cayley–Hamilton theorem and a well-known result on powers of adjacency matrices mentioned at the end of Section 3.2.1.]

13. The most economical storage scheme in terms of memory usage is the following variation of the coordinate format: Store all nonzero values a_{ij} in a real array $AA[1 : Nz]$ and the corresponding "linear array address" $(i - 1) * n + j$ in an integer array $JA[1 : Nz]$. The order in which these corresponding entries are stored is unimportant as long as they are both in the same position in their respective arrays. What are the advantages and disadvantages of this data structure? Write a short routine for performing a matrix-by-vector product in this format.

14. Write a FORTRAN-90 or C code segment to perform the matrix-by-vector product for matrices stored in Ellpack-Itpack format.

15. Write a small subroutine to perform the following operations on a sparse matrix in coordinate format, diagonal format, and CSR format.

    a. Count the number of nonzero elements in the main diagonal.

    b. Extract the diagonal whose offset is k.

    c. Add a nonzero element in position (i, j) of the matrix (this position may initially contain a zero or a nonzero element).

    d. Add a given diagonal to the matrix. What is the most convenient storage scheme for each of these operations?

16. Linked lists is another popular scheme often used for storing sparse matrices. These allow us to link together k data items (e.g., elements of a given row) in a large linear array. A starting position is given in the array, which contains the first element of the set. Then, a link to the next element in the array is provided from a LINK array.

    a. Show how to implement this scheme. A linked list is to be used for each row.

    b. What are the main advantages and disadvantages of linked lists?

    c. Write an algorithm to perform a matrix-by-vector product in this format.

Notes and References

Two good references on sparse matrix computations are the book by George and Liu [144] and the more recent volume by Duff, Erisman, and Reid [107]. These are geared toward direct solution methods and the first specializes in SPD problems. Also of interest are [220] and [226] and the early survey by Duff [106].

Sparse matrix techniques have traditionally been associated with direct solution methods. This has changed in the last decade because of the increased need to solve three-dimensional problems. The SPARSKIT library, a package for sparse matrix computations [244] is available from the author at http://www.cs.umn.edu/~saad/software. Another available software package that emphasizes object-oriented design with the goal of hiding complex data structures from users is PETSc [24].

The idea of the greedy multicoloring algorithm is known in finite element techniques (to color elements); see, e.g., Benantar and Flaherty [31]. Wu [318] presents the greedy algorithm for multicoloring vertices and uses it for successive overrelaxation–type iterations; see also [247]. The effect of multicoloring has been extensively studied by Adams [2, 3] and Poole and Ortega [227]. Interesting results regarding multicoloring in the context of finite elements based on quad-tree structures have been obtained by Benantar and Flaherty [31], who show, in particular, that with this structure a maximum of six colors is required.

Chapter 4
Basic Iterative Methods

The first iterative methods used for solving large linear systems were based on *relaxation of the coordinates*. Beginning with a given approximate solution, these methods modify the components of the approximation, one or a few at a time and in a certain order, until convergence is reached. Each of these modifications, called relaxation steps, is aimed at annihilating one or a few components of the residual vector. Now these techniques are rarely used separately. However, when combined with the more efficient methods described in later chapters, they can be quite successful. Moreover, there are a few application areas where variations of these methods are still quite popular.

4.1 Jacobi, Gauss–Seidel, and Successive Overrelaxation

This chapter begins by reviewing the basic iterative methods for solving linear systems. Given an $n \times n$ real matrix A and a real n-vector b, the problem considered is as follows: Find x belonging to $\mathbb{R}^n$ such that
$$Ax = b. \tag{4.1}$$

Equation (4.1) is a *linear system*, A is the *coefficient matrix*, b is the *right-hand side* vector, and x is the *vector of unknowns*. Most of the methods covered in this chapter involve passing from one iterate to the next by modifying one or a few components of an approximate vector solution at a time. This is natural since there are simple criteria when modifying a component in order to improve an iterate. One example is to annihilate some component(s) of the residual vector $b - Ax$. The convergence of these methods is rarely guaranteed for all matrices, but a large body of theory exists for the case where the coefficient matrix arises from the finite difference discretization of elliptic partial differential equations (PDEs).

We begin with the decomposition
$$A = D - E - F, \tag{4.2}$$

in which D is the diagonal of A, $-E$ its strict lower part, and $-F$ its strict upper part, as illustrated in Figure 4.1. It is always assumed that the diagonal entries of A are all nonzero.

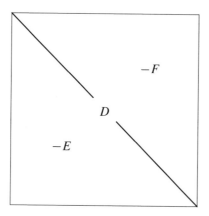

Figure 4.1. *Initial partitioning of matrix A.*

The Jacobi iteration determines the ith component of the next approximation so as to annihilate the ith component of the residual vector. In the following, $\xi_i^{(k)}$ denotes the ith component of the iterate x_k and β_i the ith component of the right-hand side b. Thus, writing

$$(b - Ax_{k+1})_i = 0, \tag{4.3}$$

in which $(y)_i$ represents the ith component of the vector y, yields

$$a_{ii}\xi_i^{(k+1)} = -\sum_{\substack{j=1\\j\neq i}}^n a_{ij}\xi_j^{(k)} + \beta_i$$

or

$$\xi_i^{(k+1)} = \frac{1}{a_{ii}}\left(\beta_i - \sum_{\substack{j=1\\j\neq i}}^n a_{ij}\xi_j^{(k)}\right), \quad i = 1, \ldots, n. \tag{4.4}$$

This is a component-wise form of the Jacobi iteration. All components of the next iterate can be grouped into the vector x_{k+1}. The above notation can be used to rewrite the Jacobi iteration (4.4) in vector form as

$$x_{k+1} = D^{-1}(E + F)x_k + D^{-1}b. \tag{4.5}$$

Similarly, the Gauss–Seidel iteration corrects the ith component of the current approximate solution, in the order $i = 1, 2, \ldots, n$, again to annihilate the ith component of the residual. However, this time the approximate solution is updated immediately after the new component is determined. The newly computed components $\xi_i^{(k)}$, $i = 1, 2, \ldots, n$, can be changed within a working vector that is redefined at each relaxation step. Thus, since the order is $i = 1, 2, \ldots$, the result at the ith step is

$$\beta_i - \sum_{j=1}^{i-1} a_{ij}\xi_j^{(k+1)} - a_{ii}\xi_i^{(k+1)} - \sum_{j=i+1}^n a_{ij}\xi_j^{(k)} = 0, \tag{4.6}$$

4.1. Jacobi, Gauss–Seidel, and Successive Overrelaxation

which leads to the iteration

$$\xi_i^{(k+1)} = \frac{1}{a_{ii}} \left(-\sum_{j=1}^{i-1} a_{ij}\xi_j^{(k+1)} - \sum_{j=i+1}^{n} a_{ij}\xi_j^{(k)} + \beta_i \right), \quad i = 1, \ldots, n. \tag{4.7}$$

The defining equation (4.6) can be written as

$$b + Ex_{k+1} - Dx_{k+1} + Fx_k = 0,$$

which leads immediately to the vector form of the Gauss–Seidel iteration

$$x_{k+1} = (D - E)^{-1} F x_k + (D - E)^{-1} b. \tag{4.8}$$

Computing the new approximation in (4.5) requires multiplying by the inverse of the diagonal matrix D. In (4.8) a triangular system must be solved with $D - E$, the lower triangular part of A. Thus, the new approximation in a Gauss–Seidel step can be determined either by solving a triangular system with the matrix $D - E$ or from the relation (4.7).

A *backward* Gauss–Seidel iteration can also be defined as

$$(D - F)x_{k+1} = Ex_k + b, \tag{4.9}$$

which is equivalent to making the coordinate corrections in the order $n, n - 1, \ldots, 1$. A symmetric Gauss–Seidel iteration consists of a forward sweep followed by a backward sweep.

The Jacobi and the Gauss–Seidel iterations are both of the form

$$Mx_{k+1} = Nx_k + b = (M - A)x_k + b, \tag{4.10}$$

in which

$$A = M - N \tag{4.11}$$

is a *splitting* of A, with $M = D$ for Jacobi, $M = D - E$ for forward Gauss–Seidel, and $M = D - F$ for backward Gauss–Seidel. An iterative method of the form (4.10) can be defined for any splitting of the form (4.11) where M is nonsingular. *Overrelaxation* is based on the splitting

$$\omega A = (D - \omega E) - (\omega F + (1 - \omega)D),$$

and the corresponding *successive overrelaxation* (SOR) method is given by the recursion

$$(D - \omega E)x_{k+1} = [\omega F + (1 - \omega)D]x_k + \omega b. \tag{4.12}$$

The above iteration corresponds to the relaxation sequence

$$\xi_i^{(k+1)} = \omega \xi_i^{GS} + (1 - \omega)\xi_i^{(k)}, \quad i = 1, 2, \ldots, n,$$

in which ξ_i^{GS} is defined by the expression on the right-hand side of (4.7). A backward SOR sweep can be defined analogously to the backward Gauss–Seidel sweep (4.9).

A symmetric SOR (SSOR) step consists of the SOR step (4.12) followed by a backward SOR step:

$$(D - \omega E)x_{k+1/2} = [\omega F + (1 - \omega)D]x_k + \omega b,$$

$$(D - \omega F)x_{k+1} = [\omega E + (1 - \omega)D]x_{k+1/2} + \omega b.$$

This gives the recurrence

$$x_{k+1} = G_\omega x_k + f_\omega,$$

where
$$G_\omega = (D - \omega F)^{-1}(\omega E + (1 - \omega)D)$$
$$\times (D - \omega E)^{-1}(\omega F + (1 - \omega)D), \quad (4.13)$$
$$f_\omega = \omega(D - \omega F)^{-1}\left(I + [\omega E + (1 - \omega)D](D - \omega E)^{-1}\right)b. \quad (4.14)$$

Observing that
$$[\omega E + (1 - \omega)D](D - \omega E)^{-1} = [-(D - \omega E) + (2 - \omega)D](D - \omega E)^{-1}$$
$$= -I + (2 - \omega)D(D - \omega E)^{-1},$$

f_ω can be rewritten as
$$f_\omega = \omega(2 - \omega)(D - \omega F)^{-1}D(D - \omega E)^{-1}b.$$

4.1.1 Block Relaxation Schemes

Block relaxation schemes are generalizations of the *point* relaxation schemes described above. They update a whole set of components at each time, typically a subvector of the solution vector, instead of only one component. The matrix A and the right-hand side and solution vectors are partitioned as follows:

$$A = \begin{pmatrix} A_{11} & A_{12} & A_{13} & \cdots & A_{1p} \\ A_{21} & A_{22} & A_{23} & \cdots & A_{2p} \\ A_{31} & A_{32} & A_{33} & \cdots & A_{3p} \\ \vdots & \vdots & \vdots & \ddots & \vdots \\ A_{p1} & A_{p2} & \cdots & \cdots & A_{pp} \end{pmatrix}, \quad x = \begin{pmatrix} \xi_1 \\ \xi_2 \\ \xi_3 \\ \vdots \\ \xi_p \end{pmatrix}, \quad b = \begin{pmatrix} \beta_1 \\ \beta_2 \\ \beta_3 \\ \vdots \\ \beta_p \end{pmatrix}, \quad (4.15)$$

in which the partitionings of b and x into subvectors β_i and ξ_i are identical and compatible with the partitioning of A. Thus, for any vector x partitioned as in (4.15),

$$(Ax)_i = \sum_{j=1}^{p} A_{ij}\xi_j,$$

in which $(y)_i$ denotes the ith component of the vector y according to the above partitioning. The diagonal blocks in A are square and assumed nonsingular.

Now define, similarly to the scalar case, the splitting
$$A = D - E - F,$$

with
$$D = \begin{pmatrix} A_{11} & & & \\ & A_{22} & & \\ & & \ddots & \\ & & & A_{pp} \end{pmatrix}, \quad (4.16)$$

$$E = -\begin{pmatrix} O & & & \\ A_{21} & O & & \\ \vdots & \vdots & \ddots & \\ A_{p1} & A_{p2} & \cdots & O \end{pmatrix}, \quad F = -\begin{pmatrix} O & A_{12} & \cdots & A_{1p} \\ & O & \cdots & A_{2p} \\ & & \ddots & \vdots \\ & & & O \end{pmatrix}.$$

4.1. Jacobi, Gauss–Seidel, and Successive Overrelaxation

With these definitions, it is easy to generalize the previous three iterative procedures defined earlier, namely, Jacobi, Gauss–Seidel, and SOR. For example, the block Jacobi iteration is now defined as a technique in which the new subvectors $\xi_i^{(k)}$ are all replaced according to

$$A_{ii}\xi_i^{(k+1)} = ((E+F)x_k)_i + \beta_i$$

or

$$\xi_i^{(k+1)} = A_{ii}^{-1}((E+F)x_k)_i + A_{ii}^{-1}\beta_i, \quad i=1,\ldots,p,$$

which leads to the same equation as before:

$$x_{k+1} = D^{-1}(E+F)x_k + D^{-1}b,$$

except that the meanings of D, E, and F have changed to their block analogues.

With finite difference approximations of PDEs, it is standard to block the variables and the matrix by partitioning along whole lines of the mesh. For example, for the two-dimensional mesh illustrated in Figure 2.5, this partitioning is

$$\xi_1 = \begin{pmatrix} u_{11} \\ u_{12} \\ u_{13} \\ u_{14} \\ u_{15} \end{pmatrix}, \quad \xi_2 = \begin{pmatrix} u_{21} \\ u_{22} \\ u_{23} \\ u_{24} \\ u_{25} \end{pmatrix}, \quad \xi_3 = \begin{pmatrix} u_{31} \\ u_{32} \\ u_{33} \\ u_{34} \\ u_{35} \end{pmatrix}.$$

This corresponds to the mesh in Figure 2.5 of Chapter 2, whose associated matrix pattern is shown in Figure 2.6. A relaxation can also be defined along the vertical instead of the horizontal lines. Techniques of this type are often known as *line relaxation* techniques.

In addition, a block can also correspond to the unknowns associated with a few consecutive lines in the plane. One such blocking is illustrated in Figure 4.2 for a 6 × 6 grid. The corresponding matrix with its block structure is shown in Figure 4.3. An important difference between this partitioning and the one corresponding to the single-line partitioning is that now the matrices A_{ii} are block tridiagonal instead of tridiagonal. As a result, solving

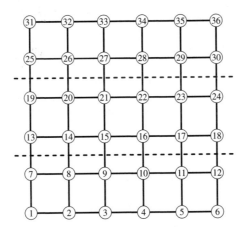

Figure 4.2. *Partitioning of a 6 × 6 square mesh into three subdomains.*

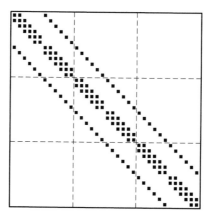

Figure 4.3. *Matrix associated with the mesh of Figure* 4.2.

linear systems with A_{ii} may be much more expensive. On the other hand, the number of iterations required to achieve convergence often decreases rapidly as the block size increases.

Finally, block techniques can be defined in more general terms. First, we use blocks that allow us to update arbitrary groups of components, and second, we allow the blocks to overlap. Since this is a form of the domain decomposition method that will be seen later, we define the approach carefully. So far, our partition has been based on an actual *set partition* of the variable set $S = \{1, 2, \ldots, n\}$ into subsets $S_1, S_2, \ldots, S_p$, with the condition that two distinct subsets are disjoint. In set theory, this is called a *partition* of S. More generally, a *set decomposition* of S removes the constraint of disjointness. In other words, it is required that the union of the subsets S_i be equal to S:

$$S_i \subseteq S, \quad \bigcup_{i=1,\ldots,p} S_i = S.$$

In the following, n_i denotes the size of S_i and the subset S_i is of the form

$$S_i = \{m_i(1), m_i(2), \ldots, m_i(n_i)\}.$$

A general block Jacobi iteration can be defined as follows. Let V_i be the $n \times n_i$ matrix

$$V_i = [e_{m_i(1)}, e_{m_i(2)}, \ldots, e_{m_i(n_i)}]$$

and let

$$W_i = [\eta_{m_i(1)} e_{m_i(1)}, \eta_{m_i(2)} e_{m_i(2)}, \ldots, \eta_{m_i(n_i)} e_{m_i(n_i)}],$$

where each e_j is the jth column of the $n \times n$ identity matrix and $\eta_{m_i(j)}$ represents a weight factor chosen so that

$$W_i^T V_i = I.$$

When there is no overlap, i.e., when the S_i's form a partition of the whole set $\{1, 2, \ldots, n\}$, then define $\eta_{m_i(j)} = 1$.

4.1. Jacobi, Gauss–Seidel, and Successive Overrelaxation

Let A_{ij} be the $n_i \times n_j$ matrix

$$A_{ij} = W_i^T A V_j$$

and define similarly the partitioned vectors

$$\xi_i = W_i^T x, \quad \beta_i = W_i^T b.$$

Note that $V_i W_i^T$ is a projector from $\mathbb{R}^n$ to the subspace K_i spanned by the columns $m_i(1), \ldots, m_i(n_i)$. In addition, we have the relation

$$x = \sum_{i=1}^{s} V_i \xi_i.$$

The n_i-dimensional vector $W_i^T x$ represents the projection $V_i W_i^T x$ of x with respect to the basis spanned by the columns of V_i. The action of V_i performs the reverse operation. That means $V_i y$ is an extension operation from a vector y in K_i (represented in the basis consisting of the columns of V_i) into a vector $V_i y$ in $\mathbb{R}^n$. The operator W_i^T is termed a *restriction operator* and V_i is a *prolongation operator*.

Each component of the Jacobi iteration can be obtained by imposing the condition that the projection of the residual in the span of S_i be zero; i.e.,

$$W_i^T \left[b - A \left(V_i W_i^T x_{k+1} + \sum_{j \neq i} V_j W_j^T x_k \right) \right] = 0.$$

Remember that $\xi_j = W_j^T x$, which can be rewritten as

$$\xi_i^{(k+1)} = \xi_i^{(k)} + A_{ii}^{-1} W_i^T (b - A x_k). \tag{4.17}$$

This leads to the following algorithm.

ALGORITHM 4.1. General Block Jacobi Iteration

1. For $k = 0, 1, \ldots$, until convergence, Do
2.   For $i = 1, 2, \ldots, p$, Do
3.   Solve $A_{ii} \delta_i = W_i^T (b - A x_k)$
4.   Set $x_{k+1} := x_k + V_i \delta_i$
5.   EndDo
6. EndDo

As was the case with the scalar algorithms, there is only a slight difference between the Jacobi and Gauss–Seidel iterations. Gauss–Seidel immediately updates the component to be corrected at step i and uses the updated approximate solution to compute the residual vector needed to correct the next component. However, the Jacobi iteration uses the same previous approximation x_k for this purpose. Therefore, the block Gauss–Seidel iteration can be defined algorithmically as follows.

ALGORITHM 4.2. General Block Gauss–Seidel Iteration

1. Until convergence, Do
2.   For $i = 1, 2, \ldots, p$, Do
3.   Solve $A_{ii}\delta_i = W_i^T(b - Ax)$
4.   Set $x := x + V_i\delta_i$
5.   EndDo
6. EndDo

From the point of view of storage, Gauss–Seidel is more economical because the new approximation can be overwritten over the same vector. Also, it typically converges faster. On the other hand, the Jacobi iteration has some appeal on parallel computers, since the second *Do* loop, corresponding to the p different blocks, can be executed in parallel. Although the point Jacobi algorithm by itself is rarely a successful technique for real-life problems, its block Jacobi variant, when using large enough overlapping blocks, can be quite attractive, especially in a parallel computing environment.

4.1.2 Iteration Matrices and Preconditioning

The Jacobi and Gauss–Seidel iterations are of the form

$$x_{k+1} = Gx_k + f, \qquad (4.18)$$

in which

$$G_{JA}(A) = I - D^{-1}A, \qquad (4.19)$$
$$G_{GS}(A) = I - (D - E)^{-1}A, \qquad (4.20)$$

for the Jacobi and Gauss–Seidel iterations, respectively. Moreover, given the matrix splitting

$$A = M - N, \qquad (4.21)$$

where A is associated with the linear system (4.1), a *linear fixed-point iteration* can be defined by the recurrence

$$x_{k+1} = M^{-1}Nx_k + M^{-1}b, \qquad (4.22)$$

which has the form (4.18) with

$$G = M^{-1}N = M^{-1}(M - A) = I - M^{-1}A, \quad f = M^{-1}b. \qquad (4.23)$$

For example, for the Jacobi iteration, $M = D$, $N = A - D$, while for the Gauss–Seidel iteration, $M = D - E$, $N = M - A = F$.

The iteration $x_{k+1} = Gx_k + f$ can be viewed as a technique for solving the system

$$(I - G)x = f.$$

Since G has the form $G = I - M^{-1}A$, this system can be rewritten as

$$M^{-1}Ax = M^{-1}b.$$

4.2. Convergence

The above system, which has the same solution as the original system, is called a *preconditioned system* and M is the *preconditioning matrix* or *preconditioner*. In other words, *a relaxation scheme is equivalent to a fixed-point iteration on a preconditioned system.*

For example, for the Jacobi, Gauss–Seidel, SOR, and SSOR iterations, these preconditioning matrices are, respectively,

$$M_{JA} = D, \tag{4.24}$$

$$M_{GS} = D - E, \tag{4.25}$$

$$M_{SOR} = \frac{1}{\omega}(D - \omega E), \tag{4.26}$$

$$M_{SSOR} = \frac{1}{\omega(2-\omega)}(D - \omega E)D^{-1}(D - \omega F). \tag{4.27}$$

Thus, the Jacobi preconditioner is simply the diagonal of A, while the Gauss–Seidel preconditioner is the lower triangular part of A. The constant coefficients in front of the matrices M_{SOR} and M_{SSOR} only have the effect of scaling the equations of the preconditioned system uniformly. Therefore, they are unimportant in the preconditioning context.

Note that the "preconditioned" system may be a full system. Indeed, there is no reason why M^{-1} should be a sparse matrix (even though M may be sparse), since the inverse of a sparse matrix is not necessarily sparse. This limits the number of techniques that can be applied to solve the preconditioned system. Most of the iterative techniques used only require matrix-by-vector products. In this case, to compute $w = M^{-1}Av$ for a given vector v, first compute $r = Av$ and then solve the system $Mw = r$:

$$r = Av,$$
$$w = M^{-1}r.$$

In some cases, it may be advantageous to exploit the splitting $A = M - N$ and compute $w = M^{-1}Av$ as $w = (I - M^{-1}N)v$ by the procedure

$$r = Nv,$$
$$w = M^{-1}r,$$
$$w := v - w.$$

The matrix N may be sparser than A and the matrix-by-vector product Nv may be less expensive than the product Av. A number of similar but somewhat more complex ideas have been exploited in the context of preconditioned iterative methods. A few of these will be examined in Chapter 9.

4.2 Convergence

All the methods seen in the previous section define a sequence of iterates of the form

$$x_{k+1} = Gx_k + f, \tag{4.28}$$

in which G is a certain *iteration matrix*. The questions addressed in this section are as follows: (a) If the iteration converges, then is the limit indeed a solution of the original system? (b) Under which conditions does the iteration converge? (c) When the iteration does converge, how fast is it?

If the above iteration converges, its limit x satisfies

$$x = Gx + f. \tag{4.29}$$

In the case where the above iteration arises from the splitting $A = M - N$, it is easy to see that the solution x to the above system is identical to that of the original system $Ax = b$. Indeed, in this case the sequence (4.28) has the form

$$x_{k+1} = M^{-1} N x_k + M^{-1} b$$

and its limit satisfies

$$Mx = Nx + b$$

or $Ax = b$. This answers question (a). Next, we focus on the other two questions.

4.2.1 General Convergence Result

If $I - G$ is nonsingular, then there is a solution x_* to (4.29). Subtracting (4.29) from (4.28) yields

$$x_{k+1} - x_* = G(x_k - x_*) = \cdots = G^{k+1}(x_0 - x_*). \tag{4.30}$$

Standard results seen in Chapter 1 imply that, if the spectral radius of the iteration matrix G is less than unity, then $x_k - x_*$ converges to zero and the iteration (4.28) converges toward the solution defined by (4.29). Conversely, the relation

$$x_{k+1} - x_k = G(x_k - x_{k-1}) = \cdots = G^k(f - (I - G)x_0)$$

shows that if the iteration converges for *any* x_0 and f, then $G^k v$ converges to zero for any vector v. As a result, $\rho(G)$ must be less than unity and the following theorem is proved.

Theorem 4.1. *Let G be a square matrix such that $\rho(G) < 1$. Then $I - G$ is nonsingular and the iteration (4.28) converges for any f and x_0. Conversely, if the iteration (4.28) converges for any f and x_0, then $\rho(G) < 1$.*

Since it is expensive to compute the spectral radius of a matrix, sufficient conditions that guarantee convergence can be useful in practice. One such sufficient condition could be obtained by utilizing the inequality $\rho(G) \leq \|G\|$ for any matrix norm.

Corollary 4.2. *Let G be a square matrix such that $\|G\| < 1$ for some matrix norm $\|\cdot\|$. Then $I - G$ is nonsingular and the iteration (4.28) converges for any initial vector x_0.*

Apart from knowing that the sequence (4.28) converges, it is also desirable to know *how fast* it converges. The error $d_k = x_k - x_*$ at step k satisfies

$$d_k = G^k d_0.$$

The matrix G can be expressed in the Jordan canonical form as $G = XJX^{-1}$. Assume for simplicity that there is only one eigenvalue of G of largest modulus and call it λ. Then

$$d_k = \lambda^k X \left(\frac{J}{\lambda}\right)^k X^{-1} d_0.$$

4.2. Convergence

A careful look at the powers of the matrix J/λ shows that all its blocks, except the block associated with the eigenvalue λ, converge to zero as k tends to infinity. Let this Jordan block be of size p and of the form

$$J_\lambda = \lambda I + E,$$

where E is nilpotent of index p; i.e., $E^p = 0$. Then, for $k \geq p$,

$$J_\lambda^k = (\lambda I + E)^k = \lambda^k (I + \lambda^{-1} E)^k = \lambda^k \left(\sum_{i=0}^{p-1} \lambda^{-i} \binom{k}{i} E^i \right).$$

If k is large enough, then for any λ the dominant term in the above sum is the last term; i.e.,

$$J_\lambda^k \approx \lambda^{k-p+1} \binom{k}{p-1} E^{p-1}.$$

Thus, the norm of $d_k = G^k d_0$ has the asymptotical form

$$\|d_k\| \approx C \times |\lambda^{k-p+1}| \binom{k}{p-1},$$

where C is some constant. The *convergence factor* of a sequence is the limit

$$\rho = \lim_{k \to \infty} \left(\frac{\|d_k\|}{\|d_0\|} \right)^{1/k}.$$

It follows from the above analysis that $\rho = \rho(G)$. The *convergence rate* τ is the (natural) logarithm of the inverse of the convergence factor:

$$\tau = -\ln \rho.$$

The above definition depends on the initial vector x_0, so it may be termed a *specific* convergence factor. A *general* convergence factor can also be defined by

$$\phi = \lim_{k \to \infty} \left(\max_{x_0 \in \mathbb{R}^n} \frac{\|d_k\|}{\|d_0\|} \right)^{1/k}.$$

This factor satisfies

$$\phi = \lim_{k \to \infty} \left(\max_{d_0 \in \mathbb{R}^n} \frac{\|G^k d_0\|}{\|d_0\|} \right)^{1/k}$$
$$= \lim_{k \to \infty} \left(\|G^k\| \right)^{1/k} = \rho(G).$$

Thus, the global asymptotic convergence factor is equal to the spectral radius of the iteration matrix G. The *general* convergence rate differs from the *specific* rate only when the initial error does not have any components in the invariant subspace associated with the dominant eigenvalue. Since it is hard to know this information in advance, the *general* convergence factor is more useful in practice.

Example 4.1. Consider the simple example of *Richardson's iteration*,

$$x_{k+1} = x_k + \alpha(b - Ax_k), \tag{4.31}$$

where α is a nonnegative scalar. This iteration can be rewritten as

$$x_{k+1} = (I - \alpha A)x_k + \alpha b. \tag{4.32}$$

Thus, the iteration matrix is $G_\alpha = I - \alpha A$ and the convergence factor is $\rho(I - \alpha A)$. Assume that the eigenvalues $\lambda_i, i = 1, \ldots, n$, are all real and such that

$$\lambda_{min} \le \lambda_i \le \lambda_{max}.$$

Then the eigenvalues μ_i of G_α are such that

$$1 - \alpha\lambda_{max} \le \mu_i \le 1 - \alpha\lambda_{min}.$$

In particular, if $\lambda_{min} < 0$ and $\lambda_{max} > 0$, at least one eigenvalue is greater than 1, and so $\rho(G_\alpha) > 1$ for any α. In this case the method will always diverge for some initial guess. Let us assume that all eigenvalues are positive; i.e., $\lambda_{min} > 0$. Then the following conditions must be satisfied in order for the method to converge:

$$1 - \alpha\lambda_{min} < 1,$$
$$1 - \alpha\lambda_{max} > -1.$$

The first condition implies that $\alpha > 0$, while the second requires that $\alpha \le 2/\lambda_{max}$. In other words, the method converges for any scalar α that satisfies

$$0 < \alpha < \frac{2}{\lambda_{max}}.$$

The next question is, What is the best value α_{opt} for the parameter α, i.e., the value of α that minimizes $\rho(G_\alpha)$? The spectral radius of G_α is

$$\rho(G_\alpha) = \max\{|1 - \alpha\lambda_{min}|, |1 - \alpha\lambda_{max}|\}.$$

This function of α is depicted in Figure 4.4. As the curve shows, the best possible α is reached at the point where the curve $|1 - \lambda_{max}\alpha|$ with positive slope crosses the curve $|1 - \lambda_{min}\alpha|$ with negative slope, i.e., when

$$-1 + \lambda_{max}\alpha = 1 - \lambda_{min}\alpha.$$

This gives

$$\alpha_{opt} = \frac{2}{\lambda_{min} + \lambda_{max}}. \tag{4.33}$$

Replacing this in one of the two curves gives the corresponding optimal spectral radius

$$\rho_{opt} = \frac{\lambda_{max} - \lambda_{min}}{\lambda_{max} + \lambda_{min}}.$$

4.2. Convergence

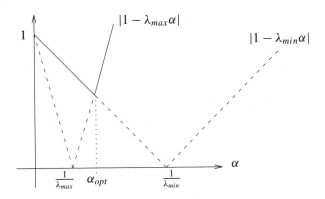

Figure 4.4. *The curve $\rho(G_\alpha)$ as a function of α.*

This expression shows the difficulty with the presence of small and large eigenvalues. The convergence rate can be extremely small for realistic problems. In addition, to achieve good convergence, eigenvalue estimates are required in order to obtain the optimal or a near-optimal α, which may cause difficulties. Finally, since λ_{max} can be very large, the curve $\rho(G_\alpha)$ can be extremely sensitive near the optimal value of α. These observations are common to many iterative methods that depend on an acceleration parameter.

4.2.2 Regular Splittings

Definition 4.3. *Let A, M, N be three given matrices satisfying $A = M - N$. The pair of matrices M, N is a regular splitting of A if M is nonsingular and M^{-1} and N are nonnegative.*

With a regular splitting, we associate the iteration

$$x_{k+1} = M^{-1}Nx_k + M^{-1}b. \tag{4.34}$$

The question is, Under which conditions does such an iteration converge? The following result, which generalizes Theorem 1.29, gives the answer.

Theorem 4.4. *Let M, N be a regular splitting of a matrix A. Then $\rho(M^{-1}N) < 1$ iff A is nonsingular and A^{-1} is nonnegative.*

Proof. Define $G = M^{-1}N$. From the fact that $\rho(G) < 1$ and the relation

$$A = M(I - G), \tag{4.35}$$

it follows that A is nonsingular. The assumptions of Theorem 1.29 are satisfied for the matrix G since $G = M^{-1}N$ is nonnegative and $\rho(G) < 1$. Therefore, $(I - G)^{-1}$ is nonnegative, as is $A^{-1} = (I - G)^{-1}M^{-1}$.

To prove the sufficient condition, assume that A is nonsingular and that its inverse is nonnegative. Since A and M are nonsingular, the relation (4.35) shows again that $I - G$ is nonsingular and, in addition,

$$\begin{aligned} A^{-1}N &= \left(M(I - M^{-1}N)\right)^{-1} N \\ &= (I - M^{-1}N)^{-1}M^{-1}N \\ &= (I - G)^{-1}G. \end{aligned} \quad (4.36)$$

Clearly, $G = M^{-1}N$ is nonnegative by the assumptions and, as a result of the Perron–Frobenius theorem, there is a nonnegative eigenvector x associated with $\rho(G)$ that is an eigenvalue such that

$$Gx = \rho(G)x.$$

From this and by virtue of (4.36), it follows that

$$A^{-1}Nx = \frac{\rho(G)}{1 - \rho(G)}x.$$

Since x and $A^{-1}N$ are nonnegative, this shows that

$$\frac{\rho(G)}{1 - \rho(G)} \geq 0,$$

which can be true only when $0 \leq \rho(G) \leq 1$. Since $I - G$ is nonsingular, then $\rho(G) \neq 1$, which implies that $\rho(G) < 1$. $\square$

This theorem establishes that the iteration (4.34) always converges if M, N is a regular splitting and A is an M-matrix.

4.2.3 Diagonally Dominant Matrices

We begin with a few standard definitions.

Definition 4.5. *A matrix A is*

- *(weakly) diagonally dominant if*

$$|a_{jj}| \geq \sum_{\substack{i=1 \\ i \neq j}}^{i=n} |a_{ij}|, \quad j = 1, \ldots, n;$$

- *strictly diagonally dominant if*

$$|a_{jj}| > \sum_{\substack{i=1 \\ i \neq j}}^{i=n} |a_{ij}|, \quad j = 1, \ldots, n;$$

- *irreducibly diagonally dominant if A is irreducible and*

$$|a_{jj}| \geq \sum_{\substack{i=1 \\ i \neq j}}^{i=n} |a_{ij}|, \quad j = 1, \ldots, n,$$

  with strict inequality for at least one j.

4.2. Convergence

Often the term diagonally dominant is used instead of *weakly* diagonally dominant.

Diagonal dominance is related to an important result in numerical linear algebra known as Gershgorin's theorem. This theorem allows rough locations for all the eigenvalues of A to be determined. In some situations, it is desirable to determine these locations in the complex plane by directly exploiting some knowledge of the entries of the matrix A. The simplest such result is the bound

$$|\lambda_i| \leq \|A\|$$

for any matrix norm. Gershgorin's theorem provides a more precise localization result.

Theorem 4.6. *(Gershgorin) Any eigenvalue λ of a matrix A is located in one of the closed discs of the complex plane centered at a_{ii} and having the radius*

$$\rho_i = \sum_{\substack{j=1 \\ j \neq i}}^{j=n} |a_{ij}|.$$

In other words,

$$\forall \lambda \in \sigma(A), \quad \exists i \quad \text{such that} \quad |\lambda - a_{ii}| \leq \sum_{\substack{j=1 \\ j \neq i}}^{j=n} |a_{ij}|. \tag{4.37}$$

Proof. Let x be an eigenvector associated with an eigenvalue λ and let m be the index of the component of largest modulus in x. Scale x so that $|\xi_m| = 1$ and $|\xi_i| \leq 1$ for $i \neq m$. Since x is an eigenvector, then

$$(\lambda - a_{mm})\xi_m = -\sum_{\substack{j=1 \\ j \neq m}}^{n} a_{mj}\xi_j,$$

which gives

$$|\lambda - a_{mm}| \leq \sum_{\substack{j=1 \\ j \neq m}}^{n} |a_{mj}||\xi_j| \leq \sum_{\substack{j=1 \\ j \neq m}}^{n} |a_{mj}| = \rho_m. \tag{4.38}$$

This completes the proof. $\square$

Since the result also holds for the transpose of A, a version of the theorem can also be formulated based on column sums instead of row sums.

The n discs defined in the theorem are called Gershgorin discs. The theorem states that the union of these n discs contains the spectrum of A. It can also be shown that, if there are m Gershgorin discs whose union S is disjoint from all other discs, then S contains exactly m eigenvalues (counted with their multiplicities). For example, when one disc is disjoint from the others, then it must contain exactly one eigenvalue.

An additional refinement, which has important consequences, concerns the particular case when A is irreducible.

Theorem 4.7. *Let A be an irreducible matrix and assume that an eigenvalue λ of A lies on the boundary of the union of the n Gershgorin discs. Then λ lies on the boundary of all Gershgorin discs.*

Proof. As in the proof of Gershgorin's theorem, let x be an eigenvector associated with λ, with $|\xi_m| = 1$ and $|\xi_i| \leq 1$ for $i \neq m$. Start from (4.38) in the proof of Gershgorin's theorem, which states that the point λ belongs to the mth disc. In addition, λ belongs to the boundary of the union of all the discs. As a result, it cannot be an interior point to the disc $D(\lambda, \rho_m)$. This implies that $|\lambda - a_{mm}| = \rho_m$. Therefore, the inequalities in (4.38) both become equalities:

$$|\lambda - a_{mm}| = \sum_{\substack{j=1 \\ j \neq m}}^{n} |a_{mj}||\xi_j| = \sum_{\substack{j=1 \\ j \neq m}}^{n} |a_{mj}| = \rho_m. \tag{4.39}$$

Let j be any integer $1 \leq j \leq n$. Since A is irreducible, its graph is connected and, therefore, there exists a path from node m to node j in the adjacency graph. Let this path be

$$m, m_1, m_2, \ldots, m_k = j.$$

By definition of an edge in the adjacency graph, $a_{m,m_1} \neq 0$. Because of the equality in (4.39), it is necessary that $|\xi_j| = 1$ for any nonzero ξ_j. Therefore, $|\xi_{m_1}|$ must be equal to one. Now repeating the argument with m replaced by m_1 shows that the following equality holds:

$$|\lambda - a_{m_1,m_1}| = \sum_{\substack{j=1 \\ j \neq m_1}}^{n} |a_{m_1,j}||\xi_j| = \sum_{\substack{j=1 \\ j \neq m_1}}^{n} |a_{m_1,j}| = \rho_{m_1}. \tag{4.40}$$

The argument can be continued showing each time that

$$|\lambda - a_{m_i,m_i}| = \rho_{m_i}, \tag{4.41}$$

which is valid for $i = 1, \ldots, k$. In the end, it will be proved that λ belongs to the boundary of the jth disc for an arbitrary j. $\square$

An immediate corollary of the Gershgorin theorem and Theorem 4.7 follows.

Corollary 4.8. *If a matrix A is strictly diagonally dominant or irreducibly diagonally dominant, then it is nonsingular.*

Proof. If a matrix is strictly diagonally dominant, then the union of the Gershgorin discs excludes the origin, so $\lambda = 0$ cannot be an eigenvalue. Assume now that it is only irreducibly diagonally dominant. Then, if it is singular, the zero eigenvalue lies on the boundary of the union of the Gershgorin discs. In this situation, according to Theorem 4.7, this eigenvalue should lie on the boundary of all the discs. This would mean that

$$|a_{jj}| = \sum_{\substack{i=1 \\ i \neq j}}^{n} |a_{ij}| \quad \text{for} \quad j = 1, \ldots, n,$$

which contradicts the assumption of irreducible diagonal dominance. $\square$

The following theorem can now be stated.

Theorem 4.9. *If A is a strictly diagonally dominant or an irreducibly diagonally dominant matrix, then the associated Jacobi and Gauss–Seidel iterations converge for any x_0.*

4.2. Convergence

Proof. We first prove the results for strictly diagonally dominant matrices. Let λ be the dominant eigenvalue of the iteration matrix $M_J = D^{-1}(E + F)$ for Jacobi and $M_G = (D - E)^{-1}F$ for Gauss–Seidel. As in the proof of Gershgorin's theorem, let x be an eigenvector associated with λ, with $|\xi_m| = 1$ and $|\xi_i| \leq 1$ for $i \neq 1$. Start from (4.38) in the proof of Gershgorin's theorem, which states that, for M_J,

$$|\lambda| \leq \sum_{\substack{j=1 \\ j \neq m}}^{n} \frac{|a_{mj}|}{|a_{mm}|} |\xi_j| \leq \sum_{\substack{j=1 \\ j \neq m}}^{n} \frac{|a_{mj}|}{|a_{mm}|} < 1.$$

This proves the result for Jacobi's method.

For the Gauss–Seidel iteration, write the mth row of the equation $Fx = \lambda(D - E)x$ in the form

$$\sum_{j<m} a_{mj}\xi_j = \lambda \left(a_{mm}\xi_m + \sum_{j>m} a_{mj}\xi_j \right),$$

which yields the inequality

$$|\lambda| \leq \frac{\sum_{j<m} |a_{mj}||\xi_j|}{|a_{mm}| - \sum_{j>m} |a_{mj}||\xi_j|} \leq \frac{\sum_{j<m} |a_{mj}|}{|a_{mm}| - \sum_{j>m} |a_{mj}|}.$$

The last term in the above inequality has the form $\sigma_2/(d - \sigma_1)$, with d, σ_1, σ_2 all nonnegative and $d - \sigma_1 - \sigma_2 > 0$. Therefore,

$$|\lambda| \leq \frac{\sigma_2}{\sigma_2 + (d - \sigma_2 - \sigma_1)} < 1.$$

In the case when the matrix is only irreducibly diagonally dominant, the above proofs only show that $\rho(M^{-1}N) \leq 1$, where $M^{-1}N$ is the iteration matrix for either Jacobi or Gauss–Seidel. A proof by contradiction will be used to show that in fact $\rho(M^{-1}N) < 1$. Assume that λ is an eigenvalue of $M^{-1}N$ with $|\lambda| = 1$. Then the matrix $M^{-1}N - \lambda I$ is singular and, as a result, $A' = N - \lambda M$ is also singular. Since $|\lambda| = 1$, it is clear that A' is also an irreducibly diagonally dominant matrix. This contradicts Corollary 4.8. □

4.2.4 Symmetric Positive Definite Matrices

It is possible to show that, when A is symmetric positive definite (SPD), then SOR will converge for any ω in the open interval $(0, 2)$ and for any initial guess x_0. In fact, the reverse is also true under certain assumptions.

Theorem 4.10. *If A is symmetric with positive diagonal elements and for $0 < \omega < 2$, SOR converges for any x_0 iff A is positive definite.*

4.2.5 Property A and Consistent Orderings

A number of properties that are related to the graph of a finite difference matrix are now defined. The first of these properties is called Property A. A matrix has Property A if its graph is *bipartite*. This means that the graph is two-colorable in the sense defined in Chapter 3: its vertices can be partitioned into two sets in such a way that no two vertices in the same set are connected by an edge. Note that, as usual, the self-connecting edges that correspond to the diagonal elements are ignored.

Definition 4.11. *A matrix has Property A if the vertices of its adjacency graph can be partitioned into two sets S_1 and S_2 so that any edge in the graph links a vertex of S_1 to a vertex of S_2.*

In other words, nodes from the first set are connected only to nodes from the second set and vice versa. This definition is illustrated in Figure 4.5.

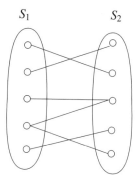

Figure 4.5. *Graph illustration of Property A.*

An alternative definition is that a matrix has Property A if it can be permuted into a matrix with the following structure:

$$A' = \begin{pmatrix} D_1 & -F \\ -E & D_2 \end{pmatrix}, \tag{4.42}$$

where D_1 and D_2 are diagonal matrices. This structure can be obtained by first labeling all the unknowns in S_1 from 1 to n_1, in which $n_1 = |S_1|$, and the rest from $n_1 + 1$ to n. Note that the Jacobi iteration matrix will have the same structure except that the D_1, D_2 blocks will be replaced by zero blocks. These Jacobi iteration matrices satisfy an important property stated in the following proposition.

Proposition 4.12. *Let B be a matrix with the following structure:*

$$B = \begin{pmatrix} O & B_{12} \\ B_{21} & O \end{pmatrix}, \tag{4.43}$$

and let L and U be the lower and upper triangular parts of B, respectively. Then the following properties hold:

1. *If μ is an eigenvalue of B, then so is $-\mu$.*

2. *The eigenvalues of the matrix*

$$B(\alpha) = \alpha L + \frac{1}{\alpha} U$$

defined for $\alpha \neq 0$ are independent of α.

Proof. The first property is shown by simply observing that, if $\begin{pmatrix} x \\ v \end{pmatrix}$ is an eigenvector associated with μ, then $\begin{pmatrix} x \\ -v \end{pmatrix}$ is an eigenvector of B associated with the eigenvalue $-\mu$.

4.2. Convergence

Consider the second property. For any α, the matrix $B(\alpha)$ is similar to B; i.e., $B(\alpha) = XBX^{-1}$, with X defined by

$$X = \begin{pmatrix} 1 & O \\ O & \alpha \end{pmatrix}.$$

This proves the desired result. □

A definition that generalizes this important property is *consistently ordered matrices*. Varga [292] calls a consistently ordered matrix one for which the eigenvalues of $B(\alpha)$ are independent of α. Another definition given by Young [321] considers a specific class of matrices that generalize this property. We will use this definition here. Unlike Property A, the consistent ordering property depends on the initial ordering of the unknowns.

Definition 4.13. *A matrix is said to be consistently ordered if the vertices of its adjacency graph can be partitioned into p sets $S_1, S_2, \ldots, S_p$ with the property that any two adjacent vertices i and j in the graph belong to two consecutive partitions S_k and $S_{k'}$, with $k' = k - 1$ if $j < i$ and $k' = k + 1$ if $j > i$.*

It is easy to show that consistently ordered matrices satisfy Property A: the first color is made up of all the partitions S_i with odd i and the second of the partitions S_i with even i.

Example 4.2. Block tridiagonal matrices of the form

$$T = \begin{pmatrix} D_1 & T_{12} & & & & \\ T_{21} & D_2 & T_{23} & & & \\ & T_{32} & D_3 & \ddots & & \\ & & \ddots & \ddots & T_{p-1,p} \\ & & & T_{p,p-1} & D_p \end{pmatrix}$$

whose diagonal blocks D_i are diagonal matrices are called T-matrices. Clearly, such matrices are consistently ordered. Note that matrices of the form (4.42) are a particular case with $p = 2$.

Consider now a general, consistently ordered matrix. By definition, there is a permutation π of $\{1, 2, \ldots, n\}$ that is the union of p disjoint subsets

$$\pi = \pi_1 \bigcup \pi_2 \cdots \bigcup \pi_p, \qquad (4.44)$$

with the property that, if $a_{ij} \neq 0$, $j \neq i$, and i belongs to π_k, then j belongs to $\pi_{k\pm 1}$ depending on whether $i < j$ or $i > j$. This permutation π can be used to permute A symmetrically. If P is the permutation matrix associated with the permutation π, then clearly

$$A' = P^T A P$$

is a T-matrix.

Not every matrix that can be symmetrically permuted into a T-matrix is consistently ordered. The important property here is that the partition $\{\pi_i\}$ *preserves the order of the indices i, j of nonzero elements*. In terms of the adjacency graph, there is a partition of the graph with the property that an oriented edge i, j from i to j always points to a set with a larger index if $j > i$ and a smaller index otherwise. In particular, a very important consequence is that edges corresponding to the lower triangular part will remain so in the permuted matrix. The same is true for the upper triangular part. Indeed, if a nonzero element in the permuted matrix is $a'_{i',j'} = a_{\pi^{-1}(i),\pi^{-1}(j)} \neq 0$ with $i' > j'$, then, by definition of the permutation, $\pi(i') > \pi(j')$ or $i = \pi(\pi^{-1}(i)) > j = \pi(\pi^{-1}(j))$. Because of the order preservation, it is necessary that $i > j$. A similar observation holds for the upper triangular part. Therefore, this results in the following proposition.

Proposition 4.14. *If a matrix A is consistently ordered, then there exists a permutation matrix P such that $P^T A P$ is a T-matrix and*

$$(P^T A P)_L = P^T A_L P, \quad (P^T A P)_U = P^T A_U P, \tag{4.45}$$

in which X_L represents the (strict) lower part of X, and X_U represents the (strict) upper part of X.

With the above property it can be shown that for consistently ordered matrices the eigenvalues of $B(\alpha)$ as defined in Proposition 4.12 are also invariant with respect to α.

Proposition 4.15. *Let B be the Jacobi iteration matrix associated with a consistently ordered matrix A and let L and U be the lower and upper triangular parts of B, respectively. Then the eigenvalues of the matrix*

$$B(\alpha) = \alpha L + \frac{1}{\alpha} U$$

defined for $\alpha \neq 0$ do not depend on α.

Proof. First transform $B(\alpha)$ into a T-matrix using the permutation π in (4.44) provided by Proposition 4.14:

$$P^T B(\alpha) P = \alpha P^T L P + \frac{1}{\alpha} P^T U P.$$

From Proposition 4.14, the lower part of $P^T B P$ is precisely $L' = P^T L P$. Similarly, the upper part is $U' = P^T U P$, the lower and upper parts of the associated T-matrix. Therefore, we only need to show that the property is true for a T-matrix.

In this case, for any α, the matrix $B(\alpha)$ is similar to B. This means that $B(\alpha) = X B X^{-1}$, with X being given by

$$X = \begin{pmatrix} 1 & & & & \\ & \alpha I & & & \\ & & \alpha^2 I & & \\ & & & \ddots & \\ & & & & \alpha^{p-1} I \end{pmatrix},$$

where the partitioning is associated with the subsets $\pi_1, \ldots, \pi_p$, respectively. □

Note that T-matrices and matrices with the structure (4.42) are two particular cases of matrices that fulfill the assumptions of the above proposition. There are a number of well-known properties related to Property A and consistent orderings. For example, it is possible to show the following:

4.2. Convergence

- Property A is invariant under symmetric permutations.
- A matrix has Property A iff there is a permutation matrix P such that $A' = P^{-1}AP$ is consistently ordered.

Consistently ordered matrices satisfy an important property that relates the eigenvalues of the corresponding SOR iteration matrices to those of the Jacobi iteration matrices. The main theorem regarding the theory for SOR is a consequence of the following result proved by Young [321]. Remember that

$$M_{SOR} = (D - \omega E)^{-1}(\omega F + (1-\omega)D)$$
$$= (I - \omega D^{-1}E)^{-1}(\omega D^{-1}F + (1-\omega)I).$$

Theorem 4.16. *Let A be a consistently ordered matrix such that $a_{ii} \neq 0$ for $i = 1, \ldots, n$ and let $\omega \neq 0$. Then, if λ is a nonzero eigenvalue of the SOR iteration matrix M_{SOR}, any scalar μ such that*

$$(\lambda + \omega - 1)^2 = \lambda \omega^2 \mu^2 \qquad (4.46)$$

is an eigenvalue of the Jacobi iteration matrix B. Conversely, if μ is an eigenvalue of the Jacobi matrix B and if a scalar λ satisfies (4.46), then λ is an eigenvalue of M_{SOR}.

Proof. Denote $D^{-1}E$ by L and $D^{-1}F$ by U, so that

$$M_{SOR} = (I - \omega L)^{-1}(\omega U + (1-\omega)I)$$

and the Jacobi iteration matrix is merely $L + U$. Writing that λ is an eigenvalue yields

$$\det\left(\lambda I - (I - \omega L)^{-1}(\omega U + (1-\omega)I)\right) = 0,$$

which is equivalent to

$$\det\left(\lambda(I - \omega L) - (\omega U + (1-\omega)I)\right) = 0$$

or

$$\det\left((\lambda + \omega - 1)I - \omega(\lambda L + U)\right) = 0.$$

Since $\omega \neq 0$, this can be rewritten as

$$\det\left(\frac{\lambda + \omega - 1}{\omega}I - (\lambda L + U)\right) = 0,$$

which means that $(\lambda + \omega - 1)/\omega$ is an eigenvalue of $\lambda L + U$. Since A is consistently ordered, the eigenvalues of $\lambda L + U$, which are equal to $\lambda^{1/2}(\lambda^{1/2}L + \lambda^{-1/2}U)$, are the same as those of $\lambda^{1/2}(L + U)$, where $L + U$ is the Jacobi iteration matrix. The proof follows immediately. $\square$

This theorem allows us to compute an optimal value for ω, which can be shown to be equal to

$$\omega_{opt} = \frac{2}{1 + \sqrt{1 - \rho(B)^2}}. \qquad (4.47)$$

A typical SOR procedure starts with some ω, for example, $\omega = 1$, then proceeds with a number of SOR steps with this ω. The convergence rate for the resulting iterates is estimated, providing an estimate for $\rho(B)$ using Theorem 4.16. A better ω is then obtained from the formula (4.47), and the iteration restarted. Further refinements of the optimal ω are calculated and retrofitted in this manner as the algorithm progresses.

4.3 Alternating Direction Methods

The alternating direction implicit (ADI) method was introduced in the mid-1950s by Peaceman and Rachford [225] specifically for solving equations arising from finite difference discretizations of elliptic and parabolic PDEs. Consider a PDE of elliptic type

$$\frac{\partial}{\partial x}\left(a(x,y)\frac{\partial u(x,y)}{\partial x}\right) + \frac{\partial}{\partial y}\left(b(x,y)\frac{\partial u(x,y)}{\partial y}\right) = f(x,y) \quad (4.48)$$

on a rectangular domain with Dirichlet boundary conditions. The equations are discretized with centered finite differences using $n+2$ points in the x direction and $m+2$ points in the y direction. This results in the system of equations

$$Hu + Vu = b, \quad (4.49)$$

in which the matrices H and V represent the three-point central difference approximations to the operators

$$\frac{\partial}{\partial x}\left(a(x,y)\frac{\partial}{\partial x}\right) \quad \text{and} \quad \frac{\partial}{\partial y}\left(b(x,y)\frac{\partial}{\partial y}\right),$$

respectively. In what follows, the same notation is used to represent the discretized version of the unknown function u.

The ADI algorithm consists of iterating by solving (4.49) in the x and y directions alternatively as follows.

ALGORITHM 4.3. Peaceman–Rachford ADI

1. *For* $k = 0, 1, \ldots$, *until convergence, Do*
2.   *Solve* $(H + \rho_k I)u_{k+\frac{1}{2}} = (\rho_k I - V)u_k + b$
3.   *Solve* $(V + \rho_k I)u_{k+1} = (\rho_k I - H)u_{k+\frac{1}{2}} + b$
4. *EndDo*

Here ρ_k, $k = 1, 2, \ldots$, is a sequence of positive acceleration parameters.

The specific case where ρ_k is chosen to be a constant ρ deserves particular attention. In this case, we can formulate the above iteration in the usual form of (4.28) with

$$G = (V + \rho I)^{-1}(H - \rho I)(H + \rho I)^{-1}(V - \rho I), \quad (4.50)$$
$$f = (V + \rho I)^{-1}\left[I - (H - \rho I)(H + \rho I)^{-1}\right]b \quad (4.51)$$

or, when $\rho > 0$, in the form (4.22), with

$$M = \frac{1}{2\rho}(H + \rho I)(V + \rho I), \quad N = \frac{1}{2\rho}(H - \rho I)(V - \rho I). \quad (4.52)$$

Note that (4.51) can be rewritten in a simpler form; see Exercise 5.

The ADI algorithm is often formulated for solving the time-dependent PDE

$$\frac{\partial u}{\partial t} = \frac{\partial}{\partial x}\left(a(x,y)\frac{\partial u}{\partial x}\right) + \frac{\partial}{\partial y}\left(b(x,y)\frac{\partial u}{\partial y}\right) \quad (4.53)$$

4.3. Alternating Direction Methods

on the domain $(x, y, t) \in \Omega \times [0, T] \equiv (0, 1) \times (0, 1) \times [0, T]$. The initial and boundary conditions are

$$u(x, y, 0) = x_0(x, y) \quad \forall (x, y) \in \Omega, \tag{4.54}$$

$$u(\bar{x}, \bar{y}, t) = g(\bar{x}, \bar{y}, t) \quad \forall (\bar{x}, \bar{y}) \in \partial\Omega, \quad t > 0, \tag{4.55}$$

where $\partial\Omega$ is the boundary of the unit square Ω. The equations are discretized with respect to the space variables x and y as before, resulting in a system of ordinary differential equations

$$\frac{du}{dt} = Hu + Vu, \tag{4.56}$$

in which the matrices H and V have been defined earlier. The ADI algorithm advances the relation (4.56) forward in time alternately in the x and y directions as follows:

$$\left(I - \frac{1}{2}\Delta t\, H\right) u_{k+\frac{1}{2}} = \left(I + \frac{1}{2}\Delta t\, V\right) u_k,$$

$$\left(I - \frac{1}{2}\Delta t\, V\right) u_{k+1} = \left(I + \frac{1}{2}\Delta t\, H\right) u_{k+\frac{1}{2}}.$$

The acceleration parameters ρ_k of Algorithm 4.3 are replaced by a natural time step.

Assuming that the mesh points are ordered by lines in the x direction, the first step of Algorithm 4.3 constitutes a set of m independent tridiagonal linear systems of size n each. However, the second step constitutes a large tridiagonal system whose three diagonals are offset by $-m$, 0, and m, respectively. This second system can also be rewritten as a set of n independent tridiagonal systems of size m each by reordering the grid points by lines, this time in the y direction. The natural (horizontal) and vertical orderings are illustrated in Figure 4.6. Whenever moving from one half-step of ADI to the next, we must implicitly work with the transpose of the matrix representing the solution on the $n \times m$ grid points. This data operation may be an expensive task on parallel machines and often it is cited as one of the drawbacks of alternating direction methods in this case.

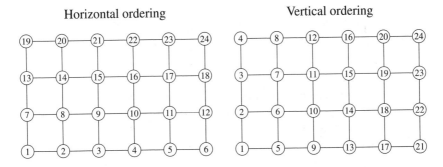

Figure 4.6. *The horizontal and vertical orderings for the unknowns in ADI.*

ADI methods were extensively studied in the 1950s and 1960s for the particular case of positive definite systems. For such systems, H and V have real eigenvalues. The following is a summary of the main results in this situation. First, when H and V are SPD, then the stationary iteration ($\rho_k = \rho > 0$ for all k) converges. For the model problem, the asymptotic rate of convergence of the stationary ADI iteration using the optimal ρ is the same as that of SSOR using the optimal ω. However, each ADI step is more expensive than one SSOR step. One of the more important results in the ADI theory is that the rate of convergence of ADI can be increased appreciably by using a cyclic sequence of parameters ρ_k. A theory for selecting the best sequence of ρ_k's is well understood in the case when H and V commute [38]. For the model problem, the parameters can be selected so that the time complexity is reduced to $O(n^2 \log n)$; for details see [225].

Exercises

1. Consider an $n \times n$ tridiagonal matrix of the form

$$T_\alpha = \begin{pmatrix} \alpha & -1 & & & & \\ -1 & \alpha & -1 & & & \\ & -1 & \alpha & -1 & & \\ & & -1 & \alpha & -1 & \\ & & & -1 & \alpha & -1 \\ & & & & -1 & \alpha \end{pmatrix}, \quad (4.57)$$

where α is a real parameter.

   a. Verify that the eigenvalues of T_α are given by

$$\lambda_j = \alpha - 2\cos(j\theta), \quad j = 1, \ldots, n,$$

where

$$\theta = \frac{\pi}{n+1},$$

and that an eigenvector associated with each λ_j is

$$q_j = [\sin(j\theta), \sin(2j\theta), \ldots, \sin(nj\theta)]^T.$$

Under what condition on α does this matrix become positive definite?

   b. Now take $\alpha = 2$. How does this matrix relate to the matrices seen in Chapter 2 for one-dimensional problems?

   (i) Will the Jacobi iteration converge for this matrix? If so, what will its convergence factor be?

   (ii) Will the Gauss–Seidel iteration converge for this matrix? If so, what will its convergence factor be?

   (iii) For which values of ω will the SOR iteration converge?

2. Prove that the iteration matrix G_ω of SSOR, as defined by (4.13), can be expressed as

$$G_\omega = I - \omega(2-\omega)(D - \omega F)^{-1} D (D - \omega E)^{-1} A.$$

Deduce the expression (4.27) for the preconditioning matrix associated with the SSOR iteration.

Exercises

3. Let A be a matrix with a positive diagonal D.

    a. Obtain an expression equivalent to that of (4.13) for G_ω but involving the matrices $S_E \equiv D^{-1/2}ED^{-1/2}$ and $S_F \equiv D^{-1/2}FD^{-1/2}$.

    b. Show that
    $$D^{1/2}G_\omega D^{-1/2} = (I - \omega S_F)^{-1}(I - \omega S_E)^{-1}(\omega S_E + (1-\omega)I)(\omega S_F + (1-\omega)I).$$

    c. Now assume that, in addition to having a positive diagonal, A is symmetric. Prove that the eigenvalues of the SSOR iteration matrix G_ω are real and nonnegative.

4. Let
$$A = \begin{pmatrix} D_1 & -F_2 & & & \\ -E_2 & D_2 & -F_3 & & \\ & -E_3 & D_3 & \ddots & \\ & & \ddots & \ddots & -F_m \\ & & & -E_m & D_m \end{pmatrix},$$
where the D_i blocks are nonsingular matrices that are not necessarily diagonal.

    a. What are the *block Jacobi* and *block Gauss–Seidel* iteration matrices?

    b. Show a result similar to that in Proposition 4.15 for the Jacobi iteration matrix.

    c. Show also that, for $\omega = 1$, (1) the block Gauss–Seidel and block Jacobi iterations either both converge or both diverge and (2) when they both converge, then the block Gauss–Seidel iteration is (asymptotically) twice as fast as the block Jacobi iteration.

5. According to formula (4.23), the f-vector in iteration (4.22) should be equal to $M^{-1}b$, where b is the right-hand side and M is given in (4.52). Yet formula (4.51) gives a different expression for f. Reconcile the two results; i.e., show that the expression (4.51) can also be rewritten as
$$f = 2\rho(V + \rho I)^{-1}(H + \rho I)^{-1}b.$$

6. Show that a matrix has Property A iff there is a permutation matrix P such that $A' = P^{-1}AP$ is consistently ordered.

7. Consider a matrix A that is consistently ordered. Show that the asymptotic convergence rate for Gauss–Seidel is double that of the Jacobi iteration.

8. A matrix of the form
$$B = \begin{pmatrix} 0 & E & 0 \\ 0 & 0 & F \\ H & 0 & 0 \end{pmatrix}$$
is called a three-cyclic matrix.

    a. What are the eigenvalues of B? (Express them in terms of eigenvalues of a certain matrix that depends on E, F, and H.)

    b. Assume that a matrix A has the form $A = D + B$, where D is a nonsingular diagonal matrix and B is three cyclic. How can the eigenvalues of the Jacobi iteration matrix be related to those of the Gauss–Seidel iteration matrix? How does the asymptotic convergence rate of the Gauss–Seidel iteration compare with that of the Jacobi iteration matrix in this case?

c. Repeat part (b) for the case when SOR replaces the Gauss–Seidel iteration.

d. Generalize the above results to p-cyclic matrices, i.e., matrices of the form

$$B = \begin{pmatrix} 0 & E_1 & & & \\ & 0 & E_2 & & \\ & & 0 & \ddots & \\ & & & 0 & E_{p-1} \\ E_p & & & & 0 \end{pmatrix}.$$

Notes and References

Two good references for the material covered in this chapter are Varga [292] and Young [321]. Although relaxation-type methods were very popular up to the 1960s, they are now mostly used as preconditioners, a topic that will be seen in detail in Chapters 9 and 10. One of the main difficulties with these methods is finding an optimal relaxation factor for general matrices. Theorem 4.7 is due to Ostrowski. For details on the use of Gershgorin's theorem in eigenvalue problems, see [245]. The original idea of the ADI method is described in [225] and those results on the optimal parameters for ADI can be found in [38]. A comprehensive text on this class of techniques can be found in [299].

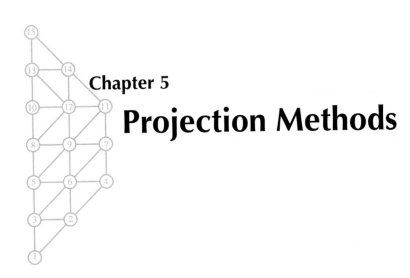

Chapter 5

Projection Methods

Most of the existing practical iterative techniques for solving large linear systems of equations utilize a projection process in one way or another. A projection process represents a *canonical* way of extracting an approximation to the solution of a linear system from a subspace. This chapter describes these techniques in a very general framework and presents some theory. The one-dimensional case is covered in detail at the end of the chapter, as it provides a good preview of the more complex projection processes to be seen in later chapters.

5.1 Basic Definitions and Algorithms

Consider the linear system
$$Ax = b, \qquad (5.1)$$
where A is an $n \times n$ real matrix. In this chapter, the same symbol A is often used to denote the matrix and the linear mapping in $\mathbb{R}^n$ that it represents. The idea of *projection techniques* is to extract an approximate solution to the above problem from a subspace of $\mathbb{R}^n$. If $\mathcal{K}$ is this subspace of *candidate approximants*, also called the *search subspace*, and if m is its dimension, then, in general, m constraints must be imposed to be able to extract such an approximation. A typical way of describing these constraints is to impose m (independent) orthogonality conditions. Specifically, the residual vector $b - Ax$ is constrained to be orthogonal to m linearly independent vectors. This defines another subspace $\mathcal{L}$ of dimension m, which will be called the *subspace of constraints* or *left subspace* for reasons that will be explained below. This simple framework is common to many different mathematical methods and is known as the Petrov–Galerkin conditions.

There are two broad classes of projection methods: *orthogonal* and *oblique*. In an orthogonal projection technique, the subspace $\mathcal{L}$ is the same as $\mathcal{K}$. In an oblique projection method, $\mathcal{L}$ is different from $\mathcal{K}$ and may be totally unrelated to it. This distinction is rather important and gives rise to different types of algorithms.

5.1.1 General Projection Methods

Let A be an $n \times n$ real matrix and $\mathcal{K}$ and $\mathcal{L}$ be two m-dimensional subspaces of $\mathbb{R}^n$. A projection technique onto the subspace $\mathcal{K}$ and orthogonal to $\mathcal{L}$ is a process that finds an approximate solution $\tilde{x}$ to (5.1) by imposing the conditions that $\tilde{x}$ belong to $\mathcal{K}$ and that the new residual vector be orthogonal to $\mathcal{L}$:

$$\text{Find } \tilde{x} \in \mathcal{K} \quad \text{such that} \quad b - A\tilde{x} \perp \mathcal{L}. \tag{5.2}$$

If we wish to exploit the knowledge of an initial guess x_0 to the solution, then the approximation must be sought in the affine space $x_0 + \mathcal{K}$ instead of the homogeneous vector space $\mathcal{K}$. This requires a slight modification to the above formulation. The approximate problem should be redefined as follows:

$$\text{Find} \quad \tilde{x} \in x_0 + \mathcal{K} \quad \text{such that} \quad b - A\tilde{x} \perp \mathcal{L}. \tag{5.3}$$

Note that, if $\tilde{x}$ is written in the form $\tilde{x} = x_0 + \delta$ and the initial residual vector r_0 is defined as

$$r_0 = b - Ax_0, \tag{5.4}$$

then the above equation becomes $b - A(x_0 + \delta) \perp \mathcal{L}$ or

$$r_0 - A\delta \perp \mathcal{L}.$$

In other words, the approximate solution can be defined as

$$\tilde{x} = x_0 + \delta, \quad \delta \in \mathcal{K}, \tag{5.5}$$

$$(r_0 - A\delta, w) = 0 \quad \forall\, w \in \mathcal{L}. \tag{5.6}$$

The orthogonality condition (5.6) imposed on the new residual $r_{new} = r_0 - A\delta$ is illustrated in Figure 5.1.

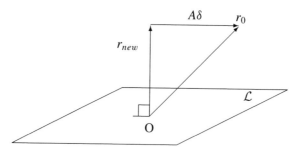

Figure 5.1. *Interpretation of the orthogonality condition.*

This is a basic projection step, in its most general form. Most standard techniques use a succession of such projections. Typically, a new projection step uses a new pair of subspaces $\mathcal{K}$ and $\mathcal{L}$ and an initial guess x_0 equal to the most recent approximation obtained from the previous projection step. Projection methods form a unifying framework for many of the well-known methods in scientific computing. In fact, virtually all of the basic iterative

5.1. Basic Definitions and Algorithms

techniques seen in the previous chapter can be considered projection techniques. Whenever an approximation is defined via m degrees of freedom (subspace $\mathcal{K}$) and m constraints (subspace $\mathcal{L}$), a projection process results.

Example 5.1. In the simplest case, an elementary Gauss–Seidel step as defined by (4.6) is nothing but a projection step with $\mathcal{K} = \mathcal{L} = \text{span}\{e_i\}$. These projection steps are cycled for $i = 1, \ldots, n$ until convergence. See Exercise 1 for an alternative way of selecting the sequence of e_i's.

Orthogonal projection methods correspond to the particular case when the two subspaces $\mathcal{L}$ and $\mathcal{K}$ are identical. The distinction is particularly important in the Hermitian case since we are guaranteed that the projected problem will be Hermitian in this situation, as will be seen shortly. In addition, a number of helpful theoretical results are true for the orthogonal case. When $\mathcal{L} = \mathcal{K}$, the Petrov–Galerkin conditions are often called the Galerkin conditions.

5.1.2 Matrix Representation

Let $V = [v_1, \ldots, v_m]$, an $n \times m$ matrix whose column vectors form a basis of $\mathcal{K}$ and, similarly, $W = [w_1, \ldots, w_m]$, an $n \times m$ matrix whose column vectors form a basis of $\mathcal{L}$. If the approximate solution is written as

$$x = x_0 + Vy,$$

then the orthogonality condition leads immediately to the following system of equations for the vector y:

$$W^T A V y = W^T r_0.$$

If the assumption is made that the $m \times m$ matrix $W^T A V$ is nonsingular, the following expression for the approximate solution $\tilde{x}$ results:

$$\tilde{x} = x_0 + V(W^T A V)^{-1} W^T r_0. \tag{5.7}$$

In many algorithms, the matrix $W^T A V$ does not have to be formed since it is available as a byproduct of the algorithm. A prototype projection technique is represented by the following algorithm.

ALGORITHM 5.1. Prototype Projection Method

1. Until convergence, Do
2.   Select a pair of subspaces $\mathcal{K}$ and $\mathcal{L}$
3.   Choose bases $V = [v_1, \ldots, v_m]$ and $W = [w_1, \ldots, w_m]$ for $\mathcal{K}$ and $\mathcal{L}$
4. $r := b - Ax$
5. $y := (W^T A V)^{-1} W^T r$
6. $x := x + Vy$
7. EndDo

The approximate solution is defined only when the matrix $W^T A V$ is nonsingular, a property that is not guaranteed to be true even when A is nonsingular.

Example 5.2. As an example, consider the matrix

$$A = \begin{pmatrix} O & I \\ I & I \end{pmatrix},$$

where I is the $m \times m$ identity matrix and O is the $m \times m$ zero matrix, and let $V = W = [e_1, e_2, \ldots, e_m]$. Although A is nonsingular, the matrix $W^T A V$ is precisely the O block in the upper-left corner of A and is therefore singular.

It can be easily verified that $W^T A V$ is nonsingular iff no vector of the subspace $A\mathcal{K}$ is orthogonal to the subspace $\mathcal{L}$. We have encountered a similar condition when defining projection operators in Chapter 1. There are two important particular cases where the nonsingularity of $W^T A V$ is guaranteed. These are discussed in the following proposition.

Proposition 5.1. *Let A, $\mathcal{L}$, and $\mathcal{K}$ satisfy either one of the two following conditions:*

(i) *A is positive definite and $\mathcal{L} = \mathcal{K}$ or*

(ii) *A is nonsingular and $\mathcal{L} = A\mathcal{K}$.*

Then the matrix $B = W^T A V$ is nonsingular for any bases V and W of $\mathcal{K}$ and $\mathcal{L}$, respectively.

Proof. Consider first case (i). Let V be any basis of $\mathcal{K}$ and W be any basis of $\mathcal{L}$. In fact, since $\mathcal{L}$ and $\mathcal{K}$ are the same, W can always be expressed as $W = VG$, where G is a nonsingular $m \times m$ matrix. Then

$$B = W^T A V = G^T V^T A V.$$

Since A is positive definite, so is $V^T A V$ (see Chapter 1), which shows that B is nonsingular.

Consider now case (ii). Let V be any basis of $\mathcal{K}$ and W be any basis of $\mathcal{L}$. Since $\mathcal{L} = A\mathcal{K}$, W can be expressed in this case as $W = AVG$, where G is a nonsingular $m \times m$ matrix. Then

$$B = W^T A V = G^T (AV)^T A V. \qquad (5.8)$$

Since A is nonsingular, the $n \times m$ matrix AV is of full rank and, as a result, $(AV)^T A V$ is nonsingular. This, along with (5.8), shows that B is nonsingular. □

Now consider the particular case where A is symmetric (real) and an orthogonal projection technique is used. In this situation, the same basis can be used for $\mathcal{L}$ and $\mathcal{K}$, which are identical subspaces, and the projected matrix, which is $B = V^T A V$, is symmetric. In addition, if the matrix A is symmetric positive definite (SPD), then so is B.

5.2 General Theory

This section gives some general theoretical results without being specific about the subspaces $\mathcal{K}$ and $\mathcal{L}$ that are used. The goal is to learn about the quality of the approximation

5.2. General Theory

obtained from a general projection process. Two main tools are used for this. The first is to exploit optimality properties of projection methods. These properties are induced from those properties of projectors seen in Section 1.12.4 of Chapter 1. The second tool consists of interpreting the projected problem with the help of projection operators in an attempt to extract residual bounds.

5.2.1 Two Optimality Results

In this section, two important optimality results will be established that are satisfied by the approximate solutions in some cases. Consider first the case when A is SPD.

Proposition 5.2. *Assume that A is SPD and $\mathcal{L} = \mathcal{K}$. Then a vector $\tilde{x}$ is the result of an (orthogonal) projection method onto $\mathcal{K}$ with the starting vector x_0 iff it minimizes the A-norm of the error over $x_0 + \mathcal{K}$, i.e., iff*

$$E(\tilde{x}) = \min_{x \in x_0 + \mathcal{K}} E(x),$$

where

$$E(x) \equiv (A(x_* - x), x_* - x)^{1/2}.$$

Proof. As was seen in Section 1.12.4, for $\tilde{x}$ to be the minimizer of $E(x)$, it is necessary and sufficient that $x_* - \tilde{x}$ be A-orthogonal to all the subspace $\mathcal{K}$. This yields

$$(A(x_* - \tilde{x}), v) = 0 \quad \forall v \in \mathcal{K}$$

or, equivalently,

$$(b - A\tilde{x}, v) = 0 \quad \forall v \in \mathcal{K},$$

which is the Galerkin condition defining an orthogonal projection process for the approximation $\tilde{x}$. □

We now take up the case when $\mathcal{L}$ is defined by $\mathcal{L} = A\mathcal{K}$.

Proposition 5.3. *Let A be an arbitrary square matrix and assume that $\mathcal{L} = A\mathcal{K}$. Then a vector $\tilde{x}$ is the result of an (oblique) projection method onto $\mathcal{K}$ orthogonally to $\mathcal{L}$ with the starting vector x_0 iff it minimizes the 2-norm of the residual vector $b - Ax$ over $x \in x_0 + \mathcal{K}$, i.e., iff*

$$R(\tilde{x}) = \min_{x \in x_0 + \mathcal{K}} R(x),$$

where $R(x) \equiv \|b - Ax\|_2$.

Proof. As was seen in Section 1.12.4, for $\tilde{x}$ to be the minimizer of $R(x)$, it is necessary and sufficient that $b - A\tilde{x}$ be orthogonal to all vectors of the form $v = Ay$, where y belongs to $\mathcal{K}$; i.e.,

$$(b - A\tilde{x}, v) = 0 \quad \forall v \in A\mathcal{K},$$

which is precisely the Petrov–Galerkin condition that defines the approximate solution $\tilde{x}$. □

It is worthwhile to point out that A need not be nonsingular in the above proposition. When A is singular there may be infinitely many vectors $\tilde{x}$ satisfying the optimality condition.

5.2.2 Interpretation in Terms of Projectors

We now return to the two important particular cases singled out in the previous section, namely, the cases $\mathcal{L} = \mathcal{K}$ and $\mathcal{L} = A\mathcal{K}$. In these cases, the result of the projection process can be interpreted easily in terms of actions of orthogonal projectors on the initial residual or initial error. Consider the second case first, as it is slightly simpler. Let r_0 be the initial residual $r_0 = b - Ax_0$ and $\tilde{r} = b - A\tilde{x}$ the residual obtained after the projection process with $\mathcal{L} = A\mathcal{K}$. Then

$$\tilde{r} = b - A(x_0 + \delta) = r_0 - A\delta. \tag{5.9}$$

In addition, δ is obtained by enforcing the condition that $r_0 - A\delta$ be orthogonal to $A\mathcal{K}$. Therefore, the vector $A\delta$ is *the orthogonal projection of the vector r_0 onto the subspace $A\mathcal{K}$*. This is illustrated in Figure 5.2. Hence, the following proposition can be stated.

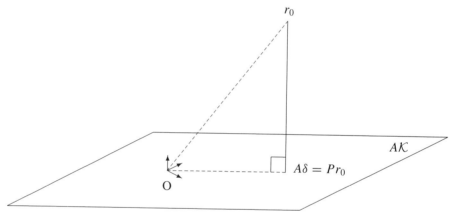

Figure 5.2. *Interpretation of the projection process for the case when $\mathcal{L} = A\mathcal{K}$.*

Proposition 5.4. *Let $\tilde{x}$ be the approximate solution obtained from a projection process onto $\mathcal{K}$ orthogonally to $\mathcal{L} = A\mathcal{K}$ and let $\tilde{r} = b - A\tilde{x}$ be the associated residual. Then*

$$\tilde{r} = (I - P)r_0, \tag{5.10}$$

where P denotes the orthogonal projector onto the subspace $A\mathcal{K}$.

A result of the proposition is that the 2-norm of the residual vector obtained after one projection step will not exceed the initial 2-norm of the residual; i.e.,

$$\|\tilde{r}\|_2 \leq \|r_0\|_2,$$

a result that has been established already. This class of methods may be termed *residual projection* methods.

Now consider the case where $\mathcal{L} = \mathcal{K}$ and A is SPD. Let $d_0 = x_* - x_0$ be the initial error, where x_* denotes the exact solution to the system, and, similarly, let $\tilde{d} = x_* - \tilde{x}$, where $\tilde{x} = x_0 + \delta$ is the approximate solution resulting from the projection step. Then (5.9) yields the relation

$$A\tilde{d} = \tilde{r} = A(d_0 - \delta),$$

5.2. General Theory

where δ is now obtained by constraining the residual vector $r_0 - A\delta$ to be orthogonal to $\mathcal{K}$:

$$(r_0 - A\delta, w) = 0 \quad \forall w \in \mathcal{K}.$$

The above condition is equivalent to

$$(A(d_0 - \delta), w) = 0 \quad \forall w \in \mathcal{K}.$$

Since A is SPD, it defines an inner product (see Section 1.11), which is usually denoted by $(\cdot, \cdot)_A$, and the above condition becomes

$$(d_0 - \delta, w)_A = 0 \quad \forall w \in \mathcal{K}.$$

The above condition is now easy to interpret: *The vector δ is the A-orthogonal projection of the initial error d_0 onto the subspace $\mathcal{K}$.*

Proposition 5.5. *Let $\tilde{x}$ be the approximate solution obtained from an orthogonal projection process onto $\mathcal{K}$ and let $\tilde{d} = x_* - \tilde{x}$ be the associated error vector. Then*

$$\tilde{d} = (I - P_A)d_0,$$

where P_A denotes the projector onto the subspace $\mathcal{K}$, which is orthogonal with respect to the A inner product.

A result of the proposition is that the A-norm of the error vector obtained after one projection step does not exceed the initial A-norm of the error; i.e.,

$$\|\tilde{d}\|_A \leq \|d_0\|_A,$$

which is expected because it is known that the A-norm of the error is minimized in $x_0 + \mathcal{K}$. This class of methods may be termed *error projection methods*.

5.2.3 General Error Bound

If no vector of the subspace $\mathcal{K}$ comes close to the exact solution x, then it is impossible to find a good approximation $\tilde{x}$ to x from $\mathcal{K}$. Therefore, the approximation obtained by any projection process based on $\mathcal{K}$ will be poor. On the other hand, if there is some vector in $\mathcal{K}$ that is a small distance ϵ away from x, then the question is as follows: How good can the approximate solution be? The purpose of this section is to try to answer this question.

Let $\mathcal{P}_\mathcal{K}$ be the orthogonal projector onto the subpace $\mathcal{K}$ and let $\mathcal{Q}_\mathcal{K}^\mathcal{L}$ be the (oblique) projector onto $\mathcal{K}$ and orthogonally to $\mathcal{L}$. These projectors are defined by

$$\mathcal{P}_\mathcal{K} x \in \mathcal{K}, \quad x - \mathcal{P}_\mathcal{K} x \perp \mathcal{K},$$
$$\mathcal{Q}_\mathcal{K}^\mathcal{L} x \in \mathcal{K}, \quad x - \mathcal{Q}_\mathcal{K}^\mathcal{L} x \perp \mathcal{L}$$

and are illustrated in Figure 5.3. The symbol A_m is used to denote the operator

$$A_m = \mathcal{Q}_\mathcal{K}^\mathcal{L} A \mathcal{P}_\mathcal{K}$$

and it is assumed, without loss of generality, that $x_0 = 0$. Then, according to the property (1.61), the approximate problem defined in (5.5)–(5.6) can be reformulated as follows: Find $\tilde{x} \in \mathcal{K}$ such that

$$\mathcal{Q}_\mathcal{K}^\mathcal{L}(b - A\tilde{x}) = 0$$

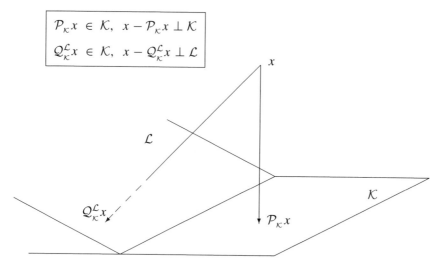

Figure 5.3. *Orthogonal and oblique projectors.*

or, equivalently,

$$A_m \tilde{x} = \mathcal{Q}_\mathcal{K}^\mathcal{L} b, \quad \tilde{x} \in \mathcal{K}.$$

Thus, an n-dimensional linear system is approximated by an m-dimensional one.

The following proposition examines what happens in the particular case when the subspace $\mathcal{K}$ is invariant under A. This is a rare occurrence in practice, but the result helps in understanding the breakdown behavior of the methods to be considered in later chapters.

Proposition 5.6. *Assume that $\mathcal{K}$ is invariant under A, $x_0 = 0$, and b belongs to $\mathcal{K}$. Then the approximate solution obtained from any (oblique or orthogonal) projection method onto $\mathcal{K}$ is exact.*

Proof. An approximate solution $\tilde{x}$ is defined by

$$\mathcal{Q}_\mathcal{K}^\mathcal{L}(b - A\tilde{x}) = 0,$$

where $\tilde{x}$ is a nonzero vector in $\mathcal{K}$. The right-hand side b is in $\mathcal{K}$, so we have $\mathcal{Q}_\mathcal{K}^\mathcal{L} b = b$. Similarly, $\tilde{x}$ belongs to $\mathcal{K}$, which is invariant under A and, therefore, $\mathcal{Q}_\mathcal{K}^\mathcal{L} A\tilde{x} = A\tilde{x}$. Then the above equation becomes

$$b - A\tilde{x} = 0,$$

showing that $\tilde{x}$ is an exact solution. □

The result can be extended trivially to the case where $x_0 \neq 0$. The required assumption in this case is that the initial residual $r_0 = b - Ax_0$ belong to the invariant subspace $\mathcal{K}$.

An important quantity for the convergence properties of projection methods is the distance $\|(I - \mathcal{P}_\mathcal{K})x_*\|_2$ of the exact solution x_* from the subspace $\mathcal{K}$. This quantity plays a key role in the analysis of projection methods. Note that the solution x_* cannot be well approximated from $\mathcal{K}$ if $\|(I - \mathcal{P}_\mathcal{K})x_*\|_2$ is not small because

$$\|\tilde{x} - x_*\|_2 \geq \|(I - \mathcal{P}_\mathcal{K})x_*\|_2.$$

The fundamental quantity $\|(I - \mathcal{P}_\mathcal{K})x_*\|_2/\|x_*\|_2$ is the *sine* of the acute angle between the solution x_* and the subspace $\mathcal{K}$. The following theorem establishes an upper bound for the residual norm of the *exact* solution with respect to the approximate operator A_m.

Theorem 5.7. *Let* $\gamma = \|\mathcal{Q}_\mathcal{K}^\mathcal{L} A(I - \mathcal{P}_\mathcal{K})\|_2$ *and assume that b is a member of $\mathcal{K}$ and $x_0 = 0$. Then the exact solution x_* of the original problem is such that*

$$\|b - A_m x_*\|_2 \leq \gamma \|(I - \mathcal{P}_\mathcal{K})x_*\|_2. \tag{5.11}$$

Proof. Since $b \in \mathcal{K}$, then

$$\begin{aligned} b - A_m x_* &= \mathcal{Q}_\mathcal{K}^\mathcal{L}(b - A\mathcal{P}_\mathcal{K} x_*) \\ &= \mathcal{Q}_\mathcal{K}^\mathcal{L}\left(Ax_* - A\mathcal{P}_\mathcal{K} x_*\right) \\ &= \mathcal{Q}_\mathcal{K}^\mathcal{L} A(x_* - \mathcal{P}_\mathcal{K} x_*) \\ &= \mathcal{Q}_\mathcal{K}^\mathcal{L} A(I - \mathcal{P}_\mathcal{K})x_*. \end{aligned}$$

Noting that $I - \mathcal{P}_\mathcal{K}$ is a projector, it follows that

$$\begin{aligned} \|b - A_m x_*\|_2 &= \|\mathcal{Q}_\mathcal{K}^\mathcal{L} A(I - \mathcal{P}_\mathcal{K})(I - \mathcal{P}_\mathcal{K})x_*\|_2 \\ &\leq \|\mathcal{Q}_\mathcal{K}^\mathcal{L} A(I - \mathcal{P}_\mathcal{K})\|_2 \|(I - \mathcal{P}_\mathcal{K})x_*\|_2, \end{aligned}$$

which completes the proof. $\square$

It is useful to consider a matrix interpretation of the theorem. We consider only the particular case of orthogonal projection methods ($\mathcal{L} = \mathcal{K}$). Assume that V is unitary, i.e., that the basis $\{v_1, \ldots, v_m\}$ is orthonormal, and that $W = V$. Observe that $b = VV^T b$. Equation (5.11) can be represented in the basis V as

$$\|b - V(V^T A V)V^T x_*\|_2 \leq \gamma \|(I - \mathcal{P}_\mathcal{K})x_*\|_2.$$

However,

$$\begin{aligned} \|b - V(V^T A V)V^T x_*\|_2 &= \|V(V^T b - (V^T A V))V^T x_*\|_2 \\ &= \|V^T b - (V^T A V)V^T x_*\|_2. \end{aligned}$$

Thus, the projection of the exact solution has a residual norm with respect to the matrix $B = V^T A V$, which is of the order of $\|(I - \mathcal{P}_\mathcal{K})x_*\|_2$.

5.3 One-Dimensional Projection Processes

This section examines simple examples provided by one-dimensional projection processes. In what follows, the vector r denotes the residual vector $r = b - Ax$ for the current approximation x. To avoid subscripts, arrow notation is used to denote *vector updates*. Thus, "$x \leftarrow x + \alpha r$" means "compute $x + \alpha r$ and overwrite the result on the current x." (This is known as a SAXPY operation.)

One-dimensional projection processes are defined when

$$\mathcal{K} = \mathrm{span}\{v\} \quad \text{and} \quad \mathcal{L} = \mathrm{span}\{w\},$$

where v and w are two vectors. In this case, the new approximation takes the form $x \leftarrow x + \alpha v$ and the Petrov–Galerkin condition $r - A\delta \perp w$ yields

$$\alpha = \frac{(r, w)}{(Av, w)}. \tag{5.12}$$

Following are three popular choices to be considered.

5.3.1 Steepest Descent

The steepest descent algorithm is defined for the case where the matrix A is SPD. It consists of taking at each step $v = r$ and $w = r$. This yields the following iterative procedure:

$$r \leftarrow b - Ax,$$
$$\alpha \leftarrow (r, r)/(Ar, r),$$
$$x \leftarrow x + \alpha v.$$

However, the above procedure requires two matrix-by-vector products, which can be reduced to only one by rearranging the computation slightly. The variation consists of computing r differently, as is shown next.

ALGORITHM 5.2. Steepest Descent Algorithm

1. Compute $r = b - Ax$ and $p = Ar$
2. Until convergence, Do
3. $\alpha \leftarrow (r, r)/(p, r)$
4. $x \leftarrow x + \alpha r$
5. $r \leftarrow r - \alpha p$
6.   Compute $p := Ar$
7. EndDo

Each step of the above iteration minimizes

$$f(x) = \|x - x_*\|_A^2 = (A(x - x_*), (x - x_*))$$

over all vectors of the form $x + \alpha d$, where d is the negative of the gradient direction $-\nabla f$. The negative of the gradient direction is *locally* the direction that yields the fastest rate of decrease for f. Next we prove that convergence is guaranteed when A is SPD. The result is a consequence of the following lemma known as the Kantorovich inequality.

Lemma 5.8. *(Kantorovich inequality)* Let B be any SPD real matrix and λ_{max}, λ_{min} its largest and smallest eigenvalues. Then

$$\frac{(Bx, x)(B^{-1}x, x)}{(x, x)^2} \leq \frac{(\lambda_{max} + \lambda_{min})^2}{4 \lambda_{max} \lambda_{min}} \quad \forall x \neq 0. \tag{5.13}$$

5.3. One-Dimensional Projection Processes

Proof. Clearly, it is equivalent to show that the result is true for any unit vector x. Since B is symmetric, it is unitarily similar to a diagonal matrix, $B = Q^T D Q$, and

$$(Bx, x)(B^{-1}x, x) = (Q^T D Q x, x)(Q^T D^{-1} Q x, x) = (DQx, Qx)(D^{-1}Qx, Qx).$$

Setting $y = Qx = (y_1, \ldots, y_n)^T$ and $\beta_i = y_i^2$, note that

$$\lambda \equiv (Dy, y) = \sum_{i=1}^{n} \beta_i \lambda_i$$

is a convex combination of the eigenvalues λ_i, $i = 1, \ldots, n$. The following relation holds:

$$(Bx, x)(B^{-1}x, x) = \lambda \psi(y), \quad \text{with} \quad \psi(y) = (D^{-1}y, y) = \sum_{i=1}^{n} \beta_i \frac{1}{\lambda_i}.$$

Noting that the function $f(\lambda) = 1/\lambda$ is convex, $\psi(y)$ is bounded from above by the linear curve that joins the points $(\lambda_1, 1/\lambda_1)$ and $(\lambda_n, 1/\lambda_n)$; i.e.,

$$\psi(y) \leq \frac{1}{\lambda_1} + \frac{1}{\lambda_n} - \frac{\lambda}{\lambda_1 \lambda_n}.$$

Therefore,

$$(Bx, x)(B^{-1}x, x) = \lambda \psi(y) \leq \lambda \left(\frac{1}{\lambda_1} + \frac{1}{\lambda_n} - \frac{\lambda}{\lambda_1 \lambda_n} \right).$$

The maximum of the right-hand side is reached for $\lambda = \frac{1}{2}(\lambda_1 + \lambda_n)$, yielding

$$(Bx, x)(B^{-1}x, x) = \lambda \psi(y) \leq \frac{(\lambda_1 + \lambda_n)^2}{4 \lambda_1 \lambda_n},$$

which gives the desired result. □

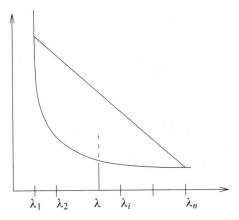

Figure 5.4. *The point $(\lambda, \psi(y))$ is a convex combination of points located on the curve $1/\lambda$. It is located in the convex set limited by the curve $1/\lambda$ and the line $1/\lambda_1 + 1/\lambda_n - \lambda/(\lambda_1 \lambda_n)$.*

This lemma helps to establish the following result regarding the convergence rate of the method.

Theorem 5.9. *Let A be an SPD matrix. Then the A-norms of the error vectors $d_k = x_* - x_k$ generated by Algorithm 5.2 satisfy the relation*

$$\|d_{k+1}\|_A \leq \frac{\lambda_{max} - \lambda_{min}}{\lambda_{max} + \lambda_{min}} \|d_k\|_A \tag{5.14}$$

and Algorithm 5.2 converges for any initial guess x_0.

Proof. Start by expanding the square of the A-norm of $d_{k+1} = d_k - \alpha_k r_k$ as

$$\|d_{k+1}\|_A^2 = (d_{k+1}, d_k - \alpha_k r_k)_A = (d_{k+1}, d_k)_A - \alpha_k (d_{k+1}, r_k)_A = (d_{k+1}, r_k).$$

The last equality is due to the orthogonality between r_k and r_{k+1}. Thus,

$$\|d_{k+1}\|_A^2 = (d_k - \alpha r_k, r_k)$$
$$= (A^{-1} r_k, r_k) - \alpha_k (r_k, r_k)$$
$$= \|d_k\|_A^2 \left(1 - \frac{(r_k, r_k)}{(r_k, A r_k)} \times \frac{(r_k, r_k)}{(r_k, A^{-1} r_k)}\right).$$

The result follows by applying the Kantorovich inequality (5.13). □

5.3.2 MR Iteration

We now assume that A is not necessarily symmetric but only positive definite, i.e., its symmetric part $A + A^T$ is SPD. Taking at each step $v = r$ and $w = Ar$ gives the following procedure:

$$r \leftarrow b - Ax,$$
$$\alpha \leftarrow (Ar, r)/(Ar, Ar),$$
$$x \leftarrow x + \alpha v.$$

This procedure can be slightly rearranged again to reduce the number of matrix-by-vector products required to only one per step, as was done for the steepest descent algorithm. This results in the following algorithm.

ALGORITHM 5.3. Minimal Residual (MR) Iteration

1. Compute $r = b - Ax$ and $p = Ar$
2. Until convergence, Do
3. $\alpha \leftarrow (Ar, r)/(p, p)$
4. $x \leftarrow x + \alpha r$
5. $r \leftarrow r - \alpha p$
6.   Compute $p := Ar$
7. EndDo

5.3. One-Dimensional Projection Processes

Here, each step minimizes $f(x) = \|b - Ax\|_2^2$ in the direction r. The iteration converges under the condition that A is positive definite, as is stated in the next theorem.

Theorem 5.10. *Let A be a real positive definite matrix and let*

$$\mu = \lambda_{min}(A + A^T)/2, \quad \sigma = \|A\|_2.$$

Then the residual vectors generated by Algorithm 5.3 satisfy the relation

$$\|r_{k+1}\|_2 \leq \left(1 - \frac{\mu^2}{\sigma^2}\right)^{1/2} \|r_k\|_2 \tag{5.15}$$

and Algorithm 5.3 converges for any initial guess x_0.

Proof. We proceed similarly to the steepest descent method, starting with the relation

$$\|r_{k+1}\|_2^2 = (r_k - \alpha_k A r_k, r_k - \alpha_k A r_k) \tag{5.16}$$
$$= (r_k - \alpha_k A r_k, r_k) - \alpha_k(r_k - \alpha_k A r_k, A r_k). \tag{5.17}$$

By construction, the new residual vector $r_k - \alpha_k A r_k$ must be orthogonal to the search direction $A r_k$ and, as a result, the second term on the right-hand side of the above equation vanishes and we obtain

$$\|r_{k+1}\|_2^2 = (r_k - \alpha_k A r_k, r_k)$$
$$= (r_k, r_k) - \alpha_k(A r_k, r_k)$$
$$= \|r_k\|_2^2 \left(1 - \frac{(A r_k, r_k)}{(r_k, r_k)} \frac{(A r_k, r_k)}{(A r_k, A r_k)}\right) \tag{5.18}$$
$$= \|r_k\|_2^2 \left(1 - \frac{(A r_k, r_k)^2}{(r_k, r_k)^2} \frac{\|r_k\|_2^2}{\|A r_k\|_2^2}\right).$$

From Theorem 1.34, it can be stated that

$$\frac{(Ax, x)}{(x, x)} \geq \mu > 0, \tag{5.19}$$

where $\mu = \lambda_{min}(A + A^T)/2$. The desired result follows immediately by using the inequality $\|A r_k\|_2 \leq \|A\|_2 \|r_k\|_2$. $\square$

There are alternative ways of obtaining inequalities that prove convergence. For example, we can start again from (5.18), use (5.19) for the term $(A r_k, r_k)/(r_k, r_k)$ and, similarly, write

$$\frac{(Ax, x)}{(Ax, Ax)} = \frac{(Ax, A^{-1}(Ax))}{(Ax, Ax)} \geq \lambda_{min}\left(\frac{A^{-1} + A^{-T}}{2}\right) > 0,$$

since A^{-1} is also positive definite. This would yield the inequality

$$\|r_{k+1}\|_2^2 \leq (1 - \mu(A)\mu(A^{-1}))\|r_k\|_2^2, \tag{5.20}$$

in which $\mu(B) = \lambda_{min}(B + B^T)/2$.

Another interesting observation is that, if we define

$$\cos \angle_k = \frac{(Ar_k, r_k)}{\|Ar_k\|_2 \|r_k\|_2},$$

then (5.18) can be rewritten as

$$\|r_{k+1}\|_2^2 = \|r_k\|_2^2 \left(1 - \frac{(Ar_k, r_k)}{(Ar_k, Ar_k)} \frac{(Ar_k, r_k)}{(r_k, r_k)}\right)$$
$$= \|r_k\|_2^2 \left(1 - \cos^2 \angle_k\right)$$
$$= \|r_k\|_2^2 \sin^2 \angle_k.$$

At each step the reduction in the residual norm is equal to the *sine* of the acute angle between r and Ar. The convergence factor is therefore bounded by

$$\rho = \max_{x \in \mathbb{R}^n,\, x \neq 0} \sin \angle(x, Ax),$$

in which $\angle(x, Ax)$ is the acute angle between x and Ax. The maximum angle $\angle(x, Ax)$ is guaranteed to be less than $\pi/2$ when A is positive definite, as the above results show.

5.3.3 Residual Norm Steepest Descent

In the residual norm steepest descent algorithm, the assumption that A is positive definite is relaxed. In fact, the only requirement is that A be a (square) nonsingular matrix. At each step the algorithm uses $v = A^T r$ and $w = Av$, giving the following sequence of operations:

$$r \leftarrow b - Ax, \, v = A^T r,$$
$$\alpha \leftarrow \|v\|_2^2 / \|Av\|_2^2, \quad (5.21)$$
$$x \leftarrow x + \alpha v.$$

However, an algorithm based on the above sequence of operations would require three matrix-by-vector products, which is three times as many as the other algorithms seen in this section. The number of matrix-by-vector operations can be reduced to two per step by computing the residual differently. This variant is as follows.

ALGORITHM 5.4. Residual Norm Steepest Descent

1. Compute $r := b - Ax$
2. Until convergence, Do
3. $\quad v := A^T r$
4. $\quad$ Compute Av and $\alpha := \|v\|_2^2 / \|Av\|_2^2$
5. $\quad x := x + \alpha v$
6. $\quad r := r - \alpha Av$
7. EndDo

Here, each step minimizes $f(x) = \|b - Ax\|_2^2$ in the direction $-\nabla f$. As it turns out, this is equivalent to the steepest descent algorithm of Section 5.3.1 applied to the normal equations $A^T A x = A^T b$. Since $A^T A$ is positive definite when A is nonsingular, then, according to Theorem 5.9, the method will converge whenever A is nonsingular.

5.4 Additive and Multiplicative Processes

We begin by considering again the block relaxation techniques seen in the previous chapter. To define these techniques, a *set decomposition* of $S = \{1, 2, \ldots, n\}$ is considered as the definition of p subsets $S_1, \ldots, S_p$ of S with

$$S_i \subseteq S, \quad \bigcup_{i=1,\ldots,p} S_i = S.$$

Denote by n_i the size of S_i and define the subset S_i as

$$S_i = \{m_i(1), m_i(2), \ldots, m_i(n_i)\}.$$

Let V_i be the $n \times n_i$ matrix

$$V_i = [e_{m_i(1)}, e_{m_i(2)}, \ldots, e_{m_i(n_i)}],$$

where each e_j is the jth column of the $n \times n$ identity matrix.

If the block Jacobi and block Gauss–Seidel algorithms, Algorithms 4.1 and 4.2, are examined carefully, it can be observed that each individual step in the main loop (lines 2–5) represents an orthogonal projection process over $K_i = \text{span}\{V_i\}$. Indeed, (4.17) is exactly (5.7) with $W = V = V_i$. This individual projection step modifies only the components corresponding to the subspace K_i. However, the general block Jacobi iteration combines these modifications, implicitly adding them together, to obtain the next iterate x_{k+1}. Borrowing from the terminology of domain decomposition techniques, this will be called an *additive projection procedure*. Generally, an additive projection procedure can be defined for any sequence of subspaces K_i, not just subspaces spanned by the columns of the identity matrix. The only requirement is that the subspaces K_i be distinct, although they are allowed to overlap.

Let a sequence of p orthogonal systems V_i be given, with the condition that $\text{span}\{V_i\} \neq \text{span}\{V_j\}$ for $i \neq j$, and define

$$A_i = V_i^T A V_i.$$

The additive projection procedure can be written as

$$y_i = A_i^{-1} V_i^T (b - Ax_k), \quad i = 1, \ldots, p,$$

$$x_{k+1} = x_k + \sum_{i=1}^{p} V_i y_i, \tag{5.22}$$

which leads to the following algorithm.

ALGORITHM 5.5. Additive Projection Procedure

1. *For $k = 0, 1, \ldots$, until convergence, Do*
2.   *For $i = 1, 2, \ldots, p$, Do*
3.   *Solve $A_i y_i = V_i^T (b - Ax_k)$*
4.   *EndDo*
5.   *Set $x_{k+1} = x_k + \sum_{i=1}^{p} V_i y_i$*
6. *EndDo*

Define $r_k = b - Ax_k$, the residual vector at step k. Then clearly

$$r_{k+1} = b - Ax_{k+1}$$
$$= b - Ax_k - \sum_{i=1}^{p} AV_i \left(V_i^T AV_i\right)^{-1} V_i^T r_k$$
$$= \left[I - \sum_{i=1}^{p} AV_i \left(V_i^T AV_i\right)^{-1} V_i^T\right] r_k.$$

Observe that each of the p operators

$$P_i = AV_i \left(V_i^T AV_i\right)^{-1} V_i^T$$

represents the projector onto the subspace spanned by AV_i and orthogonal to V_i. Often, the additive processes are used in conjunction with an acceleration parameter ω; thus (5.22) is replaced by

$$y_i = A_i^{-1} V_i^T (b - Ax_k), \quad i = 1, \ldots, p,$$
$$x_{k+1} = x_k + \omega \sum_{i=1}^{p} V_i y_i.$$

Even more generally, a different parameter ω_i can be used for each projection; i.e.,

$$y_i = A_i^{-1} V_i^T (b - Ax_k), \quad i = 1, \ldots, p,$$
$$x_{k+1} = x_k + \sum_{i=1}^{p} \omega_i V_i y_i.$$

The residual norm in this situation is given by

$$r_{k+1} = \left(I - \sum_{i=1}^{p} \omega_i P_i\right) r_k, \qquad (5.23)$$

considering the single ω parameter as a particular case. Exercise 15 gives an example of the choice of ω_i that has the effect of producing a sequence with decreasing residual norms.

We now return to the generic case, where $\omega_i = 1 \; \forall \; i$. A least-squares option can be defined by taking for each of the subproblems $L_i = AK_i$. In this situation, P_i becomes an orthogonal projector onto AK_i, since

$$P_i = AV_i \left((AV_i)^T AV_i\right)^{-1} (AV_i)^T.$$

It is interesting to note that the residual vector obtained after one outer loop is related to the previous residual by

$$r_{k+1} = \left(I - \sum_{i=1}^{p} P_i\right) r_k,$$

where the P_i's are now orthogonal projectors. In particular, in the ideal situation when the AV_i's are orthogonal to each other and the total rank of the P_i's is n, then the exact solution would be obtained in one outer step, since in this situation

$$I - \sum_{i=1}^{p} P_i = 0.$$

Thus, the maximum reduction in the residual norm is achieved when the V_i's are A-orthogonal to one another.

Similarly to the Jacobi and Gauss–Seidel iterations, what distinguishes the additive and multiplicative iterations is that the latter updates the component to be corrected at step i immediately. Then this updated approximate solution is used to compute the residual vector needed to correct the next component. The Jacobi iteration uses the same previous approximation x_k to update all the components of the solution. Thus, the analogue of the block Gauss–Seidel iteration can be defined as follows.

ALGORITHM 5.6. Multiplicative Projection Procedure

1. Until convergence, Do
2.   For $i = 1, 2, \ldots, p$, Do
3.   Solve $A_i y = V_i^T (b - Ax)$
4.   Set $x := x + V_i y$
5.   EndDo
6. EndDo

Exercises

1. Consider the linear system $Ax = b$, where A is an SPD matrix.

    a. Consider the sequence of one-dimensional projection processes with $\mathcal{K} = \mathcal{L} = \text{span}\{e_i\}$, where the sequence of indices i is selected in any fashion. Let x_{new} be a new iterate after one projection step from x and let $r = b - Ax$, $d = A^{-1}b - x$, and $d_{new} = A^{-1}b - x_{new}$. Show that

    $$(Ad_{new}, d_{new}) = (Ad, d) - (r, e_i)^2 / a_{ii}.$$

    Does this equality, as is, establish convergence of the algorithm?

    b. Assume now that i is selected at each projection step to be the index of a component of largest absolute value in the current residual vector $r = b - Ax$. Show that

    $$\|d_{new}\|_A \leq \left(1 - \frac{1}{n\kappa(A)}\right)^{1/2} \|d\|_A,$$

    in which $\kappa(A)$ is the spectral condition number of A. [Hint: Use the inequality $|e_i^T r| \geq n^{-1/2} \|r\|_2$.] Does this prove that the algorithm converges?

2. Consider the linear system $Ax = b$, where A is an SPD matrix. Consider a projection step with $\mathcal{K} = \mathcal{L} = \text{span}\{v\}$, where v is some nonzero vector. Let x_{new} be the new iterate after one projection step from x and let $d = A^{-1}b - x$ and $d_{new} = A^{-1}b - x_{new}$.

   a. Show that
   $$(Ad_{new}, d_{new}) = (Ad, d) - (r, v)^2/(Av, v).$$
   Does this equality establish convergence of the algorithm?

   b. In Gastinel's method, the vector v is selected in such a way that $(v, r) = \|r\|_1$, e.g., by defining the components of v to be $v_i = \text{sign}(e_i^T r)$, where $r = b - Ax$ is the current residual vector. Show that
   $$\|d_{new}\|_A \leq \left(1 - \frac{1}{n\kappa(A)}\right)^{1/2} \|d\|_A,$$
   in which $\kappa(A)$ is the spectral condition number of A. Does this prove that the algorithm converges?

   c. Compare the cost of one step of this method with that of cyclic Gauss–Seidel (see Example 5.1) and that of "optimal" Gauss–Seidel, where at each step $\mathcal{K} = \mathcal{L} = \text{span}\{e_i\}$ and i is a component of largest magnitude in the current residual vector.

3. In Section 5.3.3, it was shown that taking a one-dimensional projection technique with $\mathcal{K} = \text{span}\{A^T r\}$ and $\mathcal{L} = \text{span}\{AA^T r\}$ is mathematically equivalent to using the usual steepest descent algorithm applied to the normal equations $A^T A x = A^T b$. Show that an *orthogonal* projection method for $A^T A x = A^T b$ using a subspace $\mathcal{K}$ is mathematically equivalent to applying a projection method onto $\mathcal{K}$ orthogonally to $\mathcal{L} = A\mathcal{K}$ for solving the system $Ax = b$.

4. Consider the matrix
$$A = \begin{pmatrix} 1 & -6 & 0 \\ 6 & 2 & 3 \\ 0 & 3 & 2 \end{pmatrix}.$$

   a. Find a rectangle or square in the complex plane that contains all the eigenvalues of A, without computing the eigenvalues.

   b. Is the MR iteration guaranteed to converge for a linear system with the matrix A?

5. Consider the linear system
$$\begin{pmatrix} D_1 & -F \\ -E & -D_2 \end{pmatrix} \begin{pmatrix} x_1 \\ x_2 \end{pmatrix} = \begin{pmatrix} b_1 \\ b_2 \end{pmatrix},$$
in which D_1 and D_2 are both nonsingular matrices of size m each.

   a. Define an orthogonal projection method using the set of vectors $e_1, \ldots, e_m$; i.e., $\mathcal{L} = \mathcal{K} = \text{span}\{e_1, \ldots, e_m\}$. Write down the corresponding projection step (x_1 is modified into $\tilde{x}_1$). Similarly, write the projection step for the second half of the vectors, i.e., when $\mathcal{L} = \mathcal{K} = \text{span}\{e_{m+1}, \ldots, e_n\}$.

   b. Consider an iteration procedure that consists of performing the two successive half-steps described above until convergence. Show that this iteration is equivalent to a (standard) Gauss–Seidel iteration applied to the original system.

c. Now consider a similar idea in which $\mathcal{K}$ is taken to be the same as before for each half-step and $\mathcal{L} = A\mathcal{K}$. Write down the iteration procedure based on this approach. Name another technique to which it is mathematically equivalent.

6. Consider the linear system $Ax = b$, where A is an SPD matrix. We define a projection method that uses a two-dimensional space at each step. At a given step, take $\mathcal{L} = \mathcal{K} = \text{span}\{r, Ar\}$, where $r = b - Ax$ is the current residual.
    a. For a basis of $\mathcal{K}$ use the vector r and the vector p obtained by orthogonalizing Ar against r with respect to the A inner product. Give the formula for computing p (no need to normalize the resulting vector).
    b. Write the algorithm for performing the projection method described above.
    c. Will the algorithm converge for any initial guess x_0? Justify the answer. [Hint: Exploit the convergence results for one-dimensional projection techniques.]

7. Consider projection methods that update at each step the current solution with linear combinations from two directions: the current residual r and Ar.
    a. Consider an orthogonal projection method; i.e., at each step $\mathcal{L} = \mathcal{K} = \text{span}\{r, Ar\}$. Assuming that A is SPD, establish convergence of the algorithm.
    b. Consider a least-squares projection method in which at each step $\mathcal{K} = \text{span}\{r, Ar\}$ and $\mathcal{L} = A\mathcal{K}$. Assuming that A is positive definite (not necessarily symmetric), establish convergence of the algorithm.

[Hint: The convergence results for any of the one-dimensional projection techniques can be exploited.]

8. Assume that the (one-dimensional) MR iteration of Section 5.3.2 is applied to a symmetric positive definite matrix A. Will the method converge? What will the result (5.15) become in this case? Both (5.15) and (5.14) suggest a linear convergence with an estimate for the linear convergence rate given by the formulas. How do these estimated rates compare for matrices with large spectral condition numbers?

9. The least-squares Gauss–Seidel relaxation method defines a relaxation step as $x_{new} = x + \delta\, e_i$ (the same as Gauss–Seidel), but chooses δ to minimize the residual norm of x_{new}.
    a. Write down the resulting algorithm.
    b. Show that this iteration is mathematically equivalent to a Gauss–Seidel iteration applied to the normal equations $A^T A x = A^T b$.

10. Derive three types of one-dimensional projection algorithms, in the same manner as was done in Section 5.3, by replacing every occurrence of the residual vector r by a vector e_i, a column of the identity matrix.

11. Derive three types of one-dimensional projection algorithms, in the same manner as was done in Section 5.3, by replacing every occurrence of the residual vector r with a vector Ae_i, a column of the matrix A. What would be an *optimal* choice for i at each projection step? Show that the method is globally convergent in this case.

12. An MR iteration as defined in Section 5.3.2 can also be defined for an arbitrary search direction d, not necessarily related to r in any way. In this case, we still define $e = Ad$.

a. Write down the corresponding algorithm.

b. Under which condition are all iterates defined?

c. Under which condition on d does the new iterate make no progress; i.e., $\|r_{k+1}\|_2 = \|r_k\|_2$?

d. Write a general sufficient condition that must be satisfied by d at each step in order to guarantee convergence.

13. Consider the following real-valued functions of the vector variable x, where A and b are the coefficient matrix and right-hand side of a given linear system $Ax = b$ and $x_* = A^{-1}b$:

$$a(x) = \|x_* - x\|_2^2,$$
$$f(x) = \|b - Ax\|_2^2,$$
$$g(x) = \|A^T b - A^T A x\|_2^2,$$
$$h(x) = 2(b, x) - (Ax, x).$$

a. Calculate the gradients of all four functions.

b. How is the gradient of g related to that of f?

c. How is the gradient of f related to that of h when A is symmetric?

d. How does the function h relate to the A-norm of the error $x_* - x$ when A is SPD?

14. The block Gauss–Seidel iteration can be expressed as a method of successive projections. The subspace $\mathcal{K}$ used for each projection is of the form

$$\mathcal{K} = \text{span}\{e_i, e_{i+1}, \ldots, e_{i+p}\}.$$

What is $\mathcal{L}$? An uncommon alternative is to take $\mathcal{L} = A\mathcal{K}$, which amounts to solving a least-squares problem instead of a linear system. Develop algorithms for this case. What are the advantages and disadvantages of the two approaches (ignoring convergence rates)?

15. Let the scalars ω_i in the additive projection procedure satisfy the constraint

$$\sum_{i=1}^{p} \omega_i = 1. \tag{5.24}$$

It is not assumed that each ω_i is positive but only that $|\omega_i| \leq 1$ for all i. The residual vector is given by (5.23) or, equivalently,

$$r_{k+1} = \sum_{i=1}^{p} \omega_i (I - P_i) r_k.$$

a. Show that, in the least-squares case, we have $\|r_{k+1}\|_2 \leq \|r_k\|_2$ for any choice of ω_i's that satisfy the constraint (5.24).

b. We wish to choose a set of ω_i's such that the 2-norm of the residual vector r_{k+1} is minimal. Determine this set of ω_i's, assuming that the vectors $(I - P_i) r_k$ are all linearly independent.

c. The optimal ω_i's provided in part (b) require the solution of a $p \times p$ SPD linear system. Let $z_i \equiv V_i y_i$ be the *search directions* provided by each of the individual projection steps. To avoid this difficulty, a simpler strategy is used, which consists of performing p successive MR iterations along these search directions, as is described below.

$r := r_k$
For $i = 1, \ldots, p$, Do
$\quad \omega_i := (r, Az_i)/(Az_i, Az_i)$
$\quad x := x + \omega_i z_i$
$\quad r := r - \omega_i A z_i$
EndDo

Show that $\|r_{k+1}\|_2 \leq \|r_k\|_2$. Give a sufficient condition to ensure global convergence.

16. Consider the iteration $x_{k+1} = x_k + \alpha_k d_k$, where d_k is a vector called the *direction of search* and α_k is a scalar. It is assumed throughout that d_k is a nonzero vector. Consider a method that determines x_{k+1} so that the residual $\|r_{k+1}\|_2$ is the smallest possible.

    a. Determine α_k so that $\|r_{k+1}\|_2$ is minimal.
    b. Show that the residual vector r_{k+1} obtained in this manner is orthogonal to Ad_k.
    c. Show that the residual vectors satisfy the relation

    $$\|r_{k+1}\|_2 \leq \|r_k\|_2 \sin \angle(r_k, Ad_k).$$

    d. Assume that, at each step k, we have $(r_k, Ad_k) \neq 0$. Will the method always converge?
    e. Now assume that A is positive definite and select at each step $d_k \equiv r_k$. Prove that the method will converge for any initial guess x_0.

17. Consider the iteration $x_{k+1} = x_k + \alpha_k d_k$, where d_k is a vector called the *direction of search* and α_k is a scalar. It is assumed throughout that d_k is a vector selected in the form $d_k = A^T f_k$, where f_k is some nonzero vector. Let $x_* = A^{-1} b$ be the exact solution. Now consider a method that at each step k determines x_{k+1} so that the error norm $\|x_* - x_{k+1}\|_2$ is the smallest possible.

    a. Determine α_k so that $\|x_* - x_{k+1}\|_2$ is minimal and show that the error vector $e_{k+1} = x_* - x_{k+1}$ is orthogonal to d_k. The expression of α_k should not contain unknown quantities (e.g., x_* or e_k).
    b. Show that $\|e_{k+1}\|_2 \leq \|e_k\|_2 \sin \angle(e_k, d_k)$.
    c. Establish the convergence of the algorithm for any x_0, when $f_k \equiv r_k$ for all k.

Notes and References

Initially, the term *projection methods* was used mainly to describe one-dimensional techniques such as those presented in Section 5.3. An excellent account of what was done in the late 1950s and early 1960s can be found in Householder's book [172] as well as in Gastinel

[140]. For more general, including nonlinear, projection processes, a good reference is Kranoselskii and co-authors [191].

Projection techniques are present in different forms in many other areas of scientific computing and can be formulated in abstract Hilbert functional spaces. The terms *Galerkin* and *Petrov–Galerkin* are used commonly in finite element methods to describe projection methods on finite element spaces. The principles are identical to those seen in this chapter.

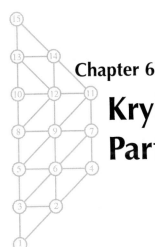

Chapter 6

Krylov Subspace Methods, Part I

The next two chapters explore a few methods that are considered currently to be among the most important iterative techniques available for solving large linear systems. These techniques are based on projection processes, both orthogonal and oblique, onto Krylov subspaces, which are subspaces spanned by vectors of the form $p(A)v$, where p is a polynomial. In short, these techniques approximate $A^{-1}b$ by $p(A)b$, where p is a "good" polynomial. This chapter covers methods derived from, or related to, the Arnoldi orthogonalization. The next chapter covers methods based on Lanczos biorthogonalization.

6.1 Introduction

Recall from the previous chapter that a general *projection method* for solving the linear system

$$Ax = b \qquad (6.1)$$

extracts an approximate solution x_m from an affine subspace $x_0 + \mathcal{K}_m$ of dimension m by imposing the Petrov–Galerkin condition

$$b - Ax_m \perp \mathcal{L}_m,$$

where $\mathcal{L}_m$ is another subspace of dimension m. Here, x_0 represents an arbitrary initial guess to the solution. A Krylov subspace method is a method for which the subspace $\mathcal{K}_m$ is the Krylov subspace

$$\mathcal{K}_m(A, r_0) = \text{span}\{r_0, Ar_0, A^2 r_0, \ldots, A^{m-1} r_0\},$$

where $r_0 = b - Ax_0$. When there is no ambiguity, $\mathcal{K}_m(A, r_0)$ will be denoted by $\mathcal{K}_m$. The different versions of Krylov subspace methods arise from different choices of the subspace $\mathcal{L}_m$ and from the ways in which the system is *preconditioned*, a topic that will be covered in detail in later chapters.

Viewed from the angle of approximation theory, it is clear that the approximations obtained from a Krylov subspace method are of the form

$$A^{-1}b \approx x_m = x_0 + q_{m-1}(A)r_0,$$

in which q_{m-1} is a certain polynomial of degree $m - 1$. In the simplest case where $x_0 = 0$, we then have

$$A^{-1}b \approx q_{m-1}(A)b.$$

In other words, $A^{-1}b$ is approximated by $q_{m-1}(A)b$.

Although all the techniques provide the same type of *polynomial* approximations, the choice of $\mathcal{L}_m$, i.e., the constraints used to build these approximations, will have an important effect on the iterative technique. Two broad choices for $\mathcal{L}_m$ give rise to the best-known techniques. The first is simply $\mathcal{L}_m = \mathcal{K}_m$ and the minimal residual (MR) variation $\mathcal{L}_m = A\mathcal{K}_m$. A few of the numerous methods in this category will be described in this chapter. The second class of methods is based on defining $\mathcal{L}_m$ to be a Krylov subspace method associated with A^T, namely, $\mathcal{L}_m = \mathcal{K}_m(A^T, r_0)$. Methods of this class will be covered in the next chapter. There are also block extensions of each of these methods, termed *block Krylov subspace methods*, which will be discussed only briefly. Note that a projection method may have several different implementations, giving rise to different algorithms that are all mathematically equivalent.

6.2 Krylov Subspaces

In this section we consider projection methods on *Krylov subspaces*, i.e., subspaces of the form

$$\mathcal{K}_m(A, v) \equiv \text{span}\{v, Av, A^2v, \ldots, A^{m-1}v\}, \tag{6.2}$$

which will be denoted simply by $\mathcal{K}_m$ if there is no ambiguity. The dimension of the subspace of approximants increases by one at each step of the approximation process. A few elementary properties of Krylov subspaces can be established. A first property is that $\mathcal{K}_m$ is the subspace of all vectors in $\mathbb{R}^n$ that can be written as $x = p(A)v$, where p is a polynomial of degree not exceeding $m - 1$. Recall that the minimal polynomial of a vector v is the nonzero monic polynomial p of lowest degree such that $p(A)v = 0$. The degree of the minimal polynomial of v with respect to A is often called the *grade of v with respect to A* or simply the grade of v if there is no ambiguity. A consequence of the Cayley–Hamilton theorem is that the grade of v does not exceed n. The following proposition is easy to prove.

Proposition 6.1. *Let μ be the grade of v. Then $\mathcal{K}_\mu$ is invariant under A and $\mathcal{K}_m = \mathcal{K}_\mu$ for all $m \geq \mu$.*

It was mentioned above that the dimension of $\mathcal{K}_m$ is nondecreasing. In fact, the following proposition determines the dimension of $\mathcal{K}_m$ in general.

Proposition 6.2. *The Krylov subspace $\mathcal{K}_m$ is of dimension m iff the grade μ of v with respect to A is not less than m; i.e.,*

$$\dim(\mathcal{K}_m) = m \quad \leftrightarrow \quad \text{grade}(v) \geq m. \tag{6.3}$$

6.3. Arnoldi's Method

Therefore,
$$\dim(\mathcal{K}_m) = \min\{m, \text{grade}(v)\}. \tag{6.4}$$

Proof. The vectors $v, Av, \ldots, A^{m-1}v$ form a basis of $\mathcal{K}_m$ iff, for any set of m scalars $\alpha_i, i = 0, \ldots, m-1$, where at least one α_i is nonzero, the linear combination $\sum_{i=0}^{m-1} \alpha_i A^i v$ is nonzero. This is equivalent to the condition that the only polynomial of degree not exceeding $m-1$ for which $p(A)v = 0$ is the zero polynomial. The equality (6.4) is a consequence of Proposition 6.1. □

Given a certain subspace X, recall that $A_{|X}$ denotes the restriction of A to X. If Q is a projector onto X, the *section of the operator* A in X is the operator from X onto itself defined by $QA_{|X}$. The following proposition characterizes the product of polynomials of A with v in terms of the section of A in $\mathcal{K}_m$.

Proposition 6.3. *Let Q_m be any projector onto $\mathcal{K}_m$ and let A_m be the section of A to $\mathcal{K}_m$; that is, $A_m = Q_m A_{|\mathcal{K}_m}$. Then, for any polynomial q of degree not exceeding $m-1$,*
$$q(A)v = q(A_m)v$$
and, for any polynomial of degree not exceeding m,
$$Q_m q(A)v = q(A_m)v.$$

Proof. First we prove that $q(A)v = q(A_m)v$ for any polynomial q of degree not exceeding $m-1$. It is sufficient to show the property for the monic polynomials $q_i(t) \equiv t^i$, $i = 0, \ldots, m-1$. The proof is by induction. The property is true for the polynomial $q_0(t) \equiv 1$. Assume that it is true for $q_i(t) \equiv t^i$:
$$q_i(A)v = q_i(A_m)v.$$
Multiplying the above equation by A on both sides yields
$$q_{i+1}(A)v = Aq_i(A_m)v.$$
If $i + 1 \leq m - 1$ the vector on the left-hand side belongs to $\mathcal{K}_m$ and, therefore, if the above equation is multiplied on both sides by Q_m, then
$$q_{i+1}(A)v = Q_m A q_i(A_m)v.$$
Looking at the right-hand side we observe that $q_i(A_m)v$ belongs to $\mathcal{K}_m$. Hence,
$$q_{i+1}(A)v = Q_m A_{|\mathcal{K}_m} q_i(A_m)v = q_{i+1}(A_m)v,$$
which proves that the property is true for $i + 1$, provided $i + 1 \leq m - 1$. For the case $i + 1 = m$, it only remains to show that $Q_m q_m(A)v = q_m(A_m)v$, which follows from $q_{m-1}(A)v = q_{m-1}(A_m)v$ by simply multiplying both sides by $Q_m A$. □

6.3 Arnoldi's Method

Arnoldi's method [9] is an orthogonal projection method onto $\mathcal{K}_m$ for general non-Hermitian matrices. The procedure was first introduced in 1951 as a means of reducing a dense matrix into Hessenberg form with a unitary transformation. In his paper, Arnoldi hinted that the eigenvalues of the Hessenberg matrix obtained from a number of steps smaller than n could provide accurate approximations to some eigenvalues of the original matrix. It was later discovered that this strategy leads to an efficient technique for approximating eigenvalues of large sparse matrices, which was then extended to the solution of large sparse linear systems of equations. The method will first be described theoretically, i.e., assuming exact arithmetic, then implementation details will be addressed.

6.3.1 The Basic Algorithm

Arnoldi's procedure is an algorithm for building an orthogonal basis of the Krylov subspace $\mathcal{K}_m$. In exact arithmetic, one variant of the algorithm is as follows.

ALGORITHM 6.1. Arnoldi

1. *Choose a vector v_1 such that $\|v_1\|_2 = 1$*
2. *For $j = 1, 2, \ldots, m$, Do*
3.   *Compute $h_{ij} = (Av_j, v_i)$ for $i = 1, 2, \ldots, j$*
4.   *Compute $w_j := Av_j - \sum_{i=1}^{j} h_{ij} v_i$*
5.   *$h_{j+1,j} = \|w_j\|_2$*
6.   *If $h_{j+1,j} = 0$ then Stop*
7.   *$v_{j+1} = w_j / h_{j+1,j}$*
8. *EndDo*

At each step, the algorithm multiplies the previous Arnoldi vector v_j by A and then orthonormalizes the resulting vector w_j against all previous v_i's by a standard Gram–Schmidt procedure. It will stop if the vector w_j computed in line 4 vanishes. This case will be examined shortly. Now a few simple properties of the algorithm are proved.

Proposition 6.4. *Assume that Algorithm 6.1 does not stop before the mth step. Then the vectors $v_1, v_2, \ldots, v_m$ form an orthonormal basis of the Krylov subspace*

$$\mathcal{K}_m = \mathrm{span}\{v_1, Av_1, \ldots, A^{m-1}v_1\}.$$

Proof. The vectors v_j, $j = 1, 2, \ldots, m$, are orthonormal by construction. That they span $\mathcal{K}_m$ follows from the fact that each vector v_j is of the form $q_{j-1}(A)v_1$, where q_{j-1} is a polynomial of degree $j - 1$. This can be shown by induction on j as follows. The result is clearly true for $j = 1$, since $v_1 = q_0(A)v_1$, with $q_0(t) \equiv 1$. Assume that the result is true for all integers not exceeding j and consider v_{j+1}. We have

$$h_{j+1,j} v_{j+1} = Av_j - \sum_{i=1}^{j} h_{ij} v_i = Aq_{j-1}(A)v_1 - \sum_{i=1}^{j} h_{ij} q_{i-1}(A)v_1, \quad (6.5)$$

which shows that v_{j+1} can be expressed as $q_j(A)v_1$, where q_j is of degree j, and completes the proof. □

Proposition 6.5. *Denote by V_m the $n \times m$ matrix with column vectors $v_1, \ldots, v_m$; by $\bar{H}_m$ the $(m+1) \times m$ Hessenberg matrix whose nonzero entries h_{ij} are defined by Algorithm*

6.3. Arnoldi's Method

6.1; and by H_m the matrix obtained from $\bar{H}_m$ by deleting its last row. Then the following relations hold:

$$AV_m = V_m H_m + w_m e_m^T \qquad (6.6)$$
$$= V_{m+1}\bar{H}_m, \qquad (6.7)$$
$$V_m^T A V_m = H_m. \qquad (6.8)$$

Proof. The relation (6.7) follows from the following equality, which is readily derived from lines 4, 5, and 7 of Algorithm 6.1:

$$Av_j = \sum_{i=1}^{j+1} h_{ij} v_i, \quad j = 1, 2, \ldots, m. \qquad (6.9)$$

Relation (6.6) is a matrix reformulation of (6.9). Relation (6.8) follows by multiplying both sides of (6.6) by V_m^T and making use of the orthonormality of $\{v_1, \ldots, v_m\}$. □

The result of the proposition is illustrated in Figure 6.1.

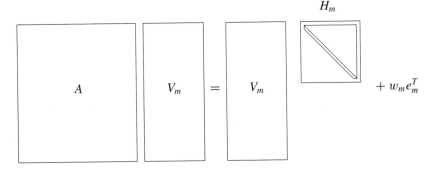

Figure 6.1. *The action of A on V_m gives $V_m H_m$ plus a rank-one matrix.*

As was noted earlier, the algorithm may break down in case the norm of w_j vanishes at a certain step j. In this case, the vector v_{j+1} cannot be computed and the algorithm stops. Still to be determined are the conditions under which this situation occurs.

Proposition 6.6. *Arnoldi's algorithm breaks down at step j (i.e., $h_{j+1,j} = 0$ in line 5 of Algorithm 6.1) iff the minimal polynomial of v_1 is of degree j. Moreover, in this case the subspace $\mathcal{K}_j$ is invariant under A.*

Proof. If the degree of the minimal polynomial is j, then w_j must be equal to zero. Indeed, otherwise v_{j+1} can be defined and, as a result, $\mathcal{K}_{j+1}$ would be of dimension $j+1$. Then Proposition 6.2 would imply that $\mu \geq j+1$, which is a contradiction. To prove the converse, assume that $w_j = 0$. Then the degree μ of the minimal polynomial of v_1 is such that $\mu \leq j$. Moreover, it is impossible that $\mu < j$. Otherwise, by the first part of this proof, the vector w_μ would be zero and the algorithm would have stopped at the earlier step number μ. The rest of the result follows from Proposition 6.1. □

A corollary of the proposition is that a projection method onto the subspace $\mathcal{K}_j$ will be exact when a breakdown occurs at step j. This result follows from Proposition 5.6 seen in Chapter 5. It is for this reason that such breakdowns are often called *lucky breakdowns*.

6.3.2 Practical Implementations

In the previous description of the Arnoldi process, exact arithmetic was assumed, mainly for simplicity. In practice, much can be gained by using the modified Gram–Schmidt (MGS) or the Householder algorithm instead of the standard Gram–Schmidt algorithm. With the MGS alternative the algorithm takes the following form.

ALGORITHM 6.2. Arnoldi MGS

1. Choose a vector v_1 of norm 1
2. For $j = 1, 2, \ldots, m$, Do
3.   Compute $w_j := A v_j$
4.   For $i = 1, \ldots, j$, Do
5. $h_{ij} = (w_j, v_i)$
6. $w_j := w_j - h_{ij} v_i$
7.   EndDo
8. $h_{j+1,j} = \|w_j\|_2$. If $h_{j+1,j} = 0$ Stop
9. $v_{j+1} = w_j / h_{j+1,j}$
10. EndDo

In exact arithmetic, this algorithm and Algorithm 6.1 are mathematically equivalent. In the presence of round-off the above formulation is much more reliable. However, there are cases where cancellations are so severe in the orthogonalization steps that even the MGS option is inadequate. In this case, two further improvements can be utilized.

The first improvement resorts to double orthogonalization. Whenever the final vector w_j obtained at the end of the main loop in the above algorithm has been computed, a test is performed to compare its norm with the norm of the initial w_j (which is $\|Av_j\|_2$). If the reduction falls below a certain threshold, indicating severe cancellation might have occurred, a second orthogonalization is made. It is known from a result by Kahan that additional orthogonalizations are superfluous (see, for example, Parlett [223]).

The second improvement is to use a different technique altogether. From the numerical point of view, one of the most reliable orthogonalization techniques is the Householder algorithm. Recall from Chapter 1 that the Householder orthogonalization uses reflection matrices of the form $P_k = I - 2 w_k w_k^T$ to transform a matrix X into upper triangular form. In the Arnoldi algorithm, the column vectors of the matrix X to be orthonormalized are not available ahead of time. Instead, the next vector is obtained as Av_j, where v_j is the current basis vector. In the Householder algorithm an orthogonal column v_i is obtained as $P_1 P_2 \cdots P_i e_i$, where $P_1, \ldots, P_i$ are the previous Householder matrices. This vector is then multiplied by A and the previous Householder transforms are applied to it. Then the next Householder transform is determined from the resulting vector. This procedure is described in the following algorithm, which was originally proposed by Walker [302].

6.3. Arnoldi's Method

ALGORITHM 6.3. Householder Arnoldi

1. Select a nonzero vector v; Set $z_1 = v$
2. For $j = 1, \ldots, m, m+1$, Do
3.   Compute the Householder unit vector w_j such that
4. $(w_j)_i = 0, i = 1, \ldots, j-1$, and
5. $(P_j z_j)_i = 0, i = j+1, \ldots, n$, where $P_j = I - 2 w_j w_j^T$
6. $h_{j-1} = P_j z_j$
7. $v_j = P_1 P_2 \cdots P_j e_j$
8.   If $j \leq m$ Compute $z_{j+1} := P_j P_{j-1} \cdots P_1 A v_j$
9. EndDo

For details regarding the determination of the Householder vector w_j in lines 3 to 5 and on its use in lines 6 to 8, see Chapter 1. Recall that the matrices P_j need not be formed explicitly. To obtain h_{j-1} from z_j in line 6, zero out all the components from positions $j+1$ through n of the n-vector z_j and change its jth component, leaving all others unchanged. Thus, the $n \times m$ matrix $[h_0, h_1, \ldots, h_m]$ will have the same structure as the matrix X_m of (1.26) in Chapter 1. By comparison with the Householder algorithm seen in Chapter 1, we can infer that the above process computes the QR factorization of the matrix $v, Av_1, Av_2, Av_3, \ldots, Av_m$. Define

$$Q_j = P_j P_{j-1} \cdots P_1. \tag{6.10}$$

The definition of z_{j+1} in line 8 of the algorithm yields the relation

$$Q_j A v_j = z_{j+1}.$$

After the next Householder transformation P_{j+1} is applied in line 6, h_j satisfies the relation

$$h_j = P_{j+1} z_{j+1} = P_{j+1} Q_j A v_j = Q_{j+1} A v_j. \tag{6.11}$$

Now observe that, since the components $j+2, \ldots, n$ of h_j are zero, then $P_i h_j = h_j$ for any $i \geq j+2$. Hence,

$$h_j = P_m P_{m-1} \cdots P_{j+2} h_j = Q_m A v_j, \quad j = 1, \ldots, m.$$

This leads to the factorization

$$Q_m [v, Av_1, Av_2, \ldots, Av_m] = [h_0, h_1, \ldots, h_m], \tag{6.12}$$

where the matrix $[h_0, \ldots, h_m]$ is $n \times (m+1)$ and upper triangular and Q_m is unitary.

It is important to relate the vectors v_i and h_i defined in this algorithm to vectors of the standard Arnoldi process. Let $\bar{H}_m$ be the $(m+1) \times m$ matrix obtained from the first $m+1$ rows of the $n \times m$ matrix $[h_1, \ldots, h_m]$. Since Q_{j+1} is unitary we have $Q_{j+1}^{-1} = Q_{j+1}^T$ and, hence, from the relation (6.11),

$$A v_j = Q_{j+1}^T \sum_{i=1}^{j+1} h_{ij} e_i = \sum_{i=1}^{j+1} h_{ij} Q_{j+1}^T e_i,$$

where each e_i is the ith column of the $n \times n$ identity matrix. Since $P_k e_i = e_i$ for $i < k$, it is not difficult to see that

$$Q_{j+1}^T e_i = P_1 \cdots P_{j+1} e_i = v_i \text{ for } i \leq j+1. \tag{6.13}$$

This yields the relation $Av_j = \sum_{i=1}^{j+1} h_{ij} v_i$ for $j = 1, \ldots, m$, which can be written in matrix form as

$$AV_m = V_{m+1} \bar{H}_m.$$

This is identical to the relation (6.7) obtained with the Gram–Schmidt or MGS implementation. The v_i's form an orthonormal basis of the Krylov subspace $\mathcal{K}_m$ and are identical to the v_i's defined by the Arnoldi process, apart from a possible sign difference.

Although the Householder algorithm is numerically more viable than the Gram–Schmidt or MGS version, it is also more expensive. The cost of each of the outer loops, corresponding to the j control variables, is dominated by lines 7 and 8. These apply the reflection matrices P_i for $i = 1, \ldots, j$ to a vector, perform the matrix-by-vector product Av_j, and then apply the matrices P_i for $i = j, j-1, \ldots, 1$ to a vector. The application of each P_i to a vector is performed as

$$(I - 2w_i w_i^T)v = v - \sigma w_i, \quad \text{with} \quad \sigma = 2w_i^T v.$$

This is essentially the result of a dot product of length $n - i + 1$ followed by a vector update of the same length, requiring a total of about $4(n - i + 1)$ operations for each application of P_i. Neglecting the last step, the number of operations due to the Householder transformations alone approximately totals

$$\sum_{j=1}^{m} \sum_{i=1}^{j} 8(n-i+1) = 8 \sum_{j=1}^{m} \left(jn - \frac{j(j-1)}{2} \right) \approx 4m^2 n - \frac{4}{3}m^3.$$

The table below shows the costs of different orthogonalization procedures. GS stands for Gram–Schmidt, MGS for modified Gram–Schmidt, MGSR for modified Gram–Schmidt with reorthogonalization, and HO for Householder.

	GS	MGS	MGSR	HO
Flops	$2m^2n$	$2m^2n$	$4m^2n$	$4m^2n - \frac{4}{3}m^3$
Storage	$(m+1)n$	$(m+1)n$	$(m+1)n$	$(m+1)n - \frac{1}{2}m^2$

The number of operations shown for MGSR corresponds to the worst-case scenario when a second orthogonalization is performed each time. In practice, the number of operations is usually closer to that of the standard MGS. Regarding storage, the vectors v_i, $i = 1, \ldots, m$, need not be saved. In the algorithms for solving linear systems, these vectors are needed at the end of the process. This issue will be covered with the Householder implementations of these algorithms. For now, assume that only the w_i's are saved. The small gain in memory usage in the Householder version can be explained by the diminishing lengths of the vectors required at each step of the Householder transformation. However, this difference is negligible relative to the whole storage requirement of the algorithm, because $m \ll n$, typically.

The Householder orthogonalization may be a reasonable choice when developing general-purpose, reliable software packages where robustness is a critical criterion. This is especially true for solving eigenvalue problems since the cost of orthogonalization is then amortized over several eigenvalue/eigenvector calculations. When solving linear systems, the MGS orthogonalization, with a reorthogonalization strategy based on a measure of the level of cancellation, is more than adequate in most cases.

6.4 Arnoldi's Method for Linear Systems

Given an initial guess x_0 to the original linear system $Ax = b$, we now consider an orthogonal *projection method*, as defined in the previous chapter, which takes $\mathcal{L} = \mathcal{K} = \mathcal{K}_m(A, r_0)$, with

$$\mathcal{K}_m(A, r_0) = \text{span}\{r_0, Ar_0, A^2 r_0, \ldots, A^{m-1} r_0\}, \tag{6.14}$$

in which $r_0 = b - Ax_0$. This method seeks an approximate solution x_m from the affine subspace $x_0 + \mathcal{K}_m$ of dimension m by imposing the Galerkin condition

$$b - Ax_m \perp \mathcal{K}_m. \tag{6.15}$$

If $v_1 = r_0 / \|r_0\|_2$ in Arnoldi's method and we set $\beta = \|r_0\|_2$, then

$$V_m^T A V_m = H_m$$

by (6.8) and

$$V_m^T r_0 = V_m^T (\beta v_1) = \beta e_1.$$

As a result, the approximate solution using the above m-dimensional subspaces is given by

$$x_m = x_0 + V_m y_m, \tag{6.16}$$
$$y_m = H_m^{-1}(\beta e_1). \tag{6.17}$$

A method based on this approach and called the full orthogonalization method (FOM) is described next. MGS is used in the Arnoldi procedure.

ALGORITHM 6.4. FOM

1. Compute $r_0 = b - Ax_0$, $\beta := \|r_0\|_2$, and $v_1 := r_0 / \beta$
2. Define the $m \times m$ matrix $H_m = \{h_{ij}\}_{i,j=1,\ldots,m}$; Set $H_m = 0$
3. For $j = 1, 2, \ldots, m$, Do
4.   Compute $w_j := Av_j$
5.   For $i = 1, \ldots, j$, Do
6. $h_{ij} = (w_j, v_i)$
7. $w_j := w_j - h_{ij} v_i$
8.   EndDo
9.   Compute $h_{j+1,j} = \|w_j\|_2$. If $h_{j+1,j} = 0$ set $m := j$ and go to 12
10.   Compute $v_{j+1} = w_j / h_{j+1,j}$
11. EndDo
12. Compute $y_m = H_m^{-1}(\beta e_1)$ and $x_m = x_0 + V_m y_m$

The above algorithm depends on a parameter m that is the dimension of the Krylov subspace. In practice it is desirable to select m in a dynamic fashion. This is possible if the residual norm of the solution x_m is available inexpensively (without having to compute x_m itself). Then the algorithm can be stopped at the appropriate step using this information. The following proposition gives a result in this direction.

Proposition 6.7. *The residual vector of the approximate solution x_m computed by the FOM algorithm is such that*

$$b - Ax_m = -h_{m+1,m} e_m^T y_m v_{m+1}$$

and, therefore,

$$\|b - Ax_m\|_2 = h_{m+1,m} |e_m^T y_m|. \tag{6.18}$$

Proof. We have the relations

$$\begin{aligned} b - Ax_m &= b - A(x_0 + V_m y_m) \\ &= r_0 - A V_m y_m \\ &= \beta v_1 - V_m H_m y_m - h_{m+1,m} e_m^T y_m v_{m+1}. \end{aligned}$$

By the definition of y_m, $H_m y_m = \beta e_1$, and so $\beta v_1 - V_m H_m y_m = 0$, from which the result follows. □

A rough estimate of the cost of each step of the algorithm is determined as follows. If $Nz(A)$ is the number of nonzero elements of A, then m steps of the Arnoldi procedure will require m matrix-by-vector products at a cost of $2m \times Nz(A)$. Each of the Gram–Schmidt steps costs approximately $4 \times j \times n$ operations, which brings the total over the m steps to approximately $2m^2 n$. Thus, on the average, a step of FOM costs approximately

$$2Nz(A) + 2mn.$$

Regarding storage, m vectors of length n are required to save the basis V_m. Additional vectors must be used to keep the current solution and right-hand side and a scratch vector is needed for the matrix-by-vector product. In addition, the Hessenberg matrix H_m must be saved. The total is therefore roughly

$$(m+3)n + \frac{m^2}{2}.$$

In most situations m is small relative to n, so this cost is dominated by the first term.

6.4.1 Variation 1: Restarted FOM

Consider now the algorithm from a practical viewpoint. As m increases, the computational cost increases at least as $O(m^2 n)$ because of the Gram–Schmidt orthogonalization. The memory cost increases as $O(mn)$. For large n this limits the largest value of m that can be used. There are two remedies. The first is to restart the algorithm periodically and the second is to "truncate" the orthogonalization in the Arnoldi algorithm. In this section we consider the first of these two options, which is described below.

6.4. Arnoldi's Method for Linear Systems

ALGORITHM 6.5. Restarted FOM (FOM(m))

1. Compute $r_0 = b - Ax_0$, $\beta = \|r_0\|_2$, and $v_1 = r_0/\beta$
2. Generate the Arnoldi basis and the matrix H_m using the Arnoldi algorithm starting with v_1
3. Compute $y_m = H_m^{-1} \beta e_1$ and $x_m = x_0 + V_m y_m$. If satisfied then Stop
4. Set $x_0 := x_m$ and go to 1

There are many possible variations of this basic scheme. One that is generally more economical in practice is based on the observation that sometimes a small m is sufficient for convergence and sometimes the largest possible m is necessary. Hence, we have the idea of averaging over different values of m. Start the algorithm with $m = 1$ and increment m by one in line 4 until a certain m_{max} is reached, after which m is reset to one or kept the same. These variations will not be considered here.

Example 6.1. Table 6.1 shows the results of applying FOM(10) with no preconditioning to three of the test problems described in Section 3.7.

Matrix	Iters	Kflops	Residual	Error
F2DA	109	4442	0.36E−03	0.67E−04
F3D	66	11664	0.87E−03	0.35E−03
ORS	300	13558	0.26E+00	0.71E−04

Table 6.1. *A test run of FOM with no preconditioning.*

The column labeled *Iters* shows the total actual number of matrix-by-vector multiplications (matvecs) required to converge. The stopping criterion used is that the 2-norm of the residual be reduced by a factor of 10^7 relative to the 2-norm of the initial residual. A maximum of 300 matvecs are allowed. *Kflops* is the total number of floating-point operations performed, in thousands. *Residual* and *Error* represent the 2-norm of the residual and error vectors, respectively. Note that the method did not succeed in solving the third problem.

6.4.2 Variation 2: IOM and DIOM

A second alternative to FOM is to truncate the Arnoldi recurrence. Specifically, an integer k is selected and the following "incomplete" orthogonalization is performed.

ALGORITHM 6.6. Incomplete Orthogonalization Process

1. For $j = 1, 2, \ldots, m$, Do
2.   Compute $w_j := Av_j$
3.   For $i = \max\{1, j-k+1\}, \ldots, j$, Do
4. $h_{i,j} = (w_j, v_i)$
5. $w_j := w_j - h_{ij} v_i$
6.   EndDo
7.   Compute $h_{j+1,j} = \|w_j\|_2$ and $v_{j+1} = w_j/h_{j+1,j}$
8. EndDo

The number of directions k against which to orthogonalize may be dictated by memory limitations. The incomplete orthogonalization method (IOM) consists of performing the above incomplete orthogonalization procedure and computing an approximate solution using the same formulas (6.16) and (6.17).

ALGORITHM 6.7. IOM Algorithm

Run a modification of Algorithm 6.4 in which the Arnoldi process in lines 3 to 11 is replaced by the incomplete orthogonalization process and every other computation remains unchanged.

It is now necessary to keep only the k previous v_i-vectors. The others are not needed in the above process and may be discarded. However, the difficulty remains that, when the solution is computed by formula (6.16), all the vectors v_i for $i = 1, 2, \ldots, m$ are required. One option is to recompute them at the end, but essentially this doubles the cost of the algorithm. Fortunately, a formula can be developed whereby the current approximate solution x_m can be updated from the previous approximation x_{m-1} and a small number of vectors that are also updated at each step. This *progressive* formulation of the solution leads to an algorithm termed *direct IOM* (DIOM), which we now derive.

The Hessenberg matrix H_m obtained from the incomplete orthogonalization process has a band structure with a bandwidth of $k+1$. For example, when $k=3$ and $m=5$, it is of the form

$$H_m = \begin{pmatrix} h_{11} & h_{12} & h_{13} & & \\ h_{21} & h_{22} & h_{23} & h_{24} & \\ & h_{32} & h_{33} & h_{34} & h_{35} \\ & & h_{43} & h_{44} & h_{45} \\ & & & h_{54} & h_{55} \end{pmatrix}. \quad (6.19)$$

The *direct* version of IOM is derived by exploiting the special structure of the LU factorization, $H_m = L_m U_m$, of the matrix H_m. Assuming no pivoting is used, the matrix L_m is unit lower bidiagonal and U_m is banded upper triangular, with k diagonals. Thus, the above matrix has a factorization of the form

$$H_m = \begin{pmatrix} 1 & & & & \\ l_{21} & 1 & & & \\ & l_{32} & 1 & & \\ & & l_{43} & 1 & \\ & & & l_{54} & 1 \end{pmatrix} \times \begin{pmatrix} u_{11} & u_{12} & u_{13} & & \\ & u_{22} & u_{23} & u_{24} & \\ & & u_{33} & u_{34} & u_{35} \\ & & & u_{44} & u_{45} \\ & & & & u_{55} \end{pmatrix}.$$

The approximate solution is then given by

$$x_m = x_0 + V_m U_m^{-1} L_m^{-1}(\beta e_1).$$

Defining

$$P_m \equiv V_m U_m^{-1}$$

and

$$z_m = L_m^{-1}(\beta e_1),$$

the approximate solution is given by

$$x_m = x_0 + P_m z_m. \quad (6.20)$$

6.4. Arnoldi's Method for Linear Systems

Because of the structure of U_m, P_m can be updated easily. Indeed, equating the last columns of the matrix relation $P_m U_m = V_m$ yields

$$\sum_{i=m-k+1}^{m} u_{im} p_i = v_m,$$

which allows the vector p_m to be computed from the previous p_i's and v_m:

$$p_m = \frac{1}{u_{mm}} \left[v_m - \sum_{i=m-k+1}^{m-1} u_{im} p_i \right].$$

In addition, because of the structure of L_m, we have the relation

$$z_m = \begin{bmatrix} z_{m-1} \\ \zeta_m \end{bmatrix},$$

in which

$$\zeta_m = -l_{m,m-1} \zeta_{m-1}.$$

From (6.20),

$$x_m = x_0 + [P_{m-1}, p_m] \begin{bmatrix} z_{m-1} \\ \zeta_m \end{bmatrix} = x_0 + P_{m-1} z_{m-1} + \zeta_m p_m.$$

Noting that $x_0 + P_{m-1} z_{m-1} = x_{m-1}$, it follows that the approximation x_m can be updated at each step by the relation

$$x_m = x_{m-1} + \zeta_m p_m, \qquad (6.21)$$

where p_m is defined above. This gives the DIOM algorithm.

ALGORITHM 6.8. DIOM

1. Choose x_0 and compute $r_0 = b - Ax_0$, $\beta := \|r_0\|_2$, $v_1 := r_0/\beta$
2. For $m = 1, 2, \ldots$, until convergence, Do
3.   Compute h_{im}, $i = \max\{1, m - k + 1\}, \ldots, m$, and v_{m+1} as in lines 2–7 of Algorithm 6.6
4.   Update the LU factorization of H_m; i.e., obtain the last column of U_m using the previous k pivots. If $u_{mm} = 0$ Stop
5. $\zeta_m = \{$ if $m = 1$ then β, else $-l_{m,m-1} \zeta_{m-1}\}$
6. $p_m = u_{mm}^{-1} \left(v_m - \sum_{i=m-k+1}^{m-1} u_{im} p_i \right)$ (for $i \leq 0$ set $u_{im} p_i \equiv 0$)
7. $x_m = x_{m-1} + \zeta_m p_m$
8. EndDo

The costs of the above algorithm as well as the IOM algorithm are the subject of Exercise 6.

Observe that (6.6) is still valid and, as a consequence, Proposition 6.7, which is based on it, still holds. That is because the orthogonality properties were not used to derive the two relations therein. A result of this is that (6.18) still holds and it is then easy to show that

$$\|b - Ax_m\|_2 = h_{m+1,m} |e_m^T y_m| = h_{m+1,m} \left| \frac{\zeta_m}{u_{mm}} \right|.$$

DIOM can also be derived by imposing the properties that are satisfied by the residual vector and the conjugate directions, i.e., the p_i's. Note that the above algorithm is based implicitly on Gaussian elimination without pivoting for the solution of the Hessenberg system $H_m y_m = \beta e_1$. This may cause a premature termination in line 4. Fortunately, an implementation is available that relies on Gaussian elimination with partial pivoting. The details of this variant can be found in [239].

Since the residual vector is a scalar multiple of v_{m+1} and since the v_i's are no longer orthogonal, IOM and DIOM are not orthogonal projection techniques. They can, however, be viewed as oblique projection techniques onto $\mathcal{K}_m$ and orthogonal to an artificially constructed subspace.

Proposition 6.8. *IOM and DIOM are mathematically equivalent to a projection process onto $\mathcal{K}_m$ and orthogonal to*

$$\mathcal{L}_m = \mathrm{span}\{z_1, z_2, \ldots, z_m\},$$

where

$$z_i = v_i - (v_i, v_{m+1})v_{m+1}, \quad i = 1, \ldots, m.$$

Proof. The proof is an immediate consequence of the fact that r_m is a multiple of v_{m+1} and, by construction, v_{m+1} is orthogonal to all z_i's defined in the proposition. □

The following simple properties can be shown:

- The residual vectors r_i, $i = 1, \ldots, m$, are *locally* orthogonal:

$$(r_j, r_i) = 0 \quad \text{for} \quad |i - j| \leq k, \quad i \neq j. \tag{6.22}$$

- The p_j's are locally A-orthogonal to the Arnoldi vectors; i.e.,

$$(Ap_j, v_i) = 0 \quad \text{for} \quad j - k + 1 < i < j. \tag{6.23}$$

- For the case $k = \infty$ (full orthogonalization), the p_j's are semiconjugate; i.e.,

$$(Ap_j, p_i) = 0 \quad \text{for} \quad i < j. \tag{6.24}$$

6.5 Generalized Minimal Residual Method

The generalized minimal residual method (GMRES) is a projection method based on taking $\mathcal{K} = \mathcal{K}_m$ and $\mathcal{L} = A\mathcal{K}_m$, in which $\mathcal{K}_m$ is the mth Krylov subspace, with $v_1 = r_0/\|r_0\|_2$. As seen in Chapter 5, such a technique minimizes the residual norm over all vectors in $x_0 + \mathcal{K}_m$. The implementation of an algorithm based on this approach is similar to that of the FOM algorithm. We first describe the basic idea and then discuss practical details and a few variations.

6.5.1 The Basic GMRES Algorithm

There are two ways to derive the algorithm. The first way exploits the optimality property and the relation (6.7). Any vector x in $x_0 + \mathcal{K}_m$ can be written as

$$x = x_0 + V_m y, \tag{6.25}$$

where y is an m-vector. Defining

$$J(y) = \|b - Ax\|_2 = \|b - A(x_0 + V_m y)\|_2, \tag{6.26}$$

6.5. Generalized Minimal Residual Method

the relation (6.7) results in

$$\begin{aligned}
b - Ax &= b - A(x_0 + V_m y) \\
&= r_0 - A V_m y \\
&= \beta v_1 - V_{m+1} \bar{H}_m y \\
&= V_{m+1}(\beta e_1 - \bar{H}_m y).
\end{aligned} \quad (6.27)$$

Since the column vectors of V_{m+1} are orthonormal, then

$$J(y) \equiv \|b - A(x_0 + V_m y)\|_2 = \|\beta e_1 - \bar{H}_m y\|_2. \quad (6.28)$$

The GMRES approximation is the unique vector of $x_0 + \mathcal{K}_m$ that minimizes (6.26). By (6.25) and (6.28), this approximation can be obtained quite simply as $x_m = x_0 + V_m y_m$, where y_m minimizes the function $J(y) = \|\beta e_1 - \bar{H}_m y\|_2$; i.e.,

$$x_m = x_0 + V_m y_m, \quad \text{where} \quad (6.29)$$
$$y_m = \text{argmin}_y \|\beta e_1 - \bar{H}_m y\|_2. \quad (6.30)$$

The minimizer y_m is inexpensive to compute since it requires the solution of an $(m+1) \times m$ least-squares problem, where m is typically small. This gives the following algorithm.

ALGORITHM 6.9. GMRES

1. Compute $r_0 = b - Ax_0$, $\beta := \|r_0\|_2$, and $v_1 := r_0/\beta$
2. For $j = 1, 2, \ldots, m$, Do
3.   Compute $w_j := Av_j$
4.   For $i = 1, \ldots, j$, Do
5. $h_{ij} := (w_j, v_i)$
6. $w_j := w_j - h_{ij} v_i$
7.   EndDo
8. $h_{j+1,j} = \|w_j\|_2$. If $h_{j+1,j} = 0$ set $m := j$ and go to 11
9. $v_{j+1} = w_j/h_{j+1,j}$
10. EndDo
11. Define the $(m+1) \times m$ Hessenberg matrix $\bar{H}_m = \{h_{ij}\}_{1 \le i \le m+1, 1 \le j \le m}$
12. Compute y_m, the minimizer of $\|\beta e_1 - \bar{H}_m y\|_2$, and $x_m = x_0 + V_m y_m$

The second way to derive the GMRES algorithm is to use (5.7) with $W_m = AV_m$. This is the subject of Exercise 5.

6.5.2 The Householder Version

The previous algorithm utilizes the MGS orthogonalization in the Arnoldi process. Section 6.3.2 described a Householder variant of the Arnoldi process that is numerically more robust than Gram–Schmidt. Here we focus on a modification of GMRES that retrofits the Householder orthogonalization. Section 6.3.2 explained how to get the v_j and the columns of $\bar{H}_{m+1}$ at each step from the Householder–Arnoldi algorithm. Since V_m and $\bar{H}_m$ are the only items needed to extract the approximate solution at the end of the GMRES process, the modification seems rather straightforward. However, this is only true if the v_i's are stored. In this case, line 12 would remain the same and the modification to the algorithm would be in lines 3–11, which are to be replaced by the Householder variant of the Arnoldi process. It was mentioned in Section 6.3.2 that it is preferable not to store the v_i's because this would double the storage requirement. In this case, a formula must be found to generate the approximate solution in line 12 using only the w_i's, i.e., the P_i's. Let

$$y_m = (\eta_1, \eta_2, \ldots, \eta_m)^T,$$

so that the solution is of the form $x_m = x_0 + \eta_1 v_1 + \cdots + \eta_m v_m$. Recall that, in the Householder variant of the Arnoldi process, each v_j is defined by

$$v_j = P_1 P_2 \cdots P_j e_j.$$

Using a Horner-like scheme, we obtain

$$\begin{aligned} x_m &= x_0 + \eta_1 P_1 e_1 + \eta_2 P_1 P_2 e_2 + \cdots + \eta_m P_1 P_2 \cdots P_m e_m \\ &= x_0 + P_1 \left(\eta_1 e_1 + P_2 \left(\eta_2 e_2 + \cdots + P_{m-1} \left(\eta_{m-1} e_{m-1} + P_m \eta_m e_m \right) \right) \right). \end{aligned}$$

Therefore, when Householder orthogonalization is used, then line 12 of the GMRES algorithm should be replaced by a step of the form

$$z := 0, \tag{6.31}$$
$$z := P_j \left(\eta_j e_j + z \right), \quad j = m, m-1, \ldots, 1, \tag{6.32}$$
$$x_m = x_0 + z. \tag{6.33}$$

The above step requires roughly as many operations as computing the last Arnoldi vector v_m. Therefore, its cost is negligible relative to the cost of the Arnoldi loop.

ALGORITHM 6.10. GMRES with Householder Orthogonalization

1. Compute $r_0 = b - Ax_0, z := r_0$
2. For $j = 1, \ldots, m, m+1$, Do
3.   Compute the Householder unit vector w_j such that
4. $(w_j)_i = 0, i = 1, \ldots, j - 1$ and
5. $(P_j z)_i = 0, i = j + 1, \ldots, n$, where $P_j = I - 2 w_j w_j^T$
6. $h_{j-1} := P_j z$; If $j = 1$ then let $\beta := e_1^T h_0$
7. $v := P_1 P_2 \cdots P_j e_j$
8.   If $j \leq m$ compute $z := P_j P_{j-1} \cdots P_1 A v$
9. EndDo
10. Define $\bar{H}_m = $ the $(m+1) \times m$ upper part of the matrix $[h_1, \ldots, h_m]$
11. Compute $y_m = \text{Argmin}_y \| \beta e_1 - \bar{H}_m y \|_2$. Let $y_m = (\eta_1, \eta_2, \ldots, \eta_m)^T$
12. $z := 0$
13. For $j = m, m-1, \ldots, 1$, Do
14. $z := P_j \left(\eta_j e_j + z \right)$
15. EndDo
16. Compute $x_m = x_0 + z$

6.5. Generalized Minimal Residual Method

Note that now only the set of w_j-vectors needs to be saved. The scalar β defined in line 6 is equal to $\pm \|r_0\|_2$. This is because $P_1 z = \beta e_1$, where β is defined by (1.25) seen in Chapter 1, which defined the first Householder transformation. As was observed earlier, the Householder factorization actually obtains the QR factorization (6.12) with $v = r_0$. We can also formulate GMRES directly from this factorization. Indeed, if $x = x_0 + V_m y_m$, then, according to this factorization, the corresponding residual norm is equal to

$$\|h_0 - \eta_1 h_1 - \eta_2 h_2 - \cdots - \eta_m h_m\|_2,$$

whose minimizer is the same as the one defined by the algorithm.

The details of implementation of the solution of the least-squares problem as well as the estimate of the residual norm are identical to those of the Gram–Schmidt versions and are discussed next.

6.5.3 Practical Implementation Issues

A clear difficulty with Algorithm 6.9 is that it does not provide the approximate solution x_m explicitly at each step. As a result, it is not easy to determine when to stop. One remedy is to compute the approximate solution x_m at regular intervals and check for convergence by a test on the residual, for example. However, there is a more elegant solution, which is related to the way in which the least-squares problem (6.30) is solved.

A common technique to solve the least-squares problem $\min \|\beta e_1 - \bar{H}_m y\|_2$ is to transform the Hessenberg matrix into upper triangular form by using plane rotations. Define the Givens rotation matrices

$$\Omega_i = \begin{pmatrix} 1 & & & & & & \\ & \ddots & & & & & \\ & & 1 & & & & \\ & & & c_i & s_i & & \\ & & & -s_i & c_i & & \\ & & & & & 1 & \\ & & & & & & \ddots \\ & & & & & & & 1 \end{pmatrix} \begin{matrix} \\ \\ \\ \leftarrow \text{row } i \\ \leftarrow \text{row } i+1 \\ \\ \\ \end{matrix} \qquad (6.34)$$

with $c_i^2 + s_i^2 = 1$. If m steps of the GMRES iteration are performed, then these matrices have dimension $(m+1) \times (m+1)$.

Multiply the Hessenberg matrix $\bar{H}_m$ and the corresponding right-hand side $\bar{g}_0 \equiv \beta e_1$ by a sequence of such matrices from the left. The coefficients s_i, c_i are selected to eliminate $h_{i+1,i}$ at each iteration. Thus, if $m = 5$, we would have

$$\bar{H}_5 = \begin{pmatrix} h_{11} & h_{12} & h_{13} & h_{14} & h_{15} \\ h_{21} & h_{22} & h_{23} & h_{24} & h_{25} \\ & h_{32} & h_{33} & h_{34} & h_{35} \\ & & h_{43} & h_{44} & h_{45} \\ & & & h_{54} & h_{55} \\ & & & & h_{65} \end{pmatrix}, \quad \bar{g}_0 = \begin{pmatrix} \beta \\ 0 \\ 0 \\ 0 \\ 0 \\ 0 \end{pmatrix}.$$

Then premultiply $\bar{H}_5$ by

$$\Omega_1 = \begin{pmatrix} c_1 & s_1 & & & \\ -s_1 & c_1 & & & \\ & & 1 & & \\ & & & 1 & \\ & & & & 1 \end{pmatrix},$$

with

$$s_1 = \frac{h_{21}}{\sqrt{h_{11}^2 + h_{21}^2}}, \quad c_1 = \frac{h_{11}}{\sqrt{h_{11}^2 + h_{21}^2}},$$

to obtain the matrix and right-hand side

$$\bar{H}_5^{(1)} = \begin{pmatrix} h_{11}^{(1)} & h_{12}^{(1)} & h_{13}^{(1)} & h_{14}^{(1)} & h_{15}^{(1)} \\ & h_{22}^{(1)} & h_{23}^{(1)} & h_{24}^{(1)} & h_{25}^{(1)} \\ & h_{32} & h_{33} & h_{34} & h_{35} \\ & & h_{43} & h_{44} & h_{45} \\ & & & h_{54} & h_{55} \\ & & & & h_{65} \end{pmatrix}, \quad \bar{g}_1 = \begin{pmatrix} c_1\beta \\ -s_1\beta \\ 0 \\ 0 \\ 0 \\ 0 \end{pmatrix}. \tag{6.35}$$

We can now premultiply the above matrix and right-hand side again by a rotation matrix Ω_2 to eliminate h_{32}. This is achieved by taking

$$s_2 = \frac{h_{32}}{\sqrt{(h_{22}^{(1)})^2 + h_{32}^2}}, \quad c_2 = \frac{h_{22}^{(1)}}{\sqrt{(h_{22}^{(1)})^2 + h_{32}^2}}.$$

This elimination process is continued until the mth rotation is applied, which transforms the problem into one involving the matrix and right-hand side

$$\bar{H}_5^{(5)} = \begin{pmatrix} h_{11}^{(5)} & h_{12}^{(5)} & h_{13}^{(5)} & h_{14}^{(5)} & h_{15}^{(5)} \\ & h_{22}^{(5)} & h_{23}^{(5)} & h_{24}^{(5)} & h_{25}^{(5)} \\ & & h_{33}^{(5)} & h_{34}^{(5)} & h_{35}^{(5)} \\ & & & h_{44}^{(5)} & h_{45}^{(5)} \\ & & & & h_{55}^{(5)} \\ & & & & 0 \end{pmatrix}, \quad \bar{g}_5 = \begin{pmatrix} \gamma_1 \\ \gamma_2 \\ \gamma_3 \\ \vdots \\ \gamma_6 \end{pmatrix}. \tag{6.36}$$

Generally, the scalars c_i and s_i of the ith rotation Ω_i are defined as

$$s_i = \frac{h_{i+1,i}}{\sqrt{(h_{ii}^{(i-1)})^2 + h_{i+1,i}^2}}, \quad c_i = \frac{h_{ii}^{(i-1)}}{\sqrt{(h_{ii}^{(i-1)})^2 + h_{i+1,i}^2}}. \tag{6.37}$$

Define Q_m as the product of matrices Ω_i:

$$Q_m = \Omega_m \Omega_{m-1} \cdots \Omega_1, \tag{6.38}$$

and let

$$\bar{R}_m = \bar{H}_m^{(m)} = Q_m \bar{H}_m, \tag{6.39}$$
$$\bar{g}_m = Q_m(\beta e_1) = (\gamma_1, \ldots, \gamma_{m+1})^T. \tag{6.40}$$

6.5. Generalized Minimal Residual Method

Since Q_m is unitary,

$$\min \|\beta e_1 - \bar{H}_m y\|_2 = \min \|\bar{g}_m - \bar{R}_m y\|_2.$$

The solution to the above least-squares problem is obtained by simply solving the triangular system resulting from deleting the last row of the matrix $\bar{R}_m$ and right-hand side $\bar{g}_m$ in (6.36). In addition, it is clear that, for the solution y_*, the residual $\|\beta e_1 - \bar{H}_m y_*\|$ is nothing but the last element of the right-hand side, i.e., the term γ_6 in the above illustration.

Proposition 6.9. *Let $m \leq n$ and Ω_i, $i = 1, \ldots, m$, be the rotation matrices used to transform $\bar{H}_m$ into an upper triangular form. Denote by $\bar{R}_m$, $\bar{g}_m = (\gamma_1, \ldots, \gamma_{m+1})^T$ the resulting matrix and right-hand side, as defined by (6.39), (6.40), and by R_m, g_m the $m \times m$ upper triangular matrix and m-dimensional vector obtained from $\bar{R}_m$, $\bar{g}_m$ by deleting their last row and component, respectively. Then we have the following properties:*

1. *The rank of AV_m is equal to the rank of R_m. In particular, if $r_{mm} = 0$, then A must be singular.*

2. *The vector y_m that minimizes $\|\beta e_1 - \bar{H}_m y\|_2$ is given by*
$$y_m = R_m^{-1} g_m.$$

3. *The residual vector at step m satisfies*
$$b - Ax_m = V_{m+1} \left(\beta e_1 - \bar{H}_m y_m\right) = V_{m+1} Q_m^T (\gamma_{m+1} e_{m+1}) \tag{6.41}$$
and, as a result,
$$\|b - Ax_m\|_2 = |\gamma_{m+1}|. \tag{6.42}$$

Proof. To prove part (1), use (6.7) to obtain the relation
$$\begin{aligned} AV_m &= V_{m+1} \bar{H}_m \\ &= V_{m+1} Q_m^T Q_m \bar{H}_m \\ &= V_{m+1} Q_m^T \bar{R}_m. \end{aligned}$$

Since $V_{m+1} Q_m^T$ is unitary, the rank of AV_m is that of $\bar{R}_m$, which equals the rank of R_m, since these two matrices differ only by a zero row (the last row of $\bar{R}_m$). If $r_{mm} = 0$, then R_m is of rank not exceeding $m - 1$ and, as a result, AV_m is also of rank not exceeding $m - 1$. Since V_m is of full rank, this means that A must be singular.

Part (2) was essentially proved before the proposition. For any vector y,
$$\begin{aligned} \|\beta e_1 - \bar{H}_m y\|_2^2 &= \|Q_m (\beta e_1 - \bar{H}_m y)\|_2^2 \\ &= \|\bar{g}_m - \bar{R}_m y\|_2^2 \\ &= |\gamma_{m+1}|^2 + \|g_m - R_m y\|_2^2. \end{aligned} \tag{6.43}$$

The minimum of the left-hand side is reached when the second term on the right-hand side of (6.43) is zero. Since R_m is nonsingular, this is achieved when $y = R_m^{-1} g_m$.

To prove part (3), we start with the definitions used for GMRES and the relation (6.27). For any $x = x_0 + V_m y$,

$$b - Ax = V_{m+1}\left(\beta e_1 - \bar{H}_m y\right)$$
$$= V_{m+1} Q_m^T Q_m \left(\beta e_1 - \bar{H}_m y\right)$$
$$= V_{m+1} Q_m^T (\bar{g}_m - \bar{R}_m y).$$

As was seen in the proof of part (2) above, the 2-norm of $\bar{g}_m - \bar{R}_m y$ is minimized when y annihilates all components of the right-hand side $\bar{g}_m$ except the last one, which is equal to γ_{m+1}. As a result,

$$b - Ax_m = V_{m+1} Q_m^T (\gamma_{m+1} e_{m+1}),$$

which is (6.41). The result (6.42) follows from the orthonormality of the column vectors of $V_{m+1} Q_m^T$. $\square$

So far we have only described a process for computing the least-squares solution y_m of (6.30). Note that this approach with plane rotations can also be used to solve the linear system (6.17) for the FOM method. The only difference is that the last rotation Ω_m must be omitted. In particular, a single program can be written to implement both algorithms using a switch for selecting the FOM or GMRES option.

It is possible to implement the above process in a progressive manner, i.e., at each step of the GMRES algorithm. This approach will allow one to obtain the residual norm at every step, with virtually no additional arithmetic operations. To illustrate this, start with (6.36); i.e., assume that the first m rotations have already been applied. Now the residual norm is available for x_5 and the stopping criterion can be applied. Assume that the test dictates that further steps be taken. One more step of the Arnoldi algorithm must be executed to get Av_6 and the 6th column of $\bar{H}_6$. This column is appended to $\bar{R}_5$, which has been augmented by a zero row to match the dimension. Then the previous rotations $\Omega_1, \Omega_2, \ldots, \Omega_5$ are applied to this last column. After this is done the following matrix and right-hand side are obtained (superscripts are now omitted from the h_{ij} entries):

$$\bar{H}_6^{(5)} = \begin{pmatrix} h_{11} & h_{12} & h_{13} & h_{14} & h_{15} & h_{16} \\ & h_{22} & h_{23} & h_{24} & h_{25} & h_{26} \\ & & h_{33} & h_{34} & h_{35} & h_{36} \\ & & & h_{44} & h_{45} & h_{46} \\ & & & & h_{55} & h_{56} \\ & & & & 0 & h_{66} \\ & & & & 0 & h_{76} \end{pmatrix}, \quad \bar{g}_6^{(5)} = \begin{pmatrix} \gamma_1 \\ \gamma_2 \\ \gamma_3 \\ \vdots \\ \gamma_6 \\ 0 \end{pmatrix}. \qquad (6.44)$$

The algorithm now continues in the same way as before. We need to premultiply the matrix by a rotation matrix Ω_6 (now of size 7×7), with

$$s_6 = \frac{h_{76}}{\sqrt{(h_{66})^2 + h_{76}^2}}, \quad c_6 = \frac{h_{66}^{(5)}}{\sqrt{(h_{66}^{(5)})^2 + h_{76}^2}}, \qquad (6.45)$$

6.5. Generalized Minimal Residual Method

to get the matrix and right-hand side

$$\bar{R}_6 = \begin{pmatrix} r_{11} & r_{12} & r_{13} & r_{14} & r_{15} & r_{16} \\ & r_{22} & r_{23} & r_{24} & r_{25} & r_{26} \\ & & r_{33} & r_{34} & r_{35} & r_{36} \\ & & & r_{44} & r_{45} & r_{46} \\ & & & & r_{55} & r_{56} \\ & & & & & r_{66} \\ & & & & & 0 \end{pmatrix}, \quad \bar{g}_6 = \begin{pmatrix} \gamma_1 \\ \gamma_2 \\ \gamma_3 \\ \vdots \\ c_6 \gamma_6 \\ -s_6 \gamma_6 \end{pmatrix}. \quad (6.46)$$

If the residual norm as given by $|\gamma_{m+1}|$ is small enough, the process must be stopped. The last rows of $\bar{R}_m$ and $\bar{g}_m$ are deleted and the resulting upper triangular system is solved to obtain y_m. Then the approximate solution $x_m = x_0 + V_m y_m$ is computed.

Note from (6.46) that the following useful relation for γ_{j+1} results:

$$\gamma_{j+1} = -s_j \gamma_j. \quad (6.47)$$

In particular, if $s_j = 0$, then the residual norm must be equal to zero, which means that the solution is exact at step j.

6.5.4 Breakdown of GMRES

If Algorithm 6.9 is examined carefully, we observe that the only possibilities for breakdown in GMRES are in the Arnoldi loop, when $w_j = 0$, i.e., when $h_{j+1,j} = 0$ at a given step j. In this situation, the algorithm stops because the next Arnoldi vector cannot be generated. However, in this situation, the residual vector is zero; i.e., the algorithm will deliver the exact solution at this step. In fact, the converse is also true: If the algorithm stops at step j with $b - Ax_j = 0$, then $h_{j+1,j} = 0$.

Proposition 6.10. *Let A be a nonsingular matrix. Then the GMRES algorithm breaks down at step j, i.e., $h_{j+1,j} = 0$, iff the approximate solution x_j is exact.*

Proof. To show the necessary condition, observe that, if $h_{j+1,j} = 0$, then $s_j = 0$. Indeed, since A is nonsingular, then $r_{jj} = h_{jj}^{(j-1)}$ is nonzero by the first part of Proposition 6.9 and (6.37) implies $s_j = 0$. Then the relations (6.42) and (6.47) imply that $r_j = 0$.

To show the sufficient condition, we use (6.47) again. Since the solution is exact at step j and not at step $j - 1$, then $s_j = 0$. From the formula (6.37), this implies that $h_{j+1,j} = 0$. □

6.5.5 Variation 1: Restarting

Similarly to the FOM algorithm of the previous section, the GMRES algorithm becomes impractical when m is large because of the growth of memory and computational requirements as m increases. These requirements are identical to those of FOM. As with FOM, there are two remedies. One is based on restarting and the other on truncating the Arnoldi orthogonalization. The straightforward restarting option is described here.

ALGORITHM 6.11. Restarted GMRES

1. Compute $r_0 = b - Ax_0$, $\beta = \|r_0\|_2$, and $v_1 = r_0/\beta$
2. Generate the Arnoldi basis and the matrix $\bar{H}_m$ using the Arnoldi algorithm starting with v_1
3. Compute y_m, which minimizes $\|\beta e_1 - \bar{H}_m y\|_2$, and $x_m = x_0 + V_m y_m$
4. If satisfied then Stop, else set $x_0 := x_m$ and go to 1

Note that the implementation tricks discussed in the previous section can be applied, providing the residual norm at each substep j without computing the approximation x_j. This enables the program to exit as soon as this norm is small enough.

A well-known difficulty with the restarted GMRES algorithm is that it can *stagnate* when the matrix is not positive definite. The full GMRES algorithm is guaranteed to converge in at most n steps, but this would be impractical if there were many steps required for convergence. A typical remedy is to use *preconditioning techniques* (see Chapters 9 and 10), whose goal is to reduce the number of steps required to converge.

Example 6.2. Table 6.2 shows the results of applying the GMRES algorithm with no preconditioning to three of the test problems described in Section 3.7.

Matrix	Iters	Kflops	Residual	Error
F2DA	95	3841	0.32E−02	0.11E−03
F3D	67	11862	0.37E−03	0.28E−03
ORS	205	9221	0.33E+00	0.68E−04

Table 6.2. *A test run of GMRES with no preconditioning.*

See Example 6.1 for the meaning of the column headers in the table. In this test, the dimension of the Krylov subspace is $m = 10$. Observe that the problem ORS, which could not be solved by FOM(10), is now solved in 205 steps.

6.5.6 Variation 2: Truncated GMRES Versions

It is possible to derive an incomplete version of the GMRES algorithm. This algorithm is called quasi-GMRES (QGMRES) for the sake of notational uniformity with other algorithms developed in the literature (some of which will be seen in the next chapter). A direct version called DQGMRES using exactly the same arguments as in Section 6.4.2 for DIOM can also be derived. We begin by defining a hypothetical QGMRES algorithm, in simple terms, by replacing the Arnoldi algorithm with Algorithm 6.6, the incomplete orthogonalization procedure.

ALGORITHM 6.12. QGMRES

Run a modification of Algorithm 6.9 in which the Arnoldi process in lines 2 to 10 is replaced by the incomplete orthogonalization process and all other computations remain unchanged.

6.5. Generalized Minimal Residual Method

Similarly to IOM, only the k previous v_i-vectors must be kept at any given step. However, this version of GMRES will potentially save computations but not storage. This is because computing the solution by formula (6.29) requires the vectors v_i for $i = 1, \ldots, m$ to be accessed. Fortunately, the approximate solution can be updated in a progressive manner, as in DIOM.

The implementation of this progressive version is quite similar to DIOM. First, note that, if $\bar{H}_m$ is banded, as, for example, when $m = 5, k = 2$:

$$\bar{H}_5 = \begin{pmatrix} h_{11} & h_{12} & & & \\ h_{21} & h_{22} & h_{23} & & \\ & h_{32} & h_{33} & h_{34} & \\ & & h_{43} & h_{44} & h_{45} \\ & & & h_{54} & h_{55} \\ & & & & h_{65} \end{pmatrix}, \quad g = \begin{pmatrix} \beta \\ 0 \\ 0 \\ 0 \\ 0 \\ 0 \end{pmatrix}, \qquad (6.48)$$

then the premultiplications by the rotation matrices Ω_i, as described in the previous section, will only introduce an additional diagonal. For the above case, the resulting least-squares system is $\bar{R}_5 y = \bar{g}_5$, with

$$\bar{R}_5 = \begin{pmatrix} r_{11} & r_{12} & r_{13} & & \\ & r_{22} & r_{23} & r_{24} & \\ & & r_{33} & r_{34} & r_{35} \\ & & & r_{44} & r_{45} \\ & & & & r_{55} \\ & & & & 0 \end{pmatrix}, \quad \bar{g}_5 = \begin{pmatrix} \gamma_1 \\ \gamma_2 \\ \gamma_3 \\ \vdots \\ \gamma_6 \end{pmatrix}. \qquad (6.49)$$

The approximate solution is given by

$$x_m = x_0 + V_m R_m^{-1} g_m,$$

where R_m and g_m are obtained by removing the last row of $\bar{R}_m$ and $\bar{g}_m$, respectively. Defining P_m as in DIOM:

$$P_m \equiv V_m R_m^{-1},$$

then

$$x_m = x_0 + P_m g_m.$$

Also note that, similarly to DIOM,

$$g_m = \begin{bmatrix} g_{m-1} \\ \gamma_m \end{bmatrix},$$

in which

$$\gamma_m = c_m \gamma_m^{(m-1)},$$

where $\gamma_m^{(m-1)}$ is the last component of the vector $\bar{g}_{m-1}$, i.e., the right-hand side before the mth rotation is applied. Thus, x_m can be updated at each step via the relation

$$x_m = x_{m-1} + \gamma_m p_m.$$

ALGORITHM 6.13. DQGMRES

1. Compute $r_0 = b - Ax_0$, $\gamma_1 := \|r_0\|_2$, and $v_1 := r_0/\gamma_1$
2. For $m = 1, 2, \ldots$, until convergence, Do
3.   Compute $h_{im}, i = \max\{1, m - k + 1\}, \ldots, m$, and v_{m+1}
         as in lines 2 to 6 of Algorithm 6.6
4.   Update the QR factorization of $\bar{H}_m$; i.e.,
5.   Apply $\Omega_i, i = m - k, \ldots, m - 1$, to the mth column of $\bar{H}_m$
6.   Compute the rotation coefficients c_m, s_m by (6.37)
7.   Apply Ω_m to $\bar{H}_m$ and $\bar{g}_m$; i.e., Compute
8. $\gamma_{m+1} := -s_m \gamma_m$
9. $\gamma_m := c_m \gamma_m$
10. $h_{mm} := c_m h_{mm} + s_m h_{m+1,m} \quad \left(= \sqrt{h_{m+1,m}^2 + h_{mm}^2}\right)$
11. $p_m = \left(v_m - \sum_{i=m-k}^{m-1} h_{im} p_i\right) / h_{mm}$
12. $x_m = x_{m-1} + \gamma_m p_m$
13.   If $|\gamma_{m+1}|$ is small enough then Stop
14. EndDo

The above algorithm does not minimize the norm of the residual vector over $x_0 + \mathcal{K}_m$. Rather, it attempts to perform an approximate minimization. Formula (6.41), which is still valid since orthogonality was not used to derive it, also yields the following equality for DQGMRES:

$$\|b - Ax_m\|_2 = \|V_{m+1} \left(\beta e_1 - \bar{H}_m y_m\right)\|_2,$$

where, as before, y_m minimizes the norm $\|\beta e_1 - \bar{H}_m y\|_2$ over all vectors y in $\mathbb{R}^m$. The norm $\|\beta e_1 - \bar{H}_m y\|_2$ is called the quasi-residual norm of the vector $x_0 + V_m y$, which is a member of $x_0 + \mathcal{K}_m$. If the v_i's were orthogonal to each other, then the quasi-residual norm and the actual residual norm would be identical and QGMRES would be equivalent to GMRES; i.e., the residual norm would be minimized over all vectors of the form $x_0 + V_m y$. Since only an incomplete orthogonalization is used, then the v_i's are only locally orthogonal and, as a result, only an approximate minimization may be obtained. Now (6.42) is no longer valid since its proof required the orthogonality of the v_i's. However, the following relation will be helpful in understanding the behavior of QGMRES:

$$b - Ax_m = V_{m+1} Q_m^T (\gamma_{m+1} e_{m+1}) \equiv \gamma_{m+1} z_{m+1}. \tag{6.50}$$

The actual residual norm is equal to the quasi-residual norm (i.e., $|\gamma_{m+1}|$) multiplied by the norm of z_{m+1}. The vector z_{m+1} is the last column of $V_{m+1} Q_m^T$, which is no longer a unitary matrix. It turns out that, in practice, $|\gamma_{m+1}|$ remains a reasonably good estimate of the actual residual norm because the v_i's are nearly orthogonal. The following inequality provides an actual upper bound on the residual norm in terms of computable quantities:

$$\|b - Ax_m\| \leq \sqrt{m - k + 1} \, |\gamma_{m+1}|. \tag{6.51}$$

Here, k is to be replaced with m when $m \leq k$. The proof of this inequality is a consequence

6.5. Generalized Minimal Residual Method

of (6.50). If the unit vector $q \equiv Q_m^T e_{m+1}$ has components $\eta_1, \eta_2, \ldots, \eta_{m+1}$, then

$$\|b - Ax_m\|_2 = |\gamma_{m+1}| \, \|V_{m+1} q\|_2$$

$$\leq |\gamma_{m+1}| \left(\left\| \sum_{i=1}^{k+1} \eta_i v_i \right\|_2 + \left\| \sum_{i=k+2}^{m+1} \eta_i v_i \right\|_2 \right)$$

$$\leq |\gamma_{m+1}| \left(\left[\sum_{i=1}^{k+1} \eta_i^2 \right]^{1/2} + \sum_{i=k+2}^{m+1} |\eta_i| \, \|v_i\|_2 \right)$$

$$\leq |\gamma_{m+1}| \left(\left[\sum_{i=1}^{k+1} \eta_i^2 \right]^{1/2} + \sqrt{m-k} \left[\sum_{i=k+2}^{m+1} \eta_i^2 \right]^{1/2} \right).$$

The orthogonality of the first $k+1$ vectors v_i was used and the last term comes from using the Cauchy–Schwarz inequality. The desired inequality follows from using the Cauchy–Schwarz inequality again in the form

$$1 \cdot a + \sqrt{m-k} \cdot b \leq \sqrt{m-k+1} \sqrt{a^2 + b^2}$$

and from the fact that the vector q is of norm unity. Thus, using $|\gamma_{m+1}|$ as a residual estimate, we would make an error of a factor of $\sqrt{m-k+1}$ at most. In general, this is an overestimate and $|\gamma_{m+1}|$ tends to give an adequate estimate for the residual norm.

It is also interesting to observe that, with a little bit more arithmetic, the exact residual vector and norm can be obtained. This is based on the observation that, according to (6.50), the residual vector is γ_{m+1} times the vector z_{m+1}, which is the last column of the matrix

$$Z_{m+1} \equiv V_{m+1} Q_m^T. \tag{6.52}$$

It is an easy exercise to see that this last column can be updated from v_{m+1} and z_m. Indeed, assuming that all the matrices related to the rotation are of size $(m+1) \times (m+1)$, the last row of Q_{m-1} is the $(m+1)$th row of the identity, so we can write

$$Z_{m+1} = [V_m, v_{m+1}] Q_{m-1}^T \Omega_m^T$$
$$= [Z_m, v_{m+1}] \Omega_m^T.$$

The result is that

$$z_{m+1} = -s_m z_m + c_m v_{m+1}. \tag{6.53}$$

The z_i's can be updated at the cost of one extra vector in memory and $4n$ operations at each step. The norm of z_{m+1} can be computed at the cost of $2n$ operations and the exact residual norm for the current approximate solution can then be obtained by multiplying this norm by $|\gamma_{m+1}|$.

Because this is a little expensive, it may be preferred to just *correct* the estimate provided by γ_{m+1} by exploiting the above recurrence relation:

$$\|z_{m+1}\|_2 \leq |s_m| \|z_m\|_2 + |c_m|.$$

If $\zeta_m \equiv \|z_m\|_2$, then the following recurrence relation holds:

$$\zeta_{m+1} \leq |s_m| \zeta_m + |c_m|. \tag{6.54}$$

The above relation is inexpensive to update, yet provides an upper bound that is sharper than (6.51); see Exercise 25.

Equation (6.53) shows an interesting relation between two successive residual vectors:

$$\begin{aligned} r_m &= \gamma_{m+1} z_{m+1} \\ &= \gamma_{m+1}[-s_m z_m + c_m v_{m+1}] \\ &= s_m^2 r_{m-1} + c_m \gamma_{m+1} v_{m+1}. \end{aligned} \tag{6.55}$$

This exploits the fact that $\gamma_{m+1} = -s_m \gamma_m$ and $r_j = \gamma_{j+1} z_{j+1}$.

Relating the DQGMRES and FOM residuals may provide some useful insight. We will denote by the superscript I all quantities related to IOM (or DIOM). For example, the mth iterate in IOM is denoted by x_m^I and its residual vector will be $r_m^I = b - Ax_m^I$. It is already known that the IOM residual is a scaled version of the vector v_{m+1} obtained by the incomplete Arnoldi process. To be more accurate, the following equality holds:

$$r_m^I = -h_{m+1,m} e_m^T y_m v_{m+1} = -h_{m+1,m} \frac{\gamma_m}{h_{mm}^{(m-1)}} v_{m+1} = \frac{h_{m+1,m}}{s_m h_{mm}^{(m-1)}} \gamma_{m+1} v_{m+1}.$$

The next relation is then obtained by observing that $h_{m+1,m}/h_{mm}^{(m)} = \tan \theta_m$. Hence,

$$\gamma_{m+1} v_{m+1} = c_m r_m^I, \tag{6.56}$$

from which it follows that

$$\rho_m^Q = |c_m| \rho_m, \tag{6.57}$$

where $\rho_m = \|r_m^I\|_2$ is the actual residual norm of the mth IOM iterate. As an important consequence of (6.56), note that (6.55) becomes

$$r_m = s_m^2 r_{m-1} + c_m^2 r_m^I. \tag{6.58}$$

Example 6.3. Table 6.3 shows the results of applying the DQGMRES algorithm with no preconditioning to three of the test problems described in Section 3.7.

Matrix	Iters	Kflops	Residual	Error
F2DA	98	7216	0.36E−02	0.13E−03
F3D	75	22798	0.64E−03	0.32E−03
ORS	300	24138	0.13E+02	0.25E−02

Table 6.3. *A test run of DQGMRES with no preconditioning.*

See Example 6.1 for the meaning of the column headers in the table. In this test the number k of directions in the recurrence is $k = 10$.

There exist several other ways to relate the quasi-minimal residual (QMR) norm to the actual MR norm provided by GMRES. The following result was proved by Freund and Nachtigal [136] for the QMR algorithm to be seen in the next chapter.

6.5. Generalized Minimal Residual Method

Theorem 6.11. *Assume that V_{m+1}, the Arnoldi basis associated with DQGMRES, is of full rank. Let r_m^Q and r_m^G be the residual norms obtained after m steps of the DQGMRES and GMRES algorithms, respectively. Then*

$$\|r_m^Q\|_2 \leq \kappa_2(V_{m+1})\|r_m^G\|_2. \tag{6.59}$$

Proof. Consider the subset of $\mathcal{K}_{m+1}$ defined by

$$\mathcal{R} = \{r : r = V_{m+1}t;\ t = \beta e_1 - \bar{H}_m y;\ y \in \mathbb{C}^m\}.$$

Denote by y_m the minimizer of $\|\beta e_1 - \bar{H}_m y\|_2$ over y and let $t_m = \beta e_1 - \bar{H}_m y_m$, $r_m = V_{m+1}t_m \equiv r_m^Q$. By assumption, V_{m+1} is of full rank and there is an $(m+1) \times (m+1)$ nonsingular matrix S such that $W_{m+1} = V_{m+1}S$ is unitary. Then, for any member of $\mathcal{R}$,

$$r = W_{m+1}S^{-1}t, \quad t = SW_{m+1}^H r,$$

and, in particular,

$$\|r_m\|_2 \leq \|S^{-1}\|_2 \|t_m\|_2. \tag{6.60}$$

Now $\|t_m\|_2$ is the minimum of the 2-norm of $\beta e_1 - \bar{H}_m y$ over all y's and, therefore,

$$\begin{aligned}
\|t_m\|_2 &= \|SW_{m+1}^H r_m\| \leq \|SW_{m+1}^H r\|_2 \quad \forall\, r \in \mathcal{R} \\
&\leq \|S\|_2 \|r\|_2 \quad \forall\, r \in \mathcal{R} \\
&\leq \|S\|_2 \|r^G\|_2. \tag{6.61}
\end{aligned}$$

The result follows from (6.60), (6.61), and the fact that $\kappa_2(V_{m+1}) = \kappa_2(S)$. $\square$

6.5.7 Relations Between FOM and GMRES

If the last row of the least-squares system in (6.44) is deleted, instead of the one in (6.46), i.e., before the last rotation Ω_6 is applied, the same approximate solution as FOM would result. Indeed, this would correspond to solving the system $H_m y = \beta e_1$ using the QR factorization. As a practical consequence a single subroutine can be written to handle both cases. This observation can also be helpful in understanding the relationships between the two algorithms.

In what follows the FOM and GMRES iterates are denoted by the superscripts F and G, respectively. The residual norm achieved at step j will be denoted by ρ_j^F for FOM and ρ_j^G for GMRES. An important consequence of (6.47) is that

$$\rho_m^G = |s_m|\rho_{m-1}^G,$$

which leads to the following equality:

$$\rho_m^G = |s_1 s_2 \cdots s_m|\beta. \tag{6.62}$$

Note that formulas (6.37) yield nonnegative s_i's, so the absolute values are not required. They are left only for generality.

Define $\bar{H}_m^{(k)}$ to be the matrix resulting from applying the first k rotations to $\bar{H}_m$ and, similarly, let $\bar{g}_m^{(k)}$ be the vector resulting from applying the first k rotations to the right-hand side βe_1. As usual $H_m^{(k)}$ is the matrix $\bar{H}_m^{(k)}$ with its last row deleted and $g_m^{(k)}$ the vector of size m obtained by removing the last component of $\bar{g}_m^{(k)}$. By formula (6.18), the residual obtained from the Arnoldi process is given by $\|r_m^F\|_2 = \|b - Ax_m^F\|_2 = h_{m+1,m}|e_m^T y_m|$. In addition, $y_m = H_m^{-1}(\beta e_1)$ can be obtained by back solving $H_m^{(m-1)} y = g^{(m-1)}$. Therefore, its last component is $e_m^T g_m^{(m-1)}/h_{mm}^{(m-1)}$. Hence,

$$\rho_m^F = h_{m+1,m}|e_m^T H_m^{-1}(\beta e_1)| = h_{m+1,m}\left|\frac{e_m^T g^{(m-1)}}{h_{mm}^{(m-1)}}\right|.$$

As before, let γ_m denote the last component of $\bar{g}_{m-1}$ or, equivalently, the mth component of $g^{(m-1)}$, i.e., before the last rotation Ω_m is applied (see (6.36) and (6.44) for an illustration). Then

$$|e_m^T g^{(m-1)}| = |s_{m-1}\gamma_m| = \cdots = |s_1 s_2 \cdots s_{m-1}\beta|.$$

Therefore, the above expression for ρ_m^F becomes

$$\rho_m^F = \frac{h_{m+1,m}}{|h_{mm}^{(m-1)}|}|s_1 s_2 \cdots s_{m-1}\beta|.$$

Now expressions (6.37) show that $h_{m+1,m}/|h_{mm}^{(m-1)}|$ is the tangent of the angle defining the mth rotation and, therefore,

$$\rho_m^F = \frac{|s_m|}{|c_m|}|s_1 s_2 \cdots s_{m-1}\beta|.$$

A comparison with (6.62) yields a revealing relation between the residuals of the FOM and GMRES algorithms; namely,

$$\rho_m^F = \frac{1}{|c_m|}\rho_m^G.$$

The trigonometric relation $1/\cos^2\theta = 1 + \tan^2\theta$ can now be invoked: $1/|c_m| = [1 + (h_{m+1,m}/h_{mm}^{(m-1)})^2]^{1/2}$. These results are summarized in the following proposition (see Brown [66]).

Proposition 6.12. *Assume that m steps of the Arnoldi process have been taken and that H_m is nonsingular. Let $\xi \equiv (Q_{m-1}\bar{H}_m)_{mm}$ and $h \equiv h_{m+1,m}$. Then the residual norms produced by the FOM and the GMRES algorithms are related by the equality*

$$\rho_m^F = \frac{1}{|c_m|}\rho_m^G = \rho_m^G\sqrt{1 + \frac{h^2}{\xi^2}}. \tag{6.63}$$

It is also possible to prove the above result by exploiting the relation (6.75); see Exercise 14.

The term ξ in the expression (6.63) is not readily available, which results in an expression that is hard to interpret practically. Another, somewhat more explicit, expression can be obtained from simply relating c_m to two consecutive residual norms of GMRES. The

6.5. Generalized Minimal Residual Method

next result, shown by Cullum and Greenbaum [92], follows immediately from the above proposition and the relation $|s_m| = \rho_m^G/\rho_{m-1}^G$, which is a consequence of (6.62).

Proposition 6.13. *Assume that m steps of the Arnoldi process have been taken and that H_m is nonsingular. Then the residual norms produced by the FOM and the GMRES algorithms are related by the equality*

$$\rho_m^F = \frac{\rho_m^G}{\sqrt{1 - (\rho_m^G/\rho_{m-1}^G)^2}}. \tag{6.64}$$

The above relation can be recast as

$$\frac{1}{(\rho_m^F)^2} + \frac{1}{(\rho_{m-1}^G)^2} = \frac{1}{(\rho_m^G)^2}. \tag{6.65}$$

Consider now these equations for $m, m-1, \ldots, 1$:

$$\frac{1}{(\rho_m^F)^2} + \frac{1}{(\rho_{m-1}^G)^2} = \frac{1}{(\rho_m^G)^2},$$

$$\frac{1}{(\rho_{m-1}^F)^2} + \frac{1}{(\rho_{m-2}^G)^2} = \frac{1}{(\rho_{m-1}^G)^2},$$

$$\cdots = \cdots$$

$$\frac{1}{(\rho_1^F)^2} + \frac{1}{(\rho_0^G)^2} = \frac{1}{(\rho_1^G)^2}.$$

Note that ρ_0^G is simply the initial residual norm and can as well be denoted by ρ_0^F. Summing the above equations yields

$$\sum_{i=0}^{m} \frac{1}{(\rho_i^F)^2} = \frac{1}{(\rho_m^G)^2}. \tag{6.66}$$

Corollary 6.14. *The residual norms produced by the FOM and the GMRES algorithms are related by the equality*

$$\rho_m^G = \frac{1}{\sqrt{\sum_{i=0}^{m} (1/\rho_i^F)^2}}. \tag{6.67}$$

The above relation establishes rigorously the intuitive fact that FOM and GMRES are never too far away from each other. It is clear that $\rho_m^G \leq \rho_m^F$. On the other hand, let $\rho_{m_*}^F$ be the smallest residual norms achieved in the first m steps of FOM. Then

$$\frac{1}{(\rho_m^G)^2} = \sum_{i=0}^{m} \frac{1}{(\rho_i^F)^2} \leq \frac{m}{(\rho_{m_*}^F)^2}.$$

An immediate consequence of this inequality is the following proposition.

Proposition 6.15. *Assume that m steps of GMRES and FOM are taken (steps in FOM with a singular H_m are skipped). Let $\rho^F_{m_*}$ be the smallest residual norm achieved by FOM in the first m steps. Then the following inequalities hold:*

$$\rho^G_m \leq \rho^F_{m_*} \leq \sqrt{m}\, \rho^G_m. \tag{6.68}$$

We now establish another interesting relation between the FOM and GMRES iterates, which will be exploited in the next chapter. A general lemma is first shown regarding the solutions of the triangular systems

$$R_m y_m = g_m$$

obtained from applying successive rotations to the Hessenberg matrices $\bar{H}_m$. As was stated before, the only difference between the y_m-vectors obtained in GMRES and Arnoldi is that the last rotation Ω_m is omitted in FOM. In other words, the R_m-matrix for the two methods differs only in its (m, m) entry, while the right-hand sides differ only in their last components.

Lemma 6.16. *Let $\tilde{R}_m$ be the $m \times m$ upper part of the matrix $Q_{m-1}\bar{H}_m$ and, as before, let R_m be the $m \times m$ upper part of the matrix $Q_m \bar{H}_m$. Similarly, let $\tilde{g}_m$ be the vector of the first m components of $Q_{m-1}(\beta e_1)$ and let g_m be the vector of the first m components of $Q_m(\beta e_1)$. Define*

$$\tilde{y}_m = \tilde{R}_m^{-1} \tilde{g}_m, \quad y_m = R_m^{-1} g_m$$

as the y-vectors obtained for m-dimensional FOM and GMRES methods, respectively. Then

$$y_m - \binom{y_{m-1}}{0} = c_m^2 \left(\tilde{y}_m - \binom{y_{m-1}}{0} \right), \tag{6.69}$$

in which c_m is the cosine used in the mth rotation Ω_m, as defined by (6.37).

Proof. The following relation holds:

$$R_m = \begin{pmatrix} R_{m-1} & z_m \\ 0 & \xi_m \end{pmatrix}, \quad \tilde{R}_m = \begin{pmatrix} R_{m-1} & z_m \\ 0 & \tilde{\xi}_m \end{pmatrix}.$$

Similarly, for the right-hand sides,

$$g_m = \begin{pmatrix} g_{m-1} \\ \gamma_m \end{pmatrix}, \quad \tilde{g}_m = \begin{pmatrix} g_{m-1} \\ \tilde{\gamma}_m \end{pmatrix},$$

with

$$\gamma_m = c_m \tilde{\gamma}_m. \tag{6.70}$$

Denoting by λ the scalar $\sqrt{\tilde{\xi}_m^2 + h_{m+1,m}^2}$ and using the definitions of s_m and c_m, we obtain

$$\xi_m = c_m \tilde{\xi}_m + s_m h_{m+1,m} = \frac{\tilde{\xi}_m^2}{\lambda} + \frac{h_{m+1,m}^2}{\lambda} = \lambda = \frac{\tilde{\xi}_m}{c_m}. \tag{6.71}$$

6.5. Generalized Minimal Residual Method

Now

$$y_m = R_m^{-1} g_m = \begin{pmatrix} R_{m-1}^{-1} & -\frac{1}{\xi_m} R_{m-1}^{-1} z_m \\ 0 & \frac{1}{\xi_m} \end{pmatrix} \begin{pmatrix} g_{m-1} \\ \gamma_m \end{pmatrix}, \quad (6.72)$$

which, upon observing that $R_{m-1}^{-1} g_{m-1} = y_{m-1}$, yields

$$y_m - \begin{pmatrix} y_{m-1} \\ 0 \end{pmatrix} = \frac{\gamma_m}{\xi_m} \begin{pmatrix} -R_{m-1}^{-1} z_m \\ 1 \end{pmatrix}. \quad (6.73)$$

Replacing y_m, ξ_m, γ_m with $\tilde{y}_m, \tilde{\xi}_m, \tilde{\gamma}_m$, respectively, in (6.72), a relation similar to (6.73) would result except that γ_m/ξ_m is replaced by $\tilde{\gamma}_m/\tilde{\xi}_m$, which, by (6.70) and (6.71), satisfies the relation

$$\frac{\gamma_m}{\xi_m} = c_m^2 \frac{\tilde{\gamma}_m}{\tilde{\xi}_m}.$$

The result follows immediately. □

If the FOM and GMRES iterates are denoted by the superscripts F and G, respectively, then the relation (6.69) implies that

$$x_m^G - x_{m-1}^G = c_m^2 \left(x_m^F - x_{m-1}^G \right)$$

or

$$x_m^G = s_m^2 x_{m-1}^G + c_m^2 x_m^F. \quad (6.74)$$

This leads to the following relation for the residual vectors obtained by the two methods:

$$r_m^G = s_m^2 r_{m-1}^G + c_m^2 r_m^F, \quad (6.75)$$

which indicates that, in general, the two residual vectors will evolve hand in hand. In particular, if $c_m = 0$, then GMRES will not progress at step m, a phenomenon known as stagnation. However, in this situation, according to the definitions (6.37) of the rotations, $h_{mm}^{(m-1)} = 0$, which implies that H_m is singular and, therefore, x_m^F is not defined. In fact, the reverse of this, a result due to Brown [66], is also true and is stated without proof in the following proposition.

Proposition 6.17. *If at any given step m, the GMRES iterates make no progress, i.e., if $x_m^G = x_{m-1}^G$, then H_m is singular and x_m^F is not defined. Conversely, if H_m is singular at step m, i.e., if FOM breaks down at step m, and A is nonsingular, then $x_m^G = x_{m-1}^G$.*

Note also that the use of Lemma 6.16 is not restricted to the GMRES–FOM pair. Some of the iterative methods defined in this chapter and the next involve a least-squares problem of the form (6.30). In such cases, the iterates of the least-squares method and those of the orthogonal residual (Galerkin) method will be related by the same equation.

6.5.8 Residual Smoothing

The previous section established strong relations between the GMRES and FOM iterates. In fact it is possible to derive the GMRES iterates from the FOM iterates by simply exploiting the relations (6.74)–(6.75), which we now rewrite as

$$x_m^G = x_{m-1}^G + c_m^2 (x_m^F - x_{m-1}^G), \qquad r_m^G = r_{m-1}^G + c_m^2 (r_m^F - r_{m-1}^G).$$

The above relations are instances of a class of algorithms derived by residual smoothing, which define a new sequence of iterates, denoted here by x_i^S, from an original sequence, denoted by x_i^O. The residual vectors of the two sequences are denoted by r_i^S and r_i^O, respectively. The new sequences are as follows:

$$x_m^S = x_{m-1}^S + \eta_m(x_m^O - x_{m-1}^S), \qquad r_m^S = r_{m-1}^S + \eta_m(r_m^O - r_{m-1}^S).$$

The parameter η_m is selected so as to make the residual r_m behave better than the original one, in the sense that large variations in the residual are dampened. In *minimal residual smoothing* (MRS) the η_m's are selected to minimize the new residual norm $\|r_m^S\|_2$. This is in essence a minimal residual projection method in the direction $x_m^O - x_{m-1}^S$ and it is achieved by selecting η_m so that the new residual r_m^S is orthogonal to $A(x_m^O - x_{m-1}^S) = -(r_m^O - r_{m-1}^S)$. Thus,

$$\eta_m = -\frac{(r_{m-1}^S, r_m^O - r_{m-1}^S)}{\|r_m^O - r_{m-1}^S\|_2^2},$$

resulting in the following algorithm.

ALGORITHM 6.14. MRS

1. $x_0^S = x_0^O, r_0^S = r_0^O$
2. For $m = 1, \ldots,$ until convergence, Do
3.   Compute x_m^O and r_m^O
4. $\eta_m = -\left(r_{m-1}^S, r_m^O - r_{m-1}^S\right)/\|r_m^O - r_{m-1}^S\|_2^2$
5. $x_m^S = x_{m-1}^S + \eta_m(x_m^O - x_{m-1}^S)$
6. $r_m^S = r_{m-1}^S + \eta_m(r_m^O - r_{m-1}^S)$
7. EndDo

In the situation when r_m^O is orthogonal to r_{m-1}^S, it is possible to show that the same relation as (6.64) (or equivalently (6.65)) is satisfied. This result is due to Weiss [306].

Lemma 6.18. *If r_m^O is orthogonal to r_{m-1}^S at each step $m \geq 1$, then the residual norms satisfy the relation*

$$\frac{1}{\|r_m^S\|_2^2} = \frac{1}{\|r_{m-1}^S\|_2^2} + \frac{1}{\|r_m^O\|_2^2} \tag{6.76}$$

and the coefficient η_m is given by

$$\eta_m = \frac{\|r_{m-1}^S\|_2^2}{\|r_{m-1}^S\|_2^2 + \|r_m^O\|_2^2}. \tag{6.77}$$

Proof. Since $r_m^O \perp r_{m-1}^S$, it follows that $(r_{m-1}^S, r_m^O - r_{m-1}^S) = -(r_{m-1}^S, r_{m-1}^S)$ and $\|r_m^O - r_{m-1}^S\|_2^2 = \|r_m^O\|_2^2 + \|r_{m-1}^S\|_2^2$. This shows (6.77). The orthogonality of r_m^S with $r_m^O - r_{m-1}^S$ implies that

$$\|r_m^S\|_2^2 = \|r_{m-1}^S\|_2^2 - \eta_m^2 \|r_m^O - r_{m-1}^S\|_2^2$$

$$= \|r_{m-1}^S\|_2^2 - \frac{\|r_{m-1}^S\|_2^4}{\|r_m^O\|_2^2 + \|r_{m-1}^S\|_2^2} = \frac{\|r_{m-1}^S\|_2^2 \|r_m^O\|_2^2}{\|r_m^O\|_2^2 + \|r_{m-1}^S\|_2^2}.$$

The result (6.76) follows by inverting both sides of the above equality. $\square$

6.5. Generalized Minimal Residual Method

The assumptions of the lemma are satisfied in particular when the residual vectors of the original algorithm are orthogonal to each other, as is the case for FOM. This can be shown by simple induction, using the fact that each new r_k^S is ultimately a linear combination of the r_i^O's for $i \leq k$. Since the relation established by the lemma is identical to that of the GMRES algorithm, it follows that the residual norms are identical, since they both satisfy (6.67). Because the approximate solutions belong to the same subspace and GMRES minimizes the residual norm, it is clear that *the resulting approximate solutions are identical*.

This result can also be shown in a different way. Induction shows that the vectors $p_j = x_j^O - x_{j-1}^S$ are $A^T A$-orthogonal; i.e., $(A p_i, A p_j) = 0$ for $i \neq j$. Then a lemma (Lemma 6.21) to be seen in Section 6.9 can be exploited to prove the same result. This is left as an exercise (Exercise 26).

The computation of the scalar η_m is likely to be subject to large errors when the residuals become small because there may be a substantial difference between the actual residual and the one computed recursively by the algorithm. One remedy is to explicitly use the directions p_j mentioned above. The formulas and the actual update will then follow closely those seen in Section 5.3, with v replaced by p_m and w by $A p_m$. Specifically, lines 5 and 6 of Algorithm 6.14 are replaced with $x_m^S = x_{m-1}^S + \eta_m p_m$ and $r_m^S = r_{m-1}^S - \eta_m A p_m$, respectively, while the coefficient η_m is computed as $\eta_m = (r_{m-1}^S, A p_m)/(A p_m, A p_m)$. Details are omitted but may be found in [323].

Lemma 6.18 yields the following equality, in which ρ_j denotes $\|r_j^O\|_2$ and τ_j denotes $\|r_j^S\|_2$:

$$r_m^S = \frac{\rho_m^2}{\rho_m^2 + \tau_{m-1}^2} r_{m-1}^S + \frac{\tau_{m-1}^2}{\rho_m^2 + \tau_{m-1}^2} r_m^O$$

$$= \frac{1}{\frac{1}{\tau_{m-1}^2} + \frac{1}{\rho_m^2}} \left[\frac{1}{\tau_{m-1}^2} r_{m-1}^S + \frac{1}{\rho_m^2} r_m^O \right]. \tag{6.78}$$

Summing up the relations (6.76) yields an expression similar to (6.66):

$$\frac{1}{\tau_j^2} = \sum_{i=0}^{j} \frac{1}{\rho_j^2}.$$

Combining this with (6.78) and using induction immediately yields the following expression, which holds under the assumptions of Lemma 6.18:

$$r_m^S = \frac{1}{\sum_{j=1}^m \frac{1}{\rho_j^2}} \sum_{j=1}^m \frac{r_j^O}{\rho_j^2}. \tag{6.79}$$

The smoothed residual is a convex combination of the residuals obtained by the original algorithm (e.g., FOM). The coefficient used for a given residual is inversely proportional to its norm squared. In other words, residual smoothing will tend to dampen wide variations in the original residuals. If the original residual moves up very high, then (6.79) or (6.78) shows that the next smoothed residual will tend to stagnate. If, on the other hand, the original residual decreases very rapidly at a given step, then the smoothed residual will be close to it. In other words, stagnation and fast convergence of the smoothed residual go hand in hand with poor convergence and fast convergence, respectively, of the original scheme.

Consider now the general situation when the residual vectors do not satisfy the conditions of Lemma 6.18. In this case the above results are not valid. However, one may ask whether or not it is possible to still select the η_m's by an alternative formula such that the nice relation (6.79) remains valid. A hint at a possible answer is provided by a look at (6.76) and (6.77). These are the only relations used to establish (6.79). This suggests computing the η_m's recursively as follows:

$$\eta_m = \frac{\tau_{m-1}^2}{\tau_{m-1}^2 + \rho_m^2}, \qquad \frac{1}{\tau_m^2} = \frac{1}{\tau_{m-1}^2} + \frac{1}{\rho_m^2}.$$

It is only when the conditions of Lemma 6.18 are satisfied that τ_k is the norm of the residuals r_k^S. What is important is that the relation (6.78) can be shown to be valid with $\|r_j^S\|_2^2$ replaced by τ_j^2. As a result, the same induction proof as before will show that (6.79) is also valid. Replacing the η_m of Algorithm 6.14 with the one defined above gives rise to an algorithm known as *quasi-minimal residual smoothing*, or QMRS.

It can easily be shown that, when applied to the sequence of iterates produced by IOM/DIOM, then QMRS will, in exact arithmetic, yield the same sequence as QGMRES/DQGMRES. The key relations are (6.47), which is still valid, and (6.57). The quasi-residual norm that replaces the actual norm r_m^S is now equal to γ_{m+1}. By (6.57), the cosine used in the mth step of QGMRES/DQGMRES satisfies $|c_m| = |\gamma_{m+1}|/\rho_m^2$. By formula (6.47), $|s_m| = |\gamma_{m+1}/\gamma_m|$. Writing $c_m^2 + s_m^2 = 1$ and using the notation $\tau_m \equiv \gamma_{m+1}$ yields

$$\frac{\gamma_{m+1}^2}{\gamma_m^2} + \frac{\gamma_{m+1}^2}{\rho_m^2} = 1 \quad \rightarrow \quad \frac{1}{\tau_m^2} = \frac{1}{\tau_{m-1}^2} + \frac{1}{\rho_m^2}.$$

This, along with the relation (6.58), shows that the residual vectors computed by QGMRES/DQGMRES obey the exact same recurrence as those defined by QMRS. QMRS is related to several other algorithms to be described in the next sections.

6.5.9 GMRES for Complex Systems

Complex linear systems arise in many important applications. Perhaps the best known of these is in solving Maxwell's equations in electromagnetics. The most common method used in this context gives rise to large dense and complex linear systems.

Adapting GMRES to the complex case is fairly straightforward. The guiding principle is that the method should minimize the 2-norm of the residual on the affine Krylov subspace. This is achieved by Algorithm 6.9, in which the inner products are now the complex inner products in $\mathbb{C}^n$, defined by (1.3) of Chapter 1. The only part requiring some attention is the solution of the least-squares problem in line 12 of the algorithm or, rather, the practical implementation using Givens rotations outlined in Section 6.5.3.

Complex Givens rotations are defined in the following way instead of (6.34):

$$\Omega_i = \begin{pmatrix} 1 & & & & & & \\ & \ddots & & & & & \\ & & 1 & & & & \\ & & & \bar{c}_i & \bar{s}_i & & \\ & & & -s_i & c_i & & \\ & & & & & 1 & \\ & & & & & & \ddots \\ & & & & & & & 1 \end{pmatrix} \begin{matrix} \\ \\ \\ \leftarrow \text{row } i \\ \leftarrow \text{row } i+1 \\ \\ \\ \end{matrix} \qquad (6.80)$$

6.6. The Symmetric Lanczos Algorithm

with $|c_i|^2 + |s_i|^2 = 1$. The description of Section 6.5.3 can be followed in the same way. In particular, the sine and cosine defined in (6.37) for the Givens rotation matrix at step i are given by

$$s_i = \frac{h_{i+1,i}}{\sqrt{|h_{ii}^{(i-1)}|^2 + h_{i+1,i}^2}}, \quad c_i = \frac{h_{ii}^{(i-1)}}{\sqrt{|h_{ii}^{(i-1)}|^2 + h_{i+1,i}^2}}. \tag{6.81}$$

A slight simplification takes place when applying the successive rotations. Since $h_{j+1,j}$ is the 2-norm of a vector, it is real (nonnegative), and so s_i is also a real (nonnegative) number while, in general, c_i is complex. The rest of the development is identical, though it is worth noting that the diagonal entries of the upper triangular matrix R are (nonnegative) real and that the scalars γ_i are real.

6.6 The Symmetric Lanczos Algorithm

The symmetric Lanczos algorithm can be viewed as a simplification of Arnoldi's method for the particular case when the matrix is symmetric. When A is symmetric, then the Hessenberg matrix H_m becomes symmetric tridiagonal. This leads to a three-term recurrence in the Arnoldi process and short-term recurrences for solution algorithms such as FOM and GMRES. On the theoretical side, there is also much more that can be said on the resulting approximation in the symmetric case.

6.6.1 The Algorithm

To introduce the Lanczos algorithm we begin by making the observation stated in the following theorem.

Theorem 6.19. *Assume that Arnoldi's method is applied to a real symmetric matrix A. Then the coefficients h_{ij} generated by the algorithm are such that*

$$h_{ij} = 0 \quad \text{for} \quad 1 \leq i < j - 1, \tag{6.82}$$

$$h_{j,j+1} = h_{j+1,j}, \quad j = 1, 2, \ldots, m. \tag{6.83}$$

In other words, the matrix H_m obtained from the Arnoldi process is tridiagonal and symmetric.

Proof. The proof is an immediate consequence of the fact that $H_m = V_m^T A V_m$ is a symmetric matrix that is also a Hessenberg matrix by construction. Therefore, H_m must be a symmetric tridiagonal matrix. □

The standard notation used to describe the Lanczos algorithm is obtained by setting

$$\alpha_j \equiv h_{jj}, \quad \beta_j \equiv h_{j-1,j},$$

and, if T_m denotes the resulting H_m-matrix, it is of the form

$$T_m = \begin{pmatrix} \alpha_1 & \beta_2 & & & \\ \beta_2 & \alpha_2 & \beta_3 & & \\ & & \ddots & & \\ & & \beta_{m-1} & \alpha_{m-1} & \beta_m \\ & & & \beta_m & \alpha_m \end{pmatrix}. \tag{6.84}$$

This leads to the following form of the MGS variant of Arnoldi's method, namely, Algorithm 6.2.

ALGORITHM 6.15. The Lanczos Algorithm

1. *Choose an initial vector v_1 of 2-norm unity. Set $\beta_1 \equiv 0$, $v_0 \equiv 0$*
2. *For $j = 1, 2, \ldots, m$, Do*
3. $w_j := Av_j - \beta_j v_{j-1}$
4. $\alpha_j := (w_j, v_j)$
5. $w_j := w_j - \alpha_j v_j$
6. $\beta_{j+1} := \|w_j\|_2$. If $\beta_{j+1} = 0$ then Stop
7. $v_{j+1} := w_j / \beta_{j+1}$
8. *EndDo*

It is rather surprising that the above simple algorithm guarantees, at least in exact arithmetic, that the vectors v_i, $i = 1, 2, \ldots$, are orthogonal. In reality, exact orthogonality of these vectors is only observed at the beginning of the process. At some point the v_i's start losing their global orthogonality rapidly. There has been much research devoted to finding ways to either recover the orthogonality or to at least diminish its effects by *partial* or *selective* orthogonalization; see Parlett [223].

The major practical differences with Arnoldi's method are that the matrix H_m is tridiagonal and, more importantly, that only three vectors must be saved, unless some form of reorthogonalization is employed.

6.6.2 Relation to Orthogonal Polynomials

In exact arithmetic, the core of Algorithm 6.15 is a relation of the form

$$\beta_{j+1} v_{j+1} = Av_j - \alpha_j v_j - \beta_j v_{j-1}.$$

This three-term recurrence relation is reminiscent of the standard three-term recurrence relation of orthogonal polynomials. In fact, there is indeed a strong relationship between the Lanczos algorithm and orthogonal polynomials. To begin, recall that, if the grade of v_1 is not less than m, then the subspace $\mathcal{K}_m$ is of dimension m and consists of all vectors of the form $q(A)v_1$, where q is a polynomial with degree(q) not exceeding $m - 1$. In this case there is even an isomorphism between $\mathcal{K}_m$ and $\mathbb{P}_{m-1}$, the space of polynomials of degree not exceeding $m - 1$, which is defined by

$$q \in \mathbb{P}_{m-1} \to x = q(A)v_1 \in \mathcal{K}_m.$$

Moreover, we can consider that the subspace $\mathbb{P}_{m-1}$ is provided with the inner product

$$\langle p, q \rangle_{v_1} = (p(A)v_1, q(A)v_1). \tag{6.85}$$

This is indeed a nondegenerate bilinear form under the assumption that m does not exceed μ, the grade of v_1. Now observe that the vectors v_i are of the form

$$v_i = q_{i-1}(A)v_1$$

and the orthogonality of the v_i's translates into the orthogonality of the polynomials with respect to the inner product (6.85).

It is known that real orthogonal polynomials satisfy a three-term recurrence. Moreover, the Lanczos procedure is nothing but the Stieltjes algorithm (see, for example, Gautschi [141]) for computing a sequence of orthogonal polynomials with respect to the inner product (6.85). It is known [245] that the characteristic polynomial of the tridiagonal matrix produced by the Lanczos algorithm minimizes the norm $\|\cdot\|_{v_1}$ over the monic polynomials. The recurrence relation between the characteristic polynomials of tridiagonal matrices also shows that the Lanczos recurrence computes the sequence of vectors $p_{T_m}(A)v_1$, where p_{T_m} is the characteristic polynomial of T_m.

6.7 The Conjugate Gradient Algorithm

The conjugate gradient (CG) algorithm is one of the best known iterative techniques for solving sparse symmetric positive definite (SPD) linear systems. Described in one sentence, the method is a realization of an orthogonal projection technique onto the Krylov subspace $\mathcal{K}_m(r_0, A)$, where r_0 is the initial residual. It is therefore mathematically equivalent to FOM. However, because A is symmetric, some simplifications resulting from the three-term Lanczos recurrence will lead to more elegant algorithms.

6.7.1 Derivation and Theory

We first derive the analogue of FOM, or Arnoldi's method, for the case when A is symmetric. Given an initial guess x_0 to the linear system $Ax = b$ and the Lanczos vectors $v_i, i = 1, \ldots, m$, together with the tridiagonal matrix T_m, the approximate solution obtained from an orthogonal projection method onto $\mathcal{K}_m$ is given by

$$x_m = x_0 + V_m y_m, \quad y_m = T_m^{-1}(\beta e_1). \tag{6.86}$$

ALGORITHM 6.16. Lanczos Method for Linear Systems

1. Compute $r_0 = b - Ax_0$, $\beta := \|r_0\|_2$, and $v_1 := r_0/\beta$
2. For $j = 1, 2, \ldots, m$, Do
3. $w_j = Av_j - \beta_j v_{j-1}$ (If $j = 1$ set $\beta_1 v_0 \equiv 0$)
4. $\alpha_j = (w_j, v_j)$
5. $w_j := w_j - \alpha_j v_j$
6. $\beta_{j+1} = \|w_j\|_2$. If $\beta_{j+1} = 0$ set $m := j$ and go to 9
7. $v_{j+1} = w_j/\beta_{j+1}$
8. EndDo
9. Set $T_m = \text{tridiag}(\beta_i, \alpha_i, \beta_{i+1})$ and $V_m = [v_1, \ldots, v_m]$
10. Compute $y_m = T_m^{-1}(\beta e_1)$ and $x_m = x_0 + V_m y_m$

Many of the results obtained from Arnoldi's method for linear systems are still valid. For example, the residual vector of the approximate solution x_m is such that

$$b - Ax_m = -\beta_{m+1} e_m^T y_m v_{m+1}. \tag{6.87}$$

The CG algorithm can be derived from the Lanczos algorithm in the same way DIOM was derived from IOM. In fact, the CG algorithm can be viewed as a variation of DIOM(2) for the case when A is symmetric. We will follow the same steps as with DIOM, except that the notation will be simplified whenever possible.

First write the LU factorization of T_m as $T_m = L_m U_m$. The matrix L_m is unit lower bidiagonal and U_m is upper bidiagonal. Thus, the factorization of T_m is of the form

$$T_m = \begin{pmatrix} 1 & & & & \\ \lambda_2 & 1 & & & \\ & \lambda_3 & 1 & & \\ & & \lambda_4 & 1 & \\ & & & \lambda_5 & 1 \end{pmatrix} \times \begin{pmatrix} \eta_1 & \beta_2 & & & \\ & \eta_2 & \beta_3 & & \\ & & \eta_3 & \beta_4 & \\ & & & \eta_4 & \beta_5 \\ & & & & \eta_5 \end{pmatrix}.$$

The approximate solution is then given by

$$x_m = x_0 + V_m U_m^{-1} L_m^{-1} (\beta e_1).$$

Letting

$$P_m \equiv V_m U_m^{-1}$$

and

$$z_m = L_m^{-1} \beta e_1,$$

then

$$x_m = x_0 + P_m z_m.$$

As for DIOM, p_m, the last column of P_m, can be computed from the previous p_i's and v_m by the simple update

$$p_m = \eta_m^{-1}[v_m - \beta_m p_{m-1}].$$

Note that β_m is a scalar computed from the Lanczos algorithm, while η_m results from the mth Gaussian elimination step on the tridiagonal matrix; i.e.,

$$\lambda_m = \frac{\beta_m}{\eta_{m-1}}, \tag{6.88}$$

$$\eta_m = \alpha_m - \lambda_m \beta_m. \tag{6.89}$$

In addition, following again what has been shown for DIOM,

$$z_m = \begin{bmatrix} z_{m-1} \\ \zeta_m \end{bmatrix},$$

in which $\zeta_m = -\lambda_m \zeta_{m-1}$. As a result, x_m can be updated at each step as

$$x_m = x_{m-1} + \zeta_m p_m,$$

where p_m is defined above.

This gives the following algorithm, which we call the direct version of the Lanczos algorithm (or D-Lanczos) for linear systems.

6.7. The Conjugate Gradient Algorithm

ALGORITHM 6.17. D-Lanczos

1. Compute $r_0 = b - Ax_0$, $\zeta_1 := \beta := \|r_0\|_2$, and $v_1 := r_0/\beta$
2. Set $\lambda_1 = \beta_1 = 0$, $p_0 = 0$
3. For $m = 1, 2, \ldots$, until convergence, Do
4.   Compute $w := Av_m - \beta_m v_{m-1}$ and $\alpha_m = (w, v_m)$
5.   If $m > 1$ then compute $\lambda_m = \frac{\beta_m}{\eta_{m-1}}$ and $\zeta_m = -\lambda_m \zeta_{m-1}$
6. $\eta_m = \alpha_m - \lambda_m \beta_m$
7. $p_m = \eta_m^{-1}(v_m - \beta_m p_{m-1})$
8. $x_m = x_{m-1} + \zeta_m p_m$
9.   If x_m has converged then Stop
10. $w := w - \alpha_m v_m$
11. $\beta_{m+1} = \|w\|_2$, $v_{m+1} = w/\beta_{m+1}$
12. EndDo

This algorithm computes the solution of the tridiagonal system $T_m y_m = \beta e_1$ progressively by using Gaussian elimination without pivoting. However, as was explained for DIOM, partial pivoting can also be implemented at the cost of having to keep an extra vector. In fact, Gaussian elimination with partial pivoting is sufficient to ensure stability for tridiagonal systems. The more complex LQ factorization has also been exploited in this context and gave rise to an algorithm known as SYMMLQ [222].

The two Algorithms 6.16 and 6.17 are mathematically equivalent; that is, they deliver the same approximate solution if they are both executable. However, since Gaussian elimination without pivoting is being used implicitly to solve the tridiagonal system $T_m y = \beta e_1$, the direct version may be more prone to breakdowns.

Observe that the residual vector for this algorithm is in the direction of v_{m+1} due to (6.87). Therefore, the residual vectors are orthogonal to each other, as in FOM. Likewise, the vectors p_i are A-orthogonal, or *conjugate*. These results are established in the next proposition.

Proposition 6.20. *Let $r_m = b - Ax_m$, $m = 0, 1, \ldots$, be the residual vectors produced by the Lanczos and the D-Lanczos algorithms (6.16 and 6.17) and p_m, $m = 0, 1, \ldots$, the auxiliary vectors produced by Algorithm 6.17. Then*

1. *each residual vector r_m is such that $r_m = \sigma_m v_{m+1}$, where σ_m is a certain scalar. As a result, the residual vectors are orthogonal to each other.*
2. *the auxiliary vectors p_i form an A-conjugate set; i.e., $(Ap_i, p_j) = 0$ for $i \neq j$.*

Proof. The first part of the proposition is an immediate consequence of the relation (6.87). For the second part, it must be proved that $P_m^T A P_m$ is a diagonal matrix, where $P_m = V_m U_m^{-1}$. This follows from

$$P_m^T A P_m = U_m^{-T} V_m^T A V_m U_m^{-1}$$
$$= U_m^{-T} T_m U_m^{-1}$$
$$= U_m^{-T} L_m.$$

Now observe that $U_m^{-T} L_m$ is a lower triangular matrix that is also symmetric, since it is equal to the symmetric matrix $P_m^T A P_m$. Therefore, it must be a diagonal matrix. □

A consequence of the above proposition is that a version of the algorithm can be derived by imposing the orthogonality and conjugacy conditions. This gives the CG algorithm, which we now derive. The vector x_{j+1} can be expressed as

$$x_{j+1} = x_j + \alpha_j p_j. \tag{6.90}$$

In order to conform with standard notation used in the literature to describe the algorithm, the indexing of the p-vectors now begins at zero instead of one, as has been done so far. This explains the difference between the above formula and formula (6.21) used for DIOM. Now the residual vectors must satisfy the recurrence

$$r_{j+1} = r_j - \alpha_j A p_j. \tag{6.91}$$

If the r_j's are to be orthogonal, then it is necessary that $(r_j - \alpha_j A p_j, r_j) = 0$ and, as a result,

$$\alpha_j = \frac{(r_j, r_j)}{(A p_j, r_j)}. \tag{6.92}$$

Also, it is known that the next search direction p_{j+1} is a linear combination of r_{j+1} and p_j and, after rescaling the p-vectors appropriately, it follows that

$$p_{j+1} = r_{j+1} + \beta_j p_j. \tag{6.93}$$

Thus, a first consequence of the above relation is that

$$(A p_j, r_j) = (A p_j, p_j - \beta_{j-1} p_{j-1}) = (A p_j, p_j),$$

because $A p_j$ is orthogonal to p_{j-1}. Then (6.92) becomes $\alpha_j = (r_j, r_j)/(A p_j, p_j)$. In addition, writing that p_{j+1} as defined by (6.93) is orthogonal to $A p_j$ yields

$$\beta_j = -\frac{(r_{j+1}, A p_j)}{(p_j, A p_j)}.$$

Note that, from (6.91),

$$A p_j = -\frac{1}{\alpha_j}(r_{j+1} - r_j) \tag{6.94}$$

and, therefore,

$$\beta_j = \frac{1}{\alpha_j} \frac{(r_{j+1}, (r_{j+1} - r_j))}{(A p_j, p_j)} = \frac{(r_{j+1}, r_{j+1})}{(r_j, r_j)}.$$

Putting these relations together gives the following algorithm.

ALGORITHM 6.18. CG

1. Compute $r_0 := b - A x_0$, $p_0 := r_0$
2. For $j = 0, 1, \ldots$, until convergence, Do
3. $\alpha_j := (r_j, r_j)/(A p_j, p_j)$
4. $x_{j+1} := x_j + \alpha_j p_j$
5. $r_{j+1} := r_j - \alpha_j A p_j$
6. $\beta_j := (r_{j+1}, r_{j+1})/(r_j, r_j)$
7. $p_{j+1} := r_{j+1} + \beta_j p_j$
8. EndDo

6.7. The Conjugate Gradient Algorithm

It is important to note that the scalars α_j, β_j in this algorithm are different from those of the Lanczos algorithm. The vectors p_j are multiples of the p_j's of Algorithm 6.17. In terms of storage, in addition to the matrix A, four vectors (x, p, Ap, and r) must be saved in Algorithm 6.18, versus five vectors (v_m, v_{m-1}, w, p, and x) in Algorithm 6.17.

6.7.2 Alternative Formulations

Algorithm 6.18 is the best known formulation of the CG algorithm. There are, however, several alternative formulations. Here, only one such formulation is shown, which can be derived once again from the Lanczos algorithm.

The residual polynomial $r_m(t)$ associated with the mth CG iterate must satisfy a three-term recurrence, implied by the three-term recurrence of the Lanczos vectors. Indeed, these vectors are just the scaled versions of the residual vectors. Therefore, we must seek a three-term recurrence of the form

$$r_{m+1}(t) = \rho_m(r_m(t) - \gamma_m t r_m(t)) + \delta_m r_{m-1}(t).$$

In addition, the consistency condition $r_m(0) = 1$ must be maintained for each m, leading to the recurrence

$$r_{m+1}(t) = \rho_m(r_m(t) - \gamma_m t r_m(t)) + (1 - \rho_m) r_{m-1}(t). \quad (6.95)$$

Observe that, if $r_m(0) = 1$ and $r_{m-1}(0) = 1$, then $r_{m+1}(0) = 1$, as desired. Translating the above relation into the sequence of residual vectors yields

$$r_{m+1} = \rho_m(r_m - \gamma_m A r_m) + (1 - \rho_m) r_{m-1}. \quad (6.96)$$

Recall that the vectors r_i are multiples of the Lanczos vectors v_i. As a result, γ_m should be the inverse of the scalar α_m of the Lanczos algorithm. In terms of the r-vectors this means

$$\gamma_m = \frac{(r_m, r_m)}{(A r_m, r_m)}.$$

Equating the inner products of both sides of (6.96) with r_{m-1} and using the orthogonality of the r-vectors gives the following expression for ρ_m, after some algebraic calculations:

$$\rho_m = \left[1 - \frac{\gamma_m}{\gamma_{m-1}} \frac{(r_m, r_m)}{(r_{m-1}, r_{m-1})} \frac{1}{\rho_{m-1}} \right]^{-1}. \quad (6.97)$$

The recurrence relation for the approximate solution vectors can be extracted from the recurrence relation for the residual vectors. This is found by starting from (6.95) and using the relation $r_m(t) = 1 - t s_{m-1}(t)$ between the solution polynomial $s_{m-1}(t)$ and the residual polynomial $r_m(t)$. Thus,

$$s_m(t) = \frac{1 - r_{m+1}(t)}{t}$$

$$= \rho_m \left(\frac{1 - r_m(t)}{t} + \gamma_m r_m(t) \right) + (1 - \rho_m) \frac{1 - r_{m-1}(t)}{t}$$

$$= \rho_m (s_{m-1}(t) + \gamma_m r_m(t)) + (1 - \rho_m) s_{m-2}(t).$$

This gives the recurrence

$$x_{m+1} = \rho_m(x_m + \gamma_m r_m) + (1 - \rho_m)x_{m-1}. \tag{6.98}$$

All that is left for the recurrence to be determined completely is to define the first two iterates. The initial iterate x_0 is given. The first vector should be of the form

$$x_1 = x_0 - \gamma_0 r_0,$$

to ensure that r_1 is orthogonal to r_0. This means that the two-term recurrence can be started with $\rho_0 = 1$ and by setting $x_{-1} \equiv 0$. Putting these relations and definitions together gives the following algorithm.

ALGORITHM 6.19. CG—Three-Term Recurrence Variant

---

1. Compute $r_0 = b - Ax_0$. Set $x_{-1} \equiv 0$ and $\rho_0 = 1$
2. For $j = 0, 1, \ldots$, until convergence, Do
3.   Compute Ar_j and $\gamma_j = \frac{(r_j, r_j)}{(Ar_j, r_j)}$
4.   If $(j > 0)$ compute $\rho_j = \left[1 - \frac{\gamma_j}{\gamma_{j-1}} \frac{(r_j, r_j)}{(r_{j-1}, r_{j-1})} \frac{1}{\rho_{j-1}}\right]^{-1}$
5.   Compute $x_{j+1} = \rho_j(x_j + \gamma_j r_j) + (1 - \rho_j)x_{j-1}$
6.   Compute $r_{j+1} = \rho_j(r_j - \gamma_j Ar_j) + (1 - \rho_j)r_{j-1}$
7. EndDo

This algorithm requires slightly more storage than the standard formulation: In addition to A, the vectors r_j, Ar_j, r_{j-1}, x_j, and x_{j-1} must be kept. It is possible to avoid keeping r_{j-1} by computing the residual r_{j+1} directly as $r_{j+1} = b - Ax_{j+1}$ in line 6 of the algorithm, but this would entail an additional matrix-by-vector product.

6.7.3 Eigenvalue Estimates from the CG Coefficients

Sometimes, it is useful to be able to obtain the tridiagonal matrix T_m related to the underlying Lanczos iteration from the coefficients of the CG Algorithm 6.18. This tridiagonal matrix can provide valuable eigenvalue information on the matrix A. For example, the largest and smallest eigenvalues of the tridiagonal matrix can approximate the smallest and largest eigenvalues of A. This could be used to compute an estimate of the condition number of A, which in turn can help provide estimates of the error norm from the residual norm. Since the Greek letters α and β have been used in both algorithms, notation must be changed. Denote by

$$T_m = \text{tridiag}[\eta_j, \delta_j, \eta_{j+1}]$$

the tridiagonal matrix (6.84) associated with the mth step of the Lanczos algorithm. We must seek expressions for the coefficients η_j, δ_j in terms of the coefficients α_j, β_j obtained from the CG algorithm. The key information regarding the correspondence between the two pairs of coefficients resides in the correspondence between the vectors generated by the two algorithms.

From (6.87) it is known that

$$r_j = \text{scalar} \times v_{j+1}. \tag{6.99}$$

6.7. The Conjugate Gradient Algorithm

As a result,
$$\delta_{j+1} = \frac{(Av_{j+1}, v_{j+1})}{(v_{j+1}, v_{j+1})} = \frac{(Ar_j, r_j)}{(r_j, r_j)}.$$

The denominator (r_j, r_j) is readily available from the coefficients of the CG algorithm, but the numerator (Ar_j, r_j) is not. The relation (6.93) can be exploited to obtain

$$r_j = p_j - \beta_{j-1} p_{j-1}, \tag{6.100}$$

which is then substituted in (Ar_j, r_j) to get

$$(Ar_j, r_j) = \big(A(p_j - \beta_{j-1} p_{j-1}), p_j - \beta_{j-1} p_{j-1}\big).$$

Note that the terms $\beta_{j-1} p_{j-1}$ are defined to be zero when $j = 0$. Because the p-vectors are A-orthogonal,

$$(Ar_j, r_j) = (Ap_j, p_j) + \beta_{j-1}^2 (Ap_{j-1}, p_{j-1}),$$

from which we finally obtain, for $j > 0$,

$$\delta_{j+1} = \frac{(Ap_j, p_j)}{(r_j, r_j)} + \beta_{j-1}^2 \frac{(Ap_{j-1}, p_{j-1})}{(r_j, r_j)} = \frac{1}{\alpha_j} + \frac{\beta_{j-1}}{\alpha_{j-1}}. \tag{6.101}$$

The above expression is only valid for $j > 0$. For $j = 0$, the second term on the right-hand side should be omitted, as was observed above. Therefore, the diagonal elements of T_m are given by

$$\delta_{j+1} = \begin{cases} \frac{1}{\alpha_j} & \text{for } j = 0, \\ \frac{1}{\alpha_j} + \frac{\beta_{j-1}}{\alpha_{j-1}} & \text{for } j > 0. \end{cases} \tag{6.102}$$

Now an expression for the codiagonal elements η_{j+1} is needed. From the definitions in the Lanczos algorithm,

$$\eta_{j+1} = (Av_j, v_{j+1}) = \frac{|(Ar_{j-1}, r_j)|}{\|r_{j-1}\|_2 \|r_j\|_2}.$$

Using (6.100) again and the relation (6.94) as well as orthogonality properties of the CG algorithm, the following sequence of equalities results:

$$(Ar_{j-1}, r_j) = (A(p_{j-1} - \beta_{j-2} p_{j-2}), r_j)$$
$$= (Ap_{j-1}, r_j) - \beta_{j-2}(Ap_{j-2}, r_j)$$
$$= \frac{-1}{\alpha_{j-1}}(r_j - r_{j-1}, r_j) + \frac{\beta_{j-2}}{\alpha_{j-2}}(r_{j-1} - r_{j-2}, r_j)$$
$$= \frac{-1}{\alpha_{j-1}}(r_j, r_j).$$

Therefore,

$$\eta_{j+1} = \frac{1}{\alpha_{j-1}} \frac{(r_j, r_j)}{\|r_{j-1}\|_2 \|r_j\|_2} = \frac{1}{\alpha_{j-1}} \frac{\|r_j\|_2}{\|r_{j-1}\|_2} = \frac{\sqrt{\beta_{j-1}}}{\alpha_{j-1}}.$$

This finally gives the general form of the m-dimensional Lanczos tridiagonal matrix in terms of the CG coefficients:

$$T_m = \begin{pmatrix} \frac{1}{\alpha_0} & \frac{\sqrt{\beta_0}}{\alpha_0} & & & \\ \frac{\sqrt{\beta_0}}{\alpha_0} & \frac{1}{\alpha_1} + \frac{\beta_0}{\alpha_0} & \frac{\sqrt{\beta_1}}{\alpha_1} & & \\ & \ddots & \ddots & & \\ & & & & \frac{\sqrt{\beta_{m-2}}}{\alpha_{m-2}} \\ & & & \frac{\sqrt{\beta_{m-2}}}{\alpha_{m-2}} & \frac{1}{\alpha_{m-1}} + \frac{\beta_{m-2}}{\alpha_{m-2}} \end{pmatrix}. \qquad (6.103)$$

6.8 The Conjugate Residual Method

In the previous section we derived the CG algorithm as a special case of FOM for SPD matrices. Similarly, a new algorithm can be derived from GMRES for the particular case where A is Hermitian. In this case, the residual vectors should be A-orthogonal, i.e., conjugate. In addition, the vectors Ap_i, $i = 0, 1, \ldots$, are orthogonal. When looking for an algorithm with the same structure as CG, but satisfying these conditions, we find the conjugate residual (CR) algorithm. Notice that the residual vectors are now conjugate to each other, hence the name of the algorithm.

ALGORITHM 6.20. CR Algorithm

1. Compute $r_0 := b - Ax_0$, $p_0 := r_0$
2. For $j = 0, 1, \ldots$, until convergence, Do
3. $\quad \alpha_j := (r_j, Ar_j)/(Ap_j, Ap_j)$
4. $\quad x_{j+1} := x_j + \alpha_j p_j$
5. $\quad r_{j+1} := r_j - \alpha_j Ap_j$
6. $\quad \beta_j := (r_{j+1}, Ar_{j+1})/(r_j, Ar_j)$
7. $\quad p_{j+1} := r_{j+1} + \beta_j p_j$
8. $\quad$ Compute $Ap_{j+1} = Ar_{j+1} + \beta_j Ap_j$
9. EndDo

Line 8 in the above algorithm computes Ap_{j+1} from Ar_{j+1} without an additional matrix-by-vector product. Five vectors of storage are needed in addition to the matrix A: x, p, Ap, r, Ar. The algorithm requires one more vector update, i.e., $2n$ more operations than the CG method and one more vector of storage. Since the two methods exhibit typically similar convergence, the CG method is often preferred.

6.9 Generalized Conjugate Residual, ORTHOMIN, and ORTHODIR

All algorithms developed in this chapter are strongly related to, as well as defined by, the choice of a basis of the Krylov subspace. The GMRES algorithm uses an orthonormal basis. In the CG algorithm, the p's are A-orthogonal, i.e., conjugate. In the CR method just described, the Ap_i's are orthogonal; i.e., the p_i's are A^TA-orthogonal. A number of algorithms can be developed using a basis of this form in the nonsymmetric case as well. The main result that is exploited in all these algorithms is the following lemma.

6.9. Generalized Conjugate Residual, ORTHOMIN, and ORTHODIR

Lemma 6.21. *Let $p_0, p_1, \ldots, p_{m-1}$ be a sequence of vectors such that each set $\{p_0, p_1, \ldots, p_{j-1}\}$ for $j \leq m$ is a basis of the Krylov subspace $\mathcal{K}_j(A, r_0)$, which is $A^T A$-orthogonal, i.e., such that*

$$(Ap_i, Ap_k) = 0 \quad \text{for } i \neq k.$$

Then the approximate solution x_m that has the smallest residual norm in the affine space $x_0 + \mathcal{K}_m(A, r_0)$ is given by

$$x_m = x_0 + \sum_{i=0}^{m-1} \frac{(r_0, Ap_i)}{(Ap_i, Ap_i)} p_i. \qquad (6.104)$$

In addition, x_m can be computed from x_{m-1} by

$$x_m = x_{m-1} + \frac{(r_{m-1}, Ap_{m-1})}{(Ap_{m-1}, Ap_{m-1})} p_{m-1}. \qquad (6.105)$$

Proof. The approximate solution and the associated residual vector can be written in the form

$$x_m = x_0 + \sum_{i=0}^{m-1} \alpha_i p_i, \quad r_m = r_0 - \sum_{i=0}^{m-1} \alpha_i Ap_i. \qquad (6.106)$$

According to the optimality result of Proposition 5.3, in order for $\|r_m\|_2$ to be minimum, the orthogonality relations

$$(r_m, Ap_i) = 0, \quad i = 0, \ldots, m-1,$$

must be enforced. Using (6.106) and the orthogonality of the Ap_i's gives immediately

$$\alpha_i = (r_0, Ap_i)/(Ap_i, Ap_i).$$

This proves the first part of the lemma. Assume now that x_{m-1} is known and that x_m must be determined. According to formula (6.104) and the fact that $p_0, \ldots, p_{m-2}$ is a basis of $\mathcal{K}_{m-1}(A, r_0)$, we can write $x_m = x_{m-1} + \alpha_{m-1} p_{m-1}$, with α_{m-1} defined above. Note that, from the second part of (6.106),

$$r_{m-1} = r_0 - \sum_{j=0}^{m-2} \alpha_j Ap_j,$$

so that

$$(r_{m-1}, Ap_{m-1}) = (r_0, Ap_{m-1}) - \sum_{j=0}^{m-2} \alpha_j (Ap_j, Ap_{m-1}) = (r_0, Ap_{m-1}),$$

exploiting, once more, the orthogonality of the vectors Ap_j, $j = 0, \ldots, m-1$. Thus,

$$\alpha_{m-1} = \frac{(r_{m-1}, Ap_{m-1})}{(Ap_{m-1}, Ap_{m-1})},$$

which proves the expression (6.105). $\square$

This lemma opens up many different ways to obtain algorithms that are mathematically equivalent to the full GMRES. The simplest option computes the next basis vector p_{m+1} as a linear combination of the current residual r_m and all previous p_i's. The approximate solution is updated by using (6.105). This is called the generalized CR (GCR) algorithm.

ALGORITHM 6.21. GCR

1. Compute $r_0 = b - Ax_0$. Set $p_0 = r_0$
2. For $j = 0, 1, \ldots$, until convergence, Do
3. $\quad \alpha_j = \frac{(r_j, Ap_j)}{(Ap_j, Ap_j)}$
4. $\quad x_{j+1} = x_j + \alpha_j p_j$
5. $\quad r_{j+1} = r_j - \alpha_j Ap_j$
6. $\quad$ Compute $\beta_{ij} = -\frac{(Ar_{j+1}, Ap_i)}{(Ap_i, Ap_i)}$ for $i = 0, 1, \ldots, j$
7. $\quad p_{j+1} = r_{j+1} + \sum_{i=0}^{j} \beta_{ij} p_i$
8. EndDo

To compute the scalars β_{ij} in the above algorithm, the vector Ar_j and the previous Ap_i's are required. In order to limit the number of matrix-by-vector products per step to one, we can proceed as follows. Follow line 5 with a computation of Ar_{j+1} and then compute Ap_{j+1} after line 7 from the relation

$$Ap_{j+1} = Ar_{j+1} + \sum_{i=0}^{j} \beta_{ij} Ap_i.$$

Both the set of p_i's and the set of Ap_i's need to be saved. This doubles the storage requirement compared to GMRES. The number of arithmetic operations per step is also roughly 50% higher than with GMRES.

The above version of GCR suffers from the same practical limitations as GMRES and FOM. A restarted version called GCR(m) can be trivially defined. Also, a truncation of the orthogonalization of the Ap_i's, similar to IOM, leads to an algorithm known as ORTHOMIN(k). Specifically, lines 6 and 7 of Algorithm 6.21 are replaced with

6a. $\quad$ Compute $\beta_{ij} = -\frac{(Ar_{j+1}, Ap_i)}{(Ap_i, Ap_i)}$ for $i = j - k + 1, \ldots, j$

7a. $\quad p_{j+1} = r_{j+1} + \sum_{i=j-k+1}^{j} \beta_{ij} p_i$

Another class of algorithms is defined by computing the next basis vector p_{j+1} as

$$p_{j+1} = Ap_j + \sum_{i=0}^{j} \beta_{ij} p_i, \tag{6.107}$$

in which, as before, the β_{ij}'s are selected to make the Ap_i's orthogonal; i.e.,

$$\beta_{ij} = -\frac{(A^2 p_j, Ap_i)}{(Ap_i, Ap_i)}.$$

The resulting algorithm is called ORTHODIR [177]. Restarted and truncated versions of ORTHODIR can also be defined.

6.10 The Faber–Manteuffel Theorem

As was seen in Section 6.6, when A is symmetric, the Arnoldi algorithm simplifies to the Lanczos procedure, which is defined through a three-term recurrence. As a consequence, FOM is mathematically equivalent to the CG algorithm in this case. Similarly, the full GMRES algorithm gives rise to the CR algorithm. It is clear that the CG-type algorithms, i.e.,

6.10. The Faber–Manteuffel Theorem

algorithms defined through short-term recurrences, are more desirable than those algorithms that require storing entire sequences of vectors, as in the GMRES process. These algorithms require less memory and fewer operations per step.

Therefore, the question is, *Is it possible to define algorithms based on optimal Krylov subspace projection that give rise to sequences involving short-term recurrences?* An optimal Krylov subspace projection means a technique that minimizes a certain norm of the error, or residual, on the Krylov subspace. Such methods can be defined from the Arnoldi process.

It is sufficient to consider the Arnoldi process. If Arnoldi's algorithm reduces to the s-term incomplete orthogonalization algorithm (Algorithm 6.6 with $k \equiv s$), i.e., if $h_{ij} = 0$ for $i < j - s + 1$, then an $(s - 1)$-term recurrence can be defined for updating the iterates, as was done in Section 6.4.2. Conversely, if the solution is updated as $x_{j+1} = x_j + \alpha_j p_j$ and p_j satisfies a short recurrence, then the residual vectors will satisfy an s-term recurrence; i.e., $h_{ij} = 0$ for $i < j - s + 1$. A similar argument can be used for the (full) GMRES algorithm when it simplifies into DQGMRES. For all purposes, it is therefore sufficient to analyze what happens to the Arnoldi process (or FOM). We start by generalizing the CG result in a simple way, by considering the DIOM algorithm.

Proposition 6.22. *Let A be a matrix such that*

$$A^T v \in \mathcal{K}_s(A, v)$$

for any vector v. Then DIOM(s) is mathematically equivalent to the FOM algorithm.

Proof. The assumption is equivalent to the statement that, for any v, there is a polynomial q_v of degree not exceeding $s - 1$ such that $A^T v = q_v(A)v$. In the Arnoldi process, the scalars h_{ij} are defined by $h_{ij} = (Av_j, v_i)$ and, therefore,

$$h_{ij} = (Av_j, v_i) = (v_j, A^T v_i) = (v_j, q_{v_i}(A)v_i). \tag{6.108}$$

Since q_{v_i} is a polynomial of degree not exceeding $s - 1$, the vector $q_{v_i}(A)v_i$ is a linear combination of the vectors $v_i, v_{i+1}, \ldots, v_{i+s-1}$. As a result, if $i < j - s + 1$, then $h_{ij} = 0$. Therefore, DIOM(s) will give the same approximate solution as FOM. ∎

In particular, if

$$A^T = q(A),$$

where q is a polynomial of degree not exceeding $s - 1$, then the result holds. However, since $Aq(A) = q(A)A$ for any polynomial q, the above relation implies that A is normal. As it turns out, the reverse is also true. That is, when A is normal, then there is a polynomial of degree not exceeding $n - 1$ such that $A^H = q(A)$. Proving this is easy because, when $A = Q\Lambda Q^H$ where Q is unitary and Λ diagonal, then $q(A) = Qq(\Lambda)Q^H$. Choosing the polynomial q so that

$$q(\lambda_j) = \bar{\lambda}_j, \ j = 1, \ldots, n,$$

results in $q(A) = Q\bar{\Lambda} Q^H = A^H$, as desired.

Let $\nu(A)$ be the smallest degree of all polynomials q such that $A^H = q(A)$. Then the following lemma due to Faber and Manteuffel [121] states an interesting relation between s and $\nu(A)$.

Lemma 6.23. *A nonsingular matrix A is such that*

$$A^H v \in \mathcal{K}_s(A, v)$$

for every vector v iff A is normal and $\nu(A) \leq s - 1$.

Proof. The sufficient condition is trivially true. To prove the necessary condition, assume that, for any vector v, $A^H v = q_v(A)v$, where q_v is a polynomial of degree not exceeding $s - 1$. Then it is easily seen that any eigenvector of A is also an eigenvector of A^H. Therefore, from Lemma 1.15, A is normal. Let μ be the degree of the minimal polynomial for A. Then, since A has μ distinct eigenvalues, there is a polynomial q of degree $\mu - 1$ such that $q(\lambda_i) = \bar{\lambda}_i$ for $i = 1, \ldots, \mu$. According to the above argument, for this q, it holds that $A^H = q(A)$ and therefore $\nu(A) \leq \mu - 1$. Now it must be shown that $\mu \leq s$. Let w be a (nonzero) vector whose grade is μ. By assumption, $A^H w \in \mathcal{K}_s(A, w)$. On the other hand, we also have $A^H w = q(A)w$. Since the vectors $w, Aw, \ldots, A^{\mu-1}w$ are linearly independent, $\mu - 1$ must not exceed $s - 1$. Otherwise, two different expressions for $A^H w$ with respect to the basis $w, Aw, \ldots, A^{\mu-1}w$ would result, which would imply that $A^H w = 0$. Since A is nonsingular, then $w = 0$, which is a contradiction. □

Proposition 6.22 gives a sufficient condition for DIOM(s) to be equivalent to FOM. According to Lemma 6.23, this condition is equivalent to A being normal and $\nu(A) \leq s - 1$. Now consider the reverse result. Faber and Manteuffel define CG(s) to be the class of all matrices such that, *for every v_1, it is true that $(Av_j, v_i) = 0$ for all i, j such that $i + s \leq j \leq \mu(v_1) - 1$*. The inner product can be different from the canonical Euclidean dot product. With this definition it is possible to show the following theorem [121], which is stated without proof.

Theorem 6.24. *$A \in CG(s)$ iff the minimal polynomial of A has degree not exceeding s or A is normal and $\nu(A) \leq s - 1$.*

It is interesting to consider the particular case where $\nu(A) \leq 1$, which is the case of the CG method. In fact, it is easy to show that in this case A either has a minimal degree not exceeding 1, is Hermitian, or is of the form

$$A = e^{i\theta}(\rho I + B),$$

where θ and ρ are real and B is skew Hermitian; i.e., $B^H = -B$. Thus, the cases in which DIOM simplifies into an (optimal) algorithm defined from a three-term recurrence are already known. The first is the CG method. The second is a version of the CG algorithm for skew-Hermitian matrices, which can be derived from the Lanczos algorithm in the same way as CG. This algorithm will be seen in Chapter 9.

6.11 Convergence Analysis

The convergence behavior of the different algorithms seen in this chapter can be analyzed by exploiting optimality properties whenever such properties exist. This is the case for the CG and the GMRES algorithms. On the other hand, the nonoptimal algorithms such as FOM, IOM, and QGMRES will be harder to analyze.

6.11. Convergence Analysis

One of the main tools in the analysis of these methods is the use of Chebyshev polynomials. These polynomials are useful both in theory, when studying convergence, and in practice, as a means of accelerating single-vector iterations or projection processes. In the following, real and complex Chebyshev polynomials are discussed separately.

6.11.1 Real Chebyshev Polynomials

The Chebyshev polynomial of the first kind of degree k is defined by

$$C_k(t) = \cos[k \cos^{-1}(t)] \quad \text{for} \quad -1 \leq t \leq 1. \tag{6.109}$$

That this is a polynomial with respect to t can be shown easily by induction from the trigonometric relation

$$\cos[(k+1)\theta] + \cos[(k-1)\theta] = 2\cos\theta \cos k\theta$$

and the fact that $C_1(t) = t$, $C_0(t) = 1$. Incidentally, this also shows the important three-term recurrence relation

$$C_{k+1}(t) = 2tC_k(t) - C_{k-1}(t).$$

The definition (6.109) can be extended to cases where $|t| > 1$ with the help of the following formula:

$$C_k(t) = \cosh[k \cosh^{-1}(t)], \quad |t| \geq 1. \tag{6.110}$$

This is readily seen by passing to complex variables and using the definition $\cos\theta = (e^{i\theta} + e^{-i\theta})/2$. As a result of (6.110), the following expression can be derived:

$$C_k(t) = \frac{1}{2}\left[\left(t + \sqrt{t^2-1}\right)^k + \left(t + \sqrt{t^2-1}\right)^{-k}\right], \tag{6.111}$$

which is valid for $|t| \geq 1$ but can also be extended to the case of $|t| < 1$. The following approximation, valid for large values of k, will sometimes be used:

$$C_k(t) \gtrsim \frac{1}{2}\left(t + \sqrt{t^2-1}\right)^k \quad \text{for} \quad |t| \geq 1. \tag{6.112}$$

In what follows we denote by $\mathbb{P}_k$ the set of all polynomials of degree k. An important result from approximation theory is the following theorem.

Theorem 6.25. *Let $[\alpha, \beta]$ be a nonempty interval in $\mathbb{R}$ and let γ be any real scalar outside the interval $[\alpha, \beta]$. Then the minimum*

$$\min_{p \in \mathbb{P}_k, \, p(\gamma)=1} \max_{t \in [\alpha, \beta]} |p(t)|$$

is reached by the polynomial

$$\hat{C}_k(t) \equiv \frac{C_k\left(1 + 2\frac{t-\beta}{\beta-\alpha}\right)}{C_k\left(1 + 2\frac{\gamma-\beta}{\beta-\alpha}\right)}. \tag{6.113}$$

For a proof, see Cheney [77]. The maximum of C_k for t in $[-1, 1]$ is 1 and a corollary of the above result is

$$\min_{p \in \mathbb{P}_k,\ p(\gamma)=1} \max_{t \in [\alpha,\beta]} |p(t)| = \frac{1}{|C_k(1 + 2\frac{\gamma-\beta}{\beta-\alpha})|} = \frac{1}{|C_k(2\frac{\gamma-\mu}{\beta-\alpha})|},$$

in which $\mu \equiv (\alpha + \beta)/2$ is the middle of the interval. The absolute values in the denominator are needed only when γ is to the left of the interval, i.e., when $\gamma \leq \alpha$. For this case, it may be more convenient to express the best polynomial as

$$\hat{C}_k(t) \equiv \frac{C_k\left(1 + 2\frac{\alpha-t}{\beta-\alpha}\right)}{C_k\left(1 + 2\frac{\alpha-\gamma}{\beta-\alpha}\right)},$$

which is obtained by exchanging the roles of α and β in (6.113).

6.11.2 Complex Chebyshev Polynomials

The standard definition of real Chebyshev polynomials given by (6.109) extends without difficulty to complex variables. First, as was seen before, when t is real and $|t| > 1$, the alternative definition, $C_k(t) = \cosh[k \cosh^{-1}(t)]$, can be used. These definitions can be unified by switching to complex variables and writing

$$C_k(z) = \cosh(k\zeta), \quad \text{where} \quad \cosh(\zeta) = z.$$

Defining the variable $w = e^\zeta$, the above formula is equivalent to

$$C_k(z) = \frac{1}{2}[w^k + w^{-k}], \quad \text{where} \quad z = \frac{1}{2}[w + w^{-1}]. \tag{6.114}$$

The above definition for Chebyshev polynomials will be used in $\mathbb{C}$. Note that the equation $\frac{1}{2}(w + w^{-1}) = z$ has two solutions w that are inverses of each other. As a result, the value of $C_k(z)$ does not depend on which of these solutions is chosen. It can be verified directly that the C_k's defined by the above equations are indeed polynomials in the z variable and that they satisfy the three-term recurrence

$$C_{k+1}(z) = 2zC_k(z) - C_{k-1}(z), \tag{6.115}$$
$$C_0(z) \equiv 1, \quad C_1(z) \equiv z.$$

As is now explained, Chebyshev polynomials are intimately related to ellipses in the complex plane. Let C_ρ be the circle of radius ρ centered at the origin. Then the so-called Joukowski mapping

$$J(w) = \frac{1}{2}[w + w^{-1}]$$

transforms C_ρ into an ellipse centered at the origin, with foci $-1, 1$; semimajor axis $\frac{1}{2}[\rho + \rho^{-1}]$; and semiminor axis $\frac{1}{2}|\rho - \rho^{-1}|$. This is illustrated in Figure 6.2.

There are two circles with the same image by the mapping $J(w)$, one with radius ρ and the other with radius ρ^{-1}. So it is sufficient to consider only those circles with radius $\rho \geq 1$. Note that the case $\rho = 1$ is a degenerate case in which the ellipse $E(0, 1, -1)$ reduces to the interval $[-1, 1]$ traveled through twice.

An important question is whether or not a generalization of the min-max result of Theorem 6.25 holds for the complex case. Here, the maximum of $|p(z)|$ is taken over

6.11. Convergence Analysis

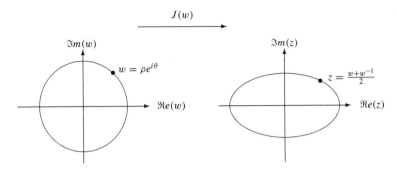

Figure 6.2. *The Joukowski mapping transforms a circle into an ellipse in the complex plane.*

the ellipse boundary and γ is some point not enclosed by the ellipse. The answer to the question is no; Chebyshev polynomials are only optimal in some cases. However, Chebyshev polynomials are asymptotically optimal, which is all that is needed in practice.

To prove the asymptotic optimality, we begin with a lemma due to Zarantonello, which deals with the particular case where the ellipse reduces to a circle. This particular case is important in itself.

Lemma 6.26. *(Zarantonello)* Let $C(0, \rho)$ be a circle centered at the origin with radius ρ and let γ be a point of $\mathbb{C}$ not enclosed by $C(0, \rho)$. Then

$$\min_{p \in \mathbb{P}_k, \ p(\gamma)=1} \max_{z \in C(0,\rho)} |p(z)| = \left(\frac{\rho}{|\gamma|}\right)^k, \tag{6.116}$$

the minimum being achieved for the polynomial $(z/\gamma)^k$.

Proof. See [231] for a proof. □

Note that, by changing variables and shifting and rescaling the polynomial, for any circle centered at c and for any scalar γ such that $|\gamma| > \rho$, the following min-max result holds:

$$\min_{p \in \mathbb{P}_k, \ p(\gamma)=1} \max_{z \in C(c,\rho)} |p(z)| = \left(\frac{\rho}{|\gamma - c|}\right)^k.$$

Now consider the case of an ellipse centered at the origin, with foci $1, -1$ and semimajor axis a, which can be considered as mapped by J from the circle $C(0, \rho)$, with the convention that $\rho \geq 1$. Denote by E_ρ such an ellipse.

Theorem 6.27. *Consider the ellipse E_ρ mapped from $C(0, \rho)$ by the mapping J and let γ be any point in the complex plane not enclosed by it. Then*

$$\frac{\rho^k}{|w_\gamma|^k} \leq \min_{p \in \mathbb{P}_k, \ p(\gamma)=1} \max_{z \in E_\rho} |p(z)| \leq \frac{\rho^k + \rho^{-k}}{|w_\gamma^k + w_\gamma^{-k}|}, \tag{6.117}$$

in which w_γ is the dominant root of the equation $J(w) = \gamma$.

Proof. We start by showing the second inequality. Any polynomial p of degree k satisfying the constraint $p(\gamma) = 1$ can be written as

$$p(z) = \frac{\sum_{j=0}^{k} \xi_j z^j}{\sum_{j=0}^{k} \xi_j \gamma^j}.$$

A point z on the ellipse is transformed by J from a certain w in $C(0, \rho)$. Similarly, let w_γ be one of the two inverse transforms of γ by the mapping, namely, the one with the largest modulus. Then p can be rewritten as

$$p(z) = \frac{\sum_{j=0}^{k} \xi_j (w^j + w^{-j})}{\sum_{j=0}^{k} \xi_j (w_\gamma^j + w_\gamma^{-j})}. \tag{6.118}$$

Consider the particular polynomial obtained by setting $\xi_k = 1$ and $\xi_j = 0$ for $j \neq k$:

$$p^*(z) = \frac{w^k + w^{-k}}{w_\gamma^k + w_\gamma^{-k}},$$

which is a scaled Chebyshev polynomial of the first kind of degree k in the variable z. It is apparent that the maximum modulus of this polynomial is reached in particular when $w = \rho e^{i\theta}$ is real, i.e., when $w = \rho$. Thus,

$$\max_{z \in E_\rho} |p^*(z)| = \frac{\rho^k + \rho^{-k}}{|w_\gamma^k + w_\gamma^{-k}|},$$

which proves the second inequality.

To prove the left inequality, we rewrite (6.118) as

$$p(z) = \left(\frac{w^{-k}}{w_\gamma^{-k}}\right) \frac{\sum_{j=0}^{k} \xi_j (w^{k+j} + w^{k-j})}{\sum_{j=0}^{k} \xi_j (w_\gamma^{k+j} + w_\gamma^{k-j})}$$

and take the modulus of $p(z)$:

$$|p(z)| = \frac{\rho^{-k}}{|w_\gamma|^{-k}} \left| \frac{\sum_{j=0}^{k} \xi_j (w^{k+j} + w^{k-j})}{\sum_{j=0}^{k} \xi_j (w_\gamma^{k+j} + w_\gamma^{k-j})} \right|.$$

The polynomial in w of degree $2k$ inside the large modulus bars on the right-hand side is such that its value at w_γ is one. By Lemma 6.26, the modulus of this polynomial over the circle $C(0, \rho)$ is not less than $(\rho/|w_\gamma|)^{2k}$; i.e., for any polynomial satisfying the constraint $p(\gamma) = 1$,

$$\max_{z \in E_\rho} |p(z)| \geq \frac{\rho^{-k}}{|w_\gamma|^{-k}} \frac{\rho^{2k}}{|w_\gamma|^{2k}} = \frac{\rho^k}{|w_\gamma|^k}.$$

This proves that the minimum over all such polynomials of the maximum modulus on the ellipse E_ρ is not less than $(\rho/|w_\gamma|)^k$. □

The difference between the left and right bounds in (6.117) tends to zero as k increases to infinity. Thus, the important point made by the theorem is that, for large k, the Chebyshev polynomial

$$p^*(z) = \frac{w^k + w^{-k}}{w_\gamma^k + w_\gamma^{-k}}, \quad \text{where} \quad z = \frac{w + w^{-1}}{2},$$

6.11. Convergence Analysis

is close to the optimal polynomial. More specifically, Chebyshev polynomials are *asymptotically* optimal.

For a more general ellipse $E(c, d, a)$ centered at c, with focal distance d and semimajor axis a, a simple change of variables shows that the near-best polynomial is given by

$$\hat{C}_k(z) = \frac{C_k\left(\frac{c-z}{d}\right)}{C_k\left(\frac{c-\gamma}{d}\right)}. \tag{6.119}$$

In addition, by examining the expression $(w^k + w^{-k})/2$ for $w = \rho e^{i\theta}$, it is easily seen that the maximum modulus of $\hat{C}_k(z)$, i.e., the infinity norm of this polynomial over the ellipse, is reached at the point $c + a$ located on the real axis. From this we get

$$\max_{z \in E(c,d,a)} |\hat{C}_k(z)| = \frac{C_k\left(\frac{a}{d}\right)}{|C_k\left(\frac{c-\gamma}{d}\right)|}.$$

Here we point out that d and a both can be purely imaginary (for an example, see Figure 6.3B). In this case a/d is real and the numerator in the above expression is always real. Using the definition of C_k we obtain the following useful expression and approximation:

$$\frac{C_k\left(\frac{a}{d}\right)}{C_k\left(\frac{c-\gamma}{d}\right)} = \frac{\left(\frac{a}{d} + \sqrt{\left(\frac{a}{d}\right)^2 - 1}\right)^k + \left(\frac{a}{d} + \sqrt{\left(\frac{a}{d}\right)^2 - 1}\right)^{-k}}{\left(\frac{c-\gamma}{d} + \sqrt{\left(\frac{c-\gamma}{d}\right)^2 - 1}\right)^k + \left(\frac{c-\gamma}{d} + \sqrt{\left(\frac{c-\gamma}{d}\right)^2 - 1}\right)^{-k}} \tag{6.120}$$

$$\approx \left(\frac{a + \sqrt{a^2 - d^2}}{c - \gamma + \sqrt{(c-\gamma)^2 - d^2}}\right)^k. \tag{6.121}$$

Finally, we note that an alternative and more detailed result has been proven by Fischer and Freund in [127].

6.11.3 Convergence of the CG Algorithm

As usual, $\|x\|_A$ denotes the norm defined by

$$\|x\|_A = (Ax, x)^{1/2}.$$

The following lemma characterizes the approximation obtained from the CG algorithm.

Lemma 6.28. *Let x_m be the approximate solution obtained from the mth step of the CG algorithm and let $d_m = x_* - x_m$, where x_* is the exact solution. Then x_m is of the form*

$$x_m = x_0 + q_m(A)r_0,$$

where q_m is a polynomial of degree $m - 1$ such that

$$\|(I - Aq_m(A))d_0\|_A = \min_{q \in \mathbb{P}_{m-1}} \|(I - Aq(A))d_0\|_A.$$

Proof. This is a consequence of the fact that x_m minimizes the A-norm of the error in the affine subspace $x_0 + \mathcal{K}_m$, a result of Proposition 5.2, and the fact that $\mathcal{K}_m$ is the set of all vectors of the form $x_0 + q(A)r_0$, where q is a polynomial of degree not exceeding $m - 1$. $\square$

From this, the following theorem can be proved.

Theorem 6.29. *Let x_m be the approximate solution obtained at the mth step of the CG algorithm and x_* be the exact solution. Define*

$$\eta = \frac{\lambda_{min}}{\lambda_{max} - \lambda_{min}}. \tag{6.122}$$

Then

$$\|x_* - x_m\|_A \leq \frac{\|x_* - x_0\|_A}{C_m(1 + 2\eta)}, \tag{6.123}$$

in which C_m is the Chebyshev polynomial of degree m of the first kind.

Proof. From the previous lemma, it is known that $\|x_* - x_m\|_A$ minimizes the A-norm of the error over polynomials $r(t)$ that take the value one at 0; i.e.,

$$\|x_* - x_m\|_A = \min_{r \in \mathbb{P}_m,\ r(0)=1} \|r(A)d_0\|_A.$$

If $\lambda_i, i = 1, \ldots, n$, are the eigenvalues of A and $\xi_i, i = 1, \ldots, n$, the components of the initial error d_0 in the eigenbasis, then

$$\|r(A)d_0\|_A^2 = \sum_{i=1}^{n} \lambda_i r(\lambda_i)^2 (\xi_i)^2 \leq \max_i (r(\lambda_i))^2 \|d_0\|_A^2$$

$$\leq \max_{\lambda \in [\lambda_{min}, \lambda_{max}]} (r(\lambda))^2 \|d_0\|_A^2.$$

Therefore,

$$\|x_* - x_m\|_A \leq \min_{r \in \mathbb{P}_m,\ r(0)=1} \max_{\lambda \in [\lambda_{min}, \lambda_{max}]} |r(\lambda)| \|d_0\|_A.$$

The result follows immediately by using the well-known result of Theorem 6.25 from approximation theory. This gives the polynomial r, which minimizes the right-hand side. $\square$

A slightly different formulation of inequality (6.123) can be derived. Using the relation

$$C_m(t) = \frac{1}{2}\left[\left(t + \sqrt{t^2 - 1}\right)^m + \left(t + \sqrt{t^2 - 1}\right)^{-m}\right]$$

$$\geq \frac{1}{2}\left(t + \sqrt{t^2 - 1}\right)^m,$$

then

$$C_m(1 + 2\eta) \geq \frac{1}{2}\left(1 + 2\eta + \sqrt{(1 + 2\eta)^2 - 1}\right)^m$$

$$\geq \frac{1}{2}\left(1 + 2\eta + 2\sqrt{\eta(\eta + 1)}\right)^m.$$

6.11. Convergence Analysis

Now notice that

$$1 + 2\eta + 2\sqrt{\eta(\eta+1)} = \left(\sqrt{\eta} + \sqrt{\eta+1}\right)^2 \tag{6.124}$$

$$= \frac{\left(\sqrt{\lambda_{min}} + \sqrt{\lambda_{max}}\right)^2}{\lambda_{max} - \lambda_{min}} \tag{6.125}$$

$$= \frac{\sqrt{\lambda_{max}} + \sqrt{\lambda_{min}}}{\sqrt{\lambda_{max}} - \sqrt{\lambda_{min}}} \tag{6.126}$$

$$= \frac{\sqrt{\kappa}+1}{\sqrt{\kappa}-1}, \tag{6.127}$$

in which κ is the spectral condition number $\kappa = \lambda_{max}/\lambda_{min}$.

Substituting this in (6.123) yields

$$\|x_* - x_m\|_A \leq 2\left[\frac{\sqrt{\kappa}-1}{\sqrt{\kappa}+1}\right]^m \|x_* - x_0\|_A. \tag{6.128}$$

This bound is similar to that of the steepest descent algorithm except that the condition number of A is now replaced with its square root.

6.11.4 Convergence of GMRES

We begin by stating a *global* convergence result. Recall that a matrix A is called positive definite if its symmetric part $(A + A^T)/2$ is SPD. This is equivalent to the property that $(Ax, x) > 0$ for all nonzero real vectors x.

Theorem 6.30. *If A is a positive definite matrix, then GMRES(m) converges for any $m \geq 1$.*

Proof. This is true because the subspace $\mathcal{K}_m$ contains the initial residual vector at each restart. Since the algorithm minimizes the residual norm in the subspace $\mathcal{K}_m$ at each outer iteration, the residual norm will be reduced by as much as the result of one step of the MR method seen in the previous chapter. Therefore, the inequality (5.15) is satisfied by residual vectors produced after each outer iteration and the method converges. □

Next we wish to establish a result similar to the one for the CG method, which would provide an upper bound on the convergence rate of the GMRES iterates. We begin with a lemma similar to Lemma 6.28.

Lemma 6.31. *Let x_m be the approximate solution obtained from the mth step of the GMRES algorithm and let $r_m = b - Ax_m$. Then x_m is of the form*

$$x_m = x_0 + q_m(A)r_0$$

and

$$\|r_m\|_2 = \|(I - Aq_m(A))r_0\|_2 = \min_{q \in \mathbb{P}_{m-1}} \|(I - Aq(A))r_0\|_2.$$

Proof. This is true because x_m minimizes the 2-norm of the residual in the affine subspace $x_0 + \mathcal{K}_m$, a result of Proposition 5.3, and because of the fact that $\mathcal{K}_m$ is the set of all vectors of the form $x_0 + q(A)r_0$, where q is a polynomial of degree not exceeding $m - 1$. □

Unfortunately, it is not possible to prove a simple result such as Theorem 6.29 unless A is normal.

Proposition 6.32. *Assume that A is a diagonalizable matrix and let $A = X \Lambda X^{-1}$, where $\Lambda = \mathrm{diag}\{\lambda_1, \lambda_2, \ldots, \lambda_n\}$ is the diagonal matrix of the eigenvalues. Define*
$$\epsilon^{(m)} = \min_{p \in \mathbb{P}_m, p(0)=1} \max_{i=1,\ldots,n} |p(\lambda_i)|.$$
Then the residual norm achieved by the mth step of GMRES satisfies the inequality
$$\|r_m\|_2 \leq \kappa_2(X) \epsilon^{(m)} \|r_0\|_2,$$
where $\kappa_2(X) \equiv \|X\|_2 \|X^{-1}\|_2$.

Proof. Let p be any polynomial of degree not exceeding m that satisfies the constraint $p(0) = 1$ and x the vector in $\mathcal{K}_m$ with which it is associated via $b - Ax = p(A) r_0$. Then
$$\|b - Ax\|_2 = \|X p(\Lambda) X^{-1} r_0\|_2 \leq \|X\|_2 \|X^{-1}\|_2 \|r_0\|_2 \|p(\Lambda)\|_2.$$
Since Λ is diagonal, observe that
$$\|p(\Lambda)\|_2 = \max_{i=1,\ldots,n} |p(\lambda_i)|.$$
Since x_m minimizes the residual norm over $x_0 + \mathcal{K}_m$, then, for any consistent polynomial p,
$$\|b - Ax_m\| \leq \|b - Ax\|_2 \leq \|X\|_2 \|X^{-1}\|_2 \|r_0\|_2 \max_{i=1,\ldots,n} |p(\lambda_i)|.$$
Now the polynomial p that minimizes the right-hand side in the above inequality can be used. This yields the desired result:
$$\|b - Ax_m\| \leq \|b - Ax\|_2 \leq \|X\|_2 \|X^{-1}\|_2 \|r_0\|_2 \epsilon^{(m)}. \qquad \square$$

The results of Section 6.11.2 on near-optimal Chebyshev polynomials in the complex plane can now be used to obtain an upper bound for $\epsilon^{(m)}$. Assume that the spectrum of A is contained in an ellipse $E(c, d, a)$ with center c, focal distance d, and semimajor axis a. In addition it is required that the origin lie outside this ellipse. The two possible cases are shown in Figure 6.3. Case (B) corresponds to the situation when d is purely imaginary; i.e., the semimajor axis is aligned with the imaginary axis.

Corollary 6.33. *Let A be a diagonalizable matrix; i.e., let $A = X \Lambda X^{-1}$, where $\Lambda = \mathrm{diag}\{\lambda_1, \lambda_2, \ldots, \lambda_n\}$ is the diagonal matrix of the eigenvalues. Assume that all the eigenvalues of A are located in the ellipse $E(c, d, a)$, which excludes the origin. Then the residual norm achieved at the mth step of GMRES satisfies the inequality*
$$\|r_m\|_2 \leq \kappa_2(X) \frac{C_m\left(\frac{a}{d}\right)}{|C_m\left(\frac{c}{d}\right)|} \|r_0\|_2.$$

Proof. All that is needed is an upper bound for the scalar $\epsilon^{(m)}$ under the assumptions. By definition,
$$\epsilon^{(m)} = \min_{p \in \mathbb{P}_m, p(0)=1} \max_{i=1,\ldots,n} |p(\lambda_i)|$$
$$\leq \min_{p \in \mathbb{P}_m, p(0)=1} \max_{\lambda \in E(c,d,a)} |p(\lambda)|.$$

6.11. Convergence Analysis

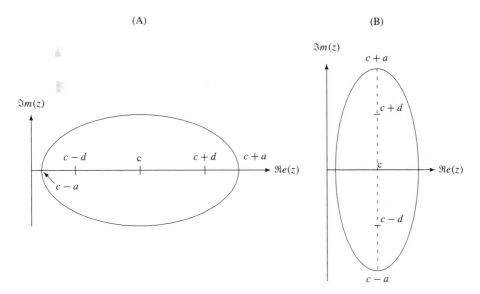

Figure 6.3. *Ellipses containing the spectrum of A. Case* (A): *real d; case* (B): *purely imaginary d.*

The second inequality is due to the fact that the maximum modulus of a complex analytical function is reached on the boundary of the domain. We can now use as a trial polynomial $\hat{C}_m$ defined by (6.119), with $\gamma = 0$:

$$\epsilon^{(m)} \leq \min_{p \in \mathbb{P}_m, p(0)=1} \max_{\lambda \in E(c,d,a)} |p(\lambda)|$$

$$\leq \max_{\lambda \in E(c,d,a)} |\hat{C}_m(\lambda)| = \frac{C_m\left(\frac{a}{d}\right)}{\left|C_m\left(\frac{c}{d}\right)\right|}.$$

This completes the proof. □

An explicit expression for the coefficient $C_m\left(\frac{a}{d}\right) / C_m\left(\frac{c}{d}\right)$ and an approximation are readily obtained from (6.120)–(6.121) by taking $\gamma = 0$:

$$\frac{C_m\left(\frac{a}{d}\right)}{C_m\left(\frac{c}{d}\right)} = \frac{\left(\frac{a}{d} + \sqrt{\left(\frac{a}{d}\right)^2 - 1}\right)^k + \left(\frac{a}{d} + \sqrt{\left(\frac{a}{d}\right)^2 - 1}\right)^{-k}}{\left(\frac{c}{d} + \sqrt{\left(\frac{c}{d}\right)^2 - 1}\right)^k + \left(\frac{c}{d} + \sqrt{\left(\frac{c}{d}\right)^2 - 1}\right)^{-k}}$$

$$\approx \left(\frac{a + \sqrt{a^2 - d^2}}{c + \sqrt{c^2 - d^2}}\right)^k.$$

Since the condition number $\kappa_2(X)$ of the matrix of eigenvectors X is typically not known and can be very large, results of the nature of the corollary are of limited practical interest. They can be useful only when it is known that the matrix is nearly normal, in which case $\kappa_2(X) \approx 1$.

6.12 Block Krylov Methods

In many circumstances, it is desirable to work with a block of vectors instead of a single vector. For example, out-of-core finite element codes are more efficient when they are programmed to exploit the presence of a block of the matrix A in fast memory, as much as possible. This can be achieved by using block generalizations of Krylov subspace methods, for which A always operates on a group of vectors instead of a single vector. We begin by describing a block version of the Arnoldi algorithm.

ALGORITHM 6.22. Block Arnoldi

1. Choose a unitary matrix V_1 of dimension $n \times p$
2. For $j = 1, 2, \ldots, m$, Do
3.   Compute $H_{ij} = V_i^T A V_j$, $i = 1, 2, \ldots, j$
4.   Compute $W_j = A V_j - \sum_{i=1}^{j} V_i H_{ij}$
5.   Compute the QR factorization of W_j: $W_j = V_{j+1} H_{j+1,j}$
6. EndDo

The above algorithm is a straightforward block analogue of Algorithm 6.1. By construction, the blocks generated by the algorithm are orthogonal blocks that are also orthogonal to each other. We use the following notation, denoting by I_k the $k \times k$ identity matrix:

$$U_m = [V_1, V_2, \ldots, V_m],$$
$$H_m = (H_{ij})_{1 \leq i,j \leq m}, \quad H_{ij} \equiv 0 \quad \text{for} \quad i > j+1,$$
$$E_m = \text{matrix of the last } p \text{ columns of } I_n.$$

Then the following analogue of the relation (6.6) is easily proved:

$$AU_m = U_m H_m + V_{m+1} H_{m+1,m} E_m^T. \tag{6.129}$$

Here, the matrix H_m is no longer Hessenberg, but band Hessenberg, meaning that it has p subdiagonals instead of only one. Note that the dimension of the subspace in which the solution is sought is not m but $m \cdot p$.

A second version of the algorithm uses a modified block Gram–Schmidt procedure instead of the simple Gram–Schmidt procedure used above. This leads to a block generalization of Algorithm 6.2, the MGS version of Arnoldi's method.

ALGORITHM 6.23. Block Arnoldi with Block MGS

1. Choose a unitary matrix V_1 of size $n \times p$
2. For $j = 1, 2, \ldots, m$, Do
3.   Compute $W_j := A V_j$
4.   For $i = 1, 2, \ldots, j$ Do
5. $H_{ij} := V_i^T W_j$
6. $W_j := W_j - V_i H_{ij}$
7.   EndDo
8.   Compute the QR decomposition $W_j = V_{j+1} H_{j+1,j}$
9. EndDo

6.12. Block Krylov Methods

Again, in practice the above algorithm is more viable than its predecessor. Finally, a third version, developed by A. Ruhe [235] for the symmetric case (block Lanczos), yields a variant that is quite similar to the original Arnoldi algorithm. Assume that the initial block of p orthonormal vectors $v_1, \ldots, v_p$ is available. The first step of the algorithm is to multiply v_1 by A and orthonormalize the resulting vector w against $v_1, \ldots, v_p$. The resulting vector is defined to be v_{p+1}. In the second step it is v_2 that is multiplied by A and orthonormalized against all available v_i's. Thus, the algorithm works similarly to Algorithm 6.2 except for a delay in the vector that is multiplied by A at each step.

ALGORITHM 6.24. Block Arnoldi—Ruhe's Variant

1. *Choose p initial orthonormal vectors $\{v_i\}_{i=1,\ldots,p}$*
2. *For $j = p, p+1, \ldots, m+p-1$, Do:*
3.   *Set $k := j - p + 1$*
4.   *Compute $w := Av_k$*
5.   *For $i = 1, 2, \ldots, j$, Do*
6. $h_{i,k} := (w, v_i)$
7. $w := w - h_{i,k} v_i$
8.   *EndDo*
9.   *Compute $h_{j+1,k} := \|w\|_2$ and $v_{j+1} := w / h_{j+1,k}$*
10. *EndDo*

Observe that the particular case $p = 1$ coincides with the usual Arnoldi process. Also, the dimension m of the subspace of approximants is no longer restricted to being a multiple of the block size p, as in the previous algorithms. The mathematical equivalence of Algorithms 6.23 and 6.24 when m is a multiple of p is straightforward to show. The advantage of the above formulation is its simplicity. A slight disadvantage is that it gives up some potential parallelism. In the original version, the columns of the matrix AV_j can be computed in parallel, whereas, in the new algorithm, they are computed in sequence. This can be remedied, however, by performing p matrix-by-vector products every p steps.

At the end of the loop consisting of lines 5 through 8 of Algorithm 6.24, the vector w satisfies the relation

$$w = Av_k - \sum_{i=1}^{j} h_{ik} v_i,$$

where k and j are related by $k = j - p + 1$. Line 9 gives $w = h_{j+1,k} v_{j+1}$, which results in

$$Av_k = \sum_{i=1}^{k+p} h_{ik} v_i.$$

As a consequence, the analogue of the relation (6.7) for Algorithm 6.24 is

$$AV_m = V_{m+p} \bar{H}_m. \tag{6.130}$$

As before, for any j the matrix V_j represents the $n \times j$ matrix with columns $v_1, \ldots, v_j$. The matrix $\bar{H}_m$ is now of size $(m+p) \times m$.

Now the block generalizations of FOM and GMRES can be defined in a straightforward way. These block algorithms can solve linear systems with multiple right-hand sides:

$$Ax^{(i)} = b^{(i)}, \quad i = 1, \ldots, p, \qquad (6.131)$$

or, in matrix form,

$$AX = B, \qquad (6.132)$$

where the columns of the $n \times p$ matrices B and X are the $b^{(i)}$'s and $x^{(i)}$'s, respectively. Given an initial block of initial guesses $x_0^{(i)}$ for $i = 1, \ldots, p$, we define R_0 as the block of initial residuals:

$$R_0 \equiv [r_0^{(1)}, r_0^{(2)}, \ldots, r_0^{(p)}],$$

where each column is $r_0^{(i)} = b^{(i)} - Ax_0^{(i)}$. It is preferable to use the unified notation derived from Algorithm 6.24. In this notation, m is not restricted to being a multiple of the block size p and the same notation is used for the v_i's as in the scalar Arnoldi algorithm. Thus, the first step of the block FOM or block GMRES algorithm is to compute the QR factorization of the block of initial residuals:

$$R_0 = [v_1, v_2, \ldots, v_p]R.$$

Here the matrix $[v_1, \ldots, v_p]$ is unitary and R is $p \times p$ upper triangular. This factorization provides the first p vectors of the block Arnoldi basis.

Each of the approximate solutions has the form

$$x^{(i)} = x_0^{(i)} + V_m y^{(i)} \qquad (6.133)$$

and, grouping these approximations $x^{(i)}$ in a block X and the $y^{(i)}$ in a block Y, we can write

$$X = X_0 + V_m Y. \qquad (6.134)$$

It is now possible to imitate what was done for the standard FOM and GMRES algorithms. The only missing link is the vector βe_1 in (6.27), which now becomes a matrix. Let E_1 be the $(m+p) \times p$ matrix whose upper $p \times p$ principal block is an identity matrix. Then the relation (6.130) results in

$$\begin{aligned}
B - AX &= B - A(X_0 + V_m Y) \\
&= R_0 - AV_m Y \\
&= [v_1, \ldots, v_p]R - V_{m+p}\bar{H}_m Y \\
&= V_{m+p}(E_1 R - \bar{H}_m Y). \qquad (6.135)
\end{aligned}$$

The vector

$$\bar{g}^{(i)} \equiv E_1 R e_i$$

is a vector of length $m + p$ whose components are zero except those from 1 to i that are extracted from the ith column of the upper triangular matrix R. The matrix $\bar{H}_m$ is an $(m+p) \times m$ matrix. The block FOM approximation would consist of deleting the last p rows of $\bar{g}^{(i)}$ and $\bar{H}_m$ and solving the resulting system:

$$H_m y^{(i)} = g^{(i)}.$$

The approximate solution $x^{(i)}$ is then computed by (6.133).

The block GMRES approximation $x^{(i)}$ is the unique vector of the form $x_0^{(i)} + V_m y^{(i)}$ that minimizes the 2-norm of the individual columns of the block residual (6.135). Since

6.12. Block Krylov Methods

the column vectors of V_{m+p} are orthonormal, then from (6.135) we get

$$\|b^{(i)} - Ax^{(i)}\|_2 = \|\bar{g}^{(i)} - \bar{H}_m y^{(i)}\|_2. \tag{6.136}$$

To minimize the residual norm, the function on the right-hand side must be minimized over $y^{(i)}$. The resulting least-squares problem is similar to the one encountered for GMRES. The only differences are on the right-hand side and in the fact that the matrix is no longer Hessenberg but band Hessenberg. Rotations can be used in a way similar to the scalar case. However, p rotations are now needed at each new step instead of only one. Thus, if $m = 6$ and $p = 2$, the matrix $\bar{H}_6$ and block right-hand side would be as follows:

$$\bar{H}_8 = \begin{pmatrix} h_{11} & h_{12} & h_{13} & h_{14} & h_{15} & h_{16} \\ h_{21} & h_{22} & h_{23} & h_{24} & h_{25} & h_{26} \\ h_{31} & h_{32} & h_{33} & h_{34} & h_{35} & h_{36} \\ & h_{42} & h_{43} & h_{44} & h_{45} & h_{46} \\ & & h_{53} & h_{54} & h_{55} & h_{56} \\ & & & h_{64} & h_{65} & h_{66} \\ & & & & h_{75} & h_{76} \\ & & & & & h_{86} \end{pmatrix}, \quad \bar{G} = \begin{pmatrix} g_{11} & g_{12} \\ & g_{22} \\ & \\ & \\ & \\ & \\ & \\ & \end{pmatrix}.$$

For each new column generated in the block Arnoldi process, p rotations are required to eliminate the elements $h_{k,j}$ for $k = j + p$ down to $k = j + 1$. This backward order is important. In the above example, a rotation is applied to eliminate $h_{3,1}$ and then a second rotation is used to eliminate the resulting $h_{2,1}$, and similarly for the second, third, ..., step. This complicates programming slightly since two-dimensional arrays must now be used to save the rotations instead of one-dimensional arrays in the scalar case. After the first column of $\bar{H}_m$ is processed, the block of right-hand sides will have a diagonal added under the diagonal of the upper triangular matrix. Specifically, the above two matrices will have the structure

$$\bar{H}_8 = \begin{pmatrix} \star & \star & \star & \star & \star & \star \\ & \star & \star & \star & \star & \star \\ & & \star & \star & \star & \star \\ & & \star & \star & \star & \star \\ & & & \star & \star & \star \\ & & & & \star & \star \\ & & & & & \star \end{pmatrix}, \quad \bar{G} = \begin{pmatrix} \star & \star \\ & \star & \star \\ & & \star \\ & & \\ & & \\ & & \\ & & \end{pmatrix},$$

where a star represents a nonzero element. After all columns are processed, the following least-squares system is obtained:

$$\bar{H}_8 = \begin{pmatrix} \star & \star & \star & \star & \star & \star \\ & \star & \star & \star & \star & \star \\ & & \star & \star & \star & \star \\ & & & \star & \star & \star \\ & & & & \star & \star \\ & & & & & \star \\ \hline & & & & & \\ & & & & & \end{pmatrix}, \quad \bar{G} = \begin{pmatrix} \star & \star \\ \star & \star \\ \star & \star \\ \star & \star \\ \star & \star \\ \star & \star \\ \star & \star \\ & \star \end{pmatrix}.$$

To obtain the least-squares solutions for each right-hand side, ignore anything below the horizontal lines in the above matrices and solve the resulting triangular systems. The residual norm of the ith system for the original problem is the 2-norm of the vector consisting of the components $m+1$ through $m+i$ in the ith column of the above block of right-hand sides.

Generally speaking, the block methods are of great practical value in applications involving linear systems with multiple right-hand sides. However, they are not as well studied from the theoretical point of view. Perhaps one of the reasons is the lack of a convincing analogue for the relationship with orthogonal polynomials established in Section 6.6.2 for the single-vector Lanczos algorithm. The block version of the Lanczos algorithm has not been covered but the generalization is straightforward.

Exercises

1. In the Householder implementation of the Arnoldi algorithm, show the following points of detail:

   a. Q_{j+1} is unitary and its inverse is Q_{j+1}^T.

   b. $Q_{j+1}^T = P_1 P_2 \cdots P_{j+1}$.

   c. $Q_{j+1}^T e_i = v_i$ for $i < j$.

   d. $Q_{j+1} A V_m = V_{m+1}[e_1, e_2, \ldots, e_{j+1}]\bar{H}_m$, where e_i is the ith column of the $n \times n$ identity matrix.

   e. The v_i's are orthonormal.

   f. The vectors $v_1, \ldots, v_j$ are equal to the Arnoldi vectors produced by the Gram–Schmidt version, except possibly for a scaling factor.

2. Rewrite the Householder implementation of the Arnoldi algorithm in more detail. In particular, define precisely the Householder vector w_j used at step j (lines 3–5).

3. Consider the Householder implementation of the Arnoldi algorithm. Give a detailed operation count of the algorithm and compare it with the Gram–Schmidt and MGS algorithms.

4. Consider a variant of the GMRES algorithm in which the Arnoldi process starts with $v_1 = Av_0/\|Av_0\|_2$, where $v_0 \equiv r_0$. The Arnoldi process is performed the same way as before to build an orthonormal system $v_1, v_2, \ldots, v_{m-1}$. Now the approximate solution is expressed in the basis $\{v_0, v_1, \ldots, v_{m-1}\}$.

   a. Show that the least-squares problem that must be solved to obtain the approximate solution is now triangular instead of Hessenberg.

   b. Show that the residual vector r_k is orthogonal to $v_1, v_2, \ldots, v_{k-1}$.

   c. Find a formula that computes the residual norm (without computing the approximate solution) and write the complete algorithm.

5. Derive the basic version of GMRES by using the standard formula (5.7) with $V = V_m$ and $W = AV_m$.

Exercises

6. Analyze the arithmetic cost, i.e., the number of operations, of Algorithms 6.7 and 6.8. Similarly analyze the memory requirements of both algorithms.

7. Derive a version of the DIOM algorithm that includes partial pivoting in the solution of the Hessenberg system.

8. Show how GMRES and FOM will converge on the linear system $Ax = b$ when

$$A = \begin{pmatrix} & & & & 1 \\ 1 & & & & \\ & 1 & & & \\ & & 1 & & \\ & & & 1 & \end{pmatrix}, \quad b = \begin{pmatrix} 1 \\ 0 \\ 0 \\ 0 \\ 0 \end{pmatrix},$$

and $x_0 = 0$.

9. Give a full proof of Proposition 6.17.

10. Let a matrix A have the form
$$A = \begin{pmatrix} I & Y \\ 0 & I \end{pmatrix}.$$
Assume that (full) GMRES is used to solve a linear system with the coefficient matrix A. What is the maximum number of steps that GMRES would require to converge?

11. Let a matrix A have the form
$$A = \begin{pmatrix} I & Y \\ 0 & S \end{pmatrix}.$$
Assume that (full) GMRES is used to solve a linear system with the coefficient matrix A. Let
$$r_0 = \begin{pmatrix} r_0^{(1)} \\ r_0^{(2)} \end{pmatrix}$$
be the initial residual vector. It is assumed that the degree of the minimal polynomial of $r_0^{(2)}$ with respect to S (i.e., its grade) is k. What is the maximum number of steps that GMRES would require to converge for this matrix? [Hint: Evaluate the sum $\sum_{i=0}^{k} \beta_i (A^{i+1} - A^i) r_0$, where $\sum_{i=0}^{k} \beta_i t^i$ is the minimal polynomial of $r_0^{(2)}$ with respect to S.]

12. Let
$$A = \begin{pmatrix} I & Y_2 & & & & \\ & I & Y_3 & & & \\ & & I & \ddots & & \\ & & & I & Y_{k-1} & \\ & & & & I & Y_k \\ & & & & & I \end{pmatrix}.$$

   a. Show that $(I - A)^k = 0$.

   b. Assume that (full) GMRES is used to solve a linear system with the coefficient matrix A. What is the maximum number of steps that GMRES would require to converge?

13. Show that, if H_m is nonsingular, i.e., the FOM iterate x_m^F is defined, and if the GMRES iterate x_m^G is such that $x_m^G = x_m^F$, then $r_m^G = r_m^F = 0$; i.e., both the GMRES and FOM solutions are exact. [Hint: use the relation (6.74) and Proposition 6.17 or Proposition 6.12.]

14. Derive the relation (6.63) from (6.75). [Hint: Use the fact that the vectors on the right-hand side of (6.75) are orthogonal.]

15. In the Householder-GMRES algorithm the approximate solution can be computed by formulas (6.31)–(6.33). What is the exact cost of this alternative (compare memory as well as arithmetic requirements)? How does it compare to the cost of keeping the v_i's?

16. An alternative to formulas (6.31)–(6.33) for accumulating the approximate solution in the Householder-GMRES algorithm without keeping the v_i's is to compute x_m as

$$x_m = x_0 + P_1 P_2 \cdots P_m y,$$

where y is a certain n-dimensional vector to be determined.

   a. What is the vector y for the above formula in order to compute the correct approximate solution x_m? [Hint: Exploit (6.13).]

   b. Write down an alternative to formulas (6.31)–(6.33) derived from this approach.

   c. Compare the cost of this approach to the cost of using (6.31)–(6.33).

17. Obtain the formula (6.97) from (6.96).

18. Show that the determinant of the matrix T_m in (6.103) is given by

$$\det(T_m) = \frac{1}{\prod_{i=0}^{m-1} \alpha_i}.$$

19. The Lanczos algorithm is more closely related to the implementation of Algorithm 6.19 of the CG algorithm. As a result, the Lanczos coefficients δ_{j+1} and η_{j+1} are easier to extract from this algorithm than from Algorithm 6.18. Obtain formulas for these coefficients from the coefficients generated by Algorithm 6.19, as was done in Section 6.7.3 for the standard CG algorithm.

20. What can be said of the Hessenberg matrix H_m when A is skew symmetric? What does this imply for the Arnoldi algorithm?

21. Consider a matrix of the form

$$A = I + \alpha B, \tag{6.137}$$

where B is skew symmetric (real), i.e., such that $B^T = -B$.

   a. Show that $(Ax, x)/(x, x) = 1$ for all nonzero x.

   b. Consider the Arnoldi process for A. Show that the resulting Hessenberg matrix will have the following tridiagonal form:

$$H_m = \begin{pmatrix} 1 & -\eta_2 & & & \\ \eta_2 & 1 & -\eta_3 & & \\ & & \ddots & & \\ & & \eta_{m-1} & 1 & -\eta_m \\ & & & \eta_m & 1 \end{pmatrix}.$$

c. Using the result of part (b), explain why the CG algorithm applied as is to a linear system with the matrix A, which is nonsymmetric, will still yield residual vectors that are orthogonal to each other.

22. Establish the three relations (6.22), (6.23), and (6.24).

23. Show that, if the rotations generated in the course of the GMRES (and DQGMRES) algorithm are such that
$$|c_m| \geq c > 0,$$
then GMRES, DQGMRES, and FOM will all converge.

24. Show the exact expression of the residual vector in the basis $v_1, v_2, \ldots, v_{m+1}$ for either GMRES or DQGMRES. [Hint: A starting point is (6.50).]

25. Prove that the inequality (6.54) is sharper than (6.51), in the sense that $\zeta_{m+1} \leq \sqrt{m-k+1}$ (for $m \geq k$). [Hint: Use the Cauchy–Schwarz inequality on (6.54).]

26. Consider the MRS algorithm (Algorithm 6.14) in the situation when the residual vectors r_j^O of the original sequence are orthogonal to each other. Show that the vectors
$$r_j^O - r_{j-1}^S = -A(x_j^O - x_{j-1}^S)$$
are orthogonal to each other. [Hint: use induction.] Then use Lemma 6.21 to conclude that the iterates of the algorithm are identical to those of ORTHOMIN and GMRES.

27. Consider the complex GMRES algorithm in Section 6.5.9. Show at least two other ways of defining complex Givens rotations (the requirement is that Ω_i be a unitary matrix, i.e., that $\Omega_i^H \Omega_i = I$). Which among the three possible choices give(s) a nonnegative real diagonal for the resulting R_m-matrix?

28. Work out the details of a Householder implementation of the GMRES algorithm for complex linear systems. (The Householder matrices are now of the form $I - 2ww^H$; a part of the practical implementation details is already available for the complex case in Section 6.5.9.)

29. Denote by S_m the unit upper triangular matrix S in the proof of Theorem 6.11, which is obtained from the Gram–Schmidt process (exact arithmetic assumed) applied to the incomplete orthogonalization basis V_m. Show that the Hessenberg matrix $\bar{H}_m^Q$ obtained in the incomplete orthogonalization process is related to the Hessenberg matrix $\bar{H}_m^G$ obtained from the (complete) Arnoldi process by
$$\bar{H}_m^G = S_{m+1}^{-1} \bar{H}_m^Q S_m.$$

Notes and References

The CG method was developed independently and in different forms by Lanczos [196] and Hestenes and Stiefel [167]. The method was essentially viewed as a direct solution technique and was abandoned early on because it did not compare well with other existing techniques. For example, in inexact arithmetic, the method does not terminate in n steps, as is predicted by the theory. This is caused by the severe loss of orthogonality of vector quantities generated by the algorithm. As a result, research on Krylov-type methods remained dormant for over two decades thereafter. This changed in the early 1970s when several researchers discovered that this loss of orthogonality did not prevent convergence. The observations were made and explained for eigenvalue problems [221, 147] as well as linear systems [230].

The early to mid 1980s saw the development of a new class of methods for solving nonsymmetric linear systems [12, 13, 177, 237, 238, 250, 297]. The works of Faber and Manteuffel [121] and Voevodin [298] showed that one could not find optimal methods that, like CG, are based on short-term recurrences. Many of the methods developed are mathematically equivalent in the sense that they realize the same projection process with different implementations.

Lemma 6.16 was proved by Freund [134] in a slightly different form. Proposition 6.12 is due to Brown [66], who proved a number of other theoretical results, including Proposition 6.17. The inequality (6.64), which can be viewed as a reformulation of Brown's result, was proved by Cullum and Greenbaum [92]. This result is equivalent to (6.67), which was shown in a very different way by Zhou and Walker [323].

The Householder version of GMRES is due to Walker [302]. The QGMRES algorithm described in Section 6.5.6 was initially described by Brown and Hindmarsh [67], and the direct version DQGMRES was discussed in [254]. The proof of Theorem 6.11 for DQGMRES is adapted from the result shown in [212] for the QMR algorithm.

Schönauer [259] seems to have been the originator of MRS methods, but Weiss [306] established much of the theory and connections with other techniques. The quasi-minimization extension of these techniques (QMRS) was developed by Zhou and Walker [323].

The nonoptimality of the Chebyshev polynomials on ellipses in the complex plane was established by Fischer and Freund [128]. Prior to this, a 1963 paper by Clayton [86] was believed to have established the optimality for the special case where the ellipse has real foci and γ is real.

Various types of block Krylov methods were considered. In addition to their attraction for solving linear systems with several right-hand sides [242, 266], one of the other motivations for these techniques is that they can also help reduce the effect of the sequential inner products in parallel environments and minimize I/O costs in out-of-core implementations. A block Lanczos algorithm was developed by Underwood [286] for the symmetric eigenvalue problem, while O'Leary discussed a block CG algorithm [214]. The block GMRES algorithm is analyzed by Simoncini and Gallopoulos [265] and in [249]. Besides the straightforward extension presented in Section 6.12, a variation was developed by Jbilou et al., in which a *global* inner product for the blocks was considered instead of the usual scalar inner product for each column [175].

Alternatives to GMRES that require fewer inner products have been proposed by Sadok [255] and Jbilou [174]. Sadok investigated a GMRES-like method based on the Hessenberg algorithm [316], while Jbilou proposed a multidimensional generalization of Gastinel's method seen in Exercise 2 of Chapter 5.

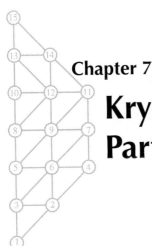

Chapter 7

Krylov Subspace Methods, Part II

The previous chapter considered a number of Krylov subspace methods that relied on some form of orthogonalization of the Krylov vectors in order to compute an approximate solution. This chapter will describe a class of Krylov subspace methods that are instead based on a biorthogonalization algorithm due to Lanczos. These are projection methods that are intrinsically nonorthogonal. They have some appealing properties, but are harder to analyze theoretically.

7.1 Lanczos Biorthogonalization

The Lanczos biorthogonalization algorithm is an extension to nonsymmetric matrices of the symmetric Lanczos algorithm seen in the previous chapter. One such extension, the Arnoldi procedure, has already been seen. However, the nonsymmetric Lanczos algorithm is quite different in concept from Arnoldi's method because it relies on biorthogonal sequences instead of orthogonal sequences.

7.1.1 The Algorithm

The algorithm proposed by Lanczos for nonsymmetric matrices builds a pair of biorthogonal bases for the two subspaces

$$\mathcal{K}_m(A, v_1) = \text{span}\{v_1, Av_1, \ldots, A^{m-1}v_1\}$$

and

$$\mathcal{K}_m(A^T, w_1) = \text{span}\{w_1, A^T w_1, \ldots, (A^T)^{m-1} w_1\}.$$

The algorithm that achieves this is the following.

ALGORITHM 7.1. The Lanczos Biorthogonalization Procedure

1. Choose two vectors v_1, w_1 such that $(v_1, w_1) = 1$
2. Set $\beta_1 = \delta_1 \equiv 0$, $w_0 = v_0 \equiv 0$
3. For $j = 1, 2, \ldots, m$, Do
4. $\quad \alpha_j = (Av_j, w_j)$
5. $\quad \hat{v}_{j+1} = Av_j - \alpha_j v_j - \beta_j v_{j-1}$
6. $\quad \hat{w}_{j+1} = A^T w_j - \alpha_j w_j - \delta_j w_{j-1}$
7. $\quad \delta_{j+1} = |(\hat{v}_{j+1}, \hat{w}_{j+1})|^{1/2}$. If $\delta_{j+1} = 0$ Stop
8. $\quad \beta_{j+1} = (\hat{v}_{j+1}, \hat{w}_{j+1})/\delta_{j+1}$
9. $\quad w_{j+1} = \hat{w}_{j+1}/\beta_{j+1}$
10. $\quad v_{j+1} = \hat{v}_{j+1}/\delta_{j+1}$
11. EndDo

Note that there are numerous ways to choose the scalars $\delta_{j+1}, \beta_{j+1}$ in lines 7 and 8. These two parameters are scaling factors for the two vectors v_{j+1} and w_{j+1} and can be selected in any manner to ensure that $(v_{j+1}, w_{j+1}) = 1$. As a result of lines 9 and 10, it is only necessary to choose two scalars $\beta_{j+1}, \delta_{j+1}$ that satisfy the equality

$$\delta_{j+1}\beta_{j+1} = (\hat{v}_{j+1}, \hat{w}_{j+1}). \tag{7.1}$$

The choice taken in the above algorithm scales the two vectors so that they are divided by two scalars with the same modulus. Both vectors can also be scaled by their 2-norms. In that case, the inner product of v_{j+1} and w_{j+1} is no longer equal to 1 and the algorithm must be modified accordingly; see Exercise 3.

Consider the case where the pair of scalars $\delta_{j+1}, \beta_{j+1}$ is *any pair* that satisfies the relation (7.1). Denote by T_m the tridiagonal matrix

$$T_m = \begin{pmatrix} \alpha_1 & \beta_2 & & & \\ \delta_2 & \alpha_2 & \beta_3 & & \\ & & \ddots & & \\ & & \delta_{m-1} & \alpha_{m-1} & \beta_m \\ & & & \delta_m & \alpha_m \end{pmatrix}. \tag{7.2}$$

If the determinations of $\beta_{j+1}, \delta_{j+1}$ of lines 7 and 8 are used, then the δ_j's are positive and $\beta_j = \pm \delta_j$.

Observe from the algorithm that the vectors v_i belong to $\mathcal{K}_m(A, v_1)$, while the w_j's are in $\mathcal{K}_m(A^T, w_1)$. In fact, the following proposition can be proved.

Proposition 7.1. *If the algorithm does not break down before step m, then the vectors $v_i, i = 1, \ldots, m$, and $w_j, j = 1, \ldots, m$, form a biorthogonal system; i.e.,*

$$(v_j, w_i) = \delta_{ij}, \quad 1 \leq i, j \leq m.$$

Moreover, $\{v_i\}_{i=1,2,\ldots,m}$ is a basis of $\mathcal{K}_m(A, v_1)$, $\{w_i\}_{i=1,2,\ldots,m}$ is a basis of $\mathcal{K}_m(A^T, w_1)$, and the following relations hold:

$$AV_m = V_m T_m + \delta_{m+1} v_{m+1} e_m^T, \tag{7.3}$$
$$A^T W_m = W_m T_m^T + \beta_{m+1} w_{m+1} e_m^T, \tag{7.4}$$
$$W_m^T A V_m = T_m. \tag{7.5}$$

7.1. Lanczos Biorthogonalization

Proof. The biorthogonality of the vectors v_i, w_i will be shown by induction. By assumption $(v_1, w_1) = 1$. Assume now that the vectors $v_1, \ldots, v_j$ and $w_1, \ldots, w_j$ are biorthogonal. Let us prove that the vectors $v_1, \ldots, v_{j+1}$ and $w_1, \ldots, w_{j+1}$ are biorthogonal.

First, we show that $(v_{j+1}, w_i) = 0$ for $i \leq j$. When $i = j$, then

$$(v_{j+1}, w_j) = \delta_{j+1}^{-1}[(Av_j, w_j) - \alpha_j(v_j, w_j) - \beta_j(v_{j-1}, w_j)].$$

The last inner product in the above expression vanishes by the induction hypothesis. The two other terms cancel each other by the definition of α_j and the fact that $(v_j, w_j) = 1$. Consider now the inner product (v_{j+1}, w_i) with $i < j$:

$$\begin{aligned}(v_{j+1}, w_i) &= \delta_{j+1}^{-1}[(Av_j, w_i) - \alpha_j(v_j, w_i) - \beta_j(v_{j-1}, w_i)] \\ &= \delta_{j+1}^{-1}[(v_j, A^T w_i) - \beta_j(v_{j-1}, w_i)] \\ &= \delta_{j+1}^{-1}[(v_j, \beta_{i+1}w_{i+1} + \alpha_i w_i + \delta_i w_{i-1}) - \beta_j(v_{j-1}, w_i)].\end{aligned}$$

For $i < j - 1$, all of the inner products in the above expression vanish by the induction hypothesis. For $i = j - 1$, the inner product is

$$\begin{aligned}(v_{j+1}, w_{j-1}) &= \delta_{j+1}^{-1}[(v_j, \beta_j w_j + \alpha_{j-1}w_{j-1} + \delta_{j-1}w_{j-2}) - \beta_j(v_{j-1}, w_{j-1})] \\ &= \delta_{j+1}^{-1}[\beta_j(v_j, w_j) - \beta_j(v_{j-1}, w_{j-1})] \\ &= 0.\end{aligned}$$

It can be proved in an identical way that $(v_i, w_{j+1}) = 0$ for $i \leq j$. Finally, by construction $(v_{j+1}, w_{j+1}) = 1$. This completes the induction proof. The proof of the matrix relations (7.3)–(7.5) is similar to that of the relations (6.6)–(6.8) in Arnoldi's method. □

The relations (7.3)–(7.5) allow us to interpret the algorithm. The matrix T_m is the projection of A obtained from an oblique projection process onto $\mathcal{K}_m(A, v_1)$ and orthogonal to $\mathcal{K}_m(A^T, w_1)$. Similarly, T_m^T represents the projection of A^T onto $\mathcal{K}_m(A^T, w_1)$ and orthogonal to $\mathcal{K}_m(A, v_1)$. Thus, an interesting new feature here is that the operators A and A^T play a dual role because similar operations are performed with them. In fact, two linear systems are solved implicitly, one with A and the other with A^T. If there are two linear systems to solve, one with A and the other with A^T, then this algorithm is suitable. Otherwise, the operations with A^T are essentially wasted. Later a number of alternative techniques developed in the literature will be introduced that avoid the use of A^T.

From a practical point of view, the Lanczos algorithm has a significant advantage over Arnoldi's method because it requires only a few vectors of storage if no reorthogonalization is performed. Specifically, six vectors of length n are needed, plus some storage for the tridiagonal matrix, no matter how large m is.

On the other hand, there are potentially more opportunities for breakdown with the nonsymmetric Lanczos method. The algorithm will break down whenever δ_{j+1} as defined in line 7 vanishes. This is examined more carefully in the next section. In practice, the difficulties are more likely to be caused by the near occurrence of this phenomenon. A look at the algorithm indicates that the Lanczos vectors may have to be scaled by small quantities when this happens. After a few steps the cumulative effect of these scalings may introduce excessive rounding errors.

Since the subspace from which the approximations are taken is identical to that of Arnoldi's method, the same bounds for the distance $\|(I - \Pi_m)u\|_2$ are valid. However, this does not mean in any way that the approximations obtained by the two methods are likely to be similar in quality. The theoretical bounds shown in Chapter 5 indicate that the norm of the projector may play a significant role.

7.1.2 Practical Implementations

There are various ways to improve the standard nonsymmetric Lanczos algorithm, which we now discuss briefly. A major concern here is the potential breakdowns or "near breakdowns" in the algorithm. There exist a number of approaches that have been developed to avoid such breakdowns. Other approaches do not attempt to eliminate the breakdown, but rather try to deal with it. The pros and cons of these strategies will be discussed after the various existing scenarios are described.

Algorithm 7.1 will abort in line 7 whenever

$$(\hat{v}_{j+1}, \hat{w}_{j+1}) = 0. \tag{7.6}$$

This can arise in two different ways. Either one of the two vectors $\hat{v}_{j+1}$ or $\hat{w}_{j+1}$ vanishes, or they are both nonzero, but their inner product is zero. The first case is the "lucky breakdown" scenario, which has been seen for symmetric matrices. Thus, if $\hat{v}_{j+1} = 0$, then span$\{V_j\}$ is invariant and, as was seen in Chapter 5, the approximate solution is exact. If $\hat{w}_{j+1} = 0$, then span$\{W_j\}$ is invariant. However, in this situation nothing can be said about the approximate solution for the linear system with A. If the algorithm is being used to solve a pair of linear systems, one with A and a *dual* system with A^T, then the approximate solution for the dual system will be exact in this case. The second scenario in which (7.6) can occur is when neither of the two vectors is zero, but their inner product is zero. Wilkinson (see [316, p. 389]) called this a *serious breakdown*. Fortunately, there are cures for this problem that allow the algorithm to continue in most cases. The corresponding modifications of the algorithm are often called *look-ahead Lanczos* algorithms. There are also rare cases of *incurable breakdowns*, which will not be discussed here (see [224] and [283]).

The main idea of look-ahead variants of the Lanczos algorithm is that the pair v_{j+2}, w_{j+2} can often be defined even though the pair v_{j+1}, w_{j+1} is not defined. The algorithm can be pursued from that iterate as before until a new breakdown is encountered. If the pair v_{j+2}, w_{j+2} cannot be defined, then the pair v_{j+3}, w_{j+3} can be tried, and so on. To better explain the idea, it is best to refer to the connection with orthogonal polynomials mentioned earlier for the symmetric case. The relationship can be extended to the nonsymmetric case by defining the bilinear form on the subspace $\mathbb{P}_{m-1}$:

$$\langle p, q \rangle = (p(A)v_1, q(A^T)w_1). \tag{7.7}$$

Unfortunately, this is now an *indefinite inner product* in general since $\langle p, p \rangle$ can be zero or even negative. Note that there is a polynomial p_j of degree j such that $\hat{v}_{j+1} = p_j(A)v_1$ and, in fact, the same polynomial intervenes in the equivalent expression of w_{j+1}. More precisely, there is a scalar γ_j such that $\hat{w}_{j+1} = \gamma_j p_j(A^T)v_1$. Similarly to the symmetric case, the nonsymmetric Lanczos algorithm attempts to compute a sequence of polynomials that are orthogonal with respect to the indefinite inner product defined above. If we define the moment matrix

$$M_k = \{\langle x^{i-1}, x^{j-1} \rangle\}_{i,j=1,\ldots,k},$$

then this process is mathematically equivalent to the computation of the factorization

$$M_k = L_k U_k$$

of the moment matrix M_k, in which U_k is upper triangular and L_k is lower triangular. Note that M_k is a Hankel matrix; i.e., its coefficients m_{ij} are constant along antidiagonals, i.e., for $i + j = constant$.

Because

$$\langle p_j, p_j \rangle = \gamma_j (p_j(A)v_1, p_j(A^T)w_1),$$

we observe that there is a serious breakdown at step j iff the indefinite norm of the polynomial p_j at step j vanishes. If this polynomial is skipped, it may still be possible to compute p_{j+1} and continue to generate the sequence. To explain this simply, consider

$$q_j(t) = x p_{j-1}(t) \quad \text{and} \quad q_{j+1}(t) = x^2 p_{j-1}(t).$$

Both q_j and q_{j+1} are orthogonal to the polynomials $p_1, \ldots, p_{j-2}$. We can define (somewhat arbitrarily) $p_j = q_j$, and then p_{j+1} can be obtained by orthogonalizing q_{j+1} against p_{j-1} and p_j. It is clear that the resulting polynomial will then be orthogonal against all polynomials of degree not exceeding j; see Exercise 5. Therefore, the algorithm can be continued from step $j + 1$ in the same manner. Exercise 5 generalizes this for the case where k polynomials are skipped rather than just one. This is a simplified description of the mechanism that underlies the various versions of look-ahead Lanczos algorithms proposed in the literature. The Parlett–Taylor–Liu implementation [224] is based on the observation that the algorithm breaks down because the pivots encountered during the LU factorization of the moment matrix M_k vanish. Then divisions by zero are avoided by performing *implicitly* a pivot with a 2×2 matrix rather than using a standard 1×1 pivot.

The drawback of look-ahead implementations is the nonnegligible added complexity. Besides the difficulty of identifying these near-breakdown situations, the matrix T_m ceases to be tridiagonal. Indeed, whenever a step is skipped, elements are introduced above the superdiagonal positions in some subsequent step. In the context of linear systems, near breakdowns are rare and their effect generally benign. Therefore, a simpler remedy, such as restarting the Lanczos procedure, may well be adequate. For eigenvalue problems, look-ahead strategies may be more justified.

7.2 The Lanczos Algorithm for Linear Systems

We present in this section a brief description of the Lanczos method for solving nonsymmetric linear systems. Consider the (single) linear system

$$Ax = b, \tag{7.8}$$

where A is $n \times n$ and nonsymmetric. Suppose that a guess x_0 to the solution is available and let its residual vector be $r_0 = b - Ax_0$. Then the Lanczos algorithm for solving (7.8) can be described as follows.

ALGORITHM 7.2. Two-Sided Lanczos Algorithm for Linear Systems

1. Compute $r_0 = b - Ax_0$ and $\beta := \|r_0\|_2$
2. Run m steps of the nonsymmetric Lanczos algorithm; i.e.,
3.   Start with $v_1 := r_0/\beta$ and any w_1 such that $(v_1, w_1) = 1$
4.   Generate the Lanczos vectors $v_1, \ldots, v_m, w_1, \ldots, w_m$ and the tridiagonal matrix T_m from Algorithm 7.1
5. Compute $y_m = T_m^{-1}(\beta e_1)$ and $x_m := x_0 + V_m y_m$

Note that it is possible to incorporate a convergence test when generating the Lanczos vectors in the second step without computing the approximate solution explicitly. This is due to the following formula, which is similar to (6.87) for the symmetric case:

$$\|b - Ax_j\|_2 = |\delta_{j+1} e_j^T y_j| \, \|v_{j+1}\|_2, \qquad (7.9)$$

and which can be proved in the same way, by using (7.3). This formula gives us the residual norm inexpensively without generating the approximate solution itself.

7.3 The Biconjugate Gradient and Quasi-Minimal Residual Algorithms

The biconjugate gradient (BCG) algorithm can be derived from Algorithm 7.1 in exactly the same way as the conjugate gradient (CG) method was derived from Algorithm 6.15. The algorithm was first proposed by Lanczos [196] in 1952 and then in a different form (CG-like version) by Fletcher [130] in 1974. Implicitly, the algorithm solves not only the original system $Ax = b$ but also a dual linear system $A^T x^* = b^*$ with A^T. This dual system is often ignored in the formulations of the algorithm.

7.3.1 The BCG Algorithm

The BCG algorithm is a projection process onto

$$\mathcal{K}_m = \mathrm{span}\{v_1, Av_1, \ldots, A^{m-1}v_1\}$$

orthogonal to

$$\mathcal{L}_m = \mathrm{span}\{w_1, A^T w_1, \ldots, (A^T)^{m-1} w_1\},$$

taking, as usual, $v_1 = r_0/\|r_0\|_2$. The vector w_1 is arbitrary, provided $(v_1, w_1) \neq 0$, but it is often chosen to be equal to v_1. If there is a dual system $A^T x^* = b^*$ to solve with A^T, then w_1 is obtained by scaling the initial residual $b^* - A^T x_0^*$.

Proceeding in the same manner as for the derivation of the CG algorithm from the symmetric Lanczos algorithm, we write the LU decomposition of T_m as

$$T_m = L_m U_m \qquad (7.10)$$

and define

$$P_m = V_m U_m^{-1}. \qquad (7.11)$$

7.3. The Biconjugate Gradient and Quasi-Minimal Residual Algorithms

The solution is then expressed as

$$\begin{aligned} x_m &= x_0 + V_m T_m^{-1}(\beta e_1) \\ &= x_0 + V_m U_m^{-1} L_m^{-1}(\beta e_1) \\ &= x_0 + P_m L_m^{-1}(\beta e_1). \end{aligned}$$

Notice that the solution x_m is updatable from x_{m-1} in a similar way to the CG algorithm. Like the CG algorithm, the vectors r_j and r_j^* are in the same direction as v_{j+1} and w_{j+1}, respectively. Hence, they form a biorthogonal sequence. Define similarly the matrix

$$P_m^* = W_m L_m^{-T}. \tag{7.12}$$

Clearly, the column vectors p_i^* of P_m^* and p_i of P_m are A-conjugate, since

$$(P_m^*)^T A P_m = L_m^{-1} W_m^T A V_m U_m^{-1} = L_m^{-1} T_m U_m^{-1} = I.$$

Utilizing this information, a CG-like algorithm can be easily derived from the Lanczos procedure.

ALGORITHM 7.3. BCG

1. Compute $r_0 := b - A x_0$. Choose r_0^* such that $(r_0, r_0^*) \neq 0$
2. Set $p_0 := r_0$, $p_0^* := r_0^*$.
3. For $j = 0, 1, \ldots,$ until convergence, Do
4. $\alpha_j := (r_j, r_j^*)/(A p_j, p_j^*)$
5. $x_{j+1} := x_j + \alpha_j p_j$
6. $r_{j+1} := r_j - \alpha_j A p_j$
7. $r_{j+1}^* := r_j^* - \alpha_j A^T p_j^*$
8. $\beta_j := (r_{j+1}, r_{j+1}^*)/(r_j, r_j^*)$
9. $p_{j+1} := r_{j+1} + \beta_j p_j$
10. $p_{j+1}^* := r_{j+1}^* + \beta_j p_j^*$
11. EndDo

If a dual system with A^T is being solved, then in line 1 r_0^* should be defined as $r_0^* = b^* - A^T x_0^*$ and the update $x_{j+1}^* := x_j^* + \alpha_j p_j^*$ to the dual approximate solution must be inserted after line 5. The vectors produced by this algorithm satisfy two orthogonality properties stated in the following proposition.

Proposition 7.2. *The vectors produced by the BCG algorithm satisfy the following orthogonality properties:*

$$(r_j, r_i^*) = 0 \quad \text{for } i \neq j, \tag{7.13}$$
$$(A p_j, p_i^*) = 0 \quad \text{for } i \neq j. \tag{7.14}$$

Proof. The proof is either by induction or by simply exploiting the relations between the vectors r_j, r_j^*, p_j, p_j^* and the vector columns of the matrices V_m, W_m, P_m, P_m^*. This is left as an exercise. □

Example 7.1. Table 7.1 shows the results of applying the BCG algorithm with no preconditioning to three of the test problems described in Section 3.7. See Example 6.1 for the meaning of the column headers in the table. Recall that Iters really represents the number of matrix-by-vector multiplications rather than the number of BCG steps.

Matrix	Iters	Kflops	Residual	Error
F2DA	163	2974	0.17E−03	0.86E−04
F3D	123	10768	0.34E−04	0.17E−03
ORS	301	6622	0.50E−01	0.37E−02

Table 7.1. *A test run of BCG without preconditioning.*

Thus, the number 163 in the first line represents 81 steps of BCG, which require 81×2 matrix-by-vector products in the iteration, and an extra one to compute the initial residual.

7.3.2 QMR Algorithm

The result of the Lanczos algorithm is a relation of the form

$$AV_m = V_{m+1}\bar{T}_m, \tag{7.15}$$

in which $\bar{T}_m$ is the $(m+1) \times m$ tridiagonal matrix

$$\bar{T}_m = \begin{pmatrix} T_m \\ \delta_{m+1}e_m^T \end{pmatrix}.$$

Now (7.15) can be exploited in the same way as was done to develop the generalized minimal residual method (GMRES). If v_1 is defined as a multiple of r_0, i.e., if $v_1 = \beta r_0$, then the residual vector associated with an approximate solution of the form

$$x = x_0 + V_m y$$

is given by

$$\begin{aligned} b - Ax &= b - A(x_0 + V_m y) \\ &= r_0 - AV_m y \\ &= \beta v_1 - V_{m+1}\bar{T}_m y \\ &= V_{m+1}(\beta e_1 - \bar{T}_m y). \end{aligned} \tag{7.16}$$

The norm of the residual vector is therefore

$$\|b - Ax\| = \|V_{m+1}(\beta e_1 - \bar{T}_m y)\|_2. \tag{7.17}$$

If the column vectors of V_{m+1} were orthonormal, then we would have $\|b - Ax\| = \|\beta e_1 - \bar{T}_m y\|_2$, as in GMRES. Therefore, a least-squares solution could be obtained from the Krylov subspace by minimizing $\|\beta e_1 - \bar{T}_m y\|_2$ over y. In the Lanczos algorithm, the v_i's are not orthonormal. However, it is still a reasonable idea to minimize the function

$$J(y) \equiv \|\beta e_1 - \bar{T}_m y\|_2$$

7.3. The Biconjugate Gradient and Quasi-Minimal Residual Algorithms

over y and compute the corresponding approximate solution $x_0 + V_m y$. The resulting solution is called the *quasi-minimal residual (QMR) approximation*. The norm $\|J(y)\|_2$ is called the quasi-residual norm for the approximation $x_0 + V_m y$.

Thus, the QMR approximation from the mth Krylov subspace is obtained as $x_m = x_0 + V_m y_m$, which minimizes the quasi-residual norm $J(y) = \|\beta e_1 - \bar{T}_m y\|_2$, i.e., just as in GMRES, except that the Arnoldi process is replaced with the Lanczos process. Because of the structure of the matrix $\bar{T}_m$, it is easy to adapt the direct quasi-GMRES (DQGMRES) algorithm (Algorithm 6.13) and obtain an efficient version of the QMR method. The algorithm is presented next.

ALGORITHM 7.4. QMR

1. Compute $r_0 = b - Ax_0$ and $\gamma_1 := \|r_0\|_2$, $w_1 := v_1 := r_0/\gamma_1$
2. For $m = 1, 2, \ldots$, until convergence, Do
3.   Compute α_m, δ_{m+1} and v_{m+1}, w_{m+1} as in Algorithm 7.1
4.   Update the QR factorization of $\bar{T}_m$; i.e.,
5.   Apply Ω_i, $i = m - 2, m - 1$, to the mth column of $\bar{T}_m$
6.   Compute the rotation coefficients c_m, s_m by (6.37)
7.   Apply rotation Ω_m to last column of $\bar{T}_m$ and to $\bar{g}_m$; i.e., compute
8. $\gamma_{m+1} := -s_m \gamma_m$
9. $\gamma_m := c_m \gamma_m$
10. $\alpha_m := c_m \alpha_m + s_m \delta_{m+1} \left(= \sqrt{\delta_{m+1}^2 + \alpha_m^2}\right)$
11. $p_m = \left(v_m - \sum_{i=m-2}^{m-1} t_{im} p_i\right) / t_{mm}$
12. $x_m = x_{m-1} + \gamma_m p_m$
13.   If $|\gamma_{m+1}|$ is small enough then Stop
14. EndDo

It is clear that the matrix T_m is not actually saved. Only the two most recent rotations need to be saved. For the remainder of this subsection, it is assumed (without loss of generality) that the v_i's are normalized to have unit 2-norms. Then the situation is similar to that of DQGMRES in that the quasi-residual norm defined by

$$\rho_m^Q = \|\beta e_1 - \bar{T}_m y_m\|_2 \equiv \min_{y \in \mathbb{R}^m} \|\beta e_1 - \bar{T}_m y\|_2$$

is usually a fairly good estimate of the actual residual norm. Following the same arguments as in Section 6.5.3 in Chapter 6, it is easily seen that

$$\rho_m^Q = |s_1 s_2 \cdots s_m| \|r_0\|_2 = |s_m| \rho_{m-1}^Q. \tag{7.18}$$

If the same notation as in Sections 6.5.3 and 6.5.7 is employed, then the actual residual $r_m = b - Ax_m$ obtained at the mth step of BCG satisfies

$$r_m = -h_{m+1,m} e_m^T y_m v_{m+1} = -h_{m+1,m} \frac{\gamma_m}{h_{mm}^{(m-1)}} v_{m+1} = \frac{h_{m+1,m}}{s_m h_{mm}^{(m-1)}} \gamma_{m+1} v_{m+1}.$$

For convenience, we have kept the notation h_{ij} used in Chapter 6 for the entries of the matrix $\bar{T}_m$. The next relation is then obtained by noticing, as in Section 6.5.7, that $h_{m+1,m}/h_{mm}^{(m)} =$

$\tan \theta_m$:
$$\gamma_{m+1} v_{m+1} = c_m r_m, \tag{7.19}$$
from which it follows that
$$\rho_m^Q = |c_m| \rho_m, \tag{7.20}$$
where $\rho_m = \|r_m\|_2$ is the actual residual norm of the mth BCG iterate.

The following proposition, which is similar to Proposition 6.9, establishes a result on the actual residual norm of the solution.

Proposition 7.3. *The residual norm of the approximate solution x_m satisfies the relation*
$$\|b - Ax_m\| \leq \|V_{m+1}\|_2 \, |s_1 s_2 \cdots s_m| \, \|r_0\|_2. \tag{7.21}$$

Proof. According to (7.16) the residual norm is given by
$$b - Ax_m = V_{m+1}[\beta e_1 - \bar{T}_m y_m] \tag{7.22}$$
and, using the same notation as in Proposition 6.9, referring to (6.43),
$$\|\beta e_1 - \bar{H}_m y\|_2^2 = |\gamma_{m+1}|^2 + \|g_m - R_m y\|_2^2,$$
in which $g_m - R_m y = 0$ by the minimization procedure. In addition, by (6.47) we have
$$\gamma_{m+1} = (-1)^m s_1 \cdots s_m \gamma_1, \quad \gamma_1 = \beta.$$
The result follows immediately using (7.22). □

A simple upper bound for $\|V_{m+1}\|_2$ can be derived from the Cauchy–Schwarz inequality:
$$\|V_{m+1}\|_2 \leq \sqrt{m+1}.$$
A comparison theorem that is similar to Theorem 6.11 can also be stated for QMR.

Theorem 7.4. *Assume that the Lanczos algorithm does not break down on or before step m and let V_{m+1} be the Lanczos basis obtained at step m. Let r_m^Q and r_m^G be the residual norms obtained after m steps of the QMR and GMRES algorithms, respectively. Then*
$$\|r_m^Q\|_2 \leq \kappa_2(V_{m+1}) \|r_m^G\|_2.$$

The proof of this theorem is essentially identical to that of Theorem 6.11. Note that V_{m+1} is now known to be of full rank, so we need not make this assumption as in Theorem 6.11.

It is not easy to analyze the QMR algorithm in terms of the exact residual norms, but the quasi-residual norms yield interesting properties. For example, an expression similar to (6.65) relates the actual BCG residual norm ρ_j to the quasi-residual norm ρ_j^Q obtained by QMR:
$$\frac{1}{\left(\rho_j^Q\right)^2} = \frac{1}{\left(\rho_{j-1}^Q\right)^2} + \frac{1}{\left(\rho_j\right)^2}. \tag{7.23}$$

7.3. The Biconjugate Gradient and Quasi-Minimal Residual Algorithms

The proof of this result is identical to that of (6.65)—it is an immediate consequence of (7.18) and (7.20). An argument similar to the one used to derive (6.67) leads to a similar conclusion:

$$\rho_m^Q = \frac{1}{\sqrt{\sum_{i=0}^{m}(1/\rho_i)^2}}. \tag{7.24}$$

The above equality underlines the smoothing property of the QMR algorithm since it shows that the quasi-residual norm is akin to a (harmonic) average of the BCG residual norms.

It is clear from (7.20) that $\rho_m^Q \leq \rho_m$. An argument similar to that used to derive Proposition 6.15 can be made. If ρ_{m_*} is the smallest residual norm achieved among those of the first m steps of BCG, then

$$\frac{1}{\left(\rho_m^Q\right)^2} = \sum_{i=0}^{m} \frac{1}{(\rho_i)^2} \leq \frac{m}{(\rho_{m_*})^2}.$$

This proves the following result.

Proposition 7.5. *Assume that m steps of QMR and BCG are taken and let ρ_{m_*} be the smallest residual norm achieved by BCG in the first m steps. Then the following inequalities hold:*

$$\rho_m^Q \leq \rho_{m_*} \leq \sqrt{m}\, \rho_m^Q. \tag{7.25}$$

The above results deal with quasi residuals instead of the actual residuals. However, it is possible to proceed as for DQGMRES (see (6.50) and (6.53)) to express the actual residual as

$$b - Ax_m^Q = \gamma_{m+1} z_{m+1}, \tag{7.26}$$

where, as before, γ_{m+1} is the last component of the right-hand side βe_1 after the m Givens rotations have been applied to it. Therefore, γ_{m+1} satisfies the recurrence (6.47) starting with $\gamma_1 = \beta$. The vector z_{m+1} can be updated by the same relation; namely,

$$z_{m+1} = -s_m z_m + c_m v_{m+1}. \tag{7.27}$$

The sequence z_{m+1} can be updated and the norm of z_{m+1} computed to yield the exact residual norm, but this entails nonnegligible additional operations ($5n$ in total). The compromise based on updating an upper bound seen for DQGMRES can be used here as well.

It is interesting to explore (7.27) further. Denote by r_m^Q the actual residual vector $b - Ax_m^Q$ obtained from QMR. Then, from (7.26), (7.27), and (6.47), it follows that

$$r_m^Q = s_m^2 r_{m-1}^Q + c_m \gamma_{m+1} v_{m+1}. \tag{7.28}$$

When combined with (7.19), the above equality leads to the following relation between the actual residuals r_m^Q produced at the mth step of QMR and the residuals r_m obtained from BCG:

$$r_m^Q = s_m^2 r_{m-1}^Q + c_m^2 r_m, \tag{7.29}$$

from which follows the same relation for the iterates:

$$x_m^Q = s_m^2 x_{m-1}^Q + c_m^2 x_m. \tag{7.30}$$

When s_m is close to zero, which corresponds to fast convergence of BCG, then QMR will be close to the BCG iterate. On the other hand, when s_m is close to one, then QMR will tend to make little progress—just as was shown by Brown [66] for the FOM/GMRES pair. A more pictorial way of stating this is that peaks of the BCG residual norms will correspond to plateaus of the QMR quasi residuals. The above relations can be rewritten as follows:

$$x_m^Q = x_{m-1}^Q + c_m^2(x_m - x_{m-1}^Q), \qquad r_m^Q = r_{m-1}^Q + c_m^2(r_m - r_{m-1}^Q). \qquad (7.31)$$

Schemes of the above general form, where now c_m^2 can be considered a parameter, are known as residual smoothing methods and were also considered in Chapter 6. The minimal residual smoothing (MRS) seen in Chapter 6 is now replaced with a *quasi-minimal residual smoothing* (QMRS). Indeed, what the above relation shows is that *it is possible to implement QMR as a QMRS algorithm*. The only missing ingredient for completing the description of the algorithm is an expression of the smoothing parameter c_m^2 in terms of quantities that do not refer to the Givens rotations. This expression can be derived from (7.20), which relates the cosine c_j to the ratio of the quasi-residual norm and the actual residual norm of BCG, and from (7.23), which allows us to compute ρ_j^Q recursively. The QMRS algorithm, developed by Zhou and Walker [323], can now be sketched.

ALGORITHM 7.5. QMRS

1. Set $r_0 = b - Ax_0$, $x_0^Q = x_0$; Set $\rho_0 = \rho_0^Q = \|r_0\|_2$
2. For $j = 1, 2, \ldots,$ Do
3.   Compute x_j and the associated residual r_j and residual norm ρ_j
4.   Compute ρ_j^Q from (7.23) and set $\eta_j = \left(\rho_j^Q/\rho_j\right)^2$
5.   Compute $\quad x_j^Q = x_{j-1}^Q + \eta_j(x_j - x_{j-1}^Q)$
6. EndDo

7.4 Transpose-Free Variants

Each step of the BCG algorithm and QMR requires a matrix-by-vector product with both A and A^T. However, observe that the vectors p_i^* and w_j generated with A^T do not contribute directly to the solution. Instead, they are used only to obtain the scalars needed in the algorithm, e.g., the scalars α_j and β_j for BCG.

The question arises as to whether or not it is possible to bypass the use of the transpose of A and still generate iterates that are related to those of the BCG algorithm. One of the motivations for this question is that, in some applications, A is available only through some approximations and not explicitly. In such situations, the transpose of A is usually not available. A simple example is when a CG-like algorithm is used in the context of Newton's iteration for solving $F(u) = 0$.

The linear system that arises at each Newton step can be solved without having to compute the Jacobian $J(u_k)$ at the current iterate u_k explicitly by using the difference formula

$$J(u_k)v = \frac{F(u_k + \epsilon v) - F(u_k)}{\epsilon}.$$

7.4. Transpose-Free Variants

This allows the action of this Jacobian to be computed on an arbitrary vector v. Unfortunately, there is no similar formula for performing operations with the transpose of $J(u_k)$.

7.4.1 CGS

The conjugate gradient squared (CGS) algorithm was developed by Sonneveld in 1984 [271], mainly to avoid using the transpose of A in the BCG and to gain faster convergence for roughly the same computational cost. The main idea is based on the following simple observation. In the BCG algorithm, the residual vector at step j can be expressed as

$$r_j = \phi_j(A)r_0, \tag{7.32}$$

where ϕ_j is a certain polynomial of degree j satisfying the constraint $\phi_j(0) = 1$. Similarly, the conjugate direction polynomial $\pi_j(t)$ is given by

$$p_j = \pi_j(A)r_0, \tag{7.33}$$

in which π_j is a polynomial of degree j. From the algorithm, observe that the directions r_j^* and p_j^* are defined through the same recurrences as r_j and p_j, in which A is replaced by A^T and, as a result,

$$r_j^* = \phi_j(A^T)r_0^*, \quad p_j^* = \pi_j(A^T)r_0^*.$$

Also, note that the scalar α_j in BCG is given by

$$\alpha_j = \frac{(\phi_j(A)r_0, \phi_j(A^T)r_0^*)}{(A\pi_j(A)r_0, \pi_j(A^T)r_0^*)} = \frac{(\phi_j^2(A)r_0, r_0^*)}{(A\pi_j^2(A)r_0, r_0^*)},$$

which indicates that, if it is possible to get a recursion for the vectors $\phi_j^2(A)r_0$ and $\pi_j^2(A)r_0$, then computing α_j and, similarly, β_j causes no problem. Hence, the idea is to seek an algorithm that would give a sequence of iterates whose residual norms r_j' satisfy

$$r_j' = \phi_j^2(A)r_0. \tag{7.34}$$

The derivation of the method relies on simple algebra only. To establish the desired recurrences for the squared polynomials, start with the recurrences that define ϕ_j and π_j, which are

$$\phi_{j+1}(t) = \phi_j(t) - \alpha_j t \pi_j(t), \tag{7.35}$$
$$\pi_{j+1}(t) = \phi_{j+1}(t) + \beta_j \pi_j(t). \tag{7.36}$$

If the above relations are squared we get

$$\phi_{j+1}^2(t) = \phi_j^2(t) - 2\alpha_j t \pi_j(t)\phi_j(t) + \alpha_j^2 t^2 \pi_j^2(t),$$
$$\pi_{j+1}^2(t) = \phi_{j+1}^2(t) + 2\beta_j \phi_{j+1}(t)\pi_j(t) + \beta_j^2 \pi_j(t)^2.$$

If it were not for the cross terms $\pi_j(t)\phi_j(t)$ and $\phi_{j+1}(t)\pi_j(t)$ on the right-hand sides, these equations would form an updatable recurrence system. The solution is to introduce one of these two cross terms, namely, $\phi_{j+1}(t)\pi_j(t)$, as a third member of the recurrence. For the other term, i.e., $\pi_j(t)\phi_j(t)$, we can exploit the relation

$$\phi_j(t)\pi_j(t) = \phi_j(t)\left(\phi_j(t) + \beta_{j-1}\pi_{j-1}(t)\right) = \phi_j^2(t) + \beta_{j-1}\phi_j(t)\pi_{j-1}(t).$$

By putting these relations together the following recurrences can be derived, in which the variable (t) is omitted where there is no ambiguity:

$$\phi_{j+1}^2 = \phi_j^2 - \alpha_j t \left(2\phi_j^2 + 2\beta_{j-1}\phi_j\pi_{j-1} - \alpha_j t\, \pi_j^2\right), \tag{7.37}$$
$$\phi_{j+1}\pi_j = \phi_j^2 + \beta_{j-1}\phi_j\pi_{j-1} - \alpha_j t\, \pi_j^2, \tag{7.38}$$
$$\pi_{j+1}^2 = \phi_{j+1}^2 + 2\beta_j\phi_{j+1}\pi_j + \beta_j^2\pi_j^2. \tag{7.39}$$

These recurrences constitute the basis of the algorithm. If we define

$$r_j = \phi_j^2(A)r_0, \tag{7.40}$$
$$p_j = \pi_j^2(A)r_0, \tag{7.41}$$
$$q_j = \phi_{j+1}(A)\pi_j(A)r_0, \tag{7.42}$$

then the above recurrences for the polynomials translate into

$$r_{j+1} = r_j - \alpha_j A\left(2r_j + 2\beta_{j-1}q_{j-1} - \alpha_j A\, p_j\right), \tag{7.43}$$
$$q_j = r_j + \beta_{j-1}q_{j-1} - \alpha_j A\, p_j, \tag{7.44}$$
$$p_{j+1} = r_{j+1} + 2\beta_j q_j + \beta_j^2 p_j. \tag{7.45}$$

It is convenient to define the auxiliary vector

$$d_j = 2r_j + 2\beta_{j-1}q_{j-1} - \alpha_j A p_j.$$

With this we obtain the following sequence of operations to compute the approximate solution, starting with $r_0 := b - Ax_0$, $p_0 := r_0$, $q_0 := 0$, $\beta_0 := 0$:

- $\alpha_j = (r_j, r_0^*)/(Ap_j, r_0^*)$,
- $d_j = 2r_j + 2\beta_{j-1}q_{j-1} - \alpha_j Ap_j$,
- $q_j = r_j + \beta_{j-1}q_{j-1} - \alpha_j Ap_j$,
- $x_{j+1} = x_j + \alpha_j d_j$,
- $r_{j+1} = r_j - \alpha_j Ad_j$,
- $\beta_j = (r_{j+1}, r_0^*)/(r_j, r_0^*)$,
- $p_{j+1} = r_{j+1} + \beta_j(2q_j + \beta_j p_j)$.

A slight simplification of the algorithm can be made by using the auxiliary vector $u_j = r_j + \beta_{j-1}q_{j-1}$. This definition leads to the relations

$$d_j = u_j + q_j,$$
$$q_j = u_j - \alpha_j Ap_j,$$
$$p_{j+1} = u_{j+1} + \beta_j(q_j + \beta_j p_j)$$

and, as a result, the vector d_j is no longer needed. The resulting algorithm is given below.

7.4. Transpose-Free Variants

ALGORITHM 7.6. CGS

1. Compute $r_0 := b - Ax_0$, r_0^* arbitrary
2. Set $p_0 := u_0 := r_0$
3. For $j = 0, 1, 2\ldots$, until convergence, Do
4. $\quad \alpha_j = (r_j, r_0^*)/(Ap_j, r_0^*)$
5. $\quad q_j = u_j - \alpha_j A p_j$
6. $\quad x_{j+1} = x_j + \alpha_j(u_j + q_j)$
7. $\quad r_{j+1} = r_j - \alpha_j A(u_j + q_j)$
8. $\quad \beta_j = (r_{j+1}, r_0^*)/(r_j, r_0^*)$
9. $\quad u_{j+1} = r_{j+1} + \beta_j q_j$
10. $\quad p_{j+1} = u_{j+1} + \beta_j(q_j + \beta_j p_j)$
11. EndDo

Observe that there are no matrix-by-vector products with the transpose of A. Instead, two matrix-by-vector products with the matrix A are now performed at each step. In general, one should expect the resulting algorithm to converge twice as fast as BCG. Therefore, what has essentially been accomplished is the replacement of the matrix-by-vector products with A^T by more useful work.

The CGS algorithm works quite well in many cases. However, one difficulty is that, since the polynomials are squared, rounding errors tend to be more damaging than in the standard BCG algorithm. In particular, very high variations of the residual vectors often cause the residual norms computed from the result of line 7 of the above algorithm to become inaccurate.

7.4.2 BICGSTAB

The CGS algorithm is based on squaring the residual polynomial, which, in cases of irregular convergence, may lead to substantial build-up of rounding errors, or possibly even overflow. The biconjugate gradient stabilized (BICGSTAB) algorithm is a variation of CGS that was developed to remedy this difficulty. Instead of seeking a method that delivers a residual vector of the form r'_j defined by (7.34), BICGSTAB produces iterates whose residual vectors are of the form

$$r'_j = \psi_j(A)\phi_j(A)r_0, \tag{7.46}$$

in which, as before, $\phi_j(t)$ is the residual polynomial associated with the BCG algorithm and $\psi_j(t)$ is a new polynomial defined recursively at each step with the goal of *stabilizing* or *smoothing* the convergence behavior of the original algorithm. Specifically, $\psi_j(t)$ is defined by the simple recurrence

$$\psi_{j+1}(t) = (1 - \omega_j t)\psi_j(t), \tag{7.47}$$

in which the scalar ω_j is to be determined. The derivation of the appropriate recurrence relations is similar to that of CGS. Ignoring the scalar coefficients at first, we start with a relation for the residual polynomial $\psi_{j+1}\phi_{j+1}$. We immediately obtain

$$\psi_{j+1}\phi_{j+1} = (1 - \omega_j t)\psi_j(t)\phi_{j+1} \tag{7.48}$$
$$= (1 - \omega_j t)\left(\psi_j\phi_j - \alpha_j t \psi_j \pi_j\right), \tag{7.49}$$

which is updatable provided a recurrence relation is found for the products $\psi_j \pi_j$. For this we write

$$\psi_j \pi_j = \psi_j(\phi_j + \beta_{j-1}\pi_{j-1}) \tag{7.50}$$
$$= \psi_j \phi_j + \beta_{j-1}(1 - \omega_{j-1}t)\psi_{j-1}\pi_{j-1}. \tag{7.51}$$

Define

$$r_j = \psi_j(A)\phi_j(A)r_0,$$
$$p_j = \psi_j(A)\pi_j(A)r_0.$$

According to the above formulas, these vectors can be updated from a double recurrence provided the scalars α_j and β_j are computable. This recurrence is

$$r_{j+1} = (I - \omega_j A)(r_j - \alpha_j A p_j), \tag{7.52}$$
$$p_{j+1} = r_{j+1} + \beta_j(I - \omega_j A)p_j.$$

Consider now the computation of the scalars needed in the recurrence. According to the original BCG algorithm, $\beta_j = \rho_{j+1}/\rho_j$, with

$$\rho_j = (\phi_j(A)r_0, \phi_j(A^T)r_0^*) = (\phi_j(A)^2 r_0, r_0^*).$$

Unfortunately, ρ_j is not computable from these formulas because none of the vectors $\phi_j(A)r_0, \phi_j(A^T)r_0^*$, and $\phi_j(A)^2 r_0$ is available. However, ρ_j can be related to the scalar

$$\tilde{\rho}_j = (\phi_j(A)r_0, \psi_j(A^T)r_0^*),$$

which is computable via

$$\tilde{\rho}_j = (\phi_j(A)r_0, \psi_j(A^T)r_0^*) = (\psi_j(A)\phi_j(A)r_0, r_0^*) = (r_j, r_0^*).$$

To relate the two scalars ρ_j and $\tilde{\rho}_j$, expand $\psi_j(A^T)r_0^*$ explicitly in the power basis to obtain

$$\tilde{\rho}_j = \left(\phi_j(A)r_0, \eta_1^{(j)}(A^T)^j r_0^* + \eta_2^{(j)}(A^T)^{j-1}r_0^* + \cdots \right).$$

Since $\phi_j(A)r_0$ is orthogonal to all vectors $(A^T)^k r_0^*$, with $k < j$, only the leading power is relevant in the expansion on the right side of the above inner product. In particular, if $\gamma_1^{(j)}$ is the leading coefficient for the polynomial $\phi_j(t)$, then

$$\tilde{\rho}_j = \left(\phi_j(A)r_0, \frac{\eta_1^{(j)}}{\gamma_1^{(j)}} \phi_j(A^T)r_0 \right) = \frac{\eta_1^{(j)}}{\gamma_1^{(j)}} \rho_j.$$

When examining the recurrence relations for ϕ_{j+1} and ψ_{j+1}, leading coefficients for these polynomials are found to satisfy the relations

$$\eta_1^{(j+1)} = -\omega_j \eta_1^{(j)}, \quad \gamma_1^{(j+1)} = -\alpha_j \gamma_1^{(j)},$$

and, as a result,

$$\frac{\tilde{\rho}_{j+1}}{\tilde{\rho}_j} = \frac{\omega_j}{\alpha_j} \frac{\rho_{j+1}}{\rho_j},$$

7.4. Transpose-Free Variants

which yields the following relation for β_j:

$$\beta_j = \frac{\tilde{\rho}_{j+1}}{\tilde{\rho}_j} \times \frac{\alpha_j}{\omega_j}. \qquad (7.53)$$

Similarly, a simple recurrence formula for α_j can be derived. By definition,

$$\alpha_j = \frac{(\phi_j(A)r_0, \phi_j(A^T)r_0^*)}{(A\pi_j(A)r_0, \pi_j(A^T)r_0^*)}$$

and, as in the previous case, the polynomials on the right sides of the inner products in both the numerator and denominator can be replaced with their leading terms. However, in this case the leading coefficients for $\phi_j(A^T)r_0^*$ and $\pi_j(A^T)r_0^*$ are identical and, therefore,

$$\alpha_j = \frac{(\phi_j(A)r_0, \phi_j(A^T)r_0^*)}{(A\pi_j(A)r_0, \phi_j(A^T)r_0^*)}$$
$$= \frac{(\phi_j(A)r_0, \psi_j(A^T)r_0^*)}{(A\pi_j(A)r_0, \psi_j(A^T)r_0^*)}$$
$$= \frac{(\psi_j(A)\phi_j(A)r_0, r_0^*)}{(A\psi_j(A)\pi_j(A)r_0, r_0^*)}.$$

Since $p_j = \psi_j(A)\pi_j(A)r_0$, this yields

$$\alpha_j = \frac{\tilde{\rho}_j}{(Ap_j, r_0^*)}. \qquad (7.54)$$

Next, the parameter ω_j must be defined. This can be thought of as an additional free parameter. One of the simplest choices, and perhaps the most natural, is to select ω_j to achieve a steepest descent step in the residual direction obtained before multiplying the residual vector by $(I - \omega_j A)$ in (7.52). In other words, ω_j is chosen to minimize the 2-norm of the vector $(I - \omega_j A)\psi_j(A)\phi_{j+1}(A)r_0$. Equation (7.52) can be rewritten as

$$r_{j+1} = (I - \omega_j A)s_j,$$

in which

$$s_j \equiv r_j - \alpha_j A p_j.$$

Then the optimal value for ω_j is given by

$$\omega_j = \frac{(As_j, s_j)}{(As_j, As_j)}. \qquad (7.55)$$

Finally, a formula is needed to update the approximate solution x_{j+1} from x_j. Equation (7.52) can be rewritten as

$$r_{j+1} = s_j - \omega_j A s_j = r_j - \alpha_j A p_j - \omega_j A s_j,$$

which yields

$$x_{j+1} = x_j + \alpha_j p_j + \omega_j s_j.$$

After putting these relations together, we obtain the final form of the BICGSTAB algorithm, due to van der Vorst [289].

ALGORITHM 7.7. BICGSTAB

1. Compute $r_0 := b - Ax_0$, r_0^* arbitrary
2. $p_0 := r_0$
3. For $j = 0, 1, \ldots,$ until convergence, Do
4. $\quad \alpha_j := (r_j, r_0^*)/(Ap_j, r_0^*)$
5. $\quad s_j := r_j - \alpha_j A p_j$
6. $\quad \omega_j := (As_j, s_j)/(As_j, As_j)$
7. $\quad x_{j+1} := x_j + \alpha_j p_j + \omega_j s_j$
8. $\quad r_{j+1} := s_j - \omega_j As_j$
9. $\quad \beta_j := \frac{(r_{j+1}, r_0^*)}{(r_j, r_0^*)} \times \frac{\alpha_j}{\omega_j}$
10. $\quad p_{j+1} := r_{j+1} + \beta_j(p_j - \omega_j A p_j)$
11. EndDo

Example 7.2. Table 7.2 shows the results of applying the BICGSTAB algorithm with no preconditioning to three of the test problems described in Section 3.7.

Matrix	Iters	Kflops	Residual	Error
F2DA	96	2048	0.14E−02	0.77E−04
F3D	64	6407	0.49E−03	0.17E−03
ORS	208	5222	0.22E+00	0.68E−04

Table 7.2. *A test run of BICGSTAB with no preconditioning.*

See Example 6.1 for the meaning of the column headers in the table. As in Example 7.1, Iters is the number of matrix-by-vector multiplications required to converge. As can be seen, it is less than with BCG. Thus, using the number of matrix-by-vector products as a criterion, BCG is more expensive than BICGSTAB in all three examples. For Problem 3, the number of matvecs exceeds the 300 limit with BCG. If the number of actual iterations is used as a criterion, then the two methods come close for the second problem (61 steps for BCG versus 64 for BICGSTAB), while BCG is slightly faster for Problem 1. Observe also that the total number of operations favors BICGSTAB. This illustrates the main weakness of BCG as well as QMR; namely, the matrix-by-vector products with transpose are essentially wasted unless a dual system with A^T must be solved simultaneously.

7.4.3 TFQMR

The transpose-free QMR (TFQMR) algorithm of Freund [134] is derived from the CGS algorithm. Observe that x_j can be updated in two half-steps in line 6 of Algorithm 7.6, namely, $x_{j+1/2} = x_j + \alpha_j u_j$ and $x_{j+1} = x_{j+1/2} + \alpha_j q_j$. This is only natural since the actual update from one iterate to the next involves two matrix-by-vector multiplications; i.e., the degree of the residual polynomial is increased by two. In order to avoid indices that are multiples of $\frac{1}{2}$, it is convenient when describing TFQMR to double all subscripts in the CGS algorithm. With this change of notation, the main steps of Algorithm 7.6 (CGS) become

$$\alpha_{2j} = (r_{2j}, r_0^*)/(Ap_{2j}, r_0^*), \qquad (7.56)$$

$$q_{2j} = u_{2j} - \alpha_{2j} A p_{2j}, \qquad (7.57)$$

7.4. Transpose-Free Variants

$$x_{2j+2} = x_{2j} + \alpha_{2j}(u_{2j} + q_{2j}), \tag{7.58}$$
$$r_{2j+2} = r_{2j} - \alpha_{2j}A(u_{2j} + q_{2j}), \tag{7.59}$$
$$\beta_{2j} = (r_{2j+2}, r_0^*)/(r_{2j}, r_0^*), \tag{7.60}$$
$$u_{2j+2} = r_{2j+2} + \beta_{2j}q_{2j}, \tag{7.61}$$
$$p_{2j+2} = u_{2j+2} + \beta_{2j}(q_{2j} + \beta p_{2j}). \tag{7.62}$$

The initialization is identical to that of Algorithm 7.6. The update of the approximate solution in (7.58) can now be split into the following two half-steps:

$$x_{2j+1} = x_{2j} + \alpha_{2j}u_{2j}, \tag{7.63}$$
$$x_{2j+2} = x_{2j+1} + \alpha_{2j}q_{2j}. \tag{7.64}$$

This can be simplified by defining the vectors u_m for odd m as $u_{2j+1} = q_{2j}$. Similarly, the sequence of α_m is defined for odd values of m as $\alpha_{2j+1} = \alpha_{2j}$. In summary,

$$\text{for } m \text{ odd define } \begin{cases} u_m \equiv q_{m-1}, \\ \alpha_m \equiv \alpha_{m-1}. \end{cases} \tag{7.65}$$

With these definitions, the relations (7.63)–(7.64) are translated into the single equation

$$x_m = x_{m-1} + \alpha_{m-1}u_{m-1},$$

which is valid whether m is even or odd. The intermediate iterates x_m, with m odd, which are now defined, do not exist in the original CGS algorithm. For even values of m the sequence x_m represents the original sequence of iterates from the CGS algorithm. It is convenient to introduce the $N \times m$ matrix

$$U_m = [u_0, \ldots, u_{m-1}]$$

and the m-dimensional vector

$$z_m = (\alpha_0, \alpha_1, \ldots, \alpha_{m-1})^T.$$

The general iterate x_m satisfies the relation

$$x_m = x_0 + U_m z_m \tag{7.66}$$
$$= x_{m-1} + \alpha_{m-1}u_{m-1}. \tag{7.67}$$

From the above equation, it is clear that the residual vectors r_m are related to the u-vectors by the relations

$$r_m = r_0 - AU_m z_m \tag{7.68}$$
$$= r_{m-1} - \alpha_{m-1}Au_{m-1}. \tag{7.69}$$

Next, a relation similar to the relation (6.7) seen for FOM and GMRES will be extracted using the matrix AU_m. As a result of (7.69), the following relation holds:

$$Au_i = \frac{1}{\alpha_i}(r_i - r_{i+1}).$$

Translated into matrix form, this relation becomes

$$AU_m = R_{m+1}\bar{B}_m, \tag{7.70}$$

where

$$R_k = [r_0, r_1, \ldots, r_{k-1}] \tag{7.71}$$

and where $\bar{B}_m$ is the $(m+1) \times m$ matrix

$$\bar{B}_m = \begin{pmatrix} 1 & 0 & \cdots & \cdots & 0 \\ -1 & 1 & & & \vdots \\ 0 & -1 & 1 & \cdots & \\ \vdots & & \ddots & \ddots & \vdots \\ \vdots & & & -1 & 1 \\ 0 & \cdots & & & -1 \end{pmatrix} \times \text{diag}\left\{\frac{1}{\alpha_0}, \frac{1}{\alpha_1}, \ldots, \frac{1}{\alpha_{m-1}}\right\}. \tag{7.72}$$

The columns of R_{m+1} can be rescaled, for example, to make each of them have a 2-norm equal to one, by multiplying R_{m+1} to the right by a diagonal matrix. Let this diagonal matrix be the inverse of the matrix

$$\Delta_{m+1} = \text{diag}[\delta_0, \delta_1, \ldots, \delta_m].$$

Then

$$AU_m = R_{m+1}\Delta_{m+1}^{-1}\Delta_{m+1}\bar{B}_m. \tag{7.73}$$

With this, (7.68) becomes

$$r_m = r_0 - AU_m z_m = R_{m+1}\left[e_1 - \bar{B}_m z_m\right] \tag{7.74}$$

$$= R_{m+1}\Delta_{m+1}^{-1}[\delta_0 e_1 - \Delta_{m+1}\bar{B}_m z_m]. \tag{7.75}$$

By analogy with the GMRES algorithm, define

$$\bar{H}_m \equiv \Delta_{m+1}\bar{B}_m.$$

Similarly, define H_m to be the matrix obtained from $\bar{H}_m$ by deleting its last row. It is easy to verify that the CGS iterates x_m (now defined for all integers $m = 0, 1, 2, \ldots$) satisfy the same definition as the full orthogonalization method (FOM); i.e.,

$$x_m = x_0 + U_m H_m^{-1}(\delta_0 e_1). \tag{7.76}$$

It is also possible to extract a GMRES-like solution from the relations (7.73) and (7.75), similar to DQGMRES. In order to minimize the residual norm over the Krylov subspace, the 2-norm of the right-hand side of (7.75) would have to be minimized, but this is not practical since the columns of $R_{m+1}\Delta_{m+1}^{-1}$ are not orthonormal as in GMRES. However, the 2-norm of $\delta_0 e_1 - \Delta_{m+1}\bar{B}_m z$ can be minimized over z, as was done for the QMR and DQGMRES algorithms.

This defines the TFQMR iterates theoretically. However, it is now necessary to find a formula for expressing the iterates in a progressive way. There are two ways to proceed. The first follows DQGMRES closely, defining the least-squares solution progressively and

7.4. Transpose-Free Variants

exploiting the structure of the matrix R_m to obtain a formula for x_m from x_{m-1}. Because of the special structure of $\bar{H}_m$, this is equivalent to using the DQGMRES algorithm with $k = 1$. The second way to proceed exploits Lemma 6.16, seen in the previous chapter. This lemma, which was shown for the FOM/GMRES pair, is also valid for the CGS/TFQMR pair. There is no fundamental difference between the two situations. Thus, the TFQMR iterates satisfy the relation

$$x_m - x_{m-1} = c_m^2 (\tilde{x}_m - x_{m-1}), \tag{7.77}$$

where the tildes are now used to denote the CGS iterate. Setting

$$d_m \equiv \frac{1}{\alpha_{m-1}} (\tilde{x}_m - x_{m-1}) = \frac{1}{c_m^2 \alpha_{m-1}} (x_m - x_{m-1}), \tag{7.78}$$

$$\eta_m \equiv c_m^2 \alpha_{m-1},$$

the above expression for x_m becomes

$$x_m = x_{m-1} + \eta_m d_m. \tag{7.79}$$

Now observe from (7.67) that the CGS iterates $\tilde{x}_m$ satisfy the relation

$$\tilde{x}_m = \tilde{x}_{m-1} + \alpha_{m-1} u_{m-1}. \tag{7.80}$$

From the above equations, a recurrence relation from d_m can be extracted. The definition of d_m and the above relations yield

$$d_m = \frac{1}{\alpha_{m-1}} (\tilde{x}_m - \tilde{x}_{m-1} + \tilde{x}_{m-1} - x_{m-1})$$

$$= u_{m-1} + \frac{1}{\alpha_{m-1}} (\tilde{x}_{m-1} - x_{m-2} - (x_{m-1} - x_{m-2}))$$

$$= u_{m-1} + \frac{1 - c_{m-1}^2}{\alpha_{m-1}} (\tilde{x}_{m-1} - x_{m-2}).$$

Therefore,

$$d_m = u_{m-1} + \frac{(1 - c_{m-1}^2)\eta_{m-1}}{c_{m-1}^2 \alpha_{m-1}} d_{m-1}.$$

The term $(1 - c_{m-1}^2)/c_{m-1}^2$ is the squared tangent of the angle used in the $(m-1)$th rotation. This tangent will be denoted by θ_{m-1}, so we have

$$\theta_m = \frac{s_m}{c_m}, \quad c_m^2 = \frac{1}{1 + \theta_m^2}, \quad d_{m+1} = u_m + \frac{\theta_m^2 \eta_m}{\alpha_m} d_m.$$

The angle used in the mth rotation or, equivalently, c_m, can be obtained by examining the matrix $\bar{H}_m$:

$$\bar{H}_m = \begin{pmatrix} \delta_0 & 0 & \cdots & & \cdots & 0 \\ -\delta_1 & \delta_1 & & & & \vdots \\ 0 & -\delta_2 & \delta_2 & \cdots & & \\ \vdots & & \ddots & \ddots & & \vdots \\ & & & & -\delta_m & \delta_m \\ 0 & \cdots & & & & -\delta_{m+1} \end{pmatrix} \times \text{diag} \left\{ \frac{1}{\alpha_i} \right\}_{i=0,\ldots,m-1}. \tag{7.81}$$

The diagonal matrix on the right-hand side scales the columns of the matrix. It is easy to see that it has no effect on the determination of the rotations. Ignoring this scaling, the above matrix becomes, after j rotations,

$$\begin{pmatrix} \star & \star & & & & & \\ & \star & \star & & & & \\ & & \ddots & \ddots & & & \\ & & & \tau_j & 0 & & \\ & & & -\delta_{j+1} & \delta_{j+1} & & \\ & & & & \ddots & \ddots & \\ & & & & & -\delta_m & \delta_m \\ & & & & & & -\delta_{m+1} \end{pmatrix}.$$

The next rotation is then determined by

$$s_{j+1} = \frac{-\delta_{j+1}}{\sqrt{\tau_j^2 + \delta_{j+1}^2}}, \quad c_{j+1} = \frac{\tau_j}{\sqrt{\tau_j^2 + \delta_{j+1}^2}}, \quad \theta_{j+1} = \frac{-\delta_{j+1}}{\tau_j}.$$

In addition, after this rotation is applied to the above matrix, the diagonal element δ_{j+1} in position $(j+1, j+1)$ is transformed into

$$\tau_{j+1} = \delta_{j+1} \times c_{j+1} = \frac{\tau_j \delta_{j+1}}{\sqrt{\tau_j^2 + \delta_{j+1}^2}} = -\tau_j s_{j+1} = -\tau_j \theta_{j+1} c_{j+1}. \tag{7.82}$$

The above relations enable us to update the direction d_m and the required quantities c_m and η_m. Since only the squares of these scalars are invoked in the update of the direction d_{m+1}, a recurrence for their absolute values is sufficient. This gives the following recurrences, which will be used in the algorithm:

$$d_{m+1} = u_m + (\theta_m^2/\alpha_m)\eta_m d_m,$$
$$\theta_{m+1} = \delta_{m+1}/\tau_m,$$
$$c_{m+1} = \left(1 + \theta_{m+1}^2\right)^{-1/2},$$
$$\tau_{m+1} = \tau_m \theta_{m+1} c_{m+1},$$
$$\eta_{m+1} = c_{m+1}^2 \alpha_m.$$

Before we write down the algorithm, a few relations must be exploited. Since the vectors r_m are no longer the actual residuals in the algorithm, we change the notation to w_m. These residual vectors can be updated by the formula

$$w_m = w_{m-1} - \alpha_{m-1} A u_{m-1}.$$

The vectors Au_i can be used to update the vectors

$$v_{2j} \equiv Ap_{2j},$$

which are needed in the CGS algorithm. Multiplying (7.62) by A results in

$$Ap_{2j} = Au_{2j} + \beta_{2j-2}(Aq_{2j-2} + \beta_j Ap_{2j-2}),$$

7.4. Transpose-Free Variants

which, upon substituting the relation

$$q_{2j} = u_{2j+1},$$

translates into

$$v_{2j} = Au_{2j} + \beta_{2j-2}(Au_{2j-1} + \beta_{2j-2}v_{2j-2}).$$

Also, observe that the recurrences in (7.57) and (7.61) for q_{2j} and u_{2j+2}, respectively, become

$$u_{2j+1} = u_{2j} - \alpha_{2j}v_{2j},$$
$$u_{2j+2} = w_{2j+2} + \beta_{2j}u_{2j+1}.$$

The first equation should be used to compute u_{m+1} when m is even, and the second when m is odd. In the following algorithm, the normalization $\delta_m = \|w_m\|_2$, which normalizes each column of R_m to have 2-norm unity, is used.

ALGORITHM 7.8. TFQMR

1. Compute $w_0 = u_0 = r_0 = b - Ax_0$, $v_0 = Au_0$, $d_0 = 0$
2. $\tau_0 = \|r_0\|_2$, $\theta_0 = \eta_0 = 0$
3. Choose r_0^* such that $\rho_0 \equiv (r_0^*, r_0) \neq 0$
4. For $m = 0, 1, 2, \ldots$, until convergence, Do
5.   If m is even then
6. $\alpha_{m+1} = \alpha_m = \rho_m/(v_m, r_0^*)$
7. $u_{m+1} = u_m - \alpha_m v_m$
8.   EndIf
9. $w_{m+1} = w_m - \alpha_m Au_m$
10. $d_{m+1} = u_m + (\theta_m^2/\alpha_m)\eta_m d_m$
11. $\theta_{m+1} = \|w_{m+1}\|_2/\tau_m$, $c_{m+1} = \left(1 + \theta_{m+1}^2\right)^{-1/2}$
12. $\tau_{m+1} = \tau_m \theta_{m+1} c_{m+1}$, $\eta_{m+1} = c_{m+1}^2 \alpha_m$
13. $x_{m+1} = x_m + \eta_{m+1} d_{m+1}$
14.   If m is odd then
15. $\rho_{m+1} = (w_{m+1}, r_0^*)$, $\beta_{m-1} = \rho_{m+1}/\rho_{m-1}$
16. $u_{m+1} = w_{m+1} + \beta_{m-1}u_m$
17. $v_{m+1} = Au_{m+1} + \beta_{m-1}(Au_m + \beta_{m-1}v_{m-1})$
18.   EndIf
19. EndDo

Notice that the quantities in the odd m loop are only defined for even values of m. The residual norm of the approximate solution x_m is not available from the above algorithm as it is described. However, good estimates can be obtained using strategies similar to those used for DQGMRES. Referring to GMRES, an interesting observation is that the recurrence (6.47) is identical to the recurrence of the scalars τ_j. In addition, these two sequences start with the same values, δ_0 for the τ's and β for the γ's. Therefore,

$$\gamma_{m+1} = \tau_m.$$

Recall that γ_{m+1} is the residual for the $(m+1) \times m$ least-squares problem

$$\min_z \|\delta_0 e_1 - \bar{H}_m z\|_2.$$

Hence, a relation similar to that for DQGMRES holds; namely,

$$\|b - Ax_m\| \le \sqrt{m+1}\,\tau_m. \tag{7.83}$$

This provides a readily computable estimate of the residual norm. Another point that should be made is that it is possible to use the scalars s_m, c_m in the recurrence instead of the pair c_m, θ_m, as was done above. In this case, the proper recurrences are

$$d_{m+1} = u_m + (s_m^2/\alpha_m)\alpha_{m-1} d_m,$$

$$s_{m+1} = \delta_{m+1}/\sqrt{\tau_m^2 + \delta_{m+1}^2},$$

$$c_{m+1} = \tau_m/\sqrt{\tau_m^2 + \delta_{m+1}^2},$$

$$\tau_{m+1} = \tau_m s_{m+1},$$

$$\eta_{m+1} = c_{m+1}^2 \alpha_m.$$

Example 7.3. Table 7.3 shows the results when the TFQMR algorithm without preconditioning is applied to three of the test problems described in Section 3.7.

Matrix	Iters	Kflops	Residual	Error
F2DA	112	2736	0.46E−04	0.68E−04
F3D	78	8772	0.52E−04	0.61E−03
ORS	252	7107	0.38E−01	0.19E−03

Table 7.3. *A test run of TFQMR with no preconditioning.*

See Example 6.1 for the meaning of the column headers in the table. As with previous examples, Iters represents the number of matrix-by-vector multiplications rather than the number of BCG steps. This number is slightly higher than that of BICGSTAB.

Using the number of matrix-by-vector products as a criterion, TFQMR is more expensive than BICGSTAB in all three cases and less expensive than BCG for all cases. If the number of actual iterations is used as a criterion, then BCG is just slightly better for Problems 1 and 2. A comparison is not possible for Problem 3, since the number of matrix-by-vector products required for convergence exceeds the limit of 300. In general, the number of steps required for convergence is similar for BICGSTAB and TFQMR. A comparison with the methods seen in the previous chapter indicates that, in many cases, GMRES will be faster if the problem is well conditioned, resulting in a moderate number of steps required to converge. If many steps (say, in the hundreds) are required, then BICGSTAB and TFQMR may perform better. If memory is not an issue, GMRES or DQGMRES, with a large number of directions, is often the most reliable choice. The issue then is one of trading robustness for memory usage. In general, a sound strategy is to focus on finding a good preconditioner rather than the best accelerator.

Exercises

1. Consider the following modification of the Lanczos algorithm, Algorithm 7.1. We replace line 6 with
$$\hat{w}_{j+1} = A^T w_j - \sum_{i=1}^{j} h_{ij} w_i,$$
where the scalars h_{ij} are arbitrary. Lines 5 and 7 through 10 remain the same but line 4, in which α_j is computed, must be changed.
   a. Show how to modify line 4 to ensure that the vector $\hat{v}_{j+1}$ is orthogonal to the vectors w_i for $i = 1, \ldots, j$.
   b. Prove that the vectors v_i and the matrix T_m do not depend on the choice of the h_{ij}'s.
   c. Consider the simplest possible choice, namely, $h_{ij} \equiv 0$ for all i, j. What are the advantages and potential difficulties with this choice?

2. Assume that the Lanczos algorithm does not break down before step m, i.e., that it is possible to generate $v_1, \ldots, v_{m+1}$. Show that V_{m+1} and W_{m+1} are both of full rank.

3. Develop a modified version of the non-Hermitian Lanczos algorithm that produces a sequence of vectors v_i, w_i such that each v_i is orthogonal to every w_j with $j \neq i$ and $\|v_i\|_2 = \|w_i\|_2 = 1$ for all i. What does the projected problem become?

4. Develop a version of the non-Hermitian Lanczos algorithm that produces a sequence of vectors v_i, w_i satisfying
$$(v_i, w_j) = \pm \delta_{ij},$$
but such that the matrix T_m is Hermitian tridiagonal. What does the projected problem become in this situation?

5. Using the notation of Section 7.1.2, prove that $q_{j+k}(t) = t^k p_j(t)$ is orthogonal to the polynomials $p_1, p_2, \ldots, p_{j-k}$, assuming that $k \leq j$. Show that, if q_{j+k} is orthogonalized against $p_1, p_2, \ldots, p_{j-k}$, the result is orthogonal to all polynomials of degree less than $j + k$. Derive a general look-ahead non-Hermitian Lanczos procedure based on this observation.

6. Consider the matrices $V_m = [v_1, \ldots, v_m]$ and $W_m = [w_1, \ldots, w_m]$ obtained from the Lanczos biorthogonalization algorithm.
   a. What are the matrix representations of the (oblique) projector onto $\mathcal{K}_m(A, v_1)$ orthogonal to the subspace $\mathcal{K}_m(A^T, w_1)$ and the projector onto $\mathcal{K}_m(A^T, w_1)$ orthogonal to the subspace $\mathcal{K}_m(A, v_1)$?
   b. Express a general condition for the existence of an oblique projector onto K orthogonal to L.
   c. How can this condition be interpreted using the Lanczos vectors and the Lanczos algorithm?

7. Show a three-term recurrence satisfied by the residual vectors r_j of the BCG algorithm. Include the first two iterates to start the recurrence. Similarly, establish a three-term recurrence for the conjugate direction vectors p_j in BCG.

8. Let $\phi_j(t)$ and $\pi_j(t)$ be the residual polynomial and the conjugate direction polynomial, respectively, for the BCG algorithm, as defined in Section 7.4.1. Let $\psi_j(t)$ be any other polynomial sequence defined from the recurrence

$$\psi_0(t) = 1, \quad \psi_1(t) = (1 - \xi_0 t)\psi_0(t),$$
$$\psi_{j+1}(t) = (1 + \eta_j - \xi_j t)\psi_j(t) - \eta_j \psi_{j-1}(t).$$

   a. Show that the polynomials ψ_j are consistent; i.e., $\psi_j(0) = 1$ for all $j \geq 0$.
   b. Show the following relations:

$$\psi_{j+1}\phi_{j+1} = \psi_j\phi_{j+1} - \eta_j(\psi_{j-1} - \psi_j)\phi_{j+1} - \xi_j t \psi_j \phi_{j+1},$$
$$\psi_j \phi_{j+1} = \psi_j \phi_j - \alpha_j t \psi_j \pi_j,$$
$$(\psi_{j-1} - \psi_j)\phi_{j+1} = \psi_{j-1}\phi_j - \psi_j \phi_{j+1} - \alpha_j t \psi_{j-1} \pi_j,$$
$$\psi_{j+1}\pi_{j+1} = \psi_{j+1}\phi_{j+1} - \beta_j \eta_j \psi_{j-1}\pi_j + \beta_j(1 + \eta_j)\psi_j \pi_j - \beta_j \xi_j t \psi_j \pi_j,$$
$$\psi_j \pi_{j+1} = \psi_j \phi_{j+1} + \beta_j \psi_j \pi_j.$$

   c. Defining

$$t_j = \psi_j(A)\phi_{j+1}(A)r_0, \quad y_j = (\psi_{j-1}(A) - \psi_j(A))\phi_{j+1}(A)r_0,$$
$$p_j = \psi_j(A)\pi_j(A)r_0, \quad s_j = \psi_{j-1}(A)\pi_j(A)r_0,$$

   show how the recurrence relations of part (b) translate for these vectors.
   d. Find a formula that allows one to update the approximation x_{j+1} from the vectors x_{j-1}, x_j and t_j, p_j, y_j, s_j defined above.
   e. Proceeding as in BICGSTAB, find formulas for generating the BCG coefficients α_j and β_j from the vectors defined in part (d).

9. Prove the expression (7.76) for the CGS approximation defined by (7.66)–(7.67). Is the relation valid for any choice of scaling Δ_{m+1}?

10. Prove that the vectors r_j and r_i^* produced by the BCG algorithm are orthogonal to each other when $i \neq j$, while the vectors p_i and p_j^* are A-orthogonal; i.e., $(Ap_j, p_i^*) = 0$ for $i \neq j$.

11. The purpose of this exercise is to develop block variants of the Lanczos algorithm. Consider a two-sided analogue of the block Arnoldi algorithm in its variant of Algorithm 6.24. Formally, the general steps that define the biorthogonalization process for $j \geq p$ are as follows:

   (i) Orthogonalize Av_{j-p+1} versus $w_1, w_2, \ldots, w_j$ (by subtracting a linear combination of $v_1, \ldots, v_j$ from Av_{j-p+1}). Call v the resulting vector.
   (ii) Orthogonalize $A^T w_{j-p+1}$ versus $v_1, v_2, \ldots, v_j$ (by subtracting a linear combination of $w_1, \ldots, w_j$ from $A^T w_{j-p+1}$). Call w the resulting vector.
   (iii) Normalize the two vectors v and w so that $(v, w) = 1$ to get v_{j+1} and w_{j+1}.

   Here, p is the block size and it is assumed that the initial blocks are biorthogonal: $(v_i, w_j) = \delta_{ij}$ for $i, j \leq p$.

a. Show that Av_{j-p+1} needs only to be orthogonalized against the $2p$ previous w_i's instead of all of them. Similarly, $A^T w_{j-p+1}$ must be orthogonalized only against the $2p$ previous v_i's.

b. Write down the algorithm completely. Show the orthogonality relations satisfied by the vectors v_i and w_j. Show also relations similar to (7.3) and (7.4).

c. We now assume that the two sets of vectors v_i and w_j have different block sizes. Call q the block size for the w's. Line (ii) of the above formal algorithm is changed as follows:

> (ii)a Orthogonalize $A^T w_{j-q+1}$ versus $v_1, v_2, \ldots, v_j$ (...). Call w the resulting vector.

The rest remains unchanged. The initial vectors are again biorthogonal: $(v_i, w_j) = \delta_{ij}$ for $i \leq p$ and $j \leq q$. Show that now Av_{j-p+1} needs only to be orthogonalized against the $q+p$ previous w_i's instead of all of them. Show a similar result for the w_j's.

d. Show how a block version of BCG and QMR can be developed based on the algorithm resulting from part (c).

Notes and References

The pioneering paper by Lanczos [196], on what is now referred to as BCG, did not receive the attention it deserved. Fletcher [130], who developed the modern version of the algorithm, mentions the 1950 Lanczos paper [194], which is devoted mostly to eigenvalue problems, but seems unaware of the second [196], which is devoted to linear systems. Likewise, the paper by Sonneveld [271], which proved for the first time that the A^T operations were not necessary, received little attention for several years (the first reference to the method [311] dates back to 1980). TFQMR (Freund and Nachtigal [136]) and BICGSTAB (van der Vorst [289]) were later developed to cure some of the numerical problems that plague CGS. Many additions to and variations of the basic BCG, BICGSTAB, and TFQMR techniques appeared; see, e.g., [63, 72, 160, 161, 259], among others. Some variations were developed to cope with the breakdown of the underlying Lanczos or BCG algorithm; see, for example, [62, 27, 135, 259, 320]. Finally, block methods of these algorithms have also been developed; see, e.g., [5].

The Lanczos-type algorithms developed for solving linear systems are rooted in the theory of orthogonal polynomials and Padé approximation. Lanczos himself certainly used this viewpoint when he wrote his breakthrough papers [194, 196] in the early 1950s. The monograph by Brezinski [59] gives an excellent coverage of the intimate relations between approximation theory and the Lanczos-type algorithms. Freund [133] establishes these relations for QMR methods. A few optimality properties for the class of methods presented in this chapter can be proved using a variable metric, i.e., an inner product that is different at each step [29]. A survey by Weiss [307] presents a framework for Krylov subspace methods explaining some of these optimality properties and the interrelationships between Krylov subspace methods. Several authors discuss a class of techniques known as residual smoothing; see, for example, [258, 323, 307, 61]. These techniques can be applied to any iterative sequence x_k to build a new sequence of iterates y_k by combining y_{k-1} with the difference $x_k - y_{k-1}$. A remarkable result shown by Zhou and Walker [323] is that the iterates of the QMR algorithm can be obtained from those of BCG as a particular case of residual smoothing.

A number of projection-type methods on Krylov subspaces, other than those seen in this chapter and the previous one, are described in [1]. The group of rank-k update methods discussed by Eirola and Nevanlinna [113] and Deuflhard et al. [100] is closely related to Krylov subspace methods. In fact, GMRES can be viewed as a particular example of these methods.

Also of interest and not covered in this book are the *vector extrapolation* techniques, which are discussed, for example, in the books by Brezinski [59] and Brezinski and Redivo Zaglia [60] and the articles [269] and [176]. Connections between these methods and Krylov subspace methods have been uncovered and are discussed by Brezinski [59] and Sidi [262].

Chapter 8
Methods Related to the Normal Equations

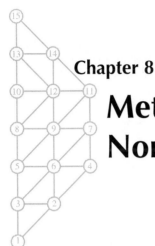

There are a number of techniques for converting a nonsymmetric linear system into a symmetric one. One such technique solves the equivalent linear system $A^T A x = A^T b$, called the normal equations. Often, this approach is avoided in practice because the coefficient matrix $A^T A$ is much worse conditioned than A. However, the normal equations approach may be adequate in some situations. Indeed, there are even applications in which it is preferred to the usual Krylov subspace techniques. This chapter covers iterative methods that are either directly or implicitly related to the normal equations.

8.1 The Normal Equations

In order to solve the linear system $Ax = b$ when A is nonsymmetric, we can solve the equivalent system

$$A^T A \, x = A^T b, \tag{8.1}$$

which is symmetric positive definite (SPD). This system is known as the system of the *normal equations* associated with the least-squares problem

$$\text{minimize} \quad \|b - Ax\|_2. \tag{8.2}$$

Note that (8.1) is typically used to solve the least-squares problem (8.2) for *overdetermined* systems, i.e., when A is a rectangular matrix of size $n \times m$, $m < n$.

A similar well-known alternative sets $x = A^T u$ and solves the following equation for u:

$$A A^T u = b. \tag{8.3}$$

Once the solution u is computed, the original unknown x can be obtained by multiplying u by A^T. However, most of the algorithms we will see do not invoke the u variable explicitly but work with the original variable x instead. The above system of equations can be used to solve *underdetermined* systems, i.e., those systems involving rectangular matrices of size $n \times m$, with $n < m$. It is related to (8.1) in the following way. Assume that $n \leq m$ and that A has full rank. Let x_* be *any* solution to the underdetermined system $Ax = b$. Then (8.3) represents the normal equations for the least-squares problem

$$\text{minimize} \quad \|x_* - A^T u\|_2. \tag{8.4}$$

Since by definition $A^T u = x$, then (8.4) will find the solution vector x that is closest to x_* in the 2-norm sense. What is interesting is that when $n < m$ there are infinitely many solutions x_* to the system $Ax = b$, but the minimizer u of (8.4) does not depend on the particular x_* used.

The system (8.1) and methods derived from it are often labeled with NR (N for "normal" and R for "residual"), while (8.3) and related techniques are labeled with NE (N for "normal" and E for "error"). If A is square and nonsingular, the coefficient matrices of these systems are both SPD, and the simpler methods for symmetric problems, such as the conjugate gradient (CG) algorithm, can be applied. Thus, CGNE denotes the CG method applied to the system (8.3), and CGNR the CG method applied to (8.1).

There are several alternative ways to formulate symmetric linear systems with the same solution as the original system. For instance, the symmetric linear system

$$\begin{pmatrix} I & A \\ A^T & O \end{pmatrix} \begin{pmatrix} r \\ x \end{pmatrix} = \begin{pmatrix} b \\ 0 \end{pmatrix}, \tag{8.5}$$

with $r = b - Ax$, arises from the standard necessary conditions satisfied by the solution of the constrained optimization problem

$$\text{minimize} \quad \frac{1}{2}\|r - b\|_2^2 \tag{8.6}$$

$$\text{subject to} \quad A^T r = 0. \tag{8.7}$$

The solution x to (8.5) is the vector of Lagrange multipliers for the above problem.

Another equivalent symmetric system is of the form

$$\begin{pmatrix} O & A \\ A^T & O \end{pmatrix} \begin{pmatrix} Ax \\ x \end{pmatrix} = \begin{pmatrix} b \\ A^T b \end{pmatrix}.$$

The eigenvalues of the coefficient matrix for this system are $\pm \sigma_i$, where σ_i is an arbitrary singular value of A. Indefinite systems of this sort are not easier to solve than the original nonsymmetric system in general. Although not obvious immediately, this approach is similar in nature to the approach (8.1) and the corresponding CG iterations applied to them should behave similarly.

A general consensus is that solving the normal equations can be an inefficient approach in the case when A is poorly conditioned. Indeed, the 2-norm condition number of $A^T A$ is given by

$$\text{Cond}_2(A^T A) = \|A^T A\|_2 \, \|(A^T A)^{-1}\|_2.$$

Now observe that $\|A^T A\|_2 = \sigma_{max}^2(A)$, where $\sigma_{max}(A)$ is the largest singular value of A, which, incidentally, is also equal to the 2-norm of A. Thus, using a similar argument for

8.2. Row Projection Methods

the inverse $(A^T A)^{-1}$ yields

$$\text{Cond}_2(A^T A) = \|A\|_2^2 \, \|A^{-1}\|_2^2 = \text{Cond}_2^2(A). \tag{8.8}$$

The 2-norm condition number of $A^T A$ is exactly the square of the condition number of A, which could cause difficulties. For example, if originally $\text{Cond}_2(A) = 10^8$, then an iterative method may be able to perform reasonably well. However, a condition number of 10^{16} can be much more difficult to handle with a standard iterative method. That is because any progress made in one step of the iterative procedure may be annihilated by the noise due to numerical errors. On the other hand, if the original matrix has a good 2-norm condition number, then the normal equations approach should not cause any serious difficulties. In the extreme case when A is unitary, i.e., when $A^H A = I$, then the normal equations are clearly the best approach (the CG method will converge in zero steps!).

8.2 Row Projection Methods

When implementing a basic relaxation scheme, such as Jacobi or successive overrelaxation (SOR), to solve the linear system

$$A^T A x = A^T b \tag{8.9}$$

or

$$A A^T u = b, \tag{8.10}$$

it is possible to exploit the fact that the matrix $A^T A$ or $A A^T$ need not be formed explicitly. As will be seen, only a row or a column of A at a time is needed at a given relaxation step. These methods are known as *row projection methods* since they are indeed projection methods on rows of A or A^T. Block row projection methods can be defined similarly.

8.2.1 Gauss–Seidel on the Normal Equations

It was stated above that, in order to use relaxation schemes on the normal equations, access to only one column of A at a time for (8.9) and one row at a time for (8.10) is needed. This is explained for (8.10) first. Starting from an approximation to the solution of (8.10), a basic relaxation-based iterative procedure modifies its components in a certain order using a succession of relaxation steps of the simple form

$$u_{new} = u + \delta_i e_i, \tag{8.11}$$

where e_i is the ith column of the identity matrix. The scalar δ_i is chosen so that the ith component of the residual vector for (8.10) becomes zero. Therefore,

$$(b - A A^T (u + \delta_i e_i), e_i) = 0, \tag{8.12}$$

which, setting $r = b - A A^T u$, yields

$$\delta_i = \frac{(r, e_i)}{\|A^T e_i\|_2^2}. \tag{8.13}$$

Denote by β_i the ith component of b. Then a basic relaxation step consists of taking

$$\delta_i = \frac{\beta_i - (A^T u, A^T e_i)}{\|A^T e_i\|_2^2}. \tag{8.14}$$

Also, (8.11) can be rewritten in terms of x variables as follows:

$$x_{new} = x + \delta_i A^T e_i. \tag{8.15}$$

The auxiliary variable u has now been removed from the scene and is replaced with the original variable $x = A^T u$.

Consider the implementation of a forward Gauss–Seidel sweep based on (8.15) and (8.13) for a general sparse matrix. The evaluation of δ_i from (8.13) requires the inner product of the current approximation $x = A^T u$ with $A^T e_i$, the ith row of A. This inner product is inexpensive to compute because $A^T e_i$ is usually sparse. If an acceleration parameter ω is used, we only need to change δ_i into $\omega \delta_i$. Therefore, a forward SOR sweep is as follows.

ALGORITHM 8.1. Forward NE-SOR Sweep

1. Choose an initial x
2. For $i = 1, 2, \ldots, n$, Do
3. $\quad \delta_i = \omega \frac{\beta_i - (A^T e_i, x)}{\|A^T e_i\|_2^2}$
4. $\quad x := x + \delta_i A^T e_i$
5. EndDo

Note that $A^T e_i$ is a vector equal to the transpose of the ith row of A. All that is needed is the row data structure for A to implement the above algorithm. Denoting by nz_i the number of nonzero elements in the ith row of A, we then see that each step of the above sweep requires $2nz_i + 2$ operations in line 3 and another $2nz_i$ operations in line 4, bringing the total to $4nz_i + 2$. The total for a whole sweep becomes $4nz + 2n$ operations, where nz represents the total number of nonzero elements of A. Twice as many operations are required for the symmetric Gauss–Seidel or the symmetric SOR (SSOR) iteration. Storage consists of the right-hand side, the vector x, and possibly an additional vector to store the 2-norms of the rows of A. A better alternative would be to rescale each row by its 2-norm at the start.

Similarly, Gauss–Seidel for (8.9) consists of a sequence of steps of the form

$$x_{new} = x + \delta_i e_i. \tag{8.16}$$

Again, the scalar δ_i is to be selected so that the ith component of the residual vector for (8.9) becomes zero, which yields

$$(A^T b - A^T A(x + \delta_i e_i), e_i) = 0. \tag{8.17}$$

With $r \equiv b - Ax$, this becomes $(A^T(r - \delta_i A e_i), e_i) = 0$, which yields

$$\delta_i = \frac{(r, A e_i)}{\|A e_i\|_2^2}. \tag{8.18}$$

Then the following algorithm is obtained.

8.2. Row Projection Methods

ALGORITHM 8.2. Forward NR-SOR Sweep

1. Choose an initial x and compute $r := b - Ax$
2. For $i = 1, 2, \ldots, n$, Do
3. $\quad \delta_i = \omega \frac{(r, Ae_i)}{\|Ae_i\|_2^2}$
4. $\quad x := x + \delta_i e_i$
5. $\quad r := r - \delta_i Ae_i$
6. EndDo

In contrast with Algorithm 8.1, the column data structure of A is now needed for the implementation instead of its row data structure. Here, the right-hand side b can be overwritten by the residual vector r, so the storage requirement is essentially the same as in the previous case. In the NE version, the scalar $\beta_i - (x, a_i)$ is just the ith component of the current residual vector $r = b - Ax$. As a result, stopping criteria can be built for both algorithms based on either the residual vector or the variation in the solution. Note that the matrices AA^T and A^TA can be dense or generally much less sparse than A, yet the cost of the above implementations depends only on the nonzero structure of A. This is a significant advantage of relaxation-type preconditioners over incomplete factorization preconditioners when using CG methods to solve the normal equations.

One question remains concerning the acceleration of the above relaxation schemes by under- or overrelaxation. If the usual acceleration parameter ω is introduced, then we only have to multiply the scalars δ_i in the previous algorithms by ω. One serious difficulty here is to determine the optimal relaxation factor. If nothing in particular is known about the matrix AA^T, then the method will converge for any ω lying strictly between 0 and 2, as was seen in Chapter 4, because the matrix is positive definite. Moreover, another unanswered question is how convergence can be affected by various reorderings of the rows. For general sparse matrices, the answer is not known.

8.2.2 Cimmino's Method

In a Jacobi iteration for the system (8.9), the components of the new iterate satisfy the following condition:
$$(A^Tb - A^TA(x + \delta_i e_i), e_i) = 0. \tag{8.19}$$

This yields
$$(b - A(x + \delta_i e_i), Ae_i) = 0 \quad \text{or} \quad (r - \delta_i Ae_i, Ae_i) = 0,$$

in which r is the old residual $b - Ax$. As a result, the i component of the new iterate x_{new} is given by
$$x_{new,i} = x_i + \delta_i e_i, \tag{8.20}$$
$$\delta_i = \frac{(r, Ae_i)}{\|Ae_i\|_2^2}. \tag{8.21}$$

Here, be aware that these equations do not result in the same approximation as that produced by Algorithm 8.2, even though the modifications are given by the same formula. Indeed, the vector x is not updated after each step and therefore the scalars δ_i are different for the two algorithms. This algorithm is usually described with an acceleration parameter ω; i.e., all δ_i's are multiplied uniformly by a certain ω. If d denotes the vector with coordinates $\delta_i, i = 1, \ldots, n$, the following algorithm results.

ALGORITHM 8.3. Cimmino-NR

1. Choose initial guess x_0. Set $x = x_0, r = b - Ax_0$
2. Until convergence, Do
3.   For $i = 1, \ldots, n$, Do
4. $\delta_i = \omega \frac{(r, Ae_i)}{\|Ae_i\|_2^2}$
5.   EndDo
6. $x := x + d$, where $d = \sum_{i=1}^{n} \delta_i e_i$
7. $r := r - Ad$
8. EndDo

Notice that all the coordinates will use the same residual vector r to compute the updates δ_i. When $\omega = 1$, each instance of the above formulas is mathematically equivalent to performing a projection step for solving $Ax = b$ with $\mathcal{K} = \text{span}\{e_i\}$ and $\mathcal{L} = A\mathcal{K}$. It is also mathematically equivalent to performing an orthogonal projection step for solving $A^T Ax = A^T b$ with $\mathcal{K} = \text{span}\{e_i\}$.

It is interesting to note that, when each column Ae_i is normalized by its 2-norm, i.e., if $\|Ae_i\|_2 = 1, i = 1, \ldots, n$, then $\delta_i = \omega(r, Ae_i) = \omega(A^T r, e_i)$. In this situation,

$$d = \omega A^T r = \omega A^T (b - Ax)$$

and the main loop of the algorithm takes the vector form

$$d := \omega A^T r,$$
$$x := x + d,$$
$$r := r - Ad.$$

Each iteration is therefore equivalent to a step of the form

$$x_{new} = x + \omega \left(A^T b - A^T Ax \right),$$

which is nothing but the Richardson iteration applied to the normal equations (8.1). In particular, as was seen in Example 4.1, convergence is guaranteed for any ω that satisfies

$$0 < \omega < \frac{2}{\lambda_{max}}, \tag{8.22}$$

where λ_{max} is the largest eigenvalue of $A^T A$. In addition, the best acceleration parameter is given by

$$\omega_{opt} = \frac{2}{\lambda_{min} + \lambda_{max}},$$

in which, similarly, λ_{min} is the smallest eigenvalue of $A^T A$. If the columns are not normalized by their 2-norms, then the procedure is equivalent to a Richardson iteration with diagonal preconditioning. The theory regarding convergence is similar but involves the

8.3. Conjugate Gradient and Normal Equations

preconditioned matrix or, equivalently, the matrix A' obtained from A by normalizing its columns.

The algorithm can be expressed in terms of projectors. Observe that the new residual satisfies

$$r_{new} = r - \sum_{i=1}^{n} \omega \frac{(r, Ae_i)}{\|Ae_i\|_2^2} Ae_i. \tag{8.23}$$

Each of the operators

$$P_i : r \longrightarrow \frac{(r, Ae_i)}{\|Ae_i\|_2^2} Ae_i \equiv P_i r \tag{8.24}$$

is an orthogonal projector onto Ae_i, the ith column of A. Hence, we can write

$$r_{new} = \left(I - \omega \sum_{i=1}^{n} P_i \right) r. \tag{8.25}$$

There are two important variations of the above scheme. First, because the point Jacobi iteration can be very slow, it may be preferable to work with sets of vectors instead. Let $\pi_1, \pi_2, \ldots, \pi_p$ be a partition of the set $\{1, 2, \ldots, n\}$ and, for each π_j, let E_j be the matrix obtained by extracting the columns of the identity matrix whose indices belong to π_j. Going back to the projection framework, define $A_i = AE_i$. If an orthogonal projection method is used onto E_j to solve (8.1), then the new iterate is given by

$$x_{new} = x + \omega \sum_{i}^{p} E_i d_i, \tag{8.26}$$

$$d_i = (E_i^T A^T A E_i)^{-1} E_i^T A^T r = (A_i^T A_i)^{-1} A_i^T r. \tag{8.27}$$

Each individual block component d_i can be obtained by solving a least-squares problem

$$\min_{d} \|r - A_i d\|_2.$$

An interpretation of this indicates that each individual substep attempts to reduce the residual as much as possible by taking linear combinations from specific columns of A_i. Similarly to the scalar iteration, we also have

$$r_{new} = \left(I - \omega \sum_{i=1}^{n} P_i \right) r,$$

where P_i now represents an orthogonal projector onto the span of A_i.

Note that $A_1, A_2, \ldots, A_p$ is a partition of the column set $\{Ae_i\}_{i=1,\ldots,n}$, which can be arbitrary. Another remark is that the original Cimmino method was formulated for rows instead of columns; i.e., it was based on (8.1) instead of (8.3). The alternative algorithm based on columns rather than rows is easy to derive.

8.3 Conjugate Gradient and Normal Equations

A popular combination to solve nonsymmetric linear systems applies the CG algorithm to solve either (8.1) or (8.3). As is shown next, the resulting algorithms can be rearranged because of the particular nature of the coefficient matrices.

8.3.1 CGNR

We begin with the CGNR algorithm applied to (8.1). Applying CG directly to the system and denoting by z_i the residual vector at step i (instead of r_i) results in the following sequence of operations:

- $\alpha_j := (z_j, z_j)/(A^T A p_j, p_j) = (z_j, z_j)/(A p_j, A p_j)$,
- $x_{j+1} := x_j + \alpha_j p_j$,
- $z_{j+1} := z_j - \alpha_j A^T A p_j$,
- $\beta_j := (z_{j+1}, z_{j+1})/(z_j, z_j)$,
- $p_{j+1} := z_{j+1} + \beta_j p_j$.

If the original residual $r_i = b - A x_i$ must be available at every step, we may compute the residual z_{i+1} in two parts: $r_{j+1} := r_j - \alpha_j A p_j$ and then $z_{i+1} = A^T r_{i+1}$, which is the residual for the normal equations (8.1). It is also convenient to introduce the vector $w_i = A p_i$. With these definitions, the algorithm can be cast in the following form.

ALGORITHM 8.4. CGNR

1. Compute $r_0 = b - A x_0$, $z_0 = A^T r_0$, $p_0 = z_0$
2. For $i = 0, \ldots,$ until convergence, Do
3. $w_i = A p_i$
4. $\alpha_i = \|z_i\|^2 / \|w_i\|_2^2$
5. $x_{i+1} = x_i + \alpha_i p_i$
6. $r_{i+1} = r_i - \alpha_i w_i$
7. $z_{i+1} = A^T r_{i+1}$
8. $\beta_i = \|z_{i+1}\|_2^2 / \|z_i\|_2^2$
9. $p_{i+1} = z_{i+1} + \beta_i p_i$
10. EndDo

In Chapter 6, the approximation x_m produced at the mth step of the CG algorithm was shown to minimize the energy norm of the error over an affine Krylov subspace. In this case, x_m minimizes the function

$$f(x) \equiv (A^T A(x_* - x), (x_* - x))$$

over all vectors x in the affine Krylov subspace

$$x_0 + \mathcal{K}_m(A^T A, A^T r_0) = x_0 + \text{span}\{A^T r_0, A^T A A^T r_0, \ldots, (A^T A)^{m-1} A^T r_0\},$$

in which $r_0 = b - A x_0$ is the initial residual with respect to the original equations $Ax = b$ and $A^T r_0$ is the residual with respect to the normal equations $A^T A x = A^T b$. However, observe that

$$f(x) = (A(x_* - x), A(x_* - x)) = \|b - Ax\|_2^2.$$

8.3. Conjugate Gradient and Normal Equations

Therefore, CGNR produces the approximate solution in the above subspace that has the smallest residual norm with respect to the original linear system $Ax = b$. The difference with the generalized minimal residual (GMRES) algorithm seen in Chapter 6 is the subspace in which the residual norm is minimized.

Example 8.1. Table 8.1 shows the results of applying the CGNR algorithm with no preconditioning to three of the test problems described in Section 3.7.

Matrix	Iters	Kflops	Residual	Error
F2DA	300	4847	0.23E+02	0.62E+00
F3D	300	23704	0.42E+00	0.15E+00
ORS	300	5981	0.30E+02	0.60E−02

Table 8.1. *A test run of CGNR with no preconditioning.*

See Example 6.1 for the meaning of the column headers in the table. The method failed to converge in less than 300 steps for all three problems. Failures of this type, characterized by very slow convergence, are rather common for CGNE and CGNR applied to problems arising from partial differential equations (PDEs). Preconditioning should improve performance somewhat but, as will be seen in Chapter 10, normal equations are also difficult to precondition.

8.3.2 CGNE

A similar reorganization of the CG algorithm is possible for the system (8.3) as well. Applying the CG algorithm directly to (8.3) and denoting by q_i the conjugate directions, the actual CG iteration for the u variable is as follows:

- $\alpha_j := (r_j, r_j)/(AA^T q_j, q_j) = (r_j, r_j)/(A^T q_j, A^T q_j)$,
- $u_{j+1} := u_j + \alpha_j q_j$,
- $r_{j+1} := r_j - \alpha_j AA^T q_j$,
- $\beta_j := (r_{j+1}, r_{j+1})/(r_j, r_j)$,
- $q_{j+1} := r_{j+1} + \beta_j q_j$.

Notice that an iteration can be written with the original variable $x_i = x_0 + A^T(u_i - u_0)$ by introducing the vector $p_i = A^T q_i$. Then the residual vectors for the vectors x_i and u_i are the same. No longer are the q_i-vectors needed because the p_i's can be obtained as $p_{j+1} := A^T r_{j+1} + \beta_j p_j$. The resulting algorithm described below, the CGNE, is also known as Craig's method.

ALGORITHM 8.5. CGNE (Craig's Method)

1. Compute $r_0 = b - Ax_0$, $p_0 = A^T r_0$
2. For $i = 0, 1, \ldots$, until convergence, Do
3. $\quad \alpha_i = (r_i, r_i)/(p_i, p_i)$
4. $\quad x_{i+1} = x_i + \alpha_i p_i$
5. $\quad r_{i+1} = r_i - \alpha_i A p_i$
6. $\quad \beta_i = (r_{i+1}, r_{i+1})/(r_i, r_i)$
7. $\quad p_{i+1} = A^T r_{i+1} + \beta_i p_i$
8. EndDo

We now explore the optimality properties of this algorithm, as was done for CGNR. The approximation u_m related to the variable x_m by $x_m = A^T u_m$ is the actual mth CG approximation for the linear system (8.3). Therefore, it minimizes the energy norm of the error on the Krylov subspace $\mathcal{K}_m$. In this case, u_m minimizes the function

$$f(u) \equiv (AA^T(u_* - u), (u_* - u))$$

over all vectors u in the affine Krylov subspace

$$u_0 + \mathcal{K}_m(AA^T, r_0) = u_0 + \text{span}\{r_0, AA^T r_0, \ldots, (AA^T)^{m-1} r_0\}.$$

Notice that $r_0 = b - AA^T u_0 = b - Ax_0$. Also, observe that

$$f(u) = (A^T(u_* - u), A^T(u_* - u)) = \|x_* - x\|_2^2,$$

where $x = A^T u$. Therefore, CGNE produces the approximate solution in the subspace

$$x_0 + A^T \mathcal{K}_m(AA^T, r_0) = x_0 + \mathcal{K}_m(A^T A, A^T r_0)$$

that has the smallest 2-norm of the error. In addition, note that the subspace $x_0 + \mathcal{K}_m(A^T A, A^T r_0)$ is identical to the subspace found for CGNR. Therefore, *the two methods find approximations from the same subspace that achieve different optimality properties: minimal residual for CGNR and minimal error for CGNE.*

8.4 Saddle-Point Problems

Now consider the equivalent system

$$\begin{pmatrix} I & A \\ A^T & O \end{pmatrix} \begin{pmatrix} r \\ x \end{pmatrix} = \begin{pmatrix} b \\ 0 \end{pmatrix},$$

with $r = b - Ax$. This system can be derived from the necessary conditions applied to the constrained least-squares problem (8.6)–(8.7). Thus, the 2-norm of $b - r = Ax$ is minimized implicitly under the constraint $A^T r = 0$. Note that A does not have to be a square matrix.

This can be extended into a more general constrained quadratic optimization problem as follows:

$$\text{minimize } f(x) \equiv \frac{1}{2}(Ax, x) - (x, b) \tag{8.28}$$

$$\text{subject to } B^T x = c. \tag{8.29}$$

8.4. Saddle-Point Problems

The necessary conditions for optimality yield the linear system

$$\begin{pmatrix} A & B \\ B^T & 0 \end{pmatrix} \begin{pmatrix} x \\ y \end{pmatrix} = \begin{pmatrix} b \\ c \end{pmatrix}, \qquad (8.30)$$

in which the names of the variables r, x are changed to x, y for notational convenience. It is assumed that the column dimension of B does not exceed its row dimension. The Lagrangian for the above optimization problem is

$$L(x, y) = \frac{1}{2}(Ax, x) - (x, b) + (y, (B^T x - c))$$

and the solution of (8.30) is the saddle point of the above Lagrangian. Optimization problems of the form (8.28)–(8.29) and the corresponding linear systems (8.30) are important and arise in many applications. Because they are intimately related to the normal equations, we discuss them briefly here.

In the context of fluid dynamics, a well-known iteration technique for solving the linear system (8.30) is Uzawa's method, which resembles a relaxed block SOR iteration.

ALGORITHM 8.6. Uzawa's Method

1. Choose x_0, y_0
2. For $k = 0, 1, \ldots,$ until convergence, Do
3. $\quad x_{k+1} = A^{-1}(b - By_k)$
4. $\quad y_{k+1} = y_k + \omega(B^T x_{k+1} - c)$
5. EndDo

The algorithm requires the solution of the linear system

$$Ax_{k+1} = b - By_k \qquad (8.31)$$

at each iteration. By substituting the result of line 3 into line 4, the x_k iterates can be eliminated to obtain the following relation for the y_k's:

$$y_{k+1} = y_k + \omega \left(B^T A^{-1}(b - By_k) - c \right),$$

which is nothing but a Richardson iteration for solving the linear system

$$B^T A^{-1} B y = B^T A^{-1} b - c. \qquad (8.32)$$

Apart from a sign, this system is the reduced system resulting from eliminating the x variable from (8.30). Convergence results can be derived from the analysis of the Richardson iteration.

Corollary 8.1. *Let A be an SPD matrix and B a matrix of full rank. Then $S = B^T A^{-1} B$ is also SPD and Uzawa's algorithm converges iff*

$$0 < \omega < \frac{2}{\lambda_{max}(S)}. \qquad (8.33)$$

In addition, the optimal convergence parameter ω is given by

$$\omega_{opt} = \frac{2}{\lambda_{min}(S) + \lambda_{max}(S)}.$$

Proof. The proof of this result is straightforward and is based on the results seen in Example 4.1. □

It is interesting to observe that, when $c = 0$ and A is SPD, then the system (8.32) can be regarded as the normal equations for minimizing the A^{-1}-norm of $b - By$. Indeed, the optimality conditions are equivalent to the orthogonality conditions

$$(b - By, Bw)_{A^{-1}} = 0 \quad \forall \, w,$$

which translate into the linear system $B^T A^{-1} By = B^T A^{-1} b$. As a consequence, the problem will tend to be easier to solve if the columns of B are almost orthogonal with respect to the A^{-1} inner product. This is true when solving the *Stokes problem*, where B represents the discretization of the gradient operator, while B^T discretizes the divergence operator, and A is the discretization of a Laplacian. In this case, if it were not for the boundary conditions, the matrix $B^T A^{-1} B$ would be the identity. This feature can be exploited in developing preconditioners for solving problems of the form (8.30). Another particular case is when A is the identity matrix and $c = 0$. Then the linear system (8.32) becomes the system of the normal equations for minimizing the 2-norm of $b - By$. These relations provide insight into understanding that the block form (8.30) is actually a form of normal equations for solving $By = b$ in the least-squares sense. However, a different inner product is used.

In Uzawa's method, a linear system at each step must be solved, namely, the system (8.31). Solving this system is equivalent to finding the minimum of the quadratic function:

$$\text{minimize} \quad f_k(x) \equiv \frac{1}{2}(Ax, x) - (x, b - By_k). \tag{8.34}$$

Apart from constants, $f_k(x)$ is the Lagrangian evaluated at the previous y iterate. The solution of (8.31), or the equivalent optimization problem (8.34), is expensive. A common alternative replaces the x variable update (8.31) by taking one step in the gradient direction for the quadratic function (8.34), usually with fixed step length ϵ. The gradient of $f_k(x)$ at the current iterate is $Ax_k - (b - By_k)$. This results in the Arrow–Hurwicz algorithm.

ALGORITHM 8.7. The Arrow–Hurwicz Algorithm

1. Select an initial guess x_0, y_0 for the system (8.30)
2. For $k = 0, 1, \ldots,$ until convergence, Do
3.   Compute $x_{k+1} = x_k + \epsilon(b - Ax_k - By_k)$
4.   Compute $y_{k+1} = y_k + \omega(B^T x_{k+1} - c)$
5. EndDo

The above algorithm is a block iteration of the form

$$\begin{pmatrix} I & 0 \\ -\omega B^T & I \end{pmatrix} \begin{pmatrix} x_{k+1} \\ y_{k+1} \end{pmatrix} = \begin{pmatrix} I - \epsilon A & -\epsilon B \\ 0 & I \end{pmatrix} \begin{pmatrix} x_k \\ y_k \end{pmatrix} + \begin{pmatrix} \epsilon b \\ -\omega c \end{pmatrix}.$$

Exercises

Uzawa's method, and many similar techniques for solving (8.30), are based on solving the reduced system (8.32). An important observation here is that the Schur complement matrix $S \equiv B^T A^{-1} B$ need not be formed explicitly. This can be useful if this reduced system is to be solved by an iterative method. The matrix A is typically factored by a Cholesky-type factorization. The linear systems with the coefficient matrix A can also be solved by a preconditioned CG method. Of course these systems must then be solved accurately.

Sometimes it is useful to *regularize* the least-squares problem (8.28) by solving the following problem in its place:

$$\text{minimize } f(x) \equiv \frac{1}{2}(Ax, x) - (x, b) + \rho(Cy, y)$$
$$\text{subject to } B^T x = c,$$

in which ρ is a scalar parameter. For example, C can be the identity matrix or the matrix $B^T B$. The matrix resulting from the Lagrange multipliers approach then becomes

$$\begin{pmatrix} A & B \\ B^T & \rho C \end{pmatrix}.$$

The new Schur complement matrix is

$$S = \rho C - B^T A^{-1} B.$$

Example 8.2. In the case where $C = B^T B$, the above matrix takes the form

$$S = B^T(\rho I - A^{-1})B.$$

Assuming that A is SPD, S is also positive definite when

$$\rho \geq \frac{1}{\lambda_{min}(A)}.$$

However, it is also *negative definite* for

$$\rho \leq \frac{1}{\lambda_{max}}(A),$$

a condition that may be easier to satisfy in practice.

Exercises

1. Derive the linear system (8.5) by expressing the standard necessary conditions for the problem (8.6)–(8.7).

2. It was stated in Section 8.2.2 that, when $\|A^T e_i\|_2 = 1$ for $i = 1, \ldots, n$, the vector d defined in Algorithm 8.3 is equal to $\omega A^T r$.
   a. What does this become in the general situation when $\|A^T e_i\|_2 \neq 1$?
   b. Is Cimmino's method still equivalent to a Richardson iteration?
   c. Show convergence results similar to those of the scaled case.

3. In Section 8.2.2, Cimmino's algorithm was derived based on the NR formulation, i.e., on (8.1). Derive an NE formulation, i.e., an algorithm based on Jacobi's method for (8.3).

4. What are the eigenvalues of the matrix (8.5)? Derive a system whose coefficient matrix has the form
$$B(\alpha) = \begin{pmatrix} 2\alpha I & A \\ A^T & O \end{pmatrix}$$
and is also equivalent to the original system $Ax = b$. What are the eigenvalues of $B(\alpha)$? Plot the spectral norm of $B(\alpha)$ as a function of α.

5. It was argued in Section 8.4 that when $c = 0$ the system (8.32) is nothing but the normal equations for minimizing the A^{-1}-norm of the residual $r = b - By$.

   a. Write the associated CGNR approach for solving this problem. Find a variant that requires only one linear system solution with the matrix A at each CG step. [Hint: Write the CG algorithm for the associated normal equations and see how the resulting procedure can be reorganized to save operations.] Find a variant that is suitable for the case where the Cholesky factorization of A is available.

   b. Derive a method for solving the equivalent system (8.30) for the case when $c = 0$ and then for the general case when $c \neq 0$. How does this technique compare with Uzawa's method?

6. Consider the linear system (8.30), in which $c = 0$ and B is of full rank. Define the matrix
$$P = I - B(B^T B)^{-1} B^T.$$

   a. Show that P is a projector. Is it an orthogonal projector? What are the range and null space of P?

   b. Show that the unknown x can be found by solving the linear system
$$PAPx = Pb, \tag{8.35}$$
   in which the coefficient matrix is singular but the system is consistent; i.e., there is a nontrivial solution because the right-hand side is in the range of the matrix (see Chapter 1).

   c. What must be done to adapt the CG algorithm for solving the above linear system (which is symmetric, but not positive definite)? In which subspace are the iterates generated from the CG algorithm applied to (8.35)?

   d. Assume that the QR factorization of the matrix B is computed. Write an algorithm based on the approach of the previous parts of this question for solving the linear system (8.30).

7. Show that Uzawa's iteration can be formulated as a fixed-point iteration associated with the splitting $C = M - N$, with
$$M = \begin{pmatrix} A & O \\ -\omega B^T & I \end{pmatrix}, \quad N = \begin{pmatrix} O & -B \\ O & I \end{pmatrix}.$$
Derive the convergence result of Corollary 8.1.

8. Show that each new vector iterate in Cimmino's method is such that
$$x_{new} = x + \omega A^{-1} \sum_i P_i r,$$
where P_i is defined by (8.24).

9. In Uzawa's method a linear system with the matrix A must be solved at each step. Assume that these systems are solved inaccurately by an iterative process. For each linear system the iterative process is applied until the norm of the residual $r_{k+1} = (b - By_k) - Ax_{k+1}$ is less than a certain threshold ϵ_{k+1}.

   a. Assume that ω is chosen so that (8.33) is satisfied and that ϵ_k converges to zero as k tends to infinity. Show that the resulting algorithm converges to the solution.
   b. Give an explicit upper bound of the error on y_k in the case when ϵ_i is chosen of the form $\epsilon = \alpha^i$, where $\alpha < 1$.

10. Assume $\|b - Ax\|_2$ is to be minimized, in which A is $n \times m$ with $n > m$. Let x_* be the minimizer and $r = b - Ax_*$. What is the minimizer of $\|(b + \alpha r) - Ax\|_2$, where α is an arbitrary scalar?

11. Consider a saddle-point linear system of the form $Ax = b$, where

$$A = \begin{pmatrix} B & C \\ C^T & 0 \end{pmatrix}, \quad x = \begin{pmatrix} u \\ p \end{pmatrix}, \quad b = \begin{pmatrix} f \\ 0 \end{pmatrix},$$

   and B is SPD. It is assumed that A is nonsingular (which is equivalent to assuming that C is of full rank).

   a. Prove that A has both negative and positive eigenvalues by showing how to select vectors $x = \binom{u}{p}$ so that $(Ax, x) > 0$ and vectors x so that $(Ax, x) < 0$.
   b. Show how to select an initial guess of the form $x_0 = \binom{u_0}{0}$ if we want its corresponding residual vector $r_0 = b - Ax_0$ to be of the form $r_0 = \binom{0}{s_0}$. What happens if we attempt to use the steepest descent algorithm with this initial guess?
   c. What happens if the minimal residual iteration is applied using the same initial guess as in part (b)?
   d. By eliminating the unknown u find a linear system $Sp = g$ that must be satisfied by the variable p. Is the coefficient matrix of this system SPD (or symmetric negative definite)?
   e. We now want to solve the linear system by the following iteration:

$$u_{k+1} = B^{-1}(f - Cp_k),$$
$$p_{k+1} = p_k + \alpha_k C^T u_{k+1}.$$

   Show that p_{k+1} is of the form $p_{k+1} = p_k + \alpha_k s_k$, where s_k is the residual relative to p_k for the reduced linear system found in part (d). How should α_k be selected if we want p_{k+1} to correspond to the iterate of steepest descent for this reduced system?

Notes and References

Methods based on the normal equations were among the first to be used for solving nonsymmetric linear systems by iterative methods [180, 85]. The work by Björk and Elfing [39] and Sameh et al. [181, 53, 52] revived these techniques by showing that they have some advantages from the implementation point of view and that they can offer good performance for a broad class of problems. In addition, they are also attractive for parallel computers. In [239], a few preconditioning ideas for normal equations were described, which will be covered in Chapter 10. It would be helpful to be able to determine whether or not it is preferable to use the normal equations approach rather than the direct equations for a given system, but this may require an eigenvalue/singular value analysis.

It is sometimes argued that the normal equations approach is *always* better than the standard approach, because it has a quality of robustness that outweighs the additional cost due to the slowness of the method in the generic elliptic case. Unfortunately, this is not true. Although variants of the Kaczmarz and Cimmino algorithms deserve a place in any robust iterative solution package, they cannot be viewed as a panacea. In *most* realistic examples arising from PDEs, the normal equations route gives rise to much slower convergence than the Krylov subspace approach for the direct equations. For ill-conditioned problems, these methods will simply fail to converge, unless a good preconditioner is available.

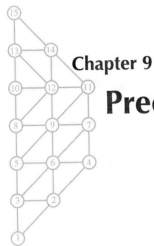

Chapter 9
Preconditioned Iterations

Although the methods seen in previous chapters are well founded theoretically, they are all likely to suffer from slow convergence for problems that arise from typical applications such as fluid dynamics and electronic device simulation. Preconditioning is a key ingredient for the success of Krylov subspace methods in these applications. This chapter discusses the preconditioned versions of the iterative methods already seen, but without being specific about the particular preconditioners used. The standard preconditioning techniques will be covered in the next chapter.

9.1 Introduction

Lack of robustness is a widely recognized weakness of iterative solvers relative to direct solvers. This drawback hampers the acceptance of iterative methods in industrial applications despite their intrinsic appeal for very large linear systems. Both the efficiency and robustness of iterative techniques can be improved by using *preconditioning*. A term introduced in Chapter 4, preconditioning is simply a means of transforming the original linear system into one with the same solution, but that is likely to be easier to solve with an iterative solver. In general, the reliability of iterative techniques, when dealing with various applications, depends much more on the quality of the preconditioner than on the particular Krylov subspace accelerators used. We will cover some of these preconditioners in detail in the next chapter. This chapter discusses the preconditioned versions of the Krylov subspace algorithms already seen, using a generic preconditioner.

To begin with, it is worthwhile to consider the options available for preconditioning a system. The first step in preconditioning is to find a preconditioning matrix M. The matrix M can be defined in many different ways but it must satisfy a few minimal requirements. From a practical point of view, the most important requirement for M is that it be inexpensive to solve linear systems $Mx = b$. This is because the preconditioned algorithms will all require a linear system solution with the matrix M at each step. Also, M should be close to A in some sense and it should clearly be nonsingular. Chapter 10 explores in detail the problem of finding preconditioners M for a given matrix S, while this chapter considers only the ways in which the preconditioner is applied to solve the original system.

Once a preconditioning matrix M is available there are three known ways of applying it. The preconditioner can be applied from the left, leading to the preconditioned system

$$M^{-1}Ax = M^{-1}b. \tag{9.1}$$

Alternatively, it can also be applied to the right:

$$AM^{-1}u = b, \quad x \equiv M^{-1}u. \tag{9.2}$$

Note that the above formulation amounts to making the change of variables $u = Mx$ and solving the system with respect to the unknown u. Finally, a common situation is when the preconditioner is available in the factored form

$$M = M_L M_R,$$

where, typically, M_L and M_R are triangular matrices. In this situation, the preconditioning can be split:

$$M_L^{-1}AM_R^{-1}u = b, \quad x \equiv M_R^{-1}u. \tag{9.3}$$

It is imperative to preserve symmetry when the original matrix is symmetric, so the split preconditioner seems mandatory in this case. However, there are other ways of preserving symmetry, or rather taking advantage of symmetry, even if M is not available in a factored form. This is discussed next for the conjugate gradient (CG) method.

9.2 Preconditioned Conjugate Gradient

Consider a matrix A that is symmetric and positive definite and assume that a preconditioner M is available. The preconditioner M is a matrix that approximates A in some yet-undefined sense. It is assumed that M is also symmetric positive definite (SPD). Then one can precondition the system in one of the three ways shown in the previous section, i.e., as in (9.1), (9.2), or (9.3). Note that the first two systems are no longer symmetric in general. The next section considers strategies for preserving symmetry. Then efficient implementations will be described for particular forms of the preconditioners.

9.2.1 Preserving Symmetry

When M is available in the form of an incomplete Cholesky factorization, i.e., when

$$M = LL^T,$$

then a simple way to preserve symmetry is to use the *split* preconditioning option (9.3), which yields the SPD matrix

$$L^{-1}AL^{-T}u = L^{-1}b, \quad x = L^{-T}u. \tag{9.4}$$

9.2. Preconditioned Conjugate Gradient

However, it is not necessary to split the preconditioner in this manner in order to preserve symmetry. Observe that $M^{-1}A$ is self-adjoint for the M inner product:

$$(x, y)_M \equiv (Mx, y) = (x, My),$$

since

$$(M^{-1}Ax, y)_M = (Ax, y) = (x, Ay) = (x, M(M^{-1}A)y) = (x, M^{-1}Ay)_M.$$

Therefore, an alternative is to replace the usual Euclidean inner product in the CG algorithm with the M inner product.

If the CG algorithm is rewritten for this new inner product, denoting by $r_j = b - Ax_j$ the original residual and by $z_j = M^{-1}r_j$ the residual for the preconditioned system, the following sequence of operations is obtained, ignoring the initial step:

1. $\alpha_j := (z_j, z_j)_M / (M^{-1}Ap_j, p_j)_M$,
2. $x_{j+1} := x_j + \alpha_j p_j$,
3. $r_{j+1} := r_j - \alpha_j Ap_j$ and $z_{j+1} := M^{-1}r_{j+1}$,
4. $\beta_j := (z_{j+1}, z_{j+1})_M / (z_j, z_j)_M$,
5. $p_{j+1} := z_{j+1} + \beta_j p_j$.

Since $(z_j, z_j)_M = (r_j, z_j)$ and $(M^{-1}Ap_j, p_j)_M = (Ap_j, p_j)$, the M inner products do not have to be computed explicitly. With this observation, the following algorithm is obtained.

ALGORITHM 9.1. Preconditioned CG (PCG)

1. Compute $r_0 := b - Ax_0$, $z_0 = M^{-1}r_0$, and $p_0 := z_0$
2. For $j = 0, 1, \ldots$, until convergence, Do
3. $\quad \alpha_j := (r_j, z_j)/(Ap_j, p_j)$
4. $\quad x_{j+1} := x_j + \alpha_j p_j$
5. $\quad r_{j+1} := r_j - \alpha_j Ap_j$
6. $\quad z_{j+1} := M^{-1}r_{j+1}$
7. $\quad \beta_j := (r_{j+1}, z_{j+1})/(r_j, z_j)$
8. $\quad p_{j+1} := z_{j+1} + \beta_j p_j$
9. EndDo

It is interesting to observe that $M^{-1}A$ is also self-adjoint with respect to the A inner product. Indeed,

$$(M^{-1}Ax, y)_A = (AM^{-1}Ax, y) = (x, AM^{-1}Ay) = (x, M^{-1}Ay)_A$$

and a similar algorithm can be written for this dot product (see Exercise 2).

In the case where M is a Cholesky product $M = LL^T$, two options are available, namely, the split preconditioning option (9.4) and the above algorithm. An immediate question arises about the iterates produced by these two options: Is one better than the other? Surprisingly, the answer is that *the iterates are identical*. To see this, start from Algorithm 9.1 and define the following auxiliary vectors and matrix from it:

$$\hat{p}_j = L^T p_j,$$
$$u_j = L^T x_j,$$
$$\hat{r}_j = L^T z_j = L^{-1} r_j,$$
$$\hat{A} = L^{-1} A L^{-T}.$$

Observe that

$$(r_j, z_j) = (r_j, L^{-T} L^{-1} r_j) = (L^{-1} r_j, L^{-1} r_j) = (\hat{r}_j, \hat{r}_j).$$

Similarly,

$$(A p_j, p_j) = (A L^{-T} \hat{p}_j, L^{-T} \hat{p}_j)(L^{-1} A L^{-T} \hat{p}_j, \hat{p}_j) = (\hat{A} \hat{p}_j, \hat{p}_j).$$

All the steps of the algorithm can be rewritten with the new variables, yielding the following sequence of operations:

1. $\alpha_j := (\hat{r}_j, \hat{r}_j)/(\hat{A}\hat{p}_j, \hat{p}_j)$,
2. $u_{j+1} := u_j + \alpha_j \hat{p}_j$,
3. $\hat{r}_{j+1} := \hat{r}_j - \alpha_j \hat{A} \hat{p}_j$,
4. $\beta_j := (\hat{r}_{j+1}, \hat{r}_{j+1})/(\hat{r}_j, \hat{r}_j)$,
5. $\hat{p}_{j+1} := \hat{r}_{j+1} + \beta_j \hat{p}_j$.

This is precisely the CG algorithm applied to the preconditioned system

$$\hat{A} u = L^{-1} b,$$

where $u = L^T x$. It is common when implementing algorithms that involve a right preconditioner to avoid the use of the u variable, since the iteration can be written with the original x variable. If the above steps are rewritten with the original x and p variables, the following algorithm results.

ALGORITHM 9.2. Split Preconditioner CG

1. Compute $r_0 := b - A x_0$, $\hat{r}_0 = L^{-1} r_0$, and $p_0 := L^{-T} \hat{r}_0$
2. For $j = 0, 1, \ldots,$ until convergence, Do
3. $\alpha_j := (\hat{r}_j, \hat{r}_j)/(A p_j, p_j)$
4. $x_{j+1} := x_j + \alpha_j p_j$
5. $\hat{r}_{j+1} := \hat{r}_j - \alpha_j L^{-1} A p_j$
6. $\beta_j := (\hat{r}_{j+1}, \hat{r}_{j+1})/(\hat{r}_j, \hat{r}_j)$
7. $p_{j+1} := L^{-T} \hat{r}_{j+1} + \beta_j p_j$
8. EndDo

9.2. Preconditioned Conjugate Gradient

The iterates x_j produced by the above algorithm and Algorithm 9.1 are identical, provided the same initial guess is used.

Consider now the right-preconditioned system (9.2). The matrix AM^{-1} is not Hermitian with either the standard inner product or the M inner product. However, it is Hermitian with respect to the M^{-1} inner product. If the CG algorithm is written with respect to the u variable for this new inner product, the following sequence of operations is obtained, ignoring again the initial step:

1. $\alpha_j := (r_j, r_j)_{M^{-1}} / (AM^{-1} p_j, p_j)_{M^{-1}}$,
2. $u_{j+1} := u_j + \alpha_j p_j$,
3. $r_{j+1} := r_j - \alpha_j AM^{-1} p_j$,
4. $\beta_j := (r_{j+1}, r_{j+1})_{M^{-1}} / (r_j, r_j)_{M^{-1}}$,
5. $p_{j+1} := r_{j+1} + \beta_j p_j$.

Recall that the u-vectors and the x-vectors are related by $x = M^{-1} u$. Since the u-vectors are not actually needed, the update for u_{j+1} in the second step can be replaced with $x_{j+1} := x_j + \alpha_j M^{-1} p_j$. Then observe that the whole algorithm can be recast in terms of $q_j = M^{-1} p_j$ and $z_j = M^{-1} r_j$:

1. $\alpha_j := (z_j, r_j) / (A q_j, q_j)$,
2. $x_{j+1} := x_j + \alpha_j q_j$,
3. $r_{j+1} := r_j - \alpha_j A q_j$, and $z_{j+1} = M^{-1} r_{j+1}$,
4. $\beta_j := (z_{j+1}, r_{j+1}) / (z_j, r_j)$,
5. $q_{j+1} := z_{j+1} + \beta_j q_j$.

Notice that the same sequence of computations is obtained as with Algorithm 9.1, the left-preconditioned CG. The implication is that *the left-preconditioned CG algorithm with the M inner product is mathematically equivalent to the right-preconditioned CG algorithm with the M^{-1} inner product.*

9.2.2 Efficient Implementations

When applying a Krylov subspace procedure to a preconditioned linear system, an operation of the form
$$v \to w = M^{-1} A v$$
or some similar operation is performed at each step. The most natural way to perform this operation is to multiply the vector v by A and then apply M^{-1} to the result. However, since A and M are related, it is sometimes possible to devise procedures that are more economical than this straightforward approach. For example, it is often the case that
$$M = A - R,$$

in which the number of nonzero elements in R is much smaller than in A. In this case, the simplest scheme would be to compute $w = M^{-1}Av$ as

$$w = M^{-1}Av = M^{-1}(M+R)v = v + M^{-1}Rv.$$

This requires that R be stored explicitly. In approximate LU factorization techniques, R is the matrix of the elements that are dropped during the incomplete factorization. An even more efficient variation of the PCG algorithm can be derived for some common forms of the preconditioner in the special situation where A is symmetric. Write A in the form

$$A = D_0 - E - E^T, \qquad (9.5)$$

in which $-E$ is the strict lower triangular part of A and D_0 its diagonal. In many cases, the preconditioner M can be written in the form

$$M = (D - E)D^{-1}(D - E^T), \qquad (9.6)$$

in which E is the same as above and D is some diagonal, not necessarily equal to D_0. For example, in the symmetric successive overrelaxation (SSOR) preconditioner with $\omega = 1$, $D \equiv D_0$. Also, for certain types of matrices, the IC(0) preconditioner can be expressed in this manner, where D can be obtained by a recurrence formula.

Eisenstat's implementation consists of applying the CG algorithm to the linear system

$$\hat{A}u = (D - E)^{-1}b, \qquad (9.7)$$

with

$$\hat{A} \equiv (D - E)^{-1}A(D - E^T)^{-1}, \quad x = (D - E^T)^{-1}u. \qquad (9.8)$$

This does not quite correspond to a preconditioning with the matrix (9.6). In order to produce the same iterates as Algorithm 9.1, the matrix $\hat{A}$ must be further preconditioned with the diagonal matrix D^{-1}. Thus, the PCG algorithm, Algorithm 9.1, is actually applied to the system (9.7), in which the preconditioning operation is $M^{-1} = D$. Alternatively, we can initially scale the rows and columns of the linear system and preconditioning to transform the diagonal to the identity; see Exercise 7.

Now note that

$$\hat{A} = (D - E)^{-1}A(D - E^T)^{-1}$$
$$= (D - E)^{-1}(D_0 - E - E^T)(D - E^T)^{-1}$$
$$= (D - E)^{-1}\left(D_0 - 2D + (D - E) + (D - E^T)\right)(D - E^T)^{-1}$$
$$\equiv (D - E)^{-1}D_1(D - E^T)^{-1} + (D - E)^{-1} + (D - E^T)^{-1},$$

in which $D_1 \equiv D_0 - 2D$. As a result,

$$\hat{A}v = (D - E)^{-1}\left[v + D_1(D - E^T)^{-1}v\right] + (D - E^T)^{-1}v.$$

Thus, the vector $w = \hat{A}v$ can be computed by the following procedure:

$z := (D - E^T)^{-1}v,$
$w := (D - E)^{-1}(v + D_1 z),$
$w := w + z.$

9.3. Preconditioned Generalized Minimal Residual

One product with the diagonal D can be saved if the matrices $D^{-1}E$ and $D^{-1}E^T$ are stored. Indeed, by setting $\hat{D}_1 = D^{-1}D_1$ and $\hat{v} = D^{-1}v$, the above procedure can be reformulated as follows.

ALGORITHM 9.3. Computation of $w = \hat{A}v$

1. $\hat{v} := D^{-1}v$
2. $z := (I - D^{-1}E^T)^{-1}\hat{v}$
3. $w := (I - D^{-1}E)^{-1}(\hat{v} + \hat{D}_1 z)$
4. $w := w + z$

Note that the matrices $D^{-1}E$ and $D^{-1}E^T$ are not the transpose of one another, so we actually need to increase the storage requirement for this formulation if these matrices are stored. However, there is a more economical variant that works with the matrix $D^{-1/2}ED^{-1/2}$ and its transpose. This is left as Exercise 8.

Denoting by $N_z(X)$ the number of nonzero elements of a sparse matrix X, the total number of operations (additions and multiplications) of this procedure is n for line 1, $2N_z(E)$ for line 2, $2N_z(E^T) + 2n$ for line 3, and n for line 4. The cost of the preconditioning operation by D^{-1}, i.e., n operations, must be added to this, yielding the total number of operations:

$$\begin{aligned} N_{op} &= n + 2N_z(E) + 2N_z(E^T) + 2n + n + n \\ &= 3n + 2(N_z(E) + N_z(E^T) + n) \\ &= 3n + 2N_z(A). \end{aligned}$$

For the straightforward approach, $2N_z(A)$ operations are needed for the product with A, $2N_z(E)$ for the forward solve, and $n + 2N_z(E^T)$ for the backward solve, giving a total of

$$2N_z(A) + 2N_z(E) + n + 2N_z(E^T) = 4N_z(A) - n.$$

Thus, Eisenstat's scheme is always more economical, when N_z is large enough, although the relative gains depend on the total number of nonzero elements in A. One disadvantage of this scheme is that it is limited to a special form of the preconditioner.

Example 9.1. For a five-point finite difference matrix, $N_z(A)$ is roughly $5n$, so that with the standard implementation $19n$ operations are performed, while with Eisenstat's implementation only $13n$ operations are performed, a savings of about $\frac{1}{3}$. However, if the other operations of the CG algorithm are included, for a total of about $10n$ operations, the relative savings become smaller. Now the original scheme will require $29n$ operations, versus $23n$ operations for Eisenstat's implementation.

9.3 Preconditioned Generalized Minimal Residual

In the case of generalized minimal residual (GMRES) or other nonsymmetric iterative solvers, the same three options for applying the preconditioning operation as for the CG (namely, left, split, and right preconditioning) are available. However, there will be one fundamental difference—the right-preconditioning versions will give rise to what is called a *flexible variant*, i.e., a variant in which the preconditioner can change at each step. This capability can be very useful in some applications.

9.3.1 Left-Preconditioned GMRES

As before, define the left-preconditioned GMRES algorithm as the GMRES algorithm applied to the system

$$M^{-1}Ax = M^{-1}b. \quad (9.9)$$

The straightforward application of GMRES to the above linear system yields the following preconditioned version of GMRES.

ALGORITHM 9.4. GMRES with Left Preconditioning

1. Compute $r_0 = M^{-1}(b - Ax_0)$, $\beta = \|r_0\|_2$, and $v_1 = r_0/\beta$
2. For $j = 1, \ldots, m$, Do
3.   Compute $w := M^{-1}Av_j$
4.   For $i = 1, \ldots, j$, Do
5. $h_{i,j} := (w, v_i)$
6. $w := w - h_{i,j}v_i$
7.   EndDo
8.   Compute $h_{j+1,j} = \|w\|_2$ and $v_{j+1} = w/h_{j+1,j}$
9. EndDo
10. Define $V_m := [v_1, \ldots, v_m]$, $\bar{H}_m = \{h_{i,j}\}_{1 \le i \le j+1, 1 \le j \le m}$
11. Compute $y_m = \mathrm{argmin}_y \|\beta e_1 - \bar{H}_m y\|_2$ and $x_m = x_0 + V_m y_m$
12. If satisfied Stop, else set $x_0 := x_m$ and go to 1

The Arnoldi loop constructs an orthogonal basis of the left-preconditioned Krylov subspace

$$\mathrm{span}\{r_0, M^{-1}Ar_0, \ldots, (M^{-1}A)^{m-1}r_0\}.$$

It uses a modified Gram–Schmidt (MGS) process, in which the new vector to be orthogonalized is obtained from the previous vector in the process. All residual vectors and their norms that are computed by the algorithm correspond to the preconditioned residuals, namely, $z_m = M^{-1}(b - Ax_m)$, instead of the original (unpreconditioned) residuals $b - Ax_m$. In addition, there is no easy access to these unpreconditioned residuals, unless they are computed explicitly, e.g., by multiplying the preconditioned residuals by M. This can cause some difficulties if a stopping criterion based on the actual residuals, instead of the preconditioned ones, is desired.

Sometimes an SPD preconditioning M for the nonsymmetric matrix A may be available. For example, if A is almost SPD, then (9.9) would not take advantage of this. It would be wiser to compute an approximate factorization to the symmetric part and use GMRES with split preconditioning. This raises the question as to whether or not a version of the preconditioned GMRES can be developed, which is similar to Algorithm 9.1, for the CG algorithm. This version would consist of using GMRES with the M inner product for the system (9.9).

9.3. Preconditioned Generalized Minimal Residual

At step j of the preconditioned GMRES algorithm, the previous v_j is multiplied by A to get a vector

$$w_j = Av_j. \tag{9.10}$$

Then this vector is preconditioned to get

$$z_j = M^{-1}w_j. \tag{9.11}$$

This vector must be M orthogonalized against all previous v_i's. If the standard Gram–Schmidt process is used, we first compute the inner products

$$h_{ij} = (z_j, v_i)_M = (Mz_j, v_i) = (w_j, v_i), \quad i = 1, \ldots, j, \tag{9.12}$$

and then modify the vector z_j into the new vector

$$\hat{z}_j := z_j - \sum_{i=1}^{j} h_{ij} v_i. \tag{9.13}$$

To complete the orthonormalization step, the final $\hat{z}_j$ must be normalized. Because of the orthogonality of $\hat{z}_j$ versus all previous v_i's, observe that

$$(\hat{z}_j, \hat{z}_j)_M = (z_j, \hat{z}_j)_M = (M^{-1}w_j, \hat{z}_j)_M = (w_j, \hat{z}_j). \tag{9.14}$$

Thus, the desired M-norm could be obtained from (9.14) and then we would set

$$h_{j+1,j} := (\hat{z}_j, w_j)^{1/2} \quad \text{and} \quad v_{j+1} = \hat{z}_j / h_{j+1,j}. \tag{9.15}$$

One serious difficulty with the above procedure is that the inner product $(\hat{z}_j, \hat{z}_j)_M$ as computed by (9.14) may be negative in the presence of round-off. There are two remedies. First, this M-norm can be computed explicitly at the expense of an additional matrix-by-vector multiplication with M. Second, the set of vectors Mv_i can be saved in order to accumulate inexpensively both the vector $\hat{z}_j$ and the vector $M\hat{z}_j$ via the relation

$$M\hat{z}_j = w_j - \sum_{i=1}^{j} h_{ij} Mv_i.$$

An MGS version of this second approach can be derived easily. The details of the algorithm are left as Exercise 13.

9.3.2 Right-Preconditioned GMRES

The right-preconditioned GMRES algorithm is based on solving

$$AM^{-1}u = b, \quad u = Mx. \tag{9.16}$$

As we now show, the new variable u never needs to be invoked explicitly. Indeed, once the initial residual $b - Ax_0 = b - AM^{-1}u_0$ is computed, all subsequent vectors of the

Krylov subspace can be obtained without any reference to the u variables. Note that u_0 is not needed at all. The initial residual for the preconditioned system can be computed from $r_0 = b - Ax_0$, which is the same as $b - AM^{-1}u_0$. In practice, it is usually x_0 that is available, not u_0. At the end, the u variable approximate solution to (9.16) is given by

$$u_m = u_0 + \sum_{i=1}^{m} v_i \eta_i,$$

with $u_0 = Mx_0$. Multiplying through by M^{-1} yields the desired approximation in terms of the x variable:

$$x_m = x_0 + M^{-1}\left[\sum_{i=1}^{m} v_i \eta_i\right].$$

Thus, one preconditioning operation is needed at the end of the outer loop, instead of at the beginning in the case of the left-preconditioned version.

ALGORITHM 9.5. GMRES with Right Preconditioning

1. Compute $r_0 = b - Ax_0$, $\beta = \|r_0\|_2$, and $v_1 = r_0/\beta$
2. For $j = 1, \ldots, m$, Do
3.   Compute $w := AM^{-1}v_j$
4.   For $i = 1, \ldots, j$, Do
5. $h_{i,j} := (w, v_i)$
6. $w := w - h_{i,j} v_i$
7.   EndDo
8.   Compute $h_{j+1,j} = \|w\|_2$ and $v_{j+1} = w/h_{j+1,j}$
9.   Define $V_m := [v_1, \ldots, v_m]$, $\bar{H}_m = \{h_{i,j}\}_{1 \leq i \leq j+1, 1 \leq j \leq m}$
10. EndDo
11. Compute $y_m = \mathrm{argmin}_y \|\beta e_1 - \bar{H}_m y\|_2$ and $x_m = x_0 + M^{-1} V_m y_m$
12. If satisfied Stop, else set $x_0 := x_m$ and go to 1

This time, the Arnoldi loop builds an orthogonal basis of the right-preconditioned Krylov subspace

$$\mathrm{span}\{r_0, AM^{-1}r_0, \ldots, (AM^{-1})^{m-1}r_0\}.$$

Note that the residual norm is now relative to the initial system, i.e., $b - Ax_m$, since the algorithm obtains the residual $b - Ax_m = b - AM^{-1}u_m$ implicitly. This is an essential difference from the left-preconditioned GMRES algorithm.

9.3.3 Split Preconditioning

In many cases, M is the result of a factorization of the form

$$M = LU.$$

Then there is the option of using GMRES on the split-preconditioned system

$$L^{-1}AU^{-1}u = L^{-1}b, \quad x = U^{-1}u.$$

9.3. Preconditioned Generalized Minimal Residual

In this situation, it is clear that we need to operate on the initial residual by L^{-1} at the start of the algorithm and by U^{-1} on the linear combination $V_m y_m$ in forming the approximate solution. The residual norm available is that of $L^{-1}(b - Ax_m)$.

A question arises about the differences among the right-, left-, and split-preconditioning options. The fact that different versions of the residuals are available in each case may affect the stopping criterion and may cause the algorithm to stop either prematurely or with delay. This can be particularly damaging in case M is very ill conditioned. The degree of symmetry, and therefore performance, can also be affected by the way in which the preconditioner is applied. For example, a split preconditioner may be much better if A is nearly symmetric. Other than these two situations, there is little difference generally among the three options. The next section establishes a theoretical connection between left- and right-preconditioned GMRES.

9.3.4 Comparison of Right and Left Preconditioning

When comparing the left-, right-, and split-preconditioning options, a first observation to make is that the spectra of the three associated operators $M^{-1}A$, AM^{-1}, and $L^{-1}AU^{-1}$ are identical. Therefore, in principle one should expect convergence to be similar, although, as is known, eigenvalues do not always govern convergence. In this section, we compare the optimality properties achieved by left- and right-preconditioned GMRES.

For the left-preconditioning option, GMRES minimizes the residual norm

$$\|M^{-1}b - M^{-1}Ax\|_2$$

among all vectors from the affine subspace

$$x_0 + \mathcal{K}_m^L = x_0 + \text{span}\{z_0, M^{-1}Az_0, \ldots, (M^{-1}A)^{m-1}z_0\}, \qquad (9.17)$$

in which z_0 is the initial preconditioned residual $z_0 = M^{-1}r_0$. Thus, the approximate solution can be expressed as

$$x_m = x_0 + M^{-1}s_{m-1}(M^{-1}A)z_0,$$

where s_{m-1} is the polynomial of degree $m - 1$ that minimizes the norm

$$\|z_0 - M^{-1}A\,s(M^{-1}A)z_0\|_2$$

among all polynomials s of degree not exceeding $m - 1$. It is also possible to express this optimality condition with respect to the original residual vector r_0. Indeed,

$$z_0 - M^{-1}A\,s(M^{-1}A)z_0 = M^{-1}[r_0 - A\,s(M^{-1}A)M^{-1}r_0].$$

A simple algebraic manipulation shows that, for any polynomial s,

$$s(M^{-1}A)M^{-1}r = M^{-1}s(AM^{-1})r, \qquad (9.18)$$

from which we obtain the relation

$$z_0 - M^{-1}As(M^{-1}A)z_0 = M^{-1}[r_0 - AM^{-1}s(AM^{-1})r_0]. \qquad (9.19)$$

Consider now the situation with the right-preconditioned GMRES. Here it is necessary to distinguish between the original x variable and the transformed variable u related to x by $x = M^{-1}u$. For the u variable, the right-preconditioned GMRES process minimizes the 2-norm of $r = b - AM^{-1}u$, where u belongs to

$$u_0 + \mathcal{K}_m^R = u_0 + \text{span}\{r_0, AM^{-1}r_0, \ldots, (AM^{-1})^{m-1}r_0\}, \qquad (9.20)$$

in which r_0 is the residual $r_0 = b - AM^{-1}u_0$. This residual is identical to the residual associated with the original x variable since $M^{-1}u_0 = x_0$. Multiplying (9.20) through to the left by M^{-1} and exploiting again (9.18), observe that the generic variable x associated with a vector of the subspace (9.20) belongs to the affine subspace

$$M^{-1}u_0 + M^{-1}\mathcal{K}_m^R = x_0 + \text{span}\{z_0, M^{-1}Az_0, \ldots, (M^{-1}A)^{m-1}z_0\}.$$

This is identical to the affine subspace (9.17) invoked in the left-preconditioned variant. In other words, for the right-preconditioned GMRES, the approximate x solution can also be expressed as

$$x_m = x_0 + s_{m-1}(AM^{-1})r_0.$$

However, now s_{m-1} is a polynomial of degree $m - 1$ that minimizes the norm

$$\|r_0 - AM^{-1} s(AM^{-1})r_0\|_2 \qquad (9.21)$$

among all polynomials s of degree not exceeding $m - 1$. What is surprising is that the two quantities that are minimized, namely, (9.19) and (9.21), differ only by a multiplication by M^{-1}. Specifically, the left-preconditioned GMRES minimizes $M^{-1}r$, whereas the right-preconditioned variant minimizes r, where r is taken over the same subspace in both cases.

Proposition 9.1. *The approximate solution obtained by left- or right-preconditioned GMRES is of the form*

$$x_m = x_0 + s_{m-1}(M^{-1}A)z_0 = x_0 + M^{-1}s_{m-1}(AM^{-1})r_0,$$

where $z_0 = M^{-1}r_0$ and s_{m-1} is a polynomial of degree $m - 1$. The polynomial s_{m-1} minimizes the residual norm $\|b - Ax_m\|_2$ in the right-preconditioning case and the preconditioned residual norm $\|M^{-1}(b - Ax_m)\|_2$ in the left-preconditioning case.

In most practical situations, the difference in the convergence behavior of the two approaches is not significant. The only exception is when M is ill conditioned, which could lead to substantial differences.

9.4 Flexible Variants

In the discussion of preconditioning techniques so far, it has been implicitly assumed that the preconditioning matrix M is fixed; i.e., it does not change from step to step. However, in some cases, no matrix M is available. Instead, the operation $M^{-1}x$ is the result of some unspecified computation, possibly another iterative process. In such cases, it may

9.4. Flexible Variants

well happen that M^{-1} is not a constant operator. The previous preconditioned iterative procedures will not converge if M is not constant. There are a number of variants of iterative procedures developed in the literature that can accommodate variations in the preconditioner, i.e., that allow the preconditioner to vary from step to step. Such iterative procedures are called *flexible* iterations. One of these iterations, a flexible variant of the GMRES algorithm, is described next.

9.4.1 FGMRES

We begin by examining the right-preconditioned GMRES algorithm. In line 11 of Algorithm 9.5 the approximate solution x_m is expressed as a linear combination of the preconditioned vectors $z_i = M^{-1} v_i, i = 1, \ldots, m$. These vectors are also computed in line 3, prior to their multiplication by A to obtain the vector w. They are all obtained by applying the same preconditioning matrix M^{-1} to the v_i's. As a result it is not necessary to save them. Instead, we only need to apply M^{-1} to the linear combination of the v_i's, i.e., to $V_m y_m$ in line 11.

Suppose now that the preconditioner could change at every step, i.e., that z_j is given by

$$z_j = M_j^{-1} v_j.$$

Then it would be natural to compute the approximate solution as

$$x_m = x_0 + Z_m y_m,$$

in which $Z_m = [z_1, \ldots, z_m]$ and y_m is computed as before, as the solution to the least-squares problem in line 11. These are the only changes that lead from the right-preconditioned algorithm to the flexible variant, described below.

ALGORITHM 9.6. Flexible GMRES (FGMRES)

1. Compute $r_0 = b - A x_0$, $\beta = \|r_0\|_2$, and $v_1 = r_0/\beta$
2. For $j = 1, \ldots, m$, Do
3.   Compute $z_j := M_j^{-1} v_j$
4.   Compute $w := A z_j$
5.   For $i = 1, \ldots, j$, Do
6. $h_{i,j} := (w, v_i)$
7. $w := w - h_{i,j} v_i$
8.   EndDo
9.   Compute $h_{j+1,j} = \|w\|_2$ and $v_{j+1} = w/h_{j+1,j}$
10.   Define $Z_m := [z_1, \ldots, z_m]$, $\bar{H}_m = \{h_{i,j}\}_{1 \le i \le j+1; 1 \le j \le m}$
11. EndDo
12. Compute $y_m = \mathrm{argmin}_y \|\beta e_1 - \bar{H}_m y\|_2$ and $x_m = x_0 + Z_m y_m$
13. If satisfied Stop, else set $x_0 \leftarrow x_m$ and go to 1

As can be seen, the main difference with the right-preconditioned version, Algorithm 9.5, is that the preconditioned vectors $z_j = M_j^{-1} v_j$ must be saved and the solution updated using these vectors. It is clear that, when $M_j = M$ for $j = 1, \ldots, m$, then this method is equivalent mathematically to Algorithm 9.5. It is important to observe that z_j can be defined in line 3 without reference to any preconditioner. That is, any given new vector z_j can be chosen. This added flexibility may cause the algorithm some problems. Indeed, z_j may be so poorly selected that a breakdown could occur, as in the worst-case scenario when z_j is zero.

One difference between FGMRES and the usual GMRES algorithm is that the action of AM_j^{-1} on a vector v of the Krylov subspace is no longer in the span of V_{m+1}. Instead, it is easy to show that

$$AZ_m = V_{m+1}\bar{H}_m, \qquad (9.22)$$

in replacement of the simpler relation $(AM^{-1})V_m = V_{m+1}\bar{H}_m$, which holds for the standard preconditioned GMRES; see (6.7). As before, H_m denotes the $m \times m$ matrix obtained from $\bar{H}_m$ by deleting its last row and $\hat{v}_{j+1}$ is the vector w that is normalized in line 9 of Algorithm 9.6 to obtain v_{j+1}. Then the following alternative formulation of (9.22) is valid, even when $h_{m+1,m} = 0$:

$$AZ_m = V_m H_m + \hat{v}_{m+1}e_m^T. \qquad (9.23)$$

An optimality property similar to the one that defines GMRES can be proved. Consider the residual vector for an arbitrary vector $z = x_0 + Z_m y$ in the affine space $x_0 + \text{span}\{Z_m\}$. This optimality property is based on the relations

$$b - Az = b - A(x_0 + Z_m y)$$
$$= r_0 - AZ_m y \qquad (9.24)$$
$$= \beta v_1 - V_{m+1}\bar{H}_m y$$
$$= V_{m+1}[\beta e_1 - \bar{H}_m y]. \qquad (9.25)$$

If $J_m(y)$ denotes the function

$$J_m(y) = \|b - A[x_0 + Z_m y]\|_2,$$

observe that, by (9.25) and the fact that V_{m+1} is unitary,

$$J_m(y) = \|\beta e_1 - \bar{H}_m y\|_2. \qquad (9.26)$$

Since the algorithm minimizes this norm over all vectors u in $\mathbb{R}^m$ to yield y_m, it is clear that the approximate solution $x_m = x_0 + Z_m y_m$ has the smallest residual norm in $x_0 + \text{span}\{Z_m\}$. Thus, the following result is proved.

Proposition 9.2. *The approximate solution x_m obtained at step m of FGMRES minimizes the residual norm $\|b - Ax_m\|_2$ over $x_0 + \text{span}\{Z_m\}$.*

Next, consider the possibility of breakdown in FGMRES. A breakdown occurs when the vector v_{j+1} cannot be computed in line 9 of Algorithm 9.6 because $h_{j+1,j} = 0$. For the standard GMRES algorithm, this is not a problem because when this happens then the approximate solution x_j is exact. The situation for FGMRES is slightly different.

Proposition 9.3. *Assume that $\beta = \|r_0\|_2 \neq 0$ and that $j - 1$ steps of FGMRES have been successfully performed, i.e., that $h_{i+1,i} \neq 0$ for $i < j$. In addition, assume that the matrix H_j is nonsingular. Then x_j is exact iff $h_{j+1,j} = 0$.*

9.4. Flexible Variants

Proof. If $h_{j+1,j} = 0$, then $AZ_j = V_j H_j$, and, as a result,

$$J_j(y) = \|\beta v_1 - AZ_j y_j\|_2 = \|\beta v_1 - V_j H_j y_j\|_2 = \|\beta e_1 - H_j y_j\|_2.$$

If H_j is nonsingular, then the above function is minimized for $y_j = H_j^{-1}(\beta e_1)$ and the corresponding minimum norm reached is zero; i.e., x_j is exact.

Conversely, if x_j is exact, then, from (9.23) and (9.24),

$$0 = b - Ax_j = V_j[\beta e_1 - H_j y_j] + \hat{v}_{j+1} e_j^T y_j. \tag{9.27}$$

We must show, by contradiction, that $\hat{v}_{j+1} = 0$. Assume that $\hat{v}_{j+1} \neq 0$. Since $\hat{v}_{j+1}$, $v_1, v_2, \ldots, v_m$ form an orthogonal system, then it follows from (9.27) that $\beta e_1 - H_j y_j = 0$ and $e_j^T y_j = 0$. The last component of y_j is equal to zero. A simple back substitution for the system $H_j y_j = \beta e_1$, starting from the last equation, will show that all components of y_j are zero. Because H_m is nonsingular, this would imply that $\beta = 0$ and contradict the assumption. □

The only difference between this result and that of Proposition 6.10 for the GMRES algorithm is that the additional assumption must be made that H_j is nonsingular since it is no longer implied by the nonsingularity of A. However, H_m is guaranteed to be nonsingular when all the z_j's are linearly independent and A is nonsingular. This is a consequence of a modification of the first part of Proposition 6.9. That same proof shows that the rank of AZ_m is equal to the rank of the matrix R_m therein. If R_m is nonsingular and $h_{m+1,m} = 0$, then H_m is also nonsingular.

A consequence of the above proposition is that, if $Az_j = v_j$ at a certain step, i.e., if the preconditioning is *exact*, then the approximation x_j will be exact provided that H_j is nonsingular. This is because $w = Az_j$ depends linearly on the previous v_i's (it is equal to v_j) and as a result the orthogonalization process will yield $\hat{v}_{j+1} = 0$.

A difficulty with the theory of the new algorithm is that general convergence results, such as those seen in earlier chapters, cannot be proved. That is because the subspace of approximants is no longer a standard Krylov subspace. However, the optimality property of Proposition 9.2 can be exploited in some specific situations. For example, if within each outer iteration *at least one* of the vectors z_j is chosen to be a steepest descent direction vector, e.g., for the function $F(x) = \|b - Ax\|_2^2$, then FGMRES is guaranteed to converge independently of m.

The additional cost of the flexible variant over the standard algorithm is only in the extra memory required to save the set of vectors $\{z_j\}_{j=1,\ldots,m}$. Yet the added advantage of *flexibility* may be worth this extra cost. A few applications can benefit from this flexibility, especially in developing robust iterative methods or preconditioners on parallel computers. Thus, *any* iterative technique can be used as a preconditioner: block SOR, SSOR, alternating direction implicit (ADI), multigrid, etc. More interestingly, iterative procedures such as GMRES, conjugate gradient normal residual (CGNR), and conjugate gradient squared (CGS) can also be used as preconditioners. Also, it may be useful to mix two or more preconditioners to solve a given problem. For example, two types of preconditioners can be applied alternately at each FGMRES step to mix the effects of local and global couplings in the partial differential equation (PDE) context.

9.4.2 DQGMRES

Recall that the direct quasi-GMRES (DQGMRES) algorithm presented in Chapter 6 uses an incomplete orthogonalization process instead of the full Arnoldi orthogonalization. At each step, the current vector is orthogonalized only against the k previous ones. The vectors thus generated are "locally" orthogonal to each other, in that $(v_i, v_j) = \delta_{ij}$ for $|i - j| < k$. The matrix $\bar{H}_m$ becomes banded and upper Hessenberg. Therefore, the approximate solution can be updated at step j from the approximate solution at step $j - 1$ via the recurrence

$$p_j = \frac{1}{r_{jj}} \left[v_j - \sum_{i=j-k+1}^{j-1} r_{ij} p_i \right], \quad x_j = x_{j-1} + \gamma_j p_j, \qquad (9.28)$$

in which the scalars γ_j and r_{ij} are obtained recursively from the Hessenberg matrix $\bar{H}_j$.

An advantage of DQGMRES is that it is also *flexible*. The principle is the same as in FGMRES. In both cases the vectors $z_j = M_j^{-1} v_j$ must be computed. In the case of FGMRES, these vectors must be saved, which requires extra storage. For DQGMRES, it can be observed that the preconditioned vectors z_j only affect the update of the vector p_j in the preconditioned version of the update formula (9.28), yielding

$$p_j = \frac{1}{r_{jj}} \left[M_j^{-1} v_j - \sum_{i=j-k+1}^{j-1} r_{ij} p_i \right].$$

As a result, $M_j^{-1} v_j$ can be discarded immediately after it is used to update p_j. The same memory locations can store this vector and the vector p_j. This contrasts with FGMRES, which requires additional vectors of storage.

9.5 Preconditioned Conjugate Gradient for the Normal Equations

There are several versions of the PCG method applied to the normal equations. Two versions come from the normal residual/normal error (NR/NE) options, and three other variations from the right-, left-, and split-preconditioning options. Here, we consider only the left-preconditioned variants.

The left-preconditioned CGNR algorithm is easily derived from Algorithm 9.1. Denote by r_j the residual for the original system, i.e., $r_j = b - Ax_j$, and by $\tilde{r}_j = A^T r_j$ the residual for the normal equations system. The preconditioned residual z_j is $z_j = M^{-1} \tilde{r}_j$. The scalar α_j in Algorithm 9.1 is now given by

$$\alpha_j = \frac{(\tilde{r}_j, z_j)}{(A^T A p_j, p_j)} = \frac{(\tilde{r}_j, z_j)}{(A p_j, A p_j)}.$$

This suggests employing the auxiliary vector $w_j = A p_j$ in the algorithm, which takes the following form.

9.5. Preconditioned Conjugate Gradient for the Normal Equations

ALGORITHM 9.7. Left-Preconditioned CGNR

1. Compute $r_0 = b - Ax_0$, $\tilde{r}_0 = A^T r_0$, $z_0 = M^{-1}\tilde{r}_0$, $p_0 = z_0$
2. For $j = 0, \ldots,$ until convergence, Do
3. $w_j = Ap_j$
4. $\alpha_j = (z_j, \tilde{r}_j)/\|w_j\|_2^2$
5. $x_{j+1} = x_j + \alpha_j p_j$
6. $r_{j+1} = r_j - \alpha_j w_j$
7. $\tilde{r}_{j+1} = A^T r_{j+1}$
8. $z_{j+1} = M^{-1}\tilde{r}_{j+1}$
9. $\beta_j = (z_{j+1}, \tilde{r}_{j+1})/(z_j, \tilde{r}_j)$
10. $p_{j+1} = z_{j+1} + \beta_j p_j$
11. EndDo

Similarly, the linear system $AA^T u = b$, with $x = A^T u$, can also be preconditioned from the left and solved with the PCG algorithm. Here it is observed that the updates of the u variable, the associated x variable, and two residuals take the form

$$\alpha_j = \frac{(r_j, z_j)}{(AA^T p_j, p_j)} = \frac{(r_j, z_j)}{(A^T p_j, A^T p_j)},$$

$$u_{j+1} = u_j + \alpha_j p_j \quad \leftrightarrow \quad x_{j+1} = x_j + \alpha_j A^T p_j,$$

$$r_{j+1} = r_j - \alpha_j AA^T p_j,$$

$$z_{j+1} = M^{-1} r_{j+1}.$$

Thus, if the algorithm for the unknown x is to be written, then the vectors $A^T p_j$ can be used instead of the vectors p_j, which are not needed. To update these vectors at the end of the algorithm the relation $p_{j+1} = z_{j+1} + \beta_{j+1} p_j$ in line 8 of Algorithm 9.1 must be multiplied through by A^T. This leads to the left-preconditioned version of CGNE, in which the notation has been changed to denote by p_j the vector $A^T p_j$ invoked in the above derivation.

ALGORITHM 9.8. Left-Preconditioned CGNE

1. Compute $r_0 = b - Ax_0$, $z_0 = M^{-1}r_0$, $p_0 = A^T z_0$
2. For $j = 0, 1, \ldots,$ until convergence, Do
3. $w_j = Ap_j$
4. $\alpha_j = (z_j, r_j)/(p_j, p_j)$
5. $x_{j+1} = x_j + \alpha_j p_j$
6. $r_{j+1} = r_j - \alpha_j w_j$
7. $z_{j+1} = M^{-1}r_{j+1}$
8. $\beta_j = (z_{j+1}, r_{j+1})/(z_j, r_j)$
9. $p_{j+1} = A^T z_{j+1} + \beta_j p_j$
10. EndDo

Not shown here are the right- and split-preconditioned versions, which are considered in Exercise 4.

9.6 The Concus, Golub, and Widlund Algorithm

When the matrix is nearly symmetric, we can think of preconditioning the system with the symmetric part of A. This gives rise to a few variants of a method known as the CGW method, from the names of the three authors Concus and Golub [88], and Widlund [312] who proposed this technique in the middle of the 1970s. Originally, the algorithm was not viewed from the angle of preconditioning. Writing $A = M - N$, with $M = \frac{1}{2}(A + A^H)$, the authors observed that the preconditioned matrix

$$M^{-1}A = I - M^{-1}N$$

is equal to the identity matrix, plus a matrix which is skew-Hermitian with respect to the M-inner product. It is not too difficult to show that the tridiagonal matrix corresponding to the Lanczos algorithm, applied to A with the M-inner product, has the form

$$T_m = \begin{pmatrix} 1 & -\eta_2 & & & \\ \eta_2 & 1 & -\eta_3 & & \\ & \cdot & \cdot & \cdot & \\ & & \eta_{m-1} & 1 & -\eta_m \\ & & & \eta_m & 1 \end{pmatrix}. \tag{9.29}$$

As a result, a three-term recurrence in the Arnoldi process is obtained, which results in a solution algorithm that resembles the standard preconditioned CG algorithm (Algorithm 9.1).

A version of the algorithm can be derived easily. The developments in Section 6.7 relating the Lanczos algorithm to the Conjugate Gradient algorithm, show that the vector x_{j+1} can be expressed as

$$x_{j+1} = x_j + \alpha_j p_j.$$

The preconditioned residual vectors must then satisfy the recurrence

$$z_{j+1} = z_j - \alpha_j M^{-1} A p_j$$

and if the z_j's are to be M-orthogonal, then we must have $(z_j - \alpha_j M^{-1}Ap_j, z_j)_M = 0$. As a result,

$$\alpha_j = \frac{(z_j, z_j)_M}{(M^{-1}Ap_j, z_j)_M} = \frac{(r_j, z_j)}{(Ap_j, z_j)}.$$

Also, the next search direction p_{j+1} is a linear combination of z_{j+1} and p_j,

$$p_{j+1} = z_{j+1} + \beta_j p_j.$$

Since $M^{-1}Ap_j$ is orthogonal to all vectors in $\mathcal{K}_{j-1}$, a first consequence is that

$$(Ap_j, z_j) = (M^{-1}Ap_j, p_j - \beta_{j-1}p_{j-1})_M = (M^{-1}Ap_j, p_j)_M = (Ap_j, p_j).$$

In addition, $M^{-1}Ap_{j+1}$ must be M-orthogonal to p_j, so that $\beta_j = -(M^{-1}Az_{j+1}, p_j)_M / (M^{-1}Ap_j, p_j)_M$. The relation $M^{-1}A = I - M^{-1}N$, the fact that $N^H = -N$, and that $(z_{j+1}, p_j)_M = 0$ yield,

$$(M^{-1}Az_{j+1}, p_j)_M = -(M^{-1}Nz_{j+1}, p_j)_M = (z_{j+1}, M^{-1}Np_j)_M = -(z_{j+1}, M^{-1}Ap_j)_M.$$

Finally, note that $M^{-1}Ap_j = -\frac{1}{\alpha_j}(z_{j+1} - z_j)$ and therefore we have (note the sign difference with the standard PCG algorithm)

$$\beta_j = -\frac{(z_{j+1}, z_{j+1})_M}{(z_j, z_j)_M} = -\frac{(z_{j+1}, r_{j+1})}{(z_j, r_j)}.$$

Exercises

1. Show that the preconditioned matrix has the same eigenvalues for all three preconditioning options (left, right, and split) described in Section 9.1.

2. Let a matrix A and its preconditioner M be SPD. Observing that $M^{-1}A$ is self-adjoint with respect to the A inner product, write an algorithm similar to Algorithm 9.1 for solving the preconditioned linear system $M^{-1}Ax = M^{-1}b$, using the A inner product. The algorithm should employ only one matrix-by-vector product per CG step.

3. In Section 9.2.1, the split-preconditioned CG algorithm, Algorithm 9.2, was derived from the PCG Algorithm 9.1. The opposite can also be done. Derive Algorithm 9.1 starting from Algorithm 9.2, providing a different proof of the equivalence of the two algorithms.

4. Six versions of the CG algorithm applied to the normal equations can be defined. Two versions come from the NR/NE options, each of which can be preconditioned from left, right, or two sides. The left-preconditioned variants have been given in Section 9.5. Describe the four other versions: right P-CGNR, right P-CGNE, split P-CGNR, split P-CGNE. Suitable inner products may be used to preserve symmetry.

5. When preconditioning the normal equations, whether the NE or NR form, two options are available in addition to the left, right, and split preconditioners. These are *centered* versions:

$$AM^{-1}A^T u = b, \quad x = M^{-1}A^T u$$

for the NE form and

$$A^T M^{-1} A x = A^T M^{-1} b$$

for the NR form. The coefficient matrices in the above systems are all symmetric. Write down the adapted versions of the CG algorithm for these options.

6. Let a matrix A and its preconditioner M be SPD. The standard result about the rate of convergence of the CG algorithm is not valid for the PCG algorithm, Algorithm 9.1. Show how to adapt this result by exploiting the M inner product. Show how to derive the same result by using the equivalence between Algorithm 9.1 and Algorithm 9.2.

7. In Eisenstat's implementation of the PCG algorithm, the operation with the diagonal D causes some difficulties when describing the algorithm. This can be avoided.

   a. Assume that the diagonal D of the preconditioning (9.6) is equal to the identity matrix. How many operations are needed to perform one step of the PCG algorithm with Eisenstat's implementation? Formulate the PCG scheme for this case carefully.

b. The rows and columns of the preconditioning matrix M can be scaled so that the matrix D of the transformed preconditioner, written in the form (9.6), is equal to the identity matrix. What scaling should be used (the resulting M should also be SPD)?

c. Assume that the same scaling of part (b) is also applied to the original matrix A. Is the resulting iteration mathematically equivalent to using Algorithm 9.1 to solve the system (9.7) preconditioned with the diagonal D?

8. In order to save operations, the two matrices $D^{-1}E$ and $D^{-1}E^T$ must be stored when computing $\hat{A}v$ by Algorithm 9.3. This exercise considers alternatives.

   a. Consider the matrix $B \equiv D\hat{A}D$. Show how to implement an algorithm similar to Algorithm 9.3 for multiplying a vector v by B. The requirement is that only ED^{-1} must be stored.

   b. The matrix B in part (a) is not the proper preconditioned version of A by the preconditioning (9.6). CG is used on an equivalent system involving B but a further preconditioning by a diagonal must be applied. Which one? How does the resulting algorithm compare in terms of cost and storage with one based on Algorithm 9.3?

   c. It was mentioned in Section 9.2.2 that $\hat{A}$ needed to be further preconditioned by D^{-1}. Consider the split-preconditioning option: CG is to be applied to the preconditioned system associated with $C = D^{1/2}\hat{A}D^{1/2}$. Defining $\hat{E} = D^{-1/2}ED^{-1/2}$, show that
   $$C = (I - \hat{E})^{-1}D_2(I - \hat{E})^{-T} + (I - \hat{E})^{-1} + (I - \hat{E})^{-T},$$
   where D_2 is a certain matrix to be determined. Then write an analogue of Algorithm 9.3 using this formulation. How does the operation count compare with that of Algorithm 9.3?

9. Assume that the number of nonzero elements of a matrix A is parameterized by $Nz(Z) = \alpha n$. How small should α be before it does not pay to use Eisenstat's implementation for the PCG algorithm? What if the matrix A is initially scaled so that D is the identity matrix?

10. Let $M = LU$ be a preconditioner for a matrix A. Show that the left-, right-, and split-preconditioned matrices all have the same eigenvalues. Does this mean that the corresponding preconditioned iterations will converge in (a) exactly the same number of steps? (b) roughly the same number of steps for any matrix? (c) roughly the same number of steps, except for ill-conditioned matrices?

11. Show that the relation (9.18) holds for any polynomial s and any vector r.

12. Write the equivalent of Algorithm 9.1 for the conjugate residual method.

13. Assume that an SPD matrix M is used to precondition GMRES for solving a nonsymmetric linear system. The main features of the preconditioned GMRES algorithm exploiting this were given in Section 9.2.1. Give a formal description of the algorithm. In particular give an MGS implementation. [Hint: The vectors Mv_i must be saved in

addition to the v_i's.] What optimality property does the approximate solution satisfy? What happens if the original matrix A is also symmetric? What is a potential advantage of the resulting algorithm?

Notes and References

The preconditioned version of CG described in Algorithm 9.1 is due to Meijerink and van der Vorst [207]. Eisenstat's implementation was developed in [114] and is often referred to as *Eisenstat's trick*. A number of other similar ideas are described in [216].

Several flexible variants of nonsymmetric Krylov subspace methods have been developed by several authors simultaneously; see, e.g., [22], [246], and [290]. There does not seem to exist a similar technique for left-preconditioned variants of the Krylov subspace methods. This is because the right-hand side $M_j^{-1}b$ of the preconditioned system now changes at each step. A rigorous flexible variant of the biconjugate gradient methods cannot be developed because the short recurrences of these algorithms rely on the preconditioned operator being constant. However, it is possible to develop an analogue of DQGMRES for quasi-minimal residual (or other quasi-minimization) methods using identical arguments; see, e.g., [281], though, as is expected, the global biorthogonality of the Lanczos basis vectors is sacrificed. Similarly, flexible variants of the CG method have been developed by sacrificing global optimality properties but by tightening the flexibility of the preconditioner in an attempt to preserve good, possibly superlinear, convergence; see [213] and [153].

The CGW algorithm can be useful in some instances, such as when the symmetric part of A can be inverted easily, e.g., using fast Poisson solvers. Otherwise, its weakness is that linear systems with the symmetric part must be solved exactly. Inner-outer variations that do not require exact solutions have been described by Golub and Overton [150].

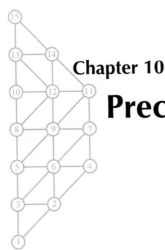

Chapter 10
Preconditioning Techniques

Finding a good preconditioner to solve a given sparse linear system is often viewed as a combination of art and science. Theoretical results are rare and some methods work surprisingly well, often despite expectations. A preconditioner can be defined as any subsidiary approximate solver that is combined with an outer iteration technique, typically one of the Krylov subspace iterations seen in previous chapters. This chapter covers some of the most successful techniques used to precondition a general sparse linear system. Note at the outset that there are virtually no limits to available options for obtaining good preconditioners. For example, preconditioners can be derived from knowledge of the original physical problems from which the linear system arises. However, a common feature of the preconditioners discussed in this chapter is that they are built from the original coefficient matrix.

10.1 Introduction

Roughly speaking, a preconditioner is any form of implicit or explicit modification of an original linear system that makes it easier to solve by a given iterative method. For example, scaling all rows of a linear system to make the diagonal elements equal to one is an explicit form of preconditioning. The resulting system can be solved by a Krylov subspace method and may require fewer steps to converge than the original system (although this is not guaranteed). As another example, solving the linear system

$$M^{-1}Ax = M^{-1}b,$$

where M^{-1} is some complicated mapping that may involve fast Fourier transforms (FFT), integral calculations, and subsidiary linear system solutions, may be another form of preconditioning. Here, it is unlikely that the matrices M and $M^{-1}A$ can be computed explicitly. Instead, the iterative processes operate with A and with M^{-1} whenever needed. In practice, the preconditioning operation M^{-1} should be inexpensive to apply to an arbitrary vector.

One of the simplest ways of defining a preconditioner is to perform an *incomplete factorization* of the original matrix A. This entails a decomposition of the form $A = LU - R$, where L and U have the same nonzero structure as the lower and upper parts of A, respectively, and R is the *residual* or *error* of the factorization. This incomplete factorization, known as ILU(0), is rather easy and inexpensive to compute. On the other hand, it often leads to a crude approximation, which may result in the Krylov subspace accelerator requiring many iterations to converge. To remedy this, several alternative incomplete factorizations have been developed by allowing more fill-in in L and U. In general, the more accurate ILU factorizations require fewer iterations to converge, but the preprocessing cost to compute the factors is higher. However, if only because of the improved robustness, these trade-offs generally favor the more accurate factorizations. This is especially true when several systems with the same matrix must be solved because the preprocessing cost can be amortized.

This chapter considers the most common preconditioners used for solving large sparse matrices and compares their performance. It begins with the simplest preconditioners (successive overrelaxation (SOR) and symmetric SOR (SSOR)) and then discusses the more accurate variants, such as ILU with threshold (ILUT).

10.2   Jacobi, Successive Overrelaxation, and Symmetric Successive Overrelaxation Preconditioners

As was seen in Chapter 4, a fixed-point iteration for solving a linear system

$$Ax = b$$

takes the general form

$$x_{k+1} = M^{-1}Nx_k + M^{-1}b, \qquad (10.1)$$

where M and N realize the splitting of A into

$$A = M - N. \qquad (10.2)$$

The above iteration is of the form

$$x_{k+1} = Gx_k + f, \qquad (10.3)$$

where $f = M^{-1}b$ and

$$\begin{aligned} G &= M^{-1}N = M^{-1}(M - A) \\ &= I - M^{-1}A. \end{aligned} \qquad (10.4)$$

Thus, for Jacobi and Gauss–Seidel, it has been shown that

$$G_{JA}(A) = I - D^{-1}A, \qquad (10.5)$$
$$G_{GS}(A) = I - (D - E)^{-1}A, \qquad (10.6)$$

where $A = D - E - F$ is the splitting defined in Chapter 4.

10.2. Jacobi, SOR, and SSOR Preconditioners

The iteration (10.3) is attempting to solve

$$(I - G)x = f, \tag{10.7}$$

which, because of the expression (10.4) for G, can be rewritten as

$$M^{-1}Ax = M^{-1}b. \tag{10.8}$$

The above system is the *preconditioned system* associated with the splitting $A = M - N$ and the iteration (10.3) is nothing but a *fixed-point iteration on this preconditioned system*. Similarly, a Krylov subspace method, e.g., generalized minimal residual (GMRES), can be used to solve (10.8), leading to a preconditioned version of the Krylov subspace method, e.g., preconditioned GMRES. The preconditioned versions of some Krylov subspace methods were discussed in the previous chapter with a generic preconditioner M. In theory, any general splitting in which M is nonsingular can be used. Ideally, M should be close to A in some sense. However, note that a linear system with the matrix M must be solved at each step of the iterative procedure. Therefore, a practical and admittedly somewhat vague requirement is that these solution steps should be inexpensive.

As was seen in Chapter 4, the SSOR preconditioner is defined by

$$M_{SSOR} = (D - \omega E)D^{-1}(D - \omega F).$$

Typically, when this matrix is used as a preconditioner, it is not necessary to choose ω as carefully as for the underlying fixed-point iteration. Taking $\omega = 1$ leads to the symmetric Gauss–Seidel (SGS) iteration

$$M_{SGS} = (D - E)D^{-1}(D - F). \tag{10.9}$$

An interesting observation is that $D - E$ is the lower part of A, including the diagonal, and $D - F$ is, similarly, the upper part of A. Thus,

$$M_{SGS} = LU,$$

with

$$L \equiv (D - E)D^{-1} = I - ED^{-1}, \quad U = D - F.$$

The matrix L is unit lower triangular and U is upper triangular. One question that may arise concerns the implementation of the preconditioning operation. To compute $w = M_{SGS}^{-1}x$, proceed as follows:

$$\text{solve} \quad (I - ED^{-1})z = x,$$
$$\text{solve} \quad (D - F)w = z.$$

A FORTRAN implementation of this preconditioning operation is illustrated in the following code for matrices stored in the modified sparse row (MSR) format described in Chapter 3.

FORTRAN CODE

```
      subroutine lusol (n,rhs,sol,luval,lucol,luptr,uptr)
      real*8 sol(n), rhs(n), luval(*)
      integer n, luptr(*), uptr(n)
c-----------------------------------------------------------------
c Performs a forward and a backward solve for an ILU or
c SSOR factorization, i.e., solves (LU) sol = rhs, where LU
c is the ILU or the SSOR factorization. For SSOR, L and U
c should contain the matrices L = I - omega E inv(D) and U
c = D - omega F, respectively, with -E = strict lower
c triangular part of A, -F = strict upper triangular part
c of A, and D = diagonal of A.
c-----------------------------------------------------------------
c PARAMETERS:
c n     = Dimension of problem
c rhs   = Right-hand side; rhs is unchanged on return
c sol   = Solution of (LU) sol = rhs.
c luval = Values of the LU matrix. L and U are stored
c         together in compressed sparse row (CSR) format.
c         The diagonal elements of U are inverted. In each
c         row, the L values are followed by the diagonal
c         element (inverted) and then the other U values.
c lucol = Column indices of corresponding elements in luval
c luptr = Contains pointers to the beginning of each row in
c         the LU matrix
c uptr  = pointer to the diagonal elements in luval, lucol
c-----------------------------------------------------------------
      integer i,k
c
c     FORWARD SOLVE. Solve   L . sol = rhs
c
      do i = 1, n
c
c     compute  sol(i) := rhs(i) - sum L(i,j) x sol(j)
c
         sol(i) = rhs(i)
         do k=luptr(i),uptr(i)-1
            sol(i) = sol(i) - luval(k)* sol(lucol(k))
         enddo
      enddo
c
c     BACKWARD SOLVE. Compute  sol := inv(U) sol
c
      do i = n, 1, -1
c
c     compute  sol(i) := sol(i) - sum U(i,j) x sol(j)
c
         do k=uptr(i)+1, luptr(i+1)-1
            sol(i) = sol(i) - luval(k)*sol(lucol(k))
         enddo
c
c     compute  sol(i) := sol(i)/ U(i,i)
c
         sol(i) = luval(uptr(i))*sol(i)
      enddo
      return
      end
```

10.3. Incomplete LU Factorization Preconditioners

As was seen above, the SSOR or SGS preconditioning matrix is of the form $M = LU$, where L and U have the same pattern as the L part and the U part of A, respectively. Here, L part means lower triangular part and, similarly, the U part is the upper triangular part. If the error matrix $A - LU$ is computed, then for SGS, for example, we would find

$$A - LU = D - E - F - (I - ED^{-1})(D - F) = -ED^{-1}F.$$

If L is restricted to have the same structure as the L part of A and U is to have the same structure as the U part of A, the question is whether or not it is possible to find L and U that yield an error that is smaller in some sense than the one above. We can, for example, try to find such an incomplete factorization in which the residual matrix $A - LU$ has zero elements in locations where A has nonzero entries. This turns out to be possible in general and yields the ILU(0) factorization to be discussed later. Generally, a pattern for L and U can be specified and L and U may be sought so that they satisfy certain conditions. This leads to the general class of incomplete factorization techniques that are discussed in the next section.

Example 10.1. Table 10.1 shows the results of applying the GMRES algorithm with SGS (SSOR with $\omega = 1$) preconditioning to the five test problems described in Section 3.7.

Matrix	Iters	Kflops	Residual	Error
F2DA	38	1986	0.76E−03	0.82E−04
F3D	20	4870	0.14E−02	0.30E−03
ORS	110	6755	0.31E+00	0.68E−04
F2DB	300	15907	0.23E+02	0.66E+00
FID	300	99070	0.26E+02	0.51E−01

Table 10.1. *A test run of GMRES with SGS preconditioning.*

See Example 6.1 for the meaning of the column headers in the table. Notice here that the method did not converge in 300 steps for the last two problems. The number of iterations for the first three problems is reduced substantially from those required by GMRES without preconditioning shown in Table 6.2. The total number of operations required is also reduced, but not proportionally because each step now costs more due to the preconditioning operation.

10.3 Incomplete LU Factorization Preconditioners

Consider a general sparse matrix A whose elements are $a_{ij}, i, j = 1, \ldots, n$. A general ILU factorization process computes a sparse lower triangular matrix L and a sparse upper triangular matrix U so that the residual matrix $R = LU - A$ satisfies certain constraints, such as having zero entries in some locations. We first describe a general ILU preconditioner geared toward M-matrices. Then we discuss the ILU(0) factorization, the simplest form of the ILU preconditioners. Finally, we show how to obtain more accurate factorizations.

10.3.1 ILU Factorizations

A general algorithm for building ILU factorizations can be derived by performing Gaussian elimination and dropping some elements in predetermined nondiagonal positions. To analyze this process and establish existence for M-matrices, the following result of Ky Fan [122] is needed.

Theorem 10.1. *Let A be an M-matrix and let A_1 be the matrix obtained from the first step of Gaussian elimination. Then A_1 is an M-matrix.*

Proof. Theorem 1.32 will be used to establish that properties 1, 2, and 3 therein are satisfied. First, consider the off-diagonal elements of A_1:

$$a_{ij}^1 = a_{ij} - \frac{a_{i1}a_{1j}}{a_{11}}.$$

Since a_{ij}, a_{i1}, a_{1j} are nonpositive and a_{11} is positive, it follows that $a_{ij}^1 \leq 0$ for $i \neq j$.

Second, the fact that A_1 is nonsingular is a trivial consequence of the following standard relation of Gaussian elimination:

$$A = L_1 A_1, \quad \text{where} \quad L_1 = \left[\frac{A_{*,1}}{a_{11}}, e_2, e_3, \ldots, e_n\right]. \tag{10.10}$$

Finally, we establish that A_1^{-1} is nonnegative by examining $A_1^{-1} e_j$ for $j = 1, \ldots, n$. For $j = 1$, it is clear that $A_1^{-1} e_1 = \frac{1}{a_{11}} e_1$ because of the structure of A_1. For the case $j \neq 1$, (10.10) can be exploited to yield

$$A_1^{-1} e_j = A^{-1} L_1^{-1} e_j = A^{-1} e_j \geq 0.$$

Therefore, all the columns of A_1^{-1} are nonnegative by assumption, which completes the proof. □

Clearly, the $(n-1) \times (n-1)$ matrix obtained from A_1 by removing its first row and first column is also an M-matrix.

Assume now that some elements are dropped from the result of Gaussian elimination outside the main diagonal. Any element that is dropped is a nonpositive entry that is transformed into a zero. Therefore, the resulting matrix $\tilde{A}_1$ is such that

$$\tilde{A}_1 = A_1 + R,$$

where the elements of R are such that $r_{ii} = 0, r_{ij} \geq 0$. Thus,

$$A_1 \leq \tilde{A}_1$$

and the off-diagonal elements of $\tilde{A}_1$ are nonpositive. Since A_1 is an M-matrix, Theorem 1.33 shows that $\tilde{A}_1$ is also an M-matrix. The process can now be repeated on the matrix $\tilde{A}(2:n, 2:n)$ and then continued until the incomplete factorization of A is obtained. The above arguments show that, at each step of this construction, we obtain an M-matrix, and that the process does not break down.

10.3. Incomplete LU Factorization Preconditioners

The elements to drop at each step have not yet been specified. This can be done statically by choosing some nonzero pattern in advance. The only restriction on the zero pattern is that it should exclude diagonal elements because this assumption was used in the above proof. Therefore, for any zero pattern set P such that

$$P \subset \{(i,j) \mid i \neq j; 1 \leq i, j \leq n\}, \tag{10.11}$$

an ILU factorization, ILU_P, can be computed as follows.

ALGORITHM 10.1. General Static Pattern ILU

0. For each $(i,j) \in P$ set $a_{ij} = 0$
1. For $k = 1, \ldots, n-1$, Do
2.   For $i = k+1, n$ and if $(i,k) \notin P$, Do
3. $a_{ik} := a_{ik}/a_{kk}$
4.   For $j = k+1, \ldots, n$ and for $(i,j) \notin P$, Do
5. $a_{ij} := a_{ij} - a_{ik} * a_{kj}$
6.   EndDo
7.   EndDo
8. EndDo

The initial step (step 0) is necessary for the case, rare in practice, when the zero pattern of A does not include the zero pattern defined by P. The *For* loop in line 4 should be interpreted as follows: For $j = k+1, \ldots, n$ and only for those indices j that are not in P execute the next line. In practice, it is wasteful to scan j from $k+1$ to n because there is an inexpensive mechanism for identifying those indices j that are in the complement of P. Using the above arguments, the following result can be proved.

Theorem 10.2. *Let A be an M-matrix and P a given zero pattern defined as in* (10.11). *Then Algorithm 10.1 does not break down and produces an incomplete factorization*

$$A = LU - R,$$

which is a regular splitting of A.

Proof. At each step of the process, we have

$$\tilde{A}_k = A_k + R_k, \quad A_k = L_k \tilde{A}_{k-1},$$

where, using O_k to denote a zero vector of dimension k and $A_{m:n,j}$ to denote the vector of components $a_{i,j}, i = m, \ldots, n$,

$$L_k = I - \frac{1}{a_{kk}^{(k)}} \begin{pmatrix} O_k \\ A(k+1:n, k) \end{pmatrix} e_k^T.$$

From this follows the relation

$$\tilde{A}_k = A_k + R_k = L_k \tilde{A}_{k-1} + R_k.$$

Applying this relation recursively, starting from $k = n - 1$ up to $k = 1$, it is found that

$$\tilde{A}_{n-1} = L_{n-1} \cdots L_1 A + L_{n-1} \cdots L_2 R_1 + \cdots + L_{n-1} R_{n-2} + R_{n-1}. \tag{10.12}$$

Now define

$$L = (L_{n-1} \cdots L_1)^{-1}, \quad U = \tilde{A}_{n-1}.$$

Then $U = L^{-1} A + S$, with

$$S = L_{n-1} \cdots L_2 R_1 + \cdots + L_{n-1} R_{n-2} + R_{n-1}.$$

Observe that, at stage k, elements are dropped only in the $(n - k) \times (n - k)$ lower part of A_k. Hence, the first k rows and columns of R_k are zero and, as a result,

$$L_{n-1} \cdots L_{k+1} R_k = L_{n-1} \cdots L_1 R_k,$$

so that S can be rewritten as

$$S = L_{n-1} \cdots L_2 (R_1 + R_2 + \cdots + R_{n-1}).$$

If R denotes the matrix

$$R = R_1 + R_2 + \cdots + R_{n-1},$$

then we obtain the factorization $A = LU - R$, where $(LU)^{-1} = U^{-1} L^{-1}$ is a nonnegative matrix and R is nonnegative. This completes the proof. □

Now consider a few practical aspects. An ILU factorization based on Algorithm 10.1 is difficult to implement because, at each step k, all rows $k + 1$ to n are being modified. However, ILU factorizations depend on which implementation of Gaussian elimination is used. Several variants of Gaussian elimination are known that depend on the order of the three loops associated with the control variables i, j, and k in the algorithm. Thus, Algorithm 10.1 is derived from what is known as the KIJ variant. In the context of ILU factorization, the variant that is most commonly used for a row-contiguous data structure is the IKJ variant, described next for dense matrices.

ALGORITHM 10.2. Gaussian Elimination—IKJ Variant

1. For $i = 2, \ldots, n$, Do
2.   For $k = 1, \ldots, i - 1$, Do
3. $a_{ik} := a_{ik}/a_{kk}$
4.   For $j = k + 1, \ldots, n$, Do
5. $a_{ij} := a_{ij} - a_{ik} * a_{kj}$
6.   EndDo
7.   EndDo
8. EndDo

The above algorithm is in place in the sense that the ith row of A can be overwritten by the ith rows of the L- and U-matrices of the factorization (since L is unit lower triangular, its diagonal entries need not be stored). Step i of the algorithm generates the ith row of L

10.3. Incomplete LU Factorization Preconditioners

and the ith row of U at the same time. The previous rows $1, 2, \ldots, i-1$ of L and U are accessed at step i but they are not modified. This is illustrated in Figure 10.1.

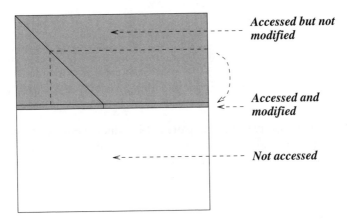

Figure 10.1. *IKJ variant of the LU factorization.*

Adapting this version for sparse matrices is easy because the rows of L and U are generated in succession. These rows can be computed one at a time and accumulated in a row-oriented data structure such as the CSR format. This constitutes an important advantage. Based on this, the general ILU factorization takes the following form.

ALGORITHM 10.3. General ILU Factorization, IKJ Version

1. For $i = 2, \ldots, n$, Do
2.   For $k = 1, \ldots, i-1$ and if $(i, k) \notin P$, Do
3. $a_{ik} := a_{ik}/a_{kk}$
4.   For $j = k+1, \ldots, n$ and if $(i, j) \notin P$, Do
5. $a_{ij} := a_{ij} - a_{ik} a_{kj}$
6.   EndDo
7.   EndDo
8. EndDo

It is not difficult to see that this more practical IKJ variant of ILU is equivalent to the KIJ version, which can be defined from Algorithm 10.1.

Proposition 10.3. *Let P be a zero pattern satisfying the condition* (10.11). *Then the ILU factors produced by the KIJ-based Algorithm* 10.1 *and the IKJ-based Algorithm* 10.3 *are identical if they can both be computed.*

Proof. Algorithm 10.3 is obtained from Algorithm 10.1 by switching the order of the loops k and i. To see that this indeed gives the same result, reformulate the first two loops of

Algorithm 10.1 as follows:

> For $k = 1, \ldots, n$, Do
>   For $i = 1, \ldots, n$, Do
>   if $k < i$ and for $(i, k) \notin P$, Do
> $ope(row(i), row(k))$
>   ......

in which $ope(row(i), row(k))$ is the operation represented by lines 3 through 6 of both Algorithm 10.1 and Algorithm 10.3. In this form, it is clear that the k and i loops can be safely permuted. Then the resulting algorithm can be reformulated to yield exactly Algorithm 10.3. □

Note that this is only true for a static pattern ILU. If the pattern is dynamically determined as the Gaussian elimination algorithm proceeds, then the patterns obtained with different versions of Gaussian elimination may be different.

It is helpful to interpret the result of one incomplete elimination step. Denoting by l_{i*}, u_{i*}, and a_{i*} the ith rows of L, U, and A, respectively, then the k loop starting at line 2 of Algorithm 10.3 can be interpreted as follows. Initially, we have $u_{i*} = a_{i*}$. Then each elimination step is an operation of the form

$$u_{i*} := u_{i*} - l_{ik} u_{k*}.$$

However, this operation is performed only on the nonzero pattern, i.e., the complement of P. This means that, in reality, the elimination step takes the form

$$u_{i*} := u_{i*} - l_{ik} u_{k*} + r_{i*}^{(k)},$$

in which $r_{ij}^{(k)}$ is zero when $(i, j) \notin P$ and equals $l_{ik} u_{kj}$ when $(i, j) \in P$. Thus, the row $r_{i*}^{(k)}$ cancels out the terms $l_{ik} u_{kj}$ that would otherwise be introduced in the zero pattern. In the end the following relation is obtained:

$$u_{i*} = a_{i*} - \sum_{k=1}^{i-1} \left(l_{ik} u_{k*} - r_{i*}^{(k)} \right).$$

Note that $l_{ik} = 0$ for $(i, k) \in P$. We now sum up all the $r_{i*}^{(k)}$'s and define

$$r_{i*} = \sum_{k=1}^{i-1} r_{i*}^{(k)}. \qquad (10.13)$$

The row r_{i*} contains the elements that fall inside the P pattern at the completion of the k loop. Using the fact that $l_{ii} = 1$, we obtain the relation

$$a_{i*} = \sum_{k=1}^{i} l_{ik} u_{k*} - r_{i*}. \qquad (10.14)$$

Therefore, the following simple property can be stated.

Proposition 10.4. *Algorithm 10.3 produces factors L and U such that*

$$A = LU - R,$$

in which $-R$ is the matrix of the elements that are dropped during the incomplete elimination

10.3. Incomplete LU Factorization Preconditioners

process. When $(i, j) \in P$, an entry r_{ij} of R is equal to the value of $-a_{ij}$ obtained at the completion of the k loop in Algorithm 10.3. Otherwise, r_{ij} is zero.

10.3.2 Zero Fill-in ILU (ILU(0))

The *ILU* factorization technique with no fill-in, denoted by ILU(0), takes the zero pattern P to be precisely the zero pattern of A. In the following, we denote by $b_{i,*}$ the ith row of a given matrix B and by $NZ(B)$ the set of pairs (i, j), $1 \leq i, j \leq n$, such that $b_{i,j} \neq 0$. The ILU(0) factorization is best illustrated by the case for which it was discovered originally, namely, for five-point and seven-point matrices related to finite difference discretization of elliptic partial differential equations (PDEs). Consider one such matrix A, as illustrated in the bottom-left corner of Figure 10.2. The A-matrix represented in this figure is a five-point matrix of size $n = 32$ corresponding to an $n_x \times n_y = 8 \times 4$ mesh. Consider now any lower triangular matrix L with the same structure as the lower part of A and any matrix U with the same structure as that of the upper part of A. Two such matrices are shown at the top of Figure 10.2. If the product LU were performed, the resulting matrix would have the pattern shown in the bottom-right part of the figure. It is impossible in general to match A with this product for any L and U. This is due to the extra diagonals in the product, namely, the diagonals with offsets $n_x - 1$ and $-n_x + 1$. The entries in these extra diagonals are called *fill-in elements*. However, if these fill-in elements are ignored, then it is possible to find L and U so that their product is equal to A in the other diagonals.

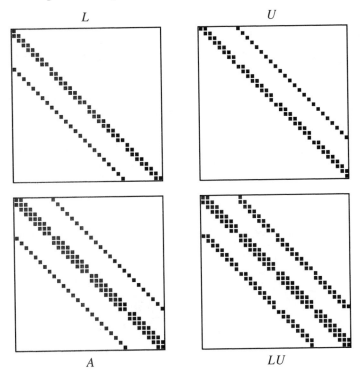

Figure 10.2. *The ILU(0) factorization for a five-point matrix.*

The ILU(0) factorization has just been defined in general terms as any pair of matrices L (unit lower triangular) and U (upper triangular) so that the elements of $A - LU$ are zero in the locations of $NZ(A)$. These constraints do not define the ILU(0) factors uniquely since there are, in general, infinitely many pairs of matrices L and U that satisfy these requirements. However, the standard ILU(0) is defined constructively using Algorithm 10.3 with the pattern P equal to the zero pattern of A.

ALGORITHM 10.4. ILU(0)

1. For $i = 2, \ldots, n$, Do
2.   For $k = 1, \ldots, i-1$ and for $(i, k) \in NZ(A)$, Do
3.   Compute $a_{ik} = a_{ik}/a_{kk}$
4.   For $j = k+1, \ldots, n$ and for $(i, j) \in NZ(A)$, Do
5.   Compute $a_{ij} := a_{ij} - a_{ik}a_{kj}$
6.   EndDo
7.   EndDo
8. EndDo

In some cases, it is possible to write the ILU(0) factorization in the form

$$M = (D - E)D^{-1}(D - F), \tag{10.15}$$

where $-E$ and $-F$ are the strict lower and strict upper triangular parts of A and D is a certain diagonal matrix, different from the diagonal of A, in general. In these cases it is sufficient to find a recursive formula for determining the elements in D. A clear advantage is that only an extra diagonal of storage is required. This form of the ILU(0) factorization is equivalent to the incomplete factorizations obtained from Algorithm 10.4 when the product of the *strict lower part* and the *strict upper part* of A consists only of diagonal elements and fill-in elements. This is true, for example, for standard five-point difference approximations to second order partial differential operators; see Exercise 4. In these instances, both the SSOR preconditioner with $\omega = 1$ and the ILU(0) preconditioner can be cast in the form (10.15), but they differ in the way the diagonal matrix D is defined. For SSOR($\omega = 1$), D is the diagonal of the matrix A itself. For ILU(0), it is defined by a recursion so that the diagonal of the product of matrices (10.15) equals the diagonal of A. By definition, together the L- and U-matrices in ILU(0) have the same number of nonzero elements as the original matrix A.

Example 10.2. Table 10.2 shows the results of applying the GMRES algorithm with ILU(0) preconditioning to the five test problems described in Section 3.7. See Example 6.1 for the meaning of the column headers in the table.

Matrix	Iters	Kflops	Residual	Error
F2DA	28	1456	0.12E−02	0.12E−03
F3D	17	4004	0.52E−03	0.30E−03
ORS	20	1228	0.18E+00	0.67E−04
F2DB	300	15907	0.23E+02	0.67E+00
FID	206	67970	0.19E+00	0.11E−03

Table 10.2. *A test run of GMRES with ILU(0) preconditioning.*

10.3. Incomplete LU Factorization Preconditioners

Observe that, for the first two problems, the gains relative to the performance of the SSOR preconditioner in Table 10.1 are rather small. For the other three problems, which are a little harder, the gains are more substantial. For the last problem, the algorithm achieves convergence in 206 steps whereas SSOR did not converge in the 300 steps allowed. The fourth problem (F2DB) is still not solvable by ILU(0) within the maximum number of steps allowed.

For the purpose of illustration, below is a sample FORTRAN code for computing the incomplete L and U factors for general sparse matrices stored in the usual CSR format. The real values of the resulting L, U factors are stored in the array *luval*, except that entries of ones of the main diagonal of the unit lower triangular matrix L are not stored. Thus, one matrix is needed to store these factors together. This matrix is denoted by L/U. Note that since the pattern of L/U is identical to that of A, the other integer arrays of the CSR representation for the LU factors are not needed. Thus, $ja(k)$, which is the column position of the element $a(k)$ in the input matrix, is also the column position of the element $luval(k)$ in the L/U matrix. The code below assumes that the nonzero elements in the input matrix A are sorted by increasing column numbers in each row.

---------------- FORTRAN CODE ----------------

```
      subroutine ilu0 (n, a, ja, ia, luval, uptr, iw, icode)
      integer n, ja(*), ia(n+1), uptr(n), iw(n)
      real*8 a(*), luval(*)
c-----------------------------------------------------------
c Set-up routine for   ILU(0)   preconditioner. This routine
c computes the L and U factors of the ILU(0) factorization
c of a general sparse matrix A stored in CSR format. Since
c L is unit triangular, the L and U factors can be stored
c as a single matrix that occupies the same storage as A.
c The ja and ia arrays are not needed for the LU matrix
c since the pattern of the LU  matrix is identical to
c that of A.
c-----------------------------------------------------------
c INPUT:
c ------
c n           = dimension of matrix
c a, ja, ia   = sparse matrix in general sparse storage format
c iw          = integer work array of length n
c OUTPUT:
c -------
c luval       = L/U matrices stored together. On return luval,
c               ja, ia is the combined CSR data structure for
c               the LU factors
c uptr        = pointer to the diagonal elements in the CSR
c               data structure luval, ja, ia
c icode       = integer indicating error code on return
c               icode = 0: normal return
c               icode = k: encountered a zero pivot at step k
c
c-----------------------------------------------------------
c     initialize work array iw to zero and luval array to a
      do 30 i = 1, ia(n+1)-1
          luval(i) = a(i)
30    continue
```

```
              do 31 i=1, n
                 iw(i) = 0
       31     continue
C----------------------- Main loop
              do 500 k = 1, n
                 j1 = ia(k)
                 j2 = ia(k+1)-1
                 do 100 j=j1, j2
                    iw(ja(j)) = j
      100     continue
                 j=j1
      150     jrow = ja(j)
C----------------------- Exit if diagonal element is reached
                 if (jrow .ge. k) goto 200
C----------------------- Compute the multiplier for jrow.
                 tl = luval(j)*luval(uptr(jrow))
                 luval(j) = tl
C----------------------- Perform linear combination
                 do 140 jj = uptr(jrow)+1, ia(jrow+1)-1
                    jw = iw(ja(jj))
                    if (jw .ne. 0) luval(jw)=luval(jw)-tl*luval(jj)
      140     continue
                 j=j+1
                 if (j .le. j2) goto 150
C----------------------- Store pointer to diagonal element
      200     uptr(k) = j
                 if (jrow .ne. k .or. luval(j) .eq. 0.0d0) goto 600
                 luval(j) = 1.0d0/luval(j)
C----------------------- Refresh all entries of iw to zero.
                 do 201 i = j1, j2
                    iw(ja(i)) = 0
      201     continue
      500     continue
C----------------------- Normal return
              icode = 0
              return
C----------------------- Error: zero pivot
      600     icode = k
              return
              end
```

10.3.3 Level of Fill and ILU(p)

The accuracy of the ILU(0) incomplete factorization may be insufficient to yield an adequate rate of convergence, as shown in Example 10.2. More accurate ILU factorizations are often more efficient as well as more reliable. These more accurate factorizations will differ from ILU(0) by allowing some fill-in. Thus, ILU(1) keeps the "first order fill-ins," a term that will be explained briefly.

To illustrate ILU(p) with the same example as before, the ILU(1) factorization results from taking P to be the zero pattern of the product LU of the factors L, U obtained from ILU(0). This pattern is shown at the bottom right of Figure 10.2. Pretend that the original matrix has this "augmented" pattern $NZ_1(A)$. In other words, the fill-in positions created in this product belong to the augmented pattern $NZ_1(A)$, but their actual values are zero. The new pattern of the matrix A is shown at the bottom left of Figure 10.3. The factors

10.3. Incomplete LU Factorization Preconditioners

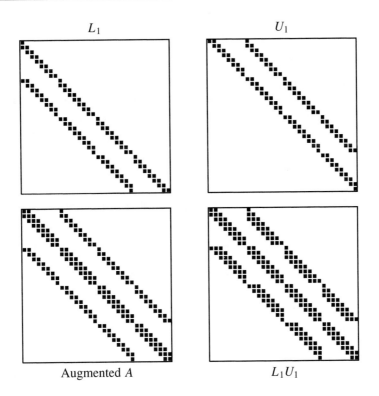

Figure 10.3. *The ILU(1) factorization for a five-point matrix.*

L_1 and U_1 of the ILU(1) factorization are obtained by performing an ILU(0) factorization on this *augmented pattern* matrix. The patterns of L_1 and U_1 are illustrated at the top of Figure 10.3. The new LU-matrix shown at the bottom right of the figure now has two additional diagonals in the lower and upper parts.

One problem with the construction defined in this illustration is that it does not extend to general sparse matrices. It can be generalized by introducing the concept of *level of fill*. A level of fill is attributed to each element that is processed by Gaussian elimination and dropping will be based on the value of the level of fill. Algorithm 10.2 will be used as a model, although any other form of Gaussian elimination can be used. The rationale is that the level of fill should be indicative of the size: the higher the level, the smaller the elements. A very simple model is employed to justify the definition: A size of ϵ^k is attributed to any element whose level of fill is k, where $\epsilon < 1$. Initially, a nonzero element has a level of fill of one (this will be changed later) and a zero element has a level of fill of ∞. An element a_{ij} is updated in line 5 of Algorithm 10.2 by the formula

$$a_{ij} = a_{ij} - a_{ik} \times a_{kj}. \tag{10.16}$$

If lev_{ij} is the current level of the element a_{ij}, then our model tells us that the size of the updated element should be

$$a_{ij} := \epsilon^{lev_{ij}} - \epsilon^{lev_{ik}} \times \epsilon^{lev_{kj}} = \epsilon^{lev_{ij}} - \epsilon^{lev_{ik} + lev_{kj}}.$$

Therefore, roughly speaking, the size of a_{ij} will be the maximum of the two sizes $\epsilon^{lev_{ij}}$ and $\epsilon^{lev_{ik}+lev_{kj}}$, so it is natural to define the new level of fill as

$$lev_{ij} := \min\{lev_{ij}, lev_{ik} + lev_{kj}\}.$$

In the common definition used in the literature, all the levels of fill are actually shifted by -1 from the definition used above. This is purely for convenience of notation and to conform with the definition used for ILU(0). Thus initially, $lev_{ij} = 0$ if $a_{ij} \neq 0$ and $lev_{ij} = \infty$ otherwise. Thereafter, define recursively

$$lev_{ij} = \min\{lev_{ij}, lev_{ik} + lev_{kj} + 1\}.$$

Definition 10.5. *The initial level of fill of an element a_{ij} of a sparse matrix A is defined by*

$$lev_{ij} = \begin{cases} 0 & \text{if } a_{ij} \neq 0 \text{ or } i = j, \\ \infty & \text{otherwise}. \end{cases}$$

Each time this element is modified in line 5 of Algorithm 10.2, its level of fill must be updated by

$$lev_{ij} = \min\{lev_{ij}, lev_{ik} + lev_{kj} + 1\}. \tag{10.17}$$

Observe that the level of fill of an element will never increase during the elimination. Thus, if $a_{ij} \neq 0$ in the original matrix A, then the element in location i, j will have a level of fill equal to zero throughout the elimination process.

An alternative way of interpreting the above definition of fill level can be drawn from the graph model of Gaussian elimination, which is a standard tool used in sparse direct solvers. Consider the adjacency graph $G(A) = (V, E)$ of the matrix A. At the completion of step $k - 1$ of Gaussian elimination, nodes $1, 2, \ldots, k - 1$ have been eliminated. Let V_{k-1} be the set of the $k - 1$ vertices that are eliminated so far and let v_i, v_j be two vertices not in V_k, i.e., such that $i, j > k$. The vertex v_i is said to be reachable from the vertex v_j through V_{k-1} if there is a path in the (original) graph $G(A)$ that connects v_i to v_j, in which all intermediate vertices are in V_{k-1}. The set of all nodes v that are reachable from u through V_{k-1} is denoted by $Reach(u, V_{k-1})$. The fill level of (i, j) at step $k - 1$ is simply the length of the shortest path through V_{k-1} between v_i and v_j, minus 1. The initial fill levels are defined, as before, to be zero when $(i, j) \in V_0$ and infinity otherwise. At the next step (k), node k will be added to V_{k-1} to get V_k. Now more paths are available and so the path lengths may be shortened by taking paths that go through the new node v_k. If we use the shifted levels (all levels are increased by one, so that $lev(i, j)$ is the actual minimum path length), then the shortest path is now the shortest of the old shortest path and new possible paths through v_k. A path through v_k is a path from i to v_k continued by a path from v_k to j. Therefore, the new path length is indeed $\min\{lev_{ij}, lev_{ik} + lev_{kj}\}$. This is illustrated in Figure 10.4.

Another useful concept in sparse direct solution methods is that of *fill path*, which is a path between two vertices i and j such that all the vertices in the path, except the endpoints i and j, are numbered less than i and j. The following result is well known in sparse direct solution methods.

Theorem 10.6. *There is a fill-in in entry (i, j) at the completion of the Gaussian elimination process iff there exists a fill path between i and j.*

10.3. Incomplete LU Factorization Preconditioners

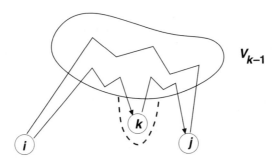

Figure 10.4. *Shortest path from i to j when k is added to V_{k-1}.*

For a proof see [144, 232]. As it turns out, a fill-in entry with level-of-fill value p corresponds to fill paths whose length is $p + 1$.

Theorem 10.7. *At the completion of the ILU process, a fill-in entry in position (i, j) has level-of-fill value p iff there exists a fill path of length $p + 1$ between i and j.*

Proof. If there is a fill path of length p, then, from what was said above on reachable sets, it is clear that $lev(a_{ij}) \leq p$. However, $lev(a_{ij})$ cannot be less than p, otherwise at some step k we would have a path between i and j that is of length less than p. Since path lengths do not increase, this would lead to a contradiction. The converse is also true. If $lev(a_{ij})$ is equal to p then at the last step k when $lev(a_{ij})$ was modified there was a path of length p between i and j. □

The above systematic definition gives rise to a natural strategy for discarding elements. In ILU(p), all fill-in elements whose level of fill does not exceed p are kept. So, using the definition of zero patterns introduced earlier, the zero pattern for ILU(p) is the set

$$P_p = \{(i, j) \mid lev_{ij} > p\},$$

where lev_{ij} is the level-of-fill value after all updates (10.17) have been performed. The case $p = 0$ coincides with the ILU(0) factorization and is consistent with the earlier definition.

Since fill levels are essentially path lengths in the graph, they are bounded from above by $\delta(G) + 1$, where the diameter $\delta(G)$ of a graph G is the maximum possible distance $d(x, y)$ between two vertices x and y of the graph:

$$\delta(G) = \max\{d(x, y) \mid x \in V, y \in V\}.$$

Recall that the distance $d(x, y)$ between vertices x and y in the graph is the length of the shortest path between x and y.

Definition 10.5 of fill levels is not the only one used in practice. An alternative definition replaces the updating formula (10.17) with

$$lev_{ij} = \min\{lev_{ij}, \max\{lev_{ik}, lev_{kj}\} + 1\}. \tag{10.18}$$

In practical implementations of the ILU(p) factorization it is common to separate the symbolic phase (where the structure of the L and U factors is determined) from the numerical factorization, when the numerical values are computed. Here a variant is described that does not separate these two phases. In the following description, a_{i*} denotes the ith row of the matrix A and a_{ij} the (i, j)th entry of A.

ALGORITHM 10.5. ILU(p)

1. For all nonzero elements a_{ij}, define $lev(a_{ij}) = 0$
2. For $i = 2, \ldots, n$, Do
3.   For each $k = 1, \ldots, i - 1$ and for $lev(a_{ik}) \leq p$, Do
4.   Compute $a_{ik} := a_{ik}/a_{kk}$
5.   Compute $a_{i*} := a_{i*} - a_{ik}a_{k*}$
6.   Update the levels of fill of the nonzero $a_{i,j}$'s using (10.17)
7.   EndDo
8.   Replace any element in row i with $lev(a_{ij}) > p$ with zero
9. EndDo

There are a number of drawbacks to the above algorithm. First, the amount of fill-in and computational work for obtaining the ILU(p) factorization is not predictable for $p > 0$. Second, the cost of updating the levels can be high. Most importantly, the level of fill-in for indefinite matrices may not be a good indicator of the size of the elements that are being dropped. Thus, the algorithm may drop large elements and result in an inaccurate incomplete factorization, in that $R = LU - A$ is not small. Experience reveals that *on the average* this will lead to a larger number of iterations to achieve convergence. The techniques that will be described in Section 10.4 have been developed to remedy these difficulties by producing incomplete factorizations with small error R and controlled fill-in.

10.3.4 Matrices with Regular Structure

Often the original matrix has a regular structure that can be exploited to formulate the ILU preconditioners in a simpler way. Historically, incomplete factorization preconditioners were developed first for such matrices, rather than for general sparse matrices. Here, we call a regularly structured matrix a matrix consisting of a small number of diagonals. As an example, consider the convection diffusion equation with Dirichlet boundary conditions

$$-\Delta u + \vec{b} \cdot \nabla u = f \text{ in } \Omega,$$
$$u = 0 \text{ on } \partial\Omega,$$

where Ω is simply a rectangle. As seen in Chapter 2, if the above problem is discretized using centered differences, a linear system is obtained whose coefficient matrix has the structure shown in Figure 10.5. In terms of the stencils seen in Chapter 4, the representation of this matrix is rather simple. Each row expresses the coupling between unknown i and unknowns $i + 1$, $i - 1$ that are in the horizontal or x direction and the unknowns $i + m$ and $i - m$ that are in the vertical or y direction. This stencil is represented in Figure 10.7. The desired L- and U-matrices in the ILU(0) factorization are shown in Figure 10.6. Now the respective stencils of these L- and U-matrices can be represented at a mesh point i, as shown in Figure 10.8. The stencil of the product LU can be obtained easily by manipulating stencils directly rather than working with the matrices they represent.

10.3. Incomplete LU Factorization Preconditioners

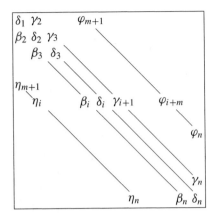

Figure 10.5. *Matrix resulting from the discretization of an elliptic problem on a rectangle.*

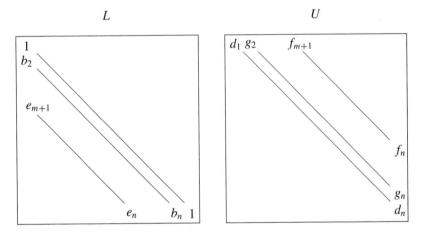

Figure 10.6. *L and U factors of the ILU(0) factorization for the five-point matrix shown in Figure 10.5.*

Indeed, the ith row of LU is obtained by performing the following operation:

$$row_i(LU) = 1 \times row_i(U) + b_i \times row_{i-1}(U) + e_i \times row_{i-m}(U).$$

This translates into a combination of the stencils associated with the rows:

$$stencil_i(LU) = 1 \times stencil_i(U) + b_i \times stencil_{i-1}(U) + e_i \times stencil_{i-m}(U),$$

in which $stencil_j(X)$ represents the stencil of the matrix X based at the mesh point labeled j. This gives the stencil for the LU-matrix represented in Figure 10.9.

302   Chapter 10. Preconditioning Techniques

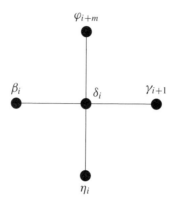

Figure 10.7. *Stencil associated with the 5-point matrix shown in Figure* 10.5.

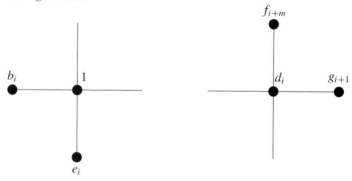

Figure 10.8. *Stencils associated with the L and U factors shown in Figure* 10.6.

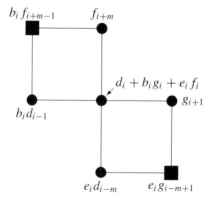

Figure 10.9. *Stencil associated with the product of the L and U factors shown in Figure* 10.6.

10.3. Incomplete LU Factorization Preconditioners

In the figures, the fill-in elements are represented by squares and all other nonzero elements of the stencil are filled circles. The ILU(0) process consists of identifying LU with A in locations where the original a_{ij}'s are nonzero. In the Gaussian elimination process, this is done from $i = 1$ to $i = n$. This provides the following equations obtained directly from comparing the stencils of LU and A (going from lowest to highest indices):

$$e_i d_{i-m} = \eta_i,$$
$$b_i d_{i-1} = \beta_i,$$
$$d_i + b_i g_i + e_i f_i = \delta_i,$$
$$g_{i+1} = \gamma_{i+1},$$
$$f_{i+m} = \varphi_{i+m}.$$

Observe that the elements g_{i+1} and f_{i+m} are identical to the corresponding elements of the A-matrix. The other values are obtained from the following recurrence:

$$e_i = \frac{\eta_i}{d_{i-m}},$$
$$b_i = \frac{\beta_i}{d_{i-1}},$$
$$d_i = \delta_i - b_i g_i - e_i f_i.$$

The above recurrence can be simplified further by making the observation that the quantities η_i/d_{i-m} and β_i/d_{i-1} need not be saved since they are scaled versions of the corresponding elements in A. With this observation, *only a recurrence for the diagonal elements d_i is needed*. This recurrence is

$$d_i = \delta_i - \frac{\beta_i \gamma_i}{d_{i-1}} - \frac{\eta_i \varphi_i}{d_{i-m}}, \quad i = 1, \ldots, n, \tag{10.19}$$

with the convention that any d_j with a nonpositive index j is replaced by 1; the entries $\beta_i, i \leq 1; \gamma_i, i \leq 1; \phi_i, i \leq m;$ and $\eta_i, i \leq m$, are zero. The factorization obtained takes the form

$$M = (D - E)D^{-1}(D - F), \tag{10.20}$$

in which $-E$ is the strict lower diagonal of A, $-F$ is the strict upper triangular part of A, and D is the diagonal obtained with the above recurrence. Note that an ILU(0) based on the IKJ version of Gaussian elimination would give the same result.

For a general sparse matrix A with irregular structure, one can also determine a preconditioner in the form (10.20) by requiring only that the diagonal elements of M match those of A (see Exercise 10). However, this will not give the same ILU factorization as the one based on the IKJ variant of Gaussian elimination seen earlier. Why the ILU(0) factorization gives rise to the same factorization as that of (10.20) is simple to understand: The product of L and U does not change the values of the existing elements in the upper part, except for the diagonal. This can also be interpreted on the adjacency graph of the matrix.

This approach can now be extended to determine the ILU(1) factorization as well as factorizations with higher levels of fill. The stencils of the L-and U-matrices in the ILU(1) factorization are the stencils of the lower and upper parts of the LU-matrix obtained from ILU(0). These are shown in Figure 10.10. In the illustration, the meaning of a given stencil is not in the usual graph theory sense. Instead, all the marked nodes at a stencil based at node i represent those nodes coupled with unknown i by an equation. Thus, all the filled circles in the picture are adjacent to the central node. Proceeding as before and combining stencils to form the stencil associated with the LU-matrix, we obtain the stencil shown in Figure 10.11.

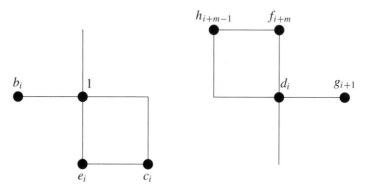

Figure 10.10. *Stencils of the L and U factors for the ILU(0) factorization of the matrix represented by the stencil of Figure 10.9.*

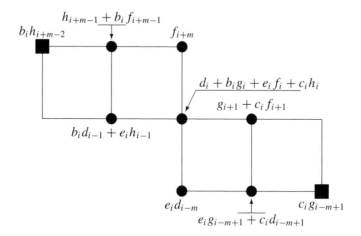

Figure 10.11. *Stencil associated with the product of the L- and U-matrices whose stencils are shown in Figure 10.10.*

As before, the fill-in elements are represented by squares and all other elements are filled circles. A typical row of the matrix associated with the above stencil has nine nonzero elements. Two of these are fill-ins, i.e., elements that fall outside the original structure of the L- and U-matrices. It is now possible to determine a recurrence relation for obtaining

10.3. Incomplete LU Factorization Preconditioners

the entries of L and U. There are seven equations in all, which, starting from the bottom, are

$$e_i d_{i-m} = \eta_i,$$
$$e_i g_{i-m+1} + c_i d_{i-m+1} = 0,$$
$$b_i d_{i-1} + e_i h_{i-1} = \beta_i,$$
$$d_i + b_i g_i + e_i f_i + c_i h_i = \delta_i,$$
$$g_{i+1} + c_i f_{i+1} = \gamma_{i+1},$$
$$h_{i+m-1} + b_i f_{i+m-1} = 0,$$
$$f_{i+m} = \varphi_{i+m}.$$

This immediately yields the following recurrence relation for the entries of the L and U factors:

$$e_i = \eta_i/d_{i-m},$$
$$c_i = -e_i g_{i-m+1}/d_{i-m+1},$$
$$b_i = (\beta_i - e_i h_{i-1})/d_{i-1},$$
$$d_i = \delta_i - b_i g_i - e_i f_i - c_i h_i,$$
$$g_{i+1} = \gamma_{i+1} - c_i f_{i+1},$$
$$h_{i+m-1} = -b_i f_{i+m-1},$$
$$f_{i+m} = \varphi_{i+m}.$$

In proceeding from the nodes of smallest index to those of largest index, we are in effect performing implicitly the IKJ version of Gaussian elimination. The result of the ILU(1) obtained in this manner is therefore identical to that obtained by using Algorithms 10.1 and 10.3.

10.3.5 MILU

In all the techniques thus far, the elements that were dropped out during the incomplete elimination process are simply discarded. There are also techniques that attempt to reduce the effect of dropping by *compensating* for the discarded entries. For example, a popular strategy is to add up all the elements that have been dropped at the completion of the k loop of Algorithm 10.3. Then this sum is subtracted from the diagonal entry in U. This *diagonal compensation* strategy gives rise to the modified ILU (MILU) factorization.

Thus, in (10.13), the final row u_{i*} obtained after completion of the k loop of Algorithm 10.3 undergoes one more modification, namely,

$$u_{ii} := u_{ii} - (r_{i*}e),$$

in which $e \equiv (1, 1, \ldots, 1)^T$. Note that r_{i*} is a row and $r_{i*}e$ is the sum of the elements in this row, i.e., its *row sum*. The above equation can be rewritten in row form as $u_{i*} := u_{i*} - (r_{i*}e)e_i^T$ and (10.14) becomes

$$a_{i*} = \sum_{k=1}^{i} l_{ik} u_{k*} + (r_{i*}e)e_i^T - r_{i*}. \tag{10.21}$$

Observe that

$$a_{i*}e = \sum_{k=1}^{i} l_{ik}u_{k*}e + (r_{i*}e)e_i^T e - r_{i*}e = \sum_{k=1}^{i-1} l_{ik}u_{k*}e = LUe.$$

This establishes that $Ae = LUe$. As a result, this strategy guarantees that the row sums of A are equal to those of LU. For PDEs, the vector of all ones represents the discretization of a constant function. This additional constraint forces the ILU factorization to be exact for constant functions in some sense. Therefore, it is not surprising that often the algorithm does well for such problems. For other problems, such as problems with discontinuous coefficients, MILU algorithms usually are not better than their ILU counterparts, in general.

Example 10.3. For regularly structured matrices there are two elements dropped at the ith step of ILU(0). These are $b_i f_{i+m-1}$ and $e_i g_{i-m+1}$, located on the northwest and southeast corners of the stencil, respectively. Thus, the row sum $r_{i,*}e$ associated with step i is

$$s_i = \frac{\beta_i \phi_{i+m-1}}{d_{i-1}} + \frac{\eta_i \gamma_{m-i+1}}{d_{i-m}}$$

and the MILU variant of the recurrence (10.19) is

$$s_i = \frac{\beta_i \phi_{i+m-1}}{d_{i-1}} + \frac{\eta_i \gamma_{m-i+1}}{d_{i-m}},$$

$$d_i = \delta_i - \frac{\beta_i \gamma_i}{d_{i-1}} - \frac{\eta_i \varphi_i}{d_{i-m}} - s_i.$$

The new ILU factorization is now such that $A = LU - R$, in which, according to (10.21), the ith row of the new remainder matrix R is given by

$$r_{i,*}^{(new)} = (r_{i*}e)e_i^T - r_{i*},$$

whose row sum is zero.

This generic idea of lumping together all the elements dropped in the elimination process and adding them to the diagonal of U can be used for *any* form of ILU factorization. In addition, there are variants of diagonal compensation in which only a fraction of the dropped elements are added to the diagonal. Thus, the term s_i in the above example would be replaced by ωs_i before being added to u_{ii}, where ω is typically between 0 and 1. Other strategies distribute the sum s_i among nonzero elements of L and U other than the diagonal.

10.4 Threshold Strategies and Incomplete LU with Threshold

Incomplete factorizations that rely on the levels of fill are blind to numerical values because elements that are dropped depend only on the structure of A. This can cause some difficulties for realistic problems that arise in many applications. A few alternative methods are available based on dropping elements in the Gaussian elimination process according to their magnitude rather than their locations. With these techniques, the zero pattern P is determined dynamically. The simplest way to obtain an incomplete factorization of this type is to take a sparse direct solver and modify it by adding lines of code that will ignore

10.4.1 The ILUT Approach

A generic ILU algorithm with threshold (ILUT) can be derived from the IKJ version of Gaussian elimination, Algorithm 10.2, by including a set of rules for dropping small elements. In what follows, *applying a dropping rule to an element* will only mean *replacing the element with zero if it satisfies a set of criteria*. A dropping rule can be applied to a whole row by applying the same rule to all the elements of the row. In the following algorithm, w is a full-length working row used to accumulate linear combinations of sparse rows in the elimination and w_k is the kth entry of this row. As usual, a_{i*} denotes the ith row of A.

ALGORITHM 10.6. ILUT

1. For $i = 1, \ldots, n$, Do
2. $w := a_{i*}$
3.   For $k = 1, \ldots, i-1$ and when $w_k \neq 0$, Do
4. $w_k := w_k / a_{kk}$
5.   Apply a dropping rule to w_k
6.   If $w_k \neq 0$ then
7. $w := w - w_k * u_{k*}$
8.   EndIf
9.   EndDo
10.   Apply a dropping rule to row w
11. $l_{i,j} := w_j$ for $j = 1, \ldots, i-1$
12. $u_{i,j} := w_j$ for $j = i, \ldots, n$
13. $w := 0$
14. EndDo

Now consider the operations involved in the above algorithm. Line 7 is a sparse update operation. A common implementation of this is to use a full vector for w and a companion pointer that points to the positions of its nonzero elements. Similarly, lines 11 and 12 are sparse vector copy operations. The vector w is filled with a few nonzero elements after the completion of each outer loop i and therefore it is necessary to zero out those elements at the end of the Gaussian elimination loop, as is done in line 13. This is a sparse *set-to-zero* operation.

ILU(0) can be viewed as a particular case of the above algorithm. The dropping rule for ILU(0) is to drop elements that are in positions not belonging to the original structure of the matrix.

In the factorization ILUT(p, τ), the following rules are used.

1. In line 5, an element w_k is dropped (i.e., replaced with zero) if it is less than the relative tolerance τ_i obtained by multiplying τ by the original norm of the ith row (e.g., the 2-norm).

2. In line 10, a dropping rule of a different type is applied. First, drop again any element in the row with a magnitude that is below the relative tolerance τ_i. Then keep only the p largest elements in the L part of the row and the p largest elements in the U part of the row in addition to the diagonal element, which is always kept.

The goal of the second dropping step is to control the number of elements per row. Roughly speaking, p can be viewed as a parameter that helps control memory usage, while τ helps reduce computational cost. There are several possible variations of the implementation of dropping step 2. For example, we can keep a number of elements equal to $nu(i) + p$ in the upper part and $nl(i) + p$ in the lower part of the row, where $nl(i)$ and $nu(i)$ are the number of nonzero elements in the L part and the U part of the ith row of A, respectively. This variant is adopted in the ILUT code used in the examples.

Note that no pivoting is performed. Partial (column) pivoting may be incorporated at little extra cost and will be discussed later. It is also possible to combine ILUT with one of the many standard reorderings, such as the nested dissection (ND) ordering or the reverse Cuthill–McKee (RCM) ordering. Reordering in the context of incomplete factorizations can also be helpful for improving robustness, *provided enough accuracy is used*. For example, when a red-black ordering is used, ILU(0) may lead to poor performance compared with the natural ordering ILU(0). On the other hand, if ILUT is used by allowing gradually more fill-in, then the performance starts improving again. In fact, in some examples, the performance of ILUT for the red-black ordering *eventually outperforms* that of ILUT for the natural ordering using the same parameters p and τ.

10.4.2 Analysis

Existence theorems for the ILUT factorization are similar to those of other incomplete factorizations. If the diagonal elements of the original matrix are positive while the off-diagonal elements are negative, then under certain conditions of diagonal dominance the matrices generated during the elimination will have the same property. If the original matrix is diagonally dominant, then the transformed matrices will also have the property of being diagonally dominant under certain conditions. These properties are analyzed in detail in this section.

The row vector w resulting from line 4 of Algorithm 10.6 will be denoted by $u_{i,*}^{k+1}$. Note that $u_{i,j}^{k+1} = 0$ for $j \leq k$. Lines 3 to 10 in the algorithm involve a sequence of operations of the form

$$l_{ik} := u_{ik}^k / u_{kk}, \qquad (10.22)$$

if $|l_{ik}|$ small enough set $l_{ik} = 0$,

else

$$u_{i,j}^{k+1} := u_{i,j}^k - l_{ik} u_{k,j} - r_{ij}^k, \quad j = k+1, \ldots, n, \qquad (10.23)$$

for $k = 1, \ldots, i-1$, in which initially $u_{i,*}^1 := a_{i,*}$ and where r_{ij}^k is an element subtracted from a fill-in element that is being dropped. It should be equal either to zero (no dropping) or to $u_{ij}^k - l_{ik} u_{kj}$ when the element $u_{i,j}^{k+1}$ is being dropped. At the end of the ith step of Gaussian elimination (outer loop in Algorithm 10.6), we obtain the ith row of U:

$$u_{i,*} \equiv u_{i-1,*}^i, \qquad (10.24)$$

10.4. Threshold Strategies and Incomplete LU with Threshold

and the following relation is satisfied:

$$a_{i,*} = \sum_{k=1}^{i} l_{k,j} u_{i,*}^{k} + r_{i,*},$$

where $r_{i,*}$ is the row containing all the fill-ins.

The existence result that will be proved is valid only for certain modifications of the basic ILUT(p, τ) strategy. We consider an ILUT strategy that uses the following modification:

- **Drop Strategy Modification.** For any $i < n$, let a_{i,j_i} be the element of largest modulus among the elements $a_{i,j}$, $j = i+1, \ldots, n$, in the original matrix. Then elements generated in position (i, j_i) during the ILUT procedure are not subject to the dropping rule.

This modification prevents elements generated in position (i, j_i) from ever being dropped. Of course, there are many alternative strategies that can lead to the same effect.

A matrix H whose entries h_{ij} satisfy the following three conditions:

$$h_{ii} > 0 \quad \text{for} \quad 1 \leq i < n \quad \text{and} \quad h_{nn} \geq 0, \qquad (10.25)$$

$$h_{ij} \leq 0 \quad \text{for} \quad i, j = 1, \ldots, n \quad \text{and} \quad i \neq j, \qquad (10.26)$$

$$\sum_{j=i+1}^{n} h_{ij} < 0 \quad \text{for} \quad 1 \leq i < n, \qquad (10.27)$$

will be referred to as an $\hat{M}$-matrix. The third condition is a requirement that there be at least one nonzero element to the right of the diagonal element in each row except the last. The row sum for the ith row is defined by

$$rs(h_{i,*}) = h_{i,*} e = \sum_{j=1}^{n} h_{i,j}.$$

A given row of an $\hat{M}$-matrix H is *diagonally dominant* if its row sum is nonnegative. An $\hat{M}$-matrix H is said to be diagonally dominant if all its rows are diagonally dominant. The following theorem is an existence result for ILUT. The underlying assumption is that an ILUT strategy is used with the modification mentioned above.

Theorem 10.8. *If the matrix A is a diagonally dominant $\hat{M}$-matrix, then the rows $u_{i,*}^k$, $k = 0, 1, 2, \ldots, i$, defined by (10.23) starting with $u_{i,*}^0 = 0$ and $u_{i,*}^1 = a_{i,*}$ satisfy the following relations for $k = 1, \ldots, l$:*

$$u_{ij}^k \leq 0, \quad j \neq i, \qquad (10.28)$$

$$rs(u_{i,*}^k) \geq rs(u_{i,*}^{k-1}) \geq 0, \qquad (10.29)$$

$$u_{ii}^k > 0 \quad \text{when} \quad i < n \quad \text{and} \quad u_{nn}^k \geq 0. \qquad (10.30)$$

Proof. The result can be proved by induction on k. It is trivially true for $k = 0$. To prove that the relation (10.28) is satisfied, start from the relation

$$u_{i,*}^{k+1} := u_{i,*}^k - l_{ik} u_{k,*} - r_{i*}^k,$$

in which $l_{ik} \leq 0$, $u_{k,j} \leq 0$. Either r_{ij}^k is zero, which yields $u_{ij}^{k+1} \leq u_{ij}^k \leq 0$, or r_{ij}^k is nonzero, which means that u_{ij}^{k+1} is being dropped, i.e., replaced by zero, and therefore again $u_{ij}^{k+1} \leq 0$.

This establishes (10.28). Note that by this argument $r_{ij}^k = 0$ except when the jth element in the row is dropped, in which case $u_{ij}^{k+1} = 0$ and $r_{ij}^k = u_{ij}^k - l_{ik}u_{k,j} \leq 0$. Therefore, $r_{ij}^k \leq 0$, always. Moreover, when an element in position (i, j) is not dropped, then

$$u_{i,j}^{k+1} := u_{i,j}^k - l_{ik}u_{k,j} \leq u_{i,j}^k$$

and, in particular, by the rule in the modification of the basic scheme described above, for $i < n$, we will always have, for $j = j_i$,

$$u_{i,j_i}^{k+1} \leq u_{i,j_i}^k, \tag{10.31}$$

in which j_i is defined in the statement of the modification.

Consider the row sum of u_{i*}^{k+1}. We have

$$rs(u_{i,*}^{k+1}) = rs(u_{i,*}^k) - l_{ik}rs(u_{k,*}) - rs(r_{i*}^k)$$
$$\geq rs(u_{i,*}^k) - l_{ik}rs(u_{k,*}) \tag{10.32}$$
$$\geq rs(u_{i,*}^k), \tag{10.33}$$

which establishes (10.29) for $k+1$.

It remains to prove (10.30). From (10.29) we have, for $i < n$,

$$u_{ii}^{k+1} \geq \sum_{j=k+1,n} -u_{i,j}^{k+1} = \sum_{j=k+1,n} |u_{i,j}^{k+1}| \tag{10.34}$$
$$\geq |u_{i,j_i}^{k+1}| \geq |u_{i,j_i}^k| \geq \cdots \tag{10.35}$$
$$\geq |u_{i,j_i}^1| = |a_{i,j_i}|. \tag{10.36}$$

Note that the inequalities in (10.35) are true because u_{i,j_i}^k is never dropped by assumption and, as a result, (10.31) applies. By the condition (10.27), which defines $\hat{M}$-matrices, $|a_{i,j_i}|$ is positive for $i < n$. Clearly, when $i = n$, we have by (10.34) $u_{nn} \geq 0$. This completes the proof. □

The theorem does not mean that the factorization is effective only when its conditions are satisfied. In practice, the preconditioner is efficient under fairly general conditions.

10.4.3 Implementation Details

A poor implementation of ILUT may well lead to an expensive factorization phase and possibly an impractical algorithm. The following is a list of the potential difficulties that may cause inefficiencies in the implementation of ILUT:

1. generation of the linear combination of rows of A (line 7 in Algorithm 10.6);

2. selection of the p largest elements in L and U;

3. need to access the elements of L in increasing order of columns (in line 3 of Algorithm 10.6).

10.4. Threshold Strategies and Incomplete LU with Threshold

For (1), the usual technique is to generate a full row and accumulate the linear combination of the previous rows in it. The row is zeroed again after the whole loop is finished using a sparse set-to-zero operation. A variation of this technique uses only a full integer array $jr(1:n)$, the values of which are zero except when there is a nonzero element. With this full row, a short real vector $w(1:maxw)$ must be maintained that contains the real values of the row as well as a corresponding short integer array $jw(1:maxw)$, which points to the column position of the real values in the row. When a nonzero element resides in position j of the row, then $jr(j)$ is set to the address k in w, jw where the nonzero element is stored. Thus, $jw(k)$ points to $jr(j)$ and $jr(j)$ points to $jw(k)$ and $w(k)$. This is illustrated in Figure 10.12.

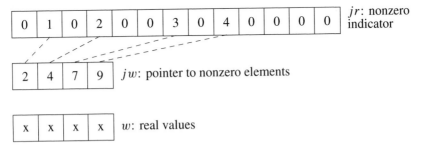

Figure 10.12. *Illustration of data structure used for the working row in ILUT.*

Note that jr holds the information on the row consisting of both the L part and the U part of the LU factorization. When the linear combinations of the rows are performed, first determine the pivot. Then, unless it is small enough to be dropped according to the dropping rule being used, proceed with the elimination. If a new element in the linear combination is not a fill-in, i.e., if $jr(j) = k \neq 0$, then update the real value $w(k)$. If it is a fill-in ($jr(j) = 0$), then append an element to the arrays w, jw and update jr accordingly.

For (2), the natural technique is to employ a heap-sort strategy. The cost of this implementation is $O(m + p \times \log_2 m)$, i.e., $O(m)$ for the heap construction and $O(\log_2 m)$ for each extraction. Another implementation is to use a modified quick-sort strategy based on the fact that sorting the array is not necessary. Only the largest p elements must be extracted. This is a *quick-split* technique to distinguish it from the full quick sort. The method consists of choosing an element, e.g., $x = w(1)$, in the array $w(1:m)$, then permuting the data so that $|w(k)| \leq |x|$ if $k \leq mid$ and $|w(k)| \geq |x|$ if $k \geq mid$, where mid is some split point. If $mid = p$, then exit. Otherwise, split *one of the left or right subarrays* recursively, depending on whether mid is smaller or larger than p. The cost of this strategy *on the average* is $O(m)$. The savings relative to the simpler bubble-sort or insertion-sort schemes are small for small values of p, but they become rather significant for large p and m.

The next implementation difficulty is that the elements in the L part of the row being built are not in increasing order of columns. Since these elements must be accessed from left to right in the elimination process, all elements in the row after those already eliminated must be scanned. The one with smallest column number is then picked as the next element to eliminate. This operation can be efficiently organized as a binary search tree, which allows easy insertions and searches. This improvement can bring substantial gains in the case when accurate factorizations are computed.

Example 10.4. Tables 10.3 and 10.4 show the results of applying GMRES(10) preconditioned with ILUT(1, 10^{-4}) and ILUT(5, 10^{-4}), respectively, to the five test problems described in Section 3.7. See Example 6.1 for the meaning of the column headers in the table. As shown, all linear systems are now solved in a relatively small number of iterations, with the exception of F2DB, which still takes 130 steps to converge with $lfil = 1$ (but only 10 with $lfil = 5$). In addition, observe a marked improvement in the operation count and error norms. Note that the operation counts shown in the column Kflops do not account for the operations required in the set-up phase to build the preconditioners. For large values of $lfil$, this may be large.

Matrix	Iters	Kflops	Residual	Error
F2DA	18	964	0.47E−03	0.41E−04
F3D	14	3414	0.11E−02	0.39E−03
ORS	6	341	0.13E+00	0.60E−04
F2DB	130	7167	0.45E−02	0.51E−03
FID	59	19112	0.19E+00	0.11E−03

Table 10.3. *A test run of GMRES(10)-ILUT(1, 10^{-4}) preconditioning.*

Matrix	Iters	Kflops	Residual	Error
F2DA	7	478	0.13E−02	0.90E−04
F3D	9	2855	0.58E−03	0.35E−03
ORS	4	270	0.92E−01	0.43E−04
F2DB	10	724	0.62E−03	0.26E−03
FID	40	14862	0.11E+00	0.11E−03

Table 10.4. *A test run of GMRES(10)-ILUT(5, 10^{-4}) preconditioning.*

If the total time to solve one linear system with A is considered, a typical curve of the total time required to solve a linear system when the $lfil$ parameter varies would look like the plot shown in Figure 10.13. As $lfil$ increases, a critical value is reached where the preprocessing time and the iteration time are equal. Beyond this critical point, the preprocessing time dominates the total time. If there are several linear systems to solve with the same matrix A, then it is advantageous to use a more accurate factorization, since the cost of the factorization will be amortized. Otherwise, a smaller value of $lfil$ will result in a more efficient, albeit also less reliable, run.

10.4.4 The ILUTP Approach

The ILUT approach may fail for many of the matrices that arise from real applications for one of the following reasons.

10.4. Threshold Strategies and Incomplete LU with Threshold

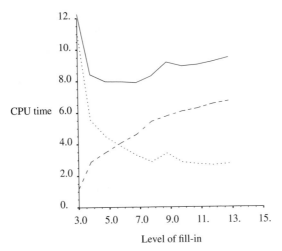

Figure 10.13. *Typical CPU time as a function of lfil. Dashed line: ILUT. Dotted line: GMRES. Solid line: total.*

1. The ILUT procedure encounters a zero pivot.

2. The ILUT procedure encounters an overflow or underflow condition because of exponential growth of the entries of the factors.

3. The ILUT preconditioner terminates normally but the incomplete factorization preconditioner that is computed is *unstable*.

An unstable ILU factorization is one for which $M^{-1} = U^{-1}L^{-1}$ has a very large norm leading to poor convergence or divergence of the outer iteration. The case (1) can be overcome to a certain degree by assigning an arbitrary nonzero value to a zero diagonal element that is encountered. Clearly, this is not a satisfactory remedy because of the loss in accuracy in the preconditioner. The ideal solution in this case is to use pivoting. However, a form of pivoting is desired that leads to an algorithm with similar cost and complexity to ILUT. Because of the data structure used in ILUT, row pivoting is not practical. Instead, column pivoting can be implemented rather easily.

Here are a few of the features that characterize the new algorithm, which is termed ILUTP (P stands for pivoting). ILUTP uses a permutation array *perm* to hold the new orderings of the variables, along with the reverse permutation array. At step i of the elimination process the largest entry in a row is selected and is defined to be the new ith variable. The two permutation arrays are then updated accordingly. The matrix elements of L and U are kept in their original numbering. However, when expanding the LU row that corresponds to the ith outer step of Gaussian elimination, the elements are loaded with respect to the new labeling, using the array *perm* for the translation. At the end of the process, there are two options. The first is to leave all elements labeled with respect to the original labeling. No additional work is required since the variables are already in this form in the algorithm, but the variables must then be permuted at each preconditioning step. The second solution is to apply the permutation to all elements of A as well as LU. This does not require applying a permutation at each step, but rather produces a permuted solution that must be permuted back at the end of the iteration phase. The complexity of the ILUTP procedure is virtually identical to that of ILUT. A few additional options can be provided. A tolerance parameter called *permtol* may be included to help determine whether or not to permute variables: A nondiagonal element a_{ij} is a candidate for a permutation only when $tol \times |a_{ij}| > |a_{ii}|$. Furthermore, pivoting may be restricted to take place only within diagonal blocks of a fixed size. The size *mbloc* of these blocks must be provided. A value of $mbloc \geq n$ indicates that there are no restrictions on the pivoting.

For difficult matrices, the following strategy seems to work well:

1. Apply a scaling to all the rows (or columns), e.g., so that their 1-norms are all equal to 1. Then apply a scaling of the columns (or rows).

2. Use a small drop tolerance (e.g., $\epsilon = 10^{-4}$ or $\epsilon = 10^{-5}$) and take a large fill-in parameter (e.g., $lfil = 20$).

3. Do not take a small value for *permtol*. Reasonable values are between 0.5 and 0.01, with 0.5 being the best in many cases.

Example 10.5. Table 10.5 shows the results of applying the GMRES algorithm with ILUTP($1, 10^{-4}$) preconditioning to the five test problems described in Section 3.7. The *permtol* parameter is set to 1.0 in this case. See Example 6.1 for the meaning of the column headers in the table. The results are identical to those of ILUT($1, 10^{-4}$) shown in Table 10.3 for the first four problems, but there is an improvement for the fifth problem.

Matrix	Iters	Kflops	Residual	Error
F2DA	18	964	0.47E−03	0.41E−04
F3D	14	3414	0.11E−02	0.39E−03
ORS	6	341	0.13E+00	0.61E−04
F2DB	130	7167	0.45E−02	0.51E−03
FID	50	16224	0.17E+00	0.18E−03

Table 10.5. *A test run of GMRES with ILUTP preconditioning.*

10.4.5 The ILUS Approach

The ILU preconditioners discussed so far are based mainly on the IKJ variant of Gaussian elimination. Different types of ILUs can be derived using other forms of Gaussian elimination. The main motivation for the version to be described next is that ILUT does not take advantage of symmetry. If A is symmetric, then the resulting $M = LU$ is nonsymmetric in general. Another motivation is that, in many applications, including computational fluid dynamics and structural engineering, the resulting matrices are stored in a *sparse skyline* (SSK) format rather than the standard CSR format.

In this format, the matrix A is decomposed as

$$A = D + L_1 + L_2^T,$$

in which D is a diagonal of A and L_1, L_2 are strictly lower triangular matrices. Then a

10.4. Threshold Strategies and Incomplete LU with Threshold

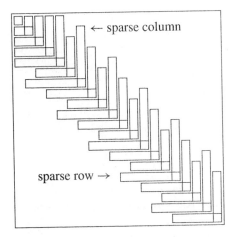

Figure 10.14. *Illustration of the SSK format.*

sparse representation of L_1 and L_2 is used in which, typically, L_1 and L_2 are stored in the CSR format and D is stored separately.

Incomplete factorization techniques may be developed for matrices in this format without having to convert them to the CSR format. Two notable advantages of this approach are (1) the savings in storage for structurally symmetric matrices and (2) the fact that the algorithm gives a symmetric preconditioner when the original matrix is symmetric.

Consider the sequence of matrices

$$A_{k+1} = \begin{pmatrix} A_k & v_k \\ w_k & \alpha_{k+1} \end{pmatrix},$$

where $A_n = A$. If A_k is nonsingular and its LDU factorization

$$A_k = L_k D_k U_k$$

is already available, then the LDU factorization of A_{k+1} is

$$A_{k+1} = \begin{pmatrix} L_k & 0 \\ y_k & 1 \end{pmatrix} \begin{pmatrix} D_k & 0 \\ 0 & d_{k+1} \end{pmatrix} \begin{pmatrix} U_k & z_k \\ 0 & 1 \end{pmatrix},$$

in which

$$z_k = D_k^{-1} L_k^{-1} v_k, \qquad (10.37)$$
$$y_k = w_k U_k^{-1} D_k^{-1}, \qquad (10.38)$$
$$d_{k+1} = \alpha_{k+1} - y_k D_k z_k. \qquad (10.39)$$

Hence, the last row/column pairs of the factorization can be obtained by solving two unit lower triangular systems and computing a scaled dot product. This can be exploited for sparse matrices provided an appropriate data structure is used to take advantage of the sparsity of the matrices L_k, U_k as well as the vectors v_k, w_k, y_k, and z_k. A convenient data structure for this is to store the row/column pairs w_k, v_k^T as a single row in sparse mode (see Figure 10.14). All these pairs are stored in sequence. The diagonal elements are stored separately. This is called the unsymmetric sparse skyline (USS) format. Each step of the ILU factorization based on this approach will consist of two approximate sparse linear system solutions and a sparse dot product. The question that arises is as follows: How can a sparse triangular system be solved inexpensively? It would seem natural to solve the triangular systems (10.37) and (10.38) exactly and then drop small terms at the end, using a numerical dropping strategy. However, the total cost of computing the ILU factorization with this strategy is $O(n^2)$ operations at least, which is not acceptable for very large problems. Since only an approximate solution is required, the first idea that comes to mind is the truncated Neumann series

$$z_k = D_k^{-1} L_k^{-1} v_k = D_k^{-1} (I + E_k + E_k^2 + \cdots + E_k^p) v_k, \qquad (10.40)$$

in which $E_k \equiv I - L_k$. In fact, by analogy with ILU(p), it is interesting to note that the powers of E_k will also tend to become smaller as p increases. A close look at the structure of $E_k^p v_k$ shows that there is indeed a strong relation between this approach and ILU(p) in the symmetric case. Now we make another important observation, namely, that the vector $E_k^j v_k$ can be computed in *sparse-sparse mode*, i.e., in terms of operations involving products of *sparse matrices with sparse vectors*. Without exploiting this, the total cost would still be $O(n^2)$. When multiplying a sparse matrix A by a sparse vector v, the operation can best be done by accumulating the linear combinations of the columns of A. A sketch of the resulting ILUS algorithm is as follows.

ALGORITHM 10.7. ILUS(ϵ, p)

1. Set $A_1 = D_1 = a_{11}$, $L_1 = U_1 = 1$
2. For $i = 1, \ldots, n - 1$, Do
3.   Compute z_k by (10.40) in sparse-sparse mode
4.   Compute y_k in a similar way
5.   Apply numerical dropping to y_k and z_k
6.   Compute d_{k+1} via (10.39)
7. EndDo

If there are only i nonzero components in the vector v and an average of ν nonzero elements per column, then the total cost per step will be $2 \times i \times \nu$ on the average. Note that the computation of d_k via (10.39) involves the inner product of two sparse vectors, which is often implemented by expanding one of the vectors into a full vector and computing the inner product of a sparse vector with this full vector. As mentioned before, in the symmetric case ILUS yields the incomplete Cholesky factorization. Here, the work can be halved since the generation of y_k is not necessary.

Also note that a simple iterative procedure such as minimal residual (MR) or GMRES(m) can be used to solve the triangular systems in sparse-sparse mode. Similar techniques will be seen in Section 10.5. Experience shows that these alternatives are not much better than the Neumann series approach [79].

10.4.6 The ILUC Approach

A notable disadvantage of the standard delayed-update IKJ factorization is that it requires access to the entries in the kth row of L in sorted order of columns. This is further complicated by the fact that the working row (denoted by w in Algorithm 10.6) is dynamically modified

10.4. Threshold Strategies and Incomplete LU with Threshold

by fill-in as the elimination proceeds. Searching for the leftmost entry in the kth row of L is usually not a problem when the fill-in allowed is small. Otherwise, when an accurate factorization is sought, it can become a significant burden and may ultimately even dominate the cost of the factorization. Sparse direct solution methods that are based on the IKJ form of Gaussian elimination obviate this difficulty by a technique known as the Gilbert–Peierls method [146]. Because of dropping, this technique cannot, however, be used as is. Another possible option is to reduce the cost of the searches through the use of clever data structures and algorithms, such as binary search trees or heaps [90].

The Crout formulation provides the most elegant solution to the problem. In fact the Crout version of Gaussian elimination has other advantages, which make it one of the most appealing ways of implementing ILU factorizations.

The Crout form of Gaussian elimination consists of computing, at step k, the entries $a_{k+1:n,k}$ (in the unit lower triangular factor L) and $a_{k,k:n}$ (in the upper triangular factor U). This is done by postponing the rank-one update in a way similar to the IKJ variant. In Figure 10.15 the parts of the factors being computed at the kth step are shown in black and those being accessed are in the shaded areas. At the kth step, all the updates of the previous steps are applied to the entries $a_{k+1:n,k}$ and $a_{k,k:n}$ and it is therefore convenient to store L by columns and U by rows.

ALGORITHM 10.8. Crout LU Factorization

1. For $k = 1 : n$, Do
2.   For $i = 1 : k - 1$ and if $a_{ki} \neq 0$, Do
3. $a_{k,k:n} = a_{k,k:n} - a_{ki} * a_{i,k:n}$
4.   EndDo
5.   For $i = 1 : k - 1$ and if $a_{ik} \neq 0$, Do
6. $a_{k+1:n,k} = a_{k+1:n,k} - a_{ik} * a_{k+1:n,i}$
7.   EndDo
8. $a_{ik} = a_{ik}/a_{kk}$ for $i = k+1, \ldots, n$
9. EndDo

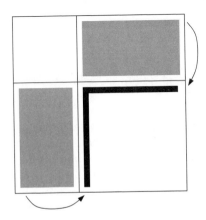

Figure 10.15. *Computational pattern of the Crout algorithm.*

The kth step of the algorithm generates the kth row of U and the kth column of L. This step is schematically represented in Figure 10.16. The above algorithm will now be adapted to the sparse case. Sparsity is taken into account and a dropping strategy is included, resulting in the following Crout version of ILU (termed ILUC).

ALGORITHM 10.9. ILUC

1.   For $k = 1 : n$, Do
2.   Initialize row z: $z_{1:k-1} = 0$, $z_{k:n} = a_{k,k:n}$
3.   For $\{i \mid 1 \leq i \leq k-1 \text{ and } l_{ki} \neq 0\}$, Do
4. $z_{k:n} = z_{k:n} - l_{ki} * u_{i,k:n}$
5.   EndDo
6.   Initialize column w: $w_{1:k} = 0$, $w_{k+1:n} = a_{k+1:n,k}$
7.   For $\{i \mid 1 \leq i \leq k-1 \text{ and } u_{ik} \neq 0\}$, Do
8. $w_{k+1:n} = w_{k+1:n} - u_{ik} * l_{k+1:n,i}$
9.   EndDo
10.   Apply a dropping rule to row z
11.   Apply a dropping rule to column w
12. $u_{k,:} = z$
13. $l_{:,k} = w/u_{kk}$, $l_{kk} = 1$
14.   EndDo

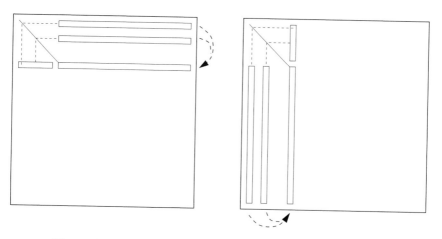

Figure 10.16. *Computing the kth row of U (left side) and the kth column of L (right side).*

Two potential sources of difficulty will require a careful and somewhat complex implementation. First, looking at lines 4 and 8, only the section $(k : n)$ of the ith row of U is required and, similarly, only the section $(k + 1 : n)$ of the ith column of L is needed. Second, line 3 requires access to the kth row of L, which is stored by columns, while line 7 requires access to the kth column of U, which is accessed by rows.

The first issue can be easily handled by keeping pointers that indicate where the relevant part of each row of U (resp. column of L) starts. An array Ufirst is used to store for

10.4. Threshold Strategies and Incomplete LU with Threshold

each row i of U the index of the first column that will be used next. If k is the current step number, this means that Ufirst(i) holds the first column index greater than k of all nonzero entries in the ith row of U. These pointers are easily updated after each elimination step, assuming that column indices (resp. row indices for L) are in increasing order.

For the second issue, consider the situation with the U factor. The problem is that the kth column of U is required for the update of L, but U is stored row-wise. An elegant solution to this problem has been known since the pioneering days of sparse direct methods [115, 144]. Before discussing this idea, consider the simpler solution of including a linked list for each column of U. These linked lists would be easy to update because the rows of U are computed one at a time. Each time a new row is computed, the nonzero entries of this row are queued to the linked lists of their corresponding columns. However, this scheme would entail nonnegligible additional storage. A clever alternative is to exploit the array Ufirst mentioned above to form incomplete linked lists of each column. Every time k is incremented the Ufirst array is updated. When Ufirst(i) is updated to point to a new nonzero with column index j, then the row index i is added to the linked list for column i. What is interesting is that, though the column structures constructed in this manner are incomplete, they *become complete as soon as they are needed*. A similar technique is used for the rows of the L factor.

In addition to avoiding searches, the ILUC has another important advantage. It enables some new dropping strategies that may be viewed as more rigorous than the standard ones seen so far. The straightforward dropping rules used in ILUT can be easily adapted for ILUC. In addition, the data structure of ILUC allows options based on estimating the norms of the inverses of L and U.

For ILU preconditioners, the error made in the inverses of the factors is more important to control than the errors in the factors themselves. This is because, when $A = LU$ and

$$\tilde{L}^{-1} = L^{-1} + X, \qquad \tilde{U}^{-1} = U^{-1} + Y,$$

then the preconditioned matrix is given by

$$\tilde{L}^{-1} A \tilde{U}^{-1} = (L^{-1} + X) A (U^{-1} + Y) = I + AY + XA + XY.$$

If the errors X and Y in the inverses of L and U are small, then the preconditioned matrix will be close to the identity matrix. On the other hand, small errors in the factors themselves may yield arbitrarily large errors in the preconditioned matrix.

Let L_k denote the matrix composed of the first k rows of L and the last $n-k$ rows of the identity matrix. Consider a term l_{jk} with $j > k$ that is dropped at step k. Then the resulting perturbed matrix $\tilde{L}_k$ differs from L_k by $l_{jk} e_j e_k^T$. Noticing that $L_k e_j = e_j$, then

$$\tilde{L}_k = L_k - l_{jk} e_j e_k^T = L_k (I - l_{jk} e_j e_k^T),$$

from which this relation between the inverses follows:

$$\tilde{L}_k^{-1} = (I - l_{jk} e_j e_k^T)^{-1} L_k^{-1} = L_k^{-1} + l_{jk} e_j e_k^T L_k^{-1}.$$

Therefore, the inverse of L_k will be perturbed by l_{jk} times the kth row of L_k^{-1}. This perturbation will affect the jth row of L_k^{-1}. Hence, using the infinity norm, for example, it is important to limit the norm of this perturbing row, which is $\|l_{jk}e_j e_k^T L_k^{-1}\|_\infty$. It follows that it is a good strategy to drop a term in L when

$$|l_{jk}|\, \|e_k^T L_k^{-1}\|_\infty < \epsilon.$$

A similar criterion can be used for the upper triangular factor U.

This strategy is not complete because the matrix L^{-1} is not available. However, standard techniques used for estimating condition numbers [149] can be adapted for estimating the norm of the kth row of L^{-1} (resp. kth column of U^{-1}). The idea is to construct a vector b, one component at a time, by following a greedy strategy to make $L^{-1}b$ large at each step. This is possible because the first $k-1$ columns of L are available at the kth step. The simplest method constructs a vector b of components $\beta_k = \pm 1$ at each step k, in such a way as to maximize the norm of the kth component of $L^{-1}b$. Since the first $k-1$ columns of L are available at step k, the kth component of the solution x is given by

$$\xi_k = \beta_k - e_k^T L_{k-1} x_{k-1}.$$

This makes the choice clear: if ξ_k is to be large in modulus, then the sign of β_k should be opposite that of $e_k^T L_{k-1} x_{k-1}$. If b is the current right-hand side at step k, then $\|e_k^T L^{-1}\|_\infty$ can be estimated by the kth component of the solution x of the system $Lx = b$:

$$\|e_k^T L^{-1}\|_\infty \approx \frac{|e_k^T L^{-1} b|}{\|b\|_\infty}.$$

Details, along with other strategies for dynamically building b, may be found in [201].

10.5 Approximate Inverse Preconditioners

The ILU factorization techniques were developed originally for M-matrices that arise from the discretization of PDEs of elliptic type, usually in one variable. For the common situation where A is indefinite, standard ILU factorizations may face several difficulties, of which the best known is the fatal breakdown due to the encounter of a zero pivot. However, there are other problems that are just as serious. Consider an incomplete factorization of the form

$$A = LU + E, \tag{10.41}$$

where E is the error. The preconditioned matrices associated with the different forms of preconditioning are similar to

$$L^{-1}AU^{-1} = I + L^{-1}EU^{-1}. \tag{10.42}$$

What is sometimes missed is the fact that the error matrix E in (10.41) is not as important as the *preconditioned* error matrix $L^{-1}EU^{-1}$ shown in (10.42). When the matrix A is diagonally dominant, then L and U are well conditioned and the size of $L^{-1}EU^{-1}$ remains confined within reasonable limits, typically with a nice clustering of its eigenvalues around the origin. On the other hand, when the original matrix is not diagonally dominant, L^{-1} or U^{-1} may have very large norms, causing the error $L^{-1}EU^{-1}$ to be very large and thus adding large perturbations to the identity matrix. It can be observed experimentally that ILU

10.5. Approximate Inverse Preconditioners

preconditioners can be very poor in these situations, which often arise when the matrices are indefinite or have large nonsymmetric parts.

One possible remedy is to try to find a preconditioner that does not require solving a linear system. For example, the original system can be preconditioned by a matrix M that is a direct approximation to the inverse of A.

10.5.1 Approximating the Inverse of a Sparse Matrix

A simple technique for finding approximate inverses of arbitrary sparse matrices is to attempt to find a sparse matrix M that minimizes the Frobenius norm of the residual matrix $I - AM$:

$$F(M) = \|I - AM\|_F^2. \tag{10.43}$$

A matrix M whose value $F(M)$ is small would be a right-approximate inverse of A. Similarly, a left-approximate inverse can be defined by using the objective function

$$\|I - MA\|_F^2. \tag{10.44}$$

Finally, a left-right pair L, U can be sought to minimize

$$\|I - LAU\|_F^2. \tag{10.45}$$

In the following, only (10.43) and (10.45) are considered. The case (10.44) is very similar to the right-preconditioner case (10.43). The objective function (10.43) decouples into the sum of the squares of the 2-norms of the individual columns of the residual matrix $I - AM$:

$$F(M) = \|I - AM\|_F^2 = \sum_{j=1}^n \|e_j - Am_j\|_2^2, \tag{10.46}$$

in which e_j and m_j are the jth columns of the identity matrix and of the matrix M, respectively. There are two different ways to proceed in order to minimize (10.46). The function (10.43) can be minimized globally as a function of the sparse matrix M, e.g., by a gradient-type method. Alternatively, the individual functions

$$f_j(m) = \|e_j - Am\|_2^2, \quad j = 1, 2, \ldots, n, \tag{10.47}$$

can be minimized. The second approach is appealing for parallel computers, although there is also parallelism to be exploited in the first approach. These two approaches will be discussed in turn.

10.5.2 Global Iteration

The *global iteration* approach consists of treating M as an unknown sparse matrix and using a descent-type method to minimize the objective function (10.43). This function is a quadratic function on the space of $n \times n$ matrices, viewed as objects in $\mathbb{R}^{n^2}$. The proper inner product on the space of matrices, with which the squared norm (10.46) is associated, is

$$\langle X, Y \rangle = \text{tr}(Y^T X). \tag{10.48}$$

In the following, an *array representation* of an n^2-vector X means the $n \times n$ matrix whose column vectors are the successive n-vectors of X.

In a descent algorithm, a new iterate M_{new} is defined by taking a step along a selected direction G, i.e.,

$$M_{new} = M + \alpha G,$$

in which α is selected to minimize the objective function $F(M_{new})$. From results seen in Chapter 5, minimizing the residual norm is equivalent to imposing the condition that $R - \alpha AG$ be orthogonal to AG with respect to the $\langle \cdot, \cdot \rangle$ inner product. Thus, the optimal α is given by

$$\alpha = \frac{\langle R, AG \rangle}{\langle AG, AG \rangle} = \frac{\mathrm{tr}(R^T AG)}{\mathrm{tr}\left((AG)^T AG\right)}. \tag{10.49}$$

The denominator may be computed as $\|AG\|_F^2$. The resulting matrix M will tend to become denser after each descent step and it is therefore essential to apply a numerical dropping strategy to the resulting M. However, the descent property of the step is now lost; i.e., it is no longer guaranteed that $F(M_{new}) \le F(M)$. An alternative would be to apply numerical dropping to the direction of search G before taking the descent step. In this case, the amount of fill-in in the matrix M cannot be controlled.

The simplest choice for the descent direction G is to take it to be equal to the residual matrix $R = I - AM$, where M is the new iterate. Except for the numerical dropping step, the corresponding descent algorithm is nothing but the MR algorithm, seen in Section 5.3.2, on the $n^2 \times n^2$ linear system $AM = I$. The global MR algorithm will have the following form.

ALGORITHM 10.10. Global MR Descent Algorithm

1. Select an initial M
2. Until convergence, Do
3.   Compute $C := AM$ and $G := I - C$
4.   Compute $\alpha = \mathrm{tr}(G^T AG)/\|C\|_F^2$
5.   Compute $M := M + \alpha G$
6.   Apply numerical dropping to M
7. EndDo

A second choice is to take G to be equal to the direction of steepest descent, i.e., the direction opposite to the gradient of the function (10.43) with respect to M. If all vectors are represented as two-dimensional $n \times n$ arrays, then the gradient can be viewed as a matrix G that satisfies the following relation for small perturbations E:

$$F(M + E) = F(M) + \langle G, E \rangle + o(\|E\|). \tag{10.50}$$

This provides a way of expressing the gradient as an operator on arrays, rather than n^2-vectors.

Proposition 10.9. *The array representation of the gradient of F with respect to M is the matrix*

$$G = -2A^T R,$$

in which R is the residual matrix $R = I - AM$.

10.5. Approximate Inverse Preconditioners

Proof. For any matrix E we have

$$\begin{aligned}
F(M+E) - F(M) &= \operatorname{tr}\left[(I - A(M+E))^T(I - A(M+E))\right] \\
&\quad - \operatorname{tr}\left[(I - AM)^T(I - AM)\right] \\
&= \operatorname{tr}\left[(R - AE)^T(R - AE) - R^TR\right] \\
&= -\operatorname{tr}\left[(AE)^TR + R^TAE - (AE)^T(AE)\right] \\
&= -2\operatorname{tr}(R^TAE) + \operatorname{tr}\left[(AE)^T(AE)\right] \\
&= -2\langle A^TR, E\rangle + \langle AE, AE\rangle.
\end{aligned}$$

Comparing this with (10.50) yields the desired result. □

Thus, the steepest descent algorithm will consist of replacing G in line 3 of Algorithm 10.10 with $G = A^TR = A^T(I - AM)$. As is expected with steepest descent techniques, the algorithm can be slow.

ALGORITHM 10.11. Global Steepest Descent Algorithm

---
1. Select an initial M
2. Until convergence, Do
3.   Compute $R = I - AM$ and $G := A^TR$
4.   Compute $\alpha = \|G\|_F^2 / \|AG\|_F^2$
5.   Compute $M := M + \alpha G$
6.   Apply numerical dropping to M
7. EndDo

In either steepest descent or MR, the G-matrix must be stored explicitly. The scalars $\|AG\|_F^2$ and $\operatorname{tr}(G^TAG)$ needed to obtain α in these algorithms can be computed from the successive columns of AG, which can be generated, used, and discarded. As a result, the matrix AG need not be stored.

10.5.3 Column-Oriented Algorithms

Column-oriented algorithms consist of minimizing the individual objective functions (10.47) separately. Each minimization can be performed by taking a sparse initial guess and solving approximately the n parallel linear subproblems

$$Am_j = e_j, \quad j = 1, 2, \ldots, n, \tag{10.51}$$

with a few steps of a nonsymmetric descent-type method, such as MR or GMRES. If these linear systems were solved (approximately) without taking advantage of sparsity, the cost of constructing the preconditioner would be of order n^2. That is because each of the n columns would require $O(n)$ operations. Such a cost would become unacceptable for large linear systems. To avoid this, the iterations must be performed in *sparse-sparse mode*, a term that was introduced in Section 10.4.5. The column m_j and the subsequent iterates in the MR algorithm must be stored and operated on as sparse vectors. The Arnoldi bases in the GMRES algorithm are now to be kept in sparse format. Inner products and vector updates involve pairs of sparse vectors.

In the following MR algorithm, n_i iterations are used to solve (10.51) approximately for each column, giving an approximation to the jth column of the inverse of A. Each initial m_j is taken from the columns of an initial guess, M_0.

ALGORITHM 10.12. Approximate Inverse (AINV) via MR Iteration

1. Start: set $M = M_0$
2. For each column $j = 1, \ldots, n$, Do
3.   Define $m_j = Me_j$
4.   For $i = 1, \ldots, n_i$, Do
5. $r_j := e_j - Am_j$
6. $\alpha_j := \frac{(r_j, Ar_j)}{(Ar_j, Ar_j)}$
7. $m_j := m_j + \alpha_j r_j$
8.   Apply numerical dropping to m_j
9.   EndDo
10. EndDo

The algorithm computes the current residual r_j and then minimizes the residual norm $\|e_j - A(m_j + \alpha r_j)\|_2$, with respect to α. The resulting column is then pruned by applying the numerical dropping step in line 8.

In the sparse implementation of MR and GMRES, the matrix-by-vector product, SAXPY, and dot product kernels now all involve sparse vectors. The matrix-by-vector product is much more efficient if the sparse matrix is stored by columns, since all the entries do not need to be traversed. Efficient codes for all these kernels may be constructed that utilize a full n-length work vector.

Columns from an initial guess M_0 for the AINV are used as the initial guesses for the iterative solution of the linear subproblems. There are two obvious choices: $M_0 = \alpha I$ and $M_0 = \alpha A^T$. The scale factor α is chosen to minimize the norm of $I - AM_0$. Thus, the initial guess is of the form $M_0 = \alpha G$, where G is either the identity or A^T. The optimal α can be computed using the formula (10.49), in which R is to be replaced with the identity, so $\alpha = \text{tr}(AG)/\text{tr}(AG(AG)^T)$. The identity initial guess is less expensive to use but $M_0 = \alpha A^T$ is sometimes a much better initial guess. For this choice, the initial preconditioned system AM_0 is symmetric positive definite (SPD).

The linear systems we need to solve when generating each column of the AINV may themselves be preconditioned with the most recent version of the preconditioning matrix M. Thus, each system (10.51) for approximating column j may be preconditioned with M'_0, where the first $j-1$ columns of M'_0 are the m_k that have already been computed, $1 \leq k < j$, and the remaining columns are the initial guesses for the m_k, $j \leq k \leq n$. Thus, *outer* iterations can be defined that sweep over the matrix, as well as *inner* iterations that compute each column. At each outer iteration, the initial guess for each column is taken to be the previous result for that column.

10.5.4 Theoretical Considerations

The first theoretical question that arises is whether or not the AINVs obtained by the approximations described earlier can be singular. It cannot be proved that M is nonsingular

10.5. Approximate Inverse Preconditioners

unless the approximation is accurate enough. This requirement may be in conflict with the requirement of keeping the approximation sparse.

Proposition 10.10. *Assume that A is nonsingular and that the residual of the AINV M satisfies the relation*
$$\|I - AM\| < 1, \tag{10.52}$$
where $\|\cdot\|$ is any consistent matrix norm. Then M is nonsingular.

Proof. The result follows immediately from the equality
$$AM = I - (I - AM) \equiv I - N. \tag{10.53}$$
Since $\|N\| < 1$, Theorem 1.11 seen in Chapter 1 implies that $I - N$ is nonsingular. $\square$

The result is true in particular for the Frobenius norm, which is consistent (see Chapter 1).

It may sometimes be the case that AM is poorly balanced and, as a result, R can be large. Then balancing AM can yield a smaller norm and possibly a less restrictive condition for the nonsingularity of M. It is easy to extend the previous result as follows. If A is nonsingular and two nonsingular diagonal matrices D_1, D_2 exist such that
$$\|I - D_1 A M D_2\| < 1, \tag{10.54}$$
where $\|\cdot\|$ is any consistent matrix norm, then M is nonsingular.

Each column is obtained independently by requiring a condition on the residual norm of the form
$$\|e_j - Am_j\| \leq \tau \tag{10.55}$$
for some vector norm $\|\cdot\|$. From a practical point of view the 2-norm is preferable since it is related to the objective function used, namely, the Frobenius norm of the residual $I - AM$. However, the 1-norm is of particular interest since it leads to a number of simple theoretical results. In the following, it is assumed that a condition of the form
$$\|e_j - Am_j\|_1 \leq \tau_j \tag{10.56}$$
is required for each column.

The above proposition does not reveal anything about the degree of sparsity of the resulting AINV M. It may well be the case that, in order to guarantee nonsingularity, M must be dense or nearly dense. In fact, in the particular case where the norm in the proposition is the 1-norm, it is known that the AINV may be *structurally dense*, in that it is always possible to find a sparse matrix A for which M will be dense if $\|I - AM\|_1 < 1$.

Next we examine the sparsity of M and prove a simple result for the case where an assumption of the form (10.56) is made.

Proposition 10.11. *Assume that M is an AINV of A computed by enforcing the condition (10.56). Let $B = A^{-1}$ and assume that a given element b_{ij} of B satisfies the inequality*
$$|b_{ij}| > \tau_j \max_{k=1,n} |b_{ik}|. \tag{10.57}$$

Then the element m_{ij} is nonzero.

Proof. From the equality $AM = I - R$ we have $M = A^{-1} - A^{-1}R$ and hence

$$m_{ij} = b_{ij} - \sum_{k=1}^{n} b_{ik} r_{kj}.$$

Therefore,

$$|m_{ij}| \geq |b_{ij}| - \sum_{k=1}^{n} |b_{ik} r_{kj}|$$

$$\geq |b_{ij}| - \max_{k=1,n} |b_{ik}| \, \|r_j\|_1$$

$$\geq |b_{ij}| - \max_{k=1,n} |b_{ik}| \tau_j.$$

Now the condition (10.57) implies that $|m_{ij}| > 0$. □

The proposition implies that, if R is small enough, then the nonzero elements of M are located in positions corresponding to the larger elements in the inverse of A. The following negative result is an immediate corollary.

Corollary 10.12. *Assume that M is an AINV of A computed by enforcing the condition (10.56) and let $\tau = \max_{j=1,\ldots,n} \tau_j$. If the nonzero elements of $B = A^{-1}$ are τ-equimodular in that*

$$|b_{ij}| > \tau \max_{k=1,n,\ l=1,n} |b_{lk}|,$$

then the nonzero sparsity pattern of M includes the nonzero sparsity pattern of A^{-1}. In particular, if A^{-1} is dense and its elements are τ-equimodular, then M is also dense.

The smaller the value of τ, the more likely the condition of the corollary will be satisfied. Another way of stating the corollary is that *accurate* and *sparse* AINVs may be computed only if the elements of the actual inverse have variations in size. Unfortunately, this is difficult to verify in advance and it is known to be true only for certain types of matrices.

10.5.5 Convergence of Self-Preconditioned MR

We now examine the convergence of the MR algorithm in the case where self-preconditioning is used, but no numerical dropping is applied. The column-oriented algorithm is considered first. Let M be the current AINV at a given substep. The self-preconditioned MR iteration for computing the jth column of the next AINV is obtained by the following sequence of operations:

1. $r_j := e_j - Am_j = e_j - AMe_j$,
2. $t_j := Mr_j$,
3. $\alpha_j := \frac{(r_j, At_j)}{(At_j, At_j)}$,
4. $m_j := m_j + \alpha_j t_j$.

Note that α_j can be written as

$$\alpha_j = \frac{(r_j, AMr_j)}{(AMr_j, AMr_j)} \equiv \frac{(r_j, Cr_j)}{(Cr_j, Cr_j)},$$

10.5. Approximate Inverse Preconditioners

where
$$C = AM$$
is the preconditioned matrix at the given substep. The subscript j is now dropped to simplify the notation. The new residual associated with the current column is given by
$$r^{new} = r - \alpha At = r - \alpha AMr \equiv r - \alpha Cr.$$
The orthogonality of the new residual against AMr can be used to obtain
$$\|r^{new}\|_2^2 = \|r\|_2^2 - \alpha^2 \|Cr\|_2^2.$$
Replacing α with its value defined above we get
$$\|r^{new}\|_2^2 = \|r\|_2^2 \left[1 - \left(\frac{(Cr, r)}{\|Cr\|_2 \|r\|_2} \right)^2 \right].$$
Thus, at each inner iteration, the residual norm for the jth column is reduced according to the formula
$$\|r^{new}\|_2 = \|r\|_2 \sin \angle(r, Cr), \tag{10.58}$$
in which $\angle(u, v)$ denotes the acute angle between the vectors u and v. Assume that each column converges. Then the preconditioned matrix C converges to the identity. As a result of this, the angle $\angle(r, Cr)$ will tend to $\angle(r, r) = 0$ and, therefore, the convergence ratio $\sin \angle(r, Cr)$ will also tend to zero, showing superlinear convergence.

Now consider (10.58) more carefully. Denote by R the residual matrix $R = I - AM$ and observe that
$$\sin \angle(r, Cr) = \min_\alpha \frac{\|r - \alpha Cr\|_2}{\|r\|_2}$$
$$\leq \frac{\|r - Cr\|_2}{\|r\|_2} \equiv \frac{\|Rr\|_2}{\|r\|_2}$$
$$\leq \|R\|_2.$$

This results in the following statement.

Proposition 10.13. *Assume that the self-preconditioned MR algorithm is employed with one inner step per iteration and no numerical dropping. Then the 2-norm of each residual $e_j - Am_j$ of the jth column is reduced by a factor of at least $\|I - AM\|_2$, where M is the AINV before the current step; i.e.,*
$$\|r_j^{new}\|_2 \leq \|I - AM\|_2 \|r_j\|_2. \tag{10.59}$$
In addition, the residual matrices $R_k = I - AM_k$ obtained after each outer iteration satisfy
$$\|R_{k+1}\|_F \leq \|R_k\|_F^2. \tag{10.60}$$

As a result, when the algorithm converges, it does so quadratically.

Proof. Inequality (10.59) was proved above. To prove quadratic convergence, first use the inequality $\|X\|_2 \leq \|X\|_F$ and (10.59) to obtain

$$\|r_j^{new}\|_2 \leq \|R_{k,j}\|_F \, \|r_j\|_2.$$

Here, the k index corresponds to the outer iteration and the j index to the column. Note that the Frobenius norm is reduced for each of the inner steps corresponding to the columns and, therefore,

$$\|R_{k,j}\|_F \leq \|R_k\|_F.$$

This yields

$$\|r_j^{new}\|_2^2 \leq \|R_k\|_F^2 \, \|r_j\|_2^2,$$

which, upon summation over j, gives

$$\|R_{k+1}\|_F \leq \|R_k\|_F^2.$$

This completes the proof. $\square$

Note that the above theorem does not prove convergence. It only states that, when the algorithm converges, it does so quadratically at the limit. In addition, the result ceases to be valid in the presence of dropping.

Consider now the case of the global iteration. When self-preconditioning is incorporated into the global MR algorithm (Algorithm 10.10), the search direction becomes $Z_k = M_k R_k$, where R_k is the current residual matrix. Then the main steps of the algorithm (without dropping) are as follows:

1. $R_k := I - AM_k$,
2. $Z_k := M_k R_k$,
3. $\alpha_k := \frac{\langle R_k, AZ_k \rangle}{\langle AZ_k, AZ_k \rangle}$,
4. $M_{k+1} := M_k + \alpha_k Z_k$.

At each step the new residual matrix R_{k+1} satisfies the relation

$$R_{k+1} = I - AM_{k+1} = I - A(M_k + \alpha_k Z_k) = R_k - \alpha_k AZ_k.$$

An important observation is that R_k is a polynomial in R_0. This is because, from the above relation,

$$R_{k+1} = R_k - \alpha_k AM_k R_k = R_k - \alpha_k (I - R_k) R_k = (1 - \alpha_k) R_k + \alpha_k R_k^2. \tag{10.61}$$

Therefore, induction shows that $R_{k+1} = p_{2^k}(R_0)$, where p_j is a polynomial of degree j. Now define the preconditioned matrices

$$B_k \equiv AM_k = I - R_k. \tag{10.62}$$

Then the following recurrence follows from (10.61):

$$B_{k+1} = B_k + \alpha_k B_k (I - B_k), \tag{10.63}$$

10.5. Approximate Inverse Preconditioners

and shows that B_{k+1} is also a polynomial of degree 2^k in B_0. In particular, *if the initial B_0 is symmetric, then so are all subsequent B_k's.* This is achieved when the initial M is a multiple of A^T, namely, if $M_0 = \alpha_0 A^T$.

Similarly to the column-oriented case, when the algorithm converges, it does so quadratically.

Proposition 10.14. *Assume that the self-preconditioned global MR algorithm is used without dropping. Then the residual matrices obtained at each iteration satisfy*

$$\|R_{k+1}\|_F \leq \|R_k^2\|_F. \tag{10.64}$$

As a result, when the algorithm converges, it does so quadratically.

Proof. Define, for any α,
$$R(\alpha) = (1-\alpha)R_k + \alpha R_k^2.$$

Recall that α_k achieves the minimum of $\|R(\alpha)\|_F$ over all α's. In particular,

$$\begin{aligned}\|R_{k+1}\|_F &= \min_\alpha \|R(\alpha)\|_F \\ &\leq \|R(1)\|_F = \|R_k^2\|_F \\ &\leq \|R_k\|_F^2.\end{aligned} \tag{10.65}$$

This proves quadratic convergence at the limit. $\square$

For further properties see Exercise 16.

10.5.6 AINVs via Bordering

A notable disadvantage of the right- or left-preconditioning method is that it is difficult to assess in advance whether or not the resulting AINV M is nonsingular. An alternative would be to seek a two-sided approximation, i.e., a pair L, U, with L lower triangular and U upper triangular, that attempts to minimize the objective function (10.45). The techniques developed in the previous sections can be exploited for this purpose.

In the factored approach, two matrices L and U that are *unit* lower and upper triangular are sought such that
$$LAU \approx D,$$
where D is some unknown diagonal matrix. When D is nonsingular and $LAU = D$, then L, U are called *inverse LU factors* of A since in this case $A^{-1} = UD^{-1}L$. Once more, the matrices are built one column or row at a time. Assume as in Section 10.4.5 that we have the sequence of matrices
$$A_{k+1} = \begin{pmatrix} A_k & v_k \\ w_k & \alpha_{k+1} \end{pmatrix},$$
in which $A_n \equiv A$. If the inverse factors L_k, U_k are available for A_k, i.e.,
$$L_k A_k U_k = D_k,$$

then the inverse factors L_{k+1}, U_{k+1} for A_{k+1} are easily obtained by writing

$$\begin{pmatrix} L_k & 0 \\ -y_k & 1 \end{pmatrix} \begin{pmatrix} A_k & v_k \\ w_k & \alpha_{k+1} \end{pmatrix} \begin{pmatrix} U_k & -z_k \\ 0 & 1 \end{pmatrix} = \begin{pmatrix} D_k & 0 \\ 0 & \delta_{k+1} \end{pmatrix}, \qquad (10.66)$$

in which z_k, y_k, and δ_{k+1} are such that

$$A_k z_k = v_k, \qquad (10.67)$$
$$y_k A_k = w_k, \qquad (10.68)$$
$$\delta_{k+1} = \alpha_{k+1} - w_k z_k = \alpha_{k+1} - y_k v_k. \qquad (10.69)$$

Note that the formula (10.69) exploits the fact that the system (10.67) is solved exactly (middle expression) or the system (10.68) is solved exactly (second expression) or both systems are solved exactly (either expression). In the realistic situation where neither of these two systems is solved exactly, then this formula should be replaced with

$$\delta_{k+1} = \alpha_{k+1} - w_k z_k - y_k v_k + y_k A_k z_k. \qquad (10.70)$$

The last row/column pairs of the approximate factored inverse can be obtained by solving two sparse systems and computing a few dot products. It is interesting to note that the only difference with the ILUS factorization seen in Section 10.4.5 is that *the coefficient matrices for these systems are not the triangular factors of A_k, but the matrix A_k itself.*

To obtain an approximate factorization, simply exploit the fact that the A_k-matrices are sparse and then employ iterative solvers in sparse-sparse mode. In this situation, formula (10.70) should be used for δ_{k+1}. The algorithm is as follows.

ALGORITHM 10.13. AINV Factors Algorithm

1. For $k = 1, \ldots, n$, Do
2.   Solve (10.67) *approximately*
3.   Solve (10.68) *approximately*
4.   Compute $\delta_{k+1} = \alpha_{k+1} - w_k z_k - y_k v_k + y_k A_k z_k$
5. EndDo

A linear system must be solved with A_k in line 2 and with A_k^T in line 3. This is a good scenario for the biconjugate gradient (BCG) algorithm or its equivalent two-sided Lanczos algorithm. In addition, the most current AINV factors can be used to precondition the linear systems to be solved in lines 2 and 3. This was termed self-preconditioning earlier. All the linear systems in the above algorithm can be solved in parallel since they are independent of one another. The diagonal D can then be obtained at the end of the process.

This approach is particularly suitable in the symmetric case. Since there is only one factor, the amount of work is halved. In addition, there is no problem with the existence in the positive definite case, as is shown in the following lemma, which states that δ_{k+1} is always greater than 0 when A is SPD, independently of the accuracy with which the system (10.67) is solved.

Lemma 10.15. *Let A be SPD. Then the scalar δ_{k+1} as computed by (10.70) is positive.*

10.5. Approximate Inverse Preconditioners

Proof. In the symmetric case, $w_k = v_k^T$. Note that δ_{k+1} as computed by formula (10.70) is the $(k+1, k+1)$th element of the matrix $L_{k+1} A_{k+1} L_{k+1}^T$. It is positive because A_{k+1} is SPD. This is independent of the accuracy for solving the system to obtain z_k. $\square$

In the general nonsymmetric case, there is no guarantee that δ_{k+1} will be nonzero, unless the systems (10.67) and (10.68) are solved accurately enough. There is no practical problem here, since δ_{k+1} is computable. The only question remaining is a theoretical one: Can δ_{k+1} be guaranteed to be nonzero if the systems are solved with enough accuracy? Intuitively, if the system is solved exactly, then the D-matrix must be nonzero since it is equal to the D-matrix of the exact inverse factors in this case. The minimal assumption to make is that each A_k is nonsingular. Let δ_{k+1}^* be the value that would be obtained if at least one of the systems (10.67) or (10.68) is solved exactly. According to (10.69), in this situation this value is given by

$$\delta_{k+1}^* = \alpha_{k+1} - w_k A_k^{-1} v_k. \tag{10.71}$$

If A_{k+1} is nonsingular, then $\delta_{k+1}^* \neq 0$. To see this refer to the defining equation (10.66) and compute the product $L_{k+1} A_{k+1} U_{k+1}$ in the general case. Let r_k and s_k be the residuals obtained for these linear systems; i.e.,

$$r_k = v_k - A_k z_k, \quad s_k = w_k - y_k A_k. \tag{10.72}$$

Then a little calculation yields

$$L_{k+1} A_{k+1} U_{k+1} = \begin{pmatrix} L_k A_k U_k & L_k r_k \\ s_k U_k & \delta_{k+1} \end{pmatrix}. \tag{10.73}$$

If one of r_k or s_k is zero, then it is clear that the term δ_{k+1} in the above relation becomes δ_{k+1}^* and it must be nonzero since the matrix on the left-hand side is nonsingular. Incidentally, this relation shows the structure of the last matrix $L_n A_n U_n \equiv LAU$. The components 1 to $j-1$ of column j consist of the vector $L_j r_j$, the components 1 to $j-1$ of row i make up the vector $s_k U_k$, and the diagonal elements are the δ_i's. Consider now the expression for δ_{k+1} from (10.70):

$$\begin{aligned}
\delta_{k+1} &= \alpha_{k+1} - w_k z_k - y_k v_k + y_k A_k z_k \\
&= \alpha_{k+1} - w_k A_k^{-1}(v_k - r_k) - (w_k - s_k) A_k^{-1} v_k + (v_k - r_k) A_k^{-1}(w_k - s_k) \\
&= \alpha_{k+1} - v_k A_k^{-1} w_k + r_k A_k^{-1} s_k \\
&= \delta_{k+1}^* + r_k A_k^{-1} s_k.
\end{aligned}$$

This perturbation formula is of second order in the sense that $|\delta_{k+1} - \delta_{k+1}^*| = O(\|r_k\| \|s_k\|)$. It guarantees that δ_{k+1} is nonzero whenever $|r_k A_k^{-1} s_k| < |\delta_{k+1}^*|$.

10.5.7 Factored Inverses via Orthogonalization: AINV

The AINV described in [34, 36] computes an approximate factorization of the form $W^T A Z = D$, where W, Z are unit upper triangular matrices and D is a diagonal. The matrices W and Z can be directly computed by performing an approximate biorthogonalization of the

Gram–Schmidt type. Indeed, when $A = LDU$ is the exact LDU factorization of A, then W will be equal to the inverse of L and we will have the equality

$$W^T A = DU,$$

which means that $W^T A$ is upper triangular. This translates into the result that any column i of W is orthogonal to the first $i - 1$ columns of A. A procedure to compute W is therefore to make the ith column of W orthogonal to the columns $1, \ldots, i - 1$ of A by subtracting multiples of the first $i - 1$ columns of W. Alternatively, columns $i + 1, \ldots, n$ of W can be made orthogonal to the first i columns of A. This will produce columns that are orthogonal to each of the columns of A. During this procedure one can drop small entries, or entries outside a certain sparsity pattern. A similar process can be applied to obtain the columns of Z. The resulting incomplete biorthogonalization process, which is sketched next, produces an approximate factored inverse.

ALGORITHM 10.14. Right-Looking Factored AINV

1. Let $p = q = (0, \ldots, 0) \in \mathbb{R}^n$, $Z = [z_1, \ldots, z_n] = I_n$, $W = [w_1, \ldots, w_n] = I_n$
2. For $k = 1, \ldots, n$,
3. $p_k = w_k^T A e_k$, $q_k = e_k^T A z_k$
4.   For $i = k + 1, \ldots, n$,
5. $p_i = \left(w_i^T A e_k\right) / p_k$, $q_i = \left(e_k^T A z_i\right) / q_k$
6.   Apply a dropping rule to p_i, q_i
7. $w_i = w_i - w_k p_i$, $z_i = z_i - z_k q_i$
8.   Apply a dropping rule to $w_{j,i}$ and $z_{j,i}$ for $j = 1, \ldots, i$
9.   EndDo
10. EndDo
11. Choose diagonal entries of D as the components of p or q

The above algorithm constitutes one of two options for computing factored AINVs via approximate orthogonalization. An alternative is based on the fact that $W^T A Z$ should become approximately diagonal. Instead of orthogonalizing W (resp. Z) with respect to the columns of A, a biorthogonalization process can be applied to force the columns of W and Z to be conjugate with respect to A. For this we must require that $e_k^T W^T A Z e_j = 0$ for all $k \neq j$, $1 \leq k, j \leq i$. The result will be a simple change to Algorithm 10.14. Specifically, the second option, which we label with a (b), replaces lines (3) and (5) with the following lines:

3b. $p_k = w_k^T A z_k$, $q_k = w_k^T A z_k$
5b. $p_i = \left(w_i^T A z_k\right) / p_k$, $q_i = \left(w_k^T A z_i\right) / q_k$

If no entries are dropped and if an LDU factorization of A exists, then $W = L^T$, $Z = U^{-1}$. A little induction proof would then show that, after step i, columns $i + 1, \ldots, n$ of W are orthogonal to columns $1, \ldots, i$ of A and likewise columns $i + 1, \ldots, n$ of Z are orthogonal to rows $1, \ldots, i$ of A. Remarkably, the computations of Z and W can be performed independently of each other for the original option represented by Algorithm 10.14.

In the original version of AINV [34, 36], dropping is performed on the vectors w_i and z_i only. Dropping entries from p_i, q_i seems not to yield as good approximations; see [34].

10.5.8 Improving a Preconditioner

After a computed ILU factorization results in an unsatisfactory convergence, it is difficult to improve it by modifying the L and U factors. One solution would be to discard this factorization and attempt to recompute a fresh one possibly with more fill-in. Clearly, this may be a wasteful process. A better alternative is to use AINV techniques. Assume a (sparse) matrix M is a preconditioner to the original matrix A, so the preconditioned matrix is

$$C = M^{-1}A.$$

A sparse matrix S is sought to approximate the inverse of $M^{-1}A$. This matrix is then to be used as a preconditioner to $M^{-1}A$. Unfortunately, the matrix C is usually dense. However, observe that all that is needed is a matrix S such that

$$AS \approx M.$$

Recall that the columns of A and M are sparse. One approach is to compute a least-squares approximation in the Frobenius norm sense. This approach was used in Section 10.5.1 when M was the identity matrix. Then the columns of S were obtained by approximately solving the linear systems $As_i \approx e_i$. The same idea can be applied here. Now the systems

$$As_i = m_i$$

must be solved instead, where m_i is the ith column of M, which is sparse. Thus, the coefficient matrix and the right-hand side are sparse, as before.

10.6 Reordering for Incomplete LU

The primary goal of reordering techniques (see Chapter 3) is to reduce fill-in during Gaussian elimination. A difficulty with such methods, whether in the context of direct or iterative solvers, is that a good ordering for reducing fill-in may lead to factors of poor numerical quality. For example, very small diagonal entries may be encountered during the process. Two types of permutations are often used to enhance ILU factorizations. First, fill-reducing symmetric permutations of the type seen in Chapter 3 have been advocated. The argument here is that, since these permutations are likely to produce fewer fill-ins, it is likely that the ILU factorizations resulting from dropping small terms will be more accurate. A second category of reorderings consists of permuting only the rows of the matrix (or its columns). These unsymmetric permutations address the other issue mentioned above, namely, avoiding poor pivots in Gaussian elimination.

10.6.1 Symmetric Permutations

The RCM ordering seen in Section 3.3.3 is among the most common techniques used to enhance the effectiveness of ILU factorizations. Recall that this reordering is designed to reduce the envelope of a matrix. Other reorderings that are geared specifically toward reducing fill-in, such as the minimum degree (MD) and multiple MD orderings, have also been advocated, though results reported in the literature are mixed. What is clear is that the results will depend on the accuracy of the ILU being computed. If ILU(0), or some low-fill, incomplete factorization is being used, then it is often reported that it is generally not a good idea to reorder the matrix. Among candidate permutations that can be applied, the RCM is the most likely to yield an improvement. As the accuracy of the preconditioner increases, i.e., as more fill-ins are allowed, then the beneficial effect of reordering becomes compelling. In many tests (see, for example, [35]), a preconditioner built on an RCM or MD reordered matrix will work while the same preconditioner built from the original ordering fails. In addition, success is often achieved with less memory than is required for the original ordering. This general observation is illustrated in the following tests.

Example 10.6. The following experiments show the performance of GMRES(20) preconditioned with ILUT for the five test problems described in Section 3.7 of Chapter 3. The first experiment uses $ILUT(5, 0.25)$. Prior to performing the ILUT factorization the coefficient matrix is reordered by three possible techniques: RCM, MD ordering (QMD), or ND. The FORTRAN codes for these three techniques are available in the book [144]. It is now important to show the amount of memory used by the factorization, which is measured here by the *fill factor*, i.e., the ratio of the number of nonzero elements required to store the LU factors to the original number of nonzero elements. This is referred to as *Fill* in the tables. Along with this measure, Table 10.6 shows the number of iterations required to reduce the initial residual by a factor of 10^{-7} with GMRES(20). Notice that reordering does not help. The RCM ordering is the best among the three orderings, with a performance that is close to that of the original ordering, but it fails on the FIDAP matrix. In many other instances we have tested, RCM does often help or its performance is close to that achieved by the original ordering. The other reorderings, MD and ND, rarely help when the factorization is inaccurate, as is the case here.

	None		RCM		QMD		ND	
Matrix	Iters	Fill	Iters	Fill	Iters	Fill	Iters	Fill
F2DA	15	1.471	16	1.448	19	1.588	20	1.592
F3D	12	1.583	13	1.391	16	1.522	15	1.527
ORS	20	0.391	20	0.391	20	0.477	20	0.480
F2DB	21	1.430	21	1.402	41	1.546	55	1.541
FID	66	1.138	300	1.131	300	0.978	300	1.032

Table 10.6. *Iteration count and fill factor for GMRES(20)-ILUT(5,0.25) with three different reordering techniques.*

We now turn to a more accurate preconditioner, namely, ILUT(10, 0.01). The results of Table 10.7 show a different picture from the one above. All reorderings are now basically helpful. A slight exception is the MD ordering, which fails on the FIDAP matrix. However, notice that this failure can be explained by the low fill factor, which is the smallest achieved by all the reorderings. What is more, good convergence of GMRES is now achieved at a lesser cost of memory.

10.6. Reordering for Incomplete LU

	None		RCM		QMD		ND	
Matrix	Iters	Fill	Iters	Fill	Iters	Fill	Iters	Fill
F2DA	7	3.382	6	3.085	8	2.456	9	2.555
F3D	8	3.438	7	3.641	11	2.383	10	2.669
ORS	9	0.708	9	0.699	9	0.779	9	0.807
F2DB	10	3.203	8	2.962	12	2.389	12	2.463
FID	197	1.798	38	1.747	300	1.388	36	1.485

Table 10.7. *Iteration count and fill factor for GMRES(20)-ILUT(10,0.01) with three different reordering techniques.*

If one ignores the fill factor it may appear that RCM is best. QMD seems to be good at reducing fill-in but results in a poor factorization. When memory cost is taken into account, the more sophisticated ND ordering is the overall winner in all cases except for the ORSIR matrix. This conclusion, namely, that reordering is most beneficial when relatively accurate factorizations are computed, is borne out by other experiments in the literature; see, for example, [35].

10.6.2 Nonsymmetric Reorderings

Nonsymmetric permutations can be applied to enhance the performance of preconditioners for matrices with extremely poor structure. Such techniques do not perform too well in other situations, such as, for example, linear systems arising from the discretization of elliptic PDEs.

The original idea on which nonsymmetric reorderings are based is to find a permutation matrix Q_π so that the matrix

$$B = Q_\pi A \tag{10.74}$$

has large entries in its diagonal. Here π is a permutation array and Q_π the corresponding permutation matrix, as defined in Section 3.3. In contrast with standard fill-reducing techniques, this is a one-sided permutation that reorders the rows of the matrix.

The first algorithm considered in this class attempts to find an ordering of the form (10.74) that guarantees that the diagonal entries of B are nonzero. In this case, the permutation matrix Q_π can be viewed from a new angle, that of bipartite transversals.

A transversal or bipartite matching is a set $\mathcal{M}$ of ordered pairs (i, j) such that $a_{ij} \neq 0$ and the column indices j and row indices i appear only once. This corresponds to selecting one nonzero diagonal element per row/column. The usual representation uses a graph whose vertices are the rows of A (squares in Figure 10.17) and columns of A (circles in Figure 10.17). There is an outgoing edge between a row i and a column j when $a_{ij} \neq 0$. A transversal is simply a subgraph of G that is *bipartite*. The transversal is *maximum* when it has maximum cardinality. For example, in Figure 10.17 the set

$$\mathcal{M} = \{(1, 2), (2, 1), (3, 3), (4, 5), (5, 6), (6, 4)\}$$

is a maximum transversal. The corresponding row permutation is $\pi = \{2, 1, 3, 6, 4, 5\}$ and the reordered matrix is shown in the bottom-middle part of the figure.

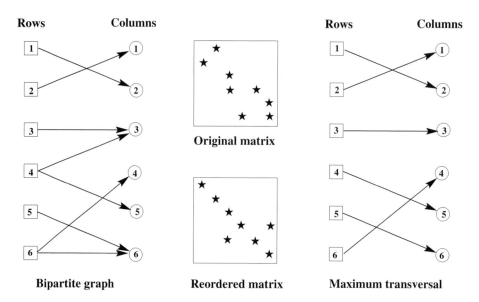

Figure 10.17. *Example of a maximum transversal. Left side: bipartite representation of matrix. Right side: maximum transversal. Middle: matrix before and after row reordering.*

When A is structurally nonsingular, it can be shown that the maximum transversal has cardinality $|\mathcal{M}| = n$. Finding the maximum transversal is a well-known problem in management sciences and has received much attention by researchers in graph theory. In particular, graph traversal algorithms based on depth first search and breadth first search have been developed to find maximum transversals.

These maximum transversal algorithms are the simplest among a class of techniques. The criterion of just finding nonzero diagonal elements to put on the diagonal is not sufficient and can be changed into one of finding a (row) permutation π so as to

$$\text{maximize} \prod_{i=1}^{n} |a_{i,\pi(i)}|. \tag{10.75}$$

A heuristic for achieving a large product of the diagonal entries is the so-called bottleneck strategy, whose goal is to maximize the smallest diagonal entry. The algorithm removes enough small elements and finds a maximum transversal of the graph. If the transversal is not of cardinality n, then the algorithm backtracks by removing fewer of the small entries and repeating the process.

Another class of algorithms solve the optimization problem (10.75) more accurately. This problem can be translated into

$$\min_{\pi} \sum_{i=1}^{n} c_{i,\pi(i)}, \qquad \text{where} \qquad c_{ij} = \begin{cases} \log\left[\,\|a_{:,j}\|_\infty \,/\, |a_{ij}|\,\right] & \text{if } a_{ij} \neq 0, \\ +\infty & \text{otherwise.} \end{cases}$$

It is known that solving this problem is equivalent to solving its dual, which can be formulated as follows:

$$\max_{u_i, u_j} \left\{ \sum_{i=1}^{n} u_i + \sum_{j=1}^{n} u_j \right\} \quad \text{subject to} \quad c_{ij} - u_i - u_j \geq 0.$$

The algorithms used to solve the above dual problem are based on graph theory techniques—in fact they can be viewed as traversal algorithms (such as depth first search) to which a cost measure is added. Details can be found in [110].

Experiments reported by Duff and Koster [110] and Benzi et al. [32] show that nonsymmetric reorderings based on the methods discussed in this section can be quite beneficial for those problems that are irregularly structured and have many zero diagonal entries. On the other hand, they do not perform as well for PDE matrices, for which symmetric orderings are often superior.

10.7 Block Preconditioners

Block preconditioning is a popular technique for block tridiagonal matrices arising from the discretization of elliptic problems. It can also be generalized to other sparse matrices. We begin with a discussion of the block tridiagonal case.

10.7.1 Block Tridiagonal Matrices

Consider a block tridiagonal matrix blocked in the form

$$A = \begin{pmatrix} D_1 & E_2 & & & \\ F_2 & D_2 & E_3 & & \\ & \ddots & \ddots & \ddots & \\ & & F_{m-1} & D_{m-1} & E_m \\ & & & F_m & D_m \end{pmatrix}. \tag{10.76}$$

One of the most popular block preconditioners used in the context of PDEs is based on this block tridiagonal form of the coefficient matrix A. Let D be the block diagonal matrix consisting of the diagonal blocks D_i, L the block strictly lower triangular matrix consisting of the subdiagonal blocks F_i, and U the block strictly upper triangular matrix consisting of the superdiagonal blocks E_i. Then the above matrix has the form

$$A = L + D + U.$$

A block ILU preconditioner is defined by

$$M = (L + \Delta)\Delta^{-1}(\Delta + U), \tag{10.77}$$

where L and U are the same as above and Δ is a block diagonal matrix whose blocks Δ_i are defined by the recurrence

$$\Delta_i = D_i - F_i \Omega_{i-1} E_i, \tag{10.78}$$

in which Ω_j is some sparse approximation to Δ_j^{-1}. Thus, to obtain a block factorization, approximations to the inverses of the blocks Δ_i must be found. This clearly will lead to difficulties if explicit inverses are used.

An important particular case is when the diagonal blocks D_i of the original matrix are tridiagonal, while the codiagonal blocks E_i and F_i are diagonal. Then a simple recurrence formula for computing the inverse of a tridiagonal matrix can be exploited. Only the tridiagonal part of the inverse must be kept in the recurrence (10.78). Thus,

$$\Delta_1 = D_1, \tag{10.79}$$
$$\Delta_i = D_i - F_i \Omega_{i-1}^{(3)} E_i, \quad i = 1, \ldots, m, \tag{10.80}$$

where $\Omega_k^{(3)}$ is the tridiagonal part of Δ_k^{-1}:

$$(\Omega_k^{(3)})_{i,j} = (\Delta_k^{-1})_{i,j} \quad \text{for} \quad |i - j| \leq 1.$$

The following theorem can be shown.

Theorem 10.16. *Let A be SPD and such that*

- $a_{ii} > 0$, $i = 1, \ldots, n$, and $a_{ij} \leq 0$ for all $j \neq i$;
- *the matrices D_i are all (strictly) diagonally dominant.*

Then each block Δ_i computed by the recurrence (10.79), (10.80) *is a symmetric M-matrix. In particular, M is also a positive definite matrix.*

We now show how the inverse of a tridiagonal matrix can be obtained. Let a tridiagonal matrix Δ of dimension l be given in the form

$$\Delta = \begin{pmatrix} \alpha_1 & -\beta_2 & & & \\ -\beta_2 & \alpha_2 & -\beta_3 & & \\ & \ddots & \ddots & \ddots & \\ & & -\beta_{l-1} & \alpha_{l-1} & -\beta_l \\ & & & -\beta_l & \alpha_l \end{pmatrix}$$

and let its Cholesky factorization be

$$\Delta = LDL^T,$$

with

$$D = \text{diag}\{\delta_i\}$$

and

$$L = \begin{pmatrix} 1 & & & & \\ -\gamma_2 & 1 & & & \\ & \ddots & \ddots & & \\ & & -\gamma_{l-1} & 1 & \\ & & & -\gamma_l & 1 \end{pmatrix}.$$

The inverse of Δ is $L^{-T}D^{-1}L^{-1}$. Start by observing that the inverse of L^T is a unit upper triangular matrix whose coefficients u_{ij} are given by

$$u_{ij} = \gamma_{i+1}\gamma_{i+2}\cdots\gamma_{j-1}\gamma_j \quad \text{for} \quad 1 \le i < j < l.$$

As a result, the jth column c_j of L^{-T} is related to the $(j-1)$th column c_{j-1} by the very simple recurrence

$$c_j = e_j + \gamma_j c_{j-1} \quad \text{for} \quad j \ge 2,$$

starting with the first column $c_1 = e_1$. The inverse of Δ becomes

$$\Delta^{-1} = L^{-T}D^{-1}L^{-1} = \sum_{j=1}^{l} \frac{1}{\delta_j} c_j c_j^T. \tag{10.81}$$

See Exercise 12 for a proof of the above equality. As noted, the recurrence formulas for computing Δ^{-1} can be unstable and lead to numerical difficulties for large values of l.

10.7.2 General Matrices

A general sparse matrix can often be put in the form (10.76), where the blocking is either natural, as provided by the physical problem, or artificial when obtained as a result of RCM ordering and some block partitioning. In such cases, a recurrence such as (10.78) can still be used to obtain a block factorization defined by (10.77). A two-level preconditioner can be defined by using sparse inverse approximate techniques to approximate Ω_i. These are sometimes termed implicit-explicit preconditioners, the implicit part referring to the block factorization and the explicit part to the AINVs used to explicitly approximate Δ_i^{-1}.

10.8 Preconditioners for the Normal Equations

When the original matrix is strongly indefinite, i.e., when it has eigenvalues spread on both sides of the imaginary axis, the usual Krylov subspace methods may fail. The CG approach applied to the normal equations may then become a good alternative. Choosing to use this alternative over the standard methods may involve inspecting the spectrum of a Hessenberg matrix obtained from a small run of an unpreconditioned GMRES algorithm.

If the normal equations approach is chosen, the question becomes how to precondition the resulting iteration. An ILU preconditioner can be computed for A and the preconditioned normal equations

$$A^T(LU)^{-T}(LU)^{-1}Ax = A^T(LU)^{-T}(LU)^{-1}b$$

can be solved. However, when A is not diagonally dominant, the ILU factorization process may encounter a zero pivot. Even when this does not happen, the resulting preconditioner may be of poor quality. An incomplete factorization routine with pivoting, such as ILUTP, may constitute a good choice. ILUTP can be used to precondition either the original equations or the normal equations shown above. This section explores a few other options available for preconditioning the normal equations.

10.8.1 Jacobi, SOR, and Variants

There are several ways to exploit the relaxation schemes for the NE seen in Chapter 8 as preconditioners for the CG method applied to either (8.1) (NR) or (8.3) (NE). Consider (8.3), for example, which requires a procedure delivering an approximation to $(AA^T)^{-1}v$ for any vector v. One such procedure is to perform one step of SSOR to solve the system $(AA^T)w = v$. Denote by M^{-1} the linear operator that transforms v into the vector resulting from this procedure. Then the usual CG method applied to (8.3) can be recast in the same form as Algorithm 8.5. This algorithm is known as CGNE/SSOR. Similarly, it is possible to incorporate the SSOR preconditioning in Algorithm 8.4, which is associated with the normal equations (8.1), by defining M^{-1} to be the linear transformation that maps a vector v into a vector w resulting from the forward sweep of Algorithm 8.2 followed by a backward sweep. We will refer to this algorithm as CGNR/SSOR.

The CGNE/SSOR and CGNR/SSOR algorithms will not break down if A is nonsingular, since then the matrices AA^T and A^TA are SPD, as are the preconditioning matrices M. There are several variations of these algorithms. The standard alternatives based on the same formulation (8.1) are either to use the preconditioner on the right, solving the system $A^TAM^{-1}y = b$, or to split the preconditioner into a forward SOR sweep on the left and a backward SOR sweep on the right of the matrix A^TA. Similar options can also be written for the normal equations (8.3), again with three different ways of preconditioning. Thus, at least six different algorithms can be defined.

10.8.2 IC(0) for the Normal Equations

The incomplete Cholesky IC(0) factorization can be used to precondition the normal equations (8.1) or (8.3). This approach may seem attractive because of the success of incomplete factorization preconditioners. However, a major problem is that the IC factorization is not guaranteed to exist for an arbitrary SPD matrix B. All the results that guarantee existence rely on some form of diagonal dominance. One of the first ideas suggested to handle this difficulty was to use an IC factorization on the "shifted" matrix $B + \alpha I$. We refer to IC(0) applied to $B = A^TA$ as ICNR(0), and likewise IC(0) applied to $B = AA^T$ as ICNE(0). Shifted variants correspond to applying IC(0) to the shifted B-matrix.

One issue often debated is how to find good values for the shift α. There is no easy and well-founded solution to this problem for irregularly structured symmetric sparse matrices. One idea is to select the smallest possible α that makes the shifted matrix diagonally dominant. However, this shift tends to be too large in general because IC(0) may exist for much smaller values of α. Another approach is to determine the smallest α for which the IC(0) factorization exists. Unfortunately, this is not a viable alternative. As is often observed, the number of steps required for convergence starts decreasing as α increases, and then increases again. The illustration shown in Figure 10.18 is from a real example using a small Laplacian matrix. This plot suggests that there is an optimal value for α, which is far from the smallest admissible one. For small α, the diagonal dominance of $B + \alpha I$ is weak and, as a result, the computed IC factorization is a poor approximation to the matrix $B(\alpha) \equiv B + \alpha I$. In other words, $B(\alpha)$ is close to the original matrix B, but the IC(0) factorization is far from $B(\alpha)$. For large α, the opposite is true. The matrix $B(\alpha)$ has a large deviation from $B(0)$, but its IC(0) factorization may be quite good. Therefore, the general shape of the curve shown in the figure is not too surprising.

10.8. Preconditioners for the Normal Equations

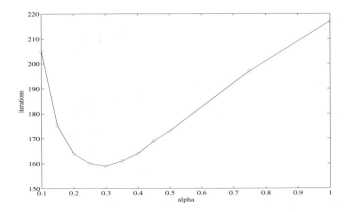

Figure 10.18. *Iteration count as a function of the shift α.*

To implement the algorithm, the matrix $B = AA^T$ need not be formed explicitly. All that is required is to be able to access one row of B at a time. This row can be computed, used, and then discarded. In the following, the ith row $e_i^T A$ of A is denoted by a_i. The algorithm is row oriented and all vectors denote row vectors. It is adapted from the ILU(0) factorization of a sparse matrix, i.e., Algorithm 10.4, but it actually computes the LDL^T factorization instead of an LU or LL^T factorization. The main difference with Algorithm 10.4 is that the loop in line 7 is now restricted to $j \leq i$ because of symmetry. If only the l_{ij} elements are stored row-wise, then the rows of $U = L^T$ that are needed in this loop are not directly available. Denote the jth row of $U = L^T$ by u_j. These rows are accessible by adding a column data structure for the L-matrix, which is updated dynamically. A linked-list data structure can be used for this purpose. With this in mind, the IC(0) algorithm will have the following structure.

ALGORITHM 10.15. Shifted ICNE(0)

1. *Initial step:* Set $d_1 := a_{11}$, $l_{11} = 1$
2. For $i = 2, 3, \ldots, n$, Do
3.   *Obtain all the nonzero inner products*
4. $l_{ij} = (a_j, a_i)$, $j = 1, 2, \ldots, i-1$, and $l_{ii} := \|a_i\|^2 + \alpha$
5.   Set $NZ(i) \equiv \{j \mid l_{ij} \neq 0\}$
6.   For $k = 1, \ldots, i-1$ and if $k \in NZ(i)$, Do
7.   Extract row $u_k = (Le_k)^T$
8.   Compute $l_{ik} := l_{ik}/d_k$
9.   For $j = k+1, \ldots, i$ and if $(i, j) \in NZ(i)$, Do
10.   Compute $l_{ik} := l_{ik} - l_{ij} u_{kj}$
11.   EndDo
12.   EndDo
13.   Set $d_i := l_{ii}$, $l_{ii} := 1$
14. EndDo

Note that initially the row u_1 in the algorithm is defined as the first row of A. All vectors in the algorithm are row vectors.

The step represented by lines 3 and 4, which computes the inner products of row number i with all previous rows, needs particular attention. If the inner products

$$a_1^T a_i,\ a_2^T a_i,\ \ldots,\ a_{i-1}^T a_i$$

are computed separately, the total cost of the incomplete factorization will be of the order of n^2 steps and the algorithm will be of little practical value. However, most of these inner products are equal to zero because of sparsity. This indicates that it may be possible to compute only the nonzero inner products at a much lower cost. Indeed, if c is the column of the $i-1$ inner products c_{ij}, then c is the product of the rectangular $(i-1) \times n$ matrix A_{i-1} whose rows are $a_1^T, \ldots, a_{i-1}^T$ with the vector a_i; i.e.,

$$c = A_{i-1} a_i. \qquad (10.82)$$

This is a sparse matrix–by–sparse vector product, which was discussed in Section 10.5. It is best performed as a linear combination of the columns of A_{i-1} that are sparse. The only difficulty with this implementation is that it requires the row data structure of both A and its transpose. A standard way to handle this problem is by building a linked-list data structure for the transpose. There is a similar problem for accessing the transpose of L, as mentioned earlier. Therefore, two linked lists are needed: one for the L-matrix and the other for the A-matrix. These linked lists avoid the storage of an additional real array for the matrices involved and simplify the process of updating the matrix A when new rows are obtained. It is important to note that these linked lists are used only in the preprocessing phase and are discarded once the incomplete factorization terminates.

10.8.3 Incomplete Gram–Schmidt and ILQ

Consider a general sparse matrix A and denote its rows by $a_1, a_2, \ldots, a_n$. The (complete) LQ factorization of A is defined by

$$A = LQ,$$

where L is a lower triangular matrix and Q is unitary; i.e., $Q^T Q = I$. The L factor in the above factorization is identical to the Cholesky factor of the matrix $B = AA^T$. Indeed, if $A = LQ$, where L is a lower triangular matrix having positive diagonal elements, then

$$B = AA^T = LQQ^T L^T = LL^T.$$

The uniqueness of the Cholesky factorization with a factor L having positive diagonal elements shows that L is equal to the Cholesky factor of B. This relationship can be exploited to obtain preconditioners for the normal equations.

Thus, there are two ways to obtain the matrix L. The first is to form the matrix B explicitly and use a sparse Cholesky factorization. This requires forming the data structure of the matrix AA^T, which may be much denser than A. However, reordering techniques can be used to reduce the amount of work required to compute L. This approach is known as *symmetric squaring*.

A second approach is to use the Gram–Schmidt process. This idea may seem undesirable at first because of its poor numerical properties when orthogonalizing a large number of vectors. However, because the rows remain very sparse in the incomplete LQ (ILQ) factorization (to be described shortly), any given row of A will be orthogonal typically to most of the previous rows of Q. As a result, the Gram–Schmidt process is much less

10.8. Preconditioners for the Normal Equations

prone to numerical difficulties. From the data structure point of view, Gram–Schmidt is optimal because it does not require allocating more space than is necessary, as is the case with approaches based on symmetric squaring. Another advantage over symmetric squaring is the simplicity of the orthogonalization process and its strong similarity to the LU factorization. At every step, a given row is combined with previous rows and then normalized. The incomplete Gram–Schmidt procedure is modeled after the following algorithm.

ALGORITHM 10.16. LQ Factorization of A

1. For $i = 1, \ldots, n$, Do
2.   Compute $l_{ij} := (a_i, q_j)$ for $j = 1, 2, \ldots, i-1$
3.   Compute $q_i := a_i - \sum_{j=1}^{i-1} l_{ij} q_j$ and $l_{ii} = \|q_i\|_2$
4.   If $l_{ii} := 0$ then Stop; else Compute $q_i := q_i / l_{ii}$
5. EndDo

If the algorithm completes, then it will result in the factorization $A = LQ$, where the rows of Q and L are the rows defined in the algorithm. To define an *incomplete* factorization, a *dropping* strategy similar to those defined for ILU factorizations must be incorporated. This can be done in very general terms as follows. Let P_L and P_Q be the chosen zero patterns for the matrices L and Q, respectively. The only restriction on P_L is that

$$P_L \subset \{(i,j) \mid i \neq j\}.$$

As for P_Q, for each row there must be at least one nonzero element; i.e.,

$$\{j \mid (i,j) \in P_Q\} \neq \{1, 2, \ldots, n\} \quad \text{for } i = 1, \ldots, n.$$

These two sets can be selected in various ways. For example, similarly to ILUT, they can be determined dynamically by using a drop strategy based on the magnitude of the elements generated. As before, x_i denotes the ith row of a matrix X and x_{ij} its (i, j)th entry.

ALGORITHM 10.17. Incomplete Gram–Schmidt

1. For $i = 1, \ldots, n$, Do
2.   Compute $l_{ij} := (a_i, q_j)$ for $j = 1, 2, \ldots, i-1$
3.   Replace l_{ij} with zero if $(i, j) \in P_L$
4.   Compute $q_i := a_i - \sum_{j=1}^{i-1} l_{ij} q_j$
5.   Replace each q_{ij}, $j = 1, \ldots, n$, with zero if $(i, j) \in P_Q$
6. $l_{ii} := \|q_i\|_2$
7.   If $l_{ii} = 0$ then Stop; else Compute $q_i := q_i / l_{ii}$
8. EndDo

We recognize in line 2 the same practical problem encountered in the previous section for IC(0) for the normal equations. It can be handled in the same manner. Thus, the row structures of A, L, and Q are needed, as well as a linked list for the column structure of Q.

After the ith step is performed, the following relation holds:

$$q_i = l_{ii} q_i + r_i = a_i - \sum_{j=1}^{j-1} l_{ij} q_j$$

or
$$a_i = \sum_{j=1}^{j} l_{ij} q_j + r_i, \qquad (10.83)$$
where r_i is the row of elements that have been dropped from the row q_i in line 5. The above equation translates into
$$A = LQ + R, \qquad (10.84)$$
where R is the matrix whose ith row is r_i and the notation for L and Q is as before.

The case where the elements in Q are not dropped, i.e., the case when P_Q is the empty set, is of particular interest. Indeed, in this situation, $R = 0$ and we have the exact relation $A = LQ$. However, Q is not unitary in general because elements are dropped from L. If at a given step $l_{ii} = 0$, then (10.83) implies that a_i is a linear combination of the rows $q_1, \ldots, q_{j-1}$. Each of these q_k is, inductively, a linear combination of $a_1, \ldots, a_k$. Therefore, a_i is a linear combination of the previous rows, $a_1, \ldots, a_{i-1}$, which cannot be true if A is nonsingular. As a result, the following proposition can be stated.

Proposition 10.17. *If A is nonsingular and $P_Q = \emptyset$, then Algorithm 10.17 completes and computes an ILQ factorization $A = LQ$, in which Q is nonsingular and L is a lower triangular matrix with positive elements.*

A major problem with the decomposition (10.84) is that the matrix Q is not orthogonal in general. In fact, nothing guarantees that it is even nonsingular unless Q is not dropped or the dropping strategy is made tight enough.

Because the matrix L of the *complete* LQ factorization of A is identical to the Cholesky factor of B, one might wonder why the IC(0) factorization of B does not always exist while the ILQ factorization seems to always exist. In fact, the relationship between ILQ and ICNE, i.e., IC for $B = AA^T$, can lead to a more rigorous way of choosing a good pattern for ICNE, as is explained next.

We turn our attention to modified Gram–Schmidt (MGS). The only difference is that the row q_j is updated immediately after an inner product is computed. The algorithm is described without dropping for Q for simplicity.

ALGORITHM 10.18. Incomplete MGS

1. For $i = 1, \ldots, n$, Do
2. $q_i := a_i$
3.   For $j = 1, \ldots, i - 1$, Do
4.   Compute $l_{ij} := \begin{cases} 0 & \text{if } (i, j) \in P_L \\ (q_i, q_j) & \text{otherwise} \end{cases}$
5.   Compute $q_i := q_i - l_{ij} q_j$
6.   EndDo
7. $l_{ii} := \|q_i\|_2$
8.   If $l_{ii} = 0$ then Stop; else Compute $q_i := q_i / l_{ii}$
9. EndDo

When A is nonsingular, the same result as before is obtained if no dropping is used on Q, namely, that the factorization will exist and be exact in that $A = LQ$. Regarding the

Exercises

implementation, if the zero pattern P_L is known in advance, the computation of the inner products in line 4 does not pose a particular problem. Without any dropping in Q, this algorithm may be too costly in terms of storage. It is interesting to see that this algorithm has a connection with ICNE, the IC applied to the matrix AA^T. The following result is stated without proof.

Theorem 10.18. *Let A be an $n \times m$ matrix and let $B = AA^T$. Consider a zero pattern set P_L such that, for any $1 \leq i, j, k \leq n$, with $i < j$ and $i < k$, the following holds:*

$$(i, j) \in P_L \text{ and } (i, k) \notin P_L \rightarrow (j, k) \in P_L.$$

Then the matrix L obtained from Algorithm 10.18 with the zero pattern set P_L is identical to the L factor that would be obtained from the IC factorization applied to B with the zero pattern set P_L.

For a proof, see [303]. This result shows how a zero pattern can be defined that guarantees the existence of an IC factorization on AA^T.

Exercises

1. Assume that A is the SPD matrix arising from the five-point finite difference discretization of the Laplacian on a given mesh. We reorder the matrix using the red-black ordering and obtain

    $$B = \begin{pmatrix} D_1 & E \\ E^T & D_2 \end{pmatrix}.$$

    We then form the IC factorization of this matrix.

    a. Show the fill-in pattern for the IC(0) factorization for a matrix of size $n = 12$ associated with a 4×3 mesh.
    b. Show the nodes associated with these fill-ins on the five-point stencil in the finite difference mesh.
    c. Give an approximate count of the total number of fill-ins when the original mesh is square, with the same number of mesh points in each direction. How does this compare with the natural ordering? Any conclusions?

2. Consider a 6×6 tridiagonal nonsingular matrix A.

    a. What can be said about its ILU(0) factorization (when it exists)?
    b. Suppose that the matrix is permuted (symmetrically, i.e., both rows and columns) using the permutation

    $$\pi = [1, 3, 5, 2, 4, 6].$$

    (i) Show the pattern of the permuted matrix.
    (ii) Show the locations of the fill-in elements in the ILU(0) factorization.
    (iii) Show the pattern of the ILU(1) factorization as well as the fill-ins generated.

(iv) Show the level of fill of each element at the end of the ILU(1) process (including the fill-ins).

(v) What can be said of the ILU(2) factorization for this permuted matrix?

3. Assume that A is the matrix arising from the five-point finite difference discretization of an elliptic operator on a given mesh. We reorder the original linear system using the red-black ordering and obtain

$$\begin{pmatrix} D_1 & E \\ F & D_2 \end{pmatrix} \begin{pmatrix} x_1 \\ x_2 \end{pmatrix} = \begin{pmatrix} b_1 \\ b_2 \end{pmatrix}.$$

   a. Show how to obtain a system (called the *reduced system*) that involves the variable x_2 only.

   b. Show that this reduced system is also a sparse matrix. Show the stencil associated with the reduced system matrix on the original finite difference mesh and give a graph theory interpretation of the reduction process. What is the maximum number of nonzero elements in each row of the reduced system?

4. It was stated in Section 10.3.2 that for some specific matrices the ILU(0) factorization of A can be put in the form

$$M = (D - E)D^{-1}(D - F),$$

in which $-E$ and $-F$ are the strict lower and upper parts of A, respectively.

   a. Characterize these matrices carefully and give an interpretation with respect to their adjacency graphs.

   b. Verify that this is true for standard five-point matrices associated with any domain Ω.

   c. Is it true for nine-point matrices?

   d. Is it true for the higher level ILU factorizations?

5. Let A be a pentadiagonal matrix having diagonals in offset positions $-m, -1, 0, 1, m$. The coefficients in these diagonals are all constants: a for the main diagonal and -1 for all others. It is assumed that $a \geq \sqrt{8}$. Consider the ILU(0) factorization of A as given in the form (10.20). The elements d_i of the diagonal D are determined by a recurrence of the form (10.19).

   a. Show that $\frac{a}{2} < d_i \leq a$ for $i = 1, \ldots, n$.

   b. Show that d_i is a decreasing sequence. [Hint: Use induction.]

   c. Prove that the formal (infinite) sequence defined by the recurrence converges. What is its limit?

6. Consider a matrix A that is split in the form $A = D_0 - E - F$, where D_0 is a block diagonal matrix whose block diagonal entries are the same as those of A and where $-E$ is strictly lower triangular and $-F$ is strictly upper triangular. In some cases the block form of the ILU(0) factorization can be put in the form (Section 10.3.2)

$$M = (D - E)D^{-1}(D - F).$$

The block entries of D can be defined by a simple matrix recurrence. Find this recurrence relation. The algorithm may be expressed in terms of the block entries of the matrix A.

Exercises

7. Generalize the formulas developed at the end of Section 10.7.1, for the inverses of symmetric tridiagonal matrices, to the nonsymmetric case.

8. Develop recurrence relations for IC with no fill-in (IC(0)) for five-point matrices, similar to those seen in Section 10.3.4 for ILU(0). Repeat for IC(1).

9. What becomes of the formulas seen in Section 10.3.4 in the case of a seven-point matrix (for three-dimensional problems)? In particular, can the ILU(0) factorization be cast in the form (10.20) in which $-E$ is the strict lower diagonal of A, $-F$ is the strict upper triangular part of A, and D is a certain diagonal?

10. Consider an arbitrary matrix A that is split in the usual manner as $A = D_0 - E - F$, in which $-E$ and $-F$ are the strict lower and upper parts of A, respectively, and define, for any diagonal matrix D, the approximate factorization of A given by

    $$M = (D - E)D^{-1}(D - F).$$

    Show how a diagonal D can be determined such that A and M have the same diagonal elements. Find a recurrence relation for the elements of D. Consider now the symmetric case and assume that the matrix D that is positive can be found. Write M in the form

    $$M = (D^{1/2} - ED^{-1/2})(D^{1/2} - ED^{-1/2})^T \equiv L_1 L_1^T.$$

    What is the relation between this matrix and the matrix of the SSOR(ω) preconditioning, in the particular case when $D^{-1/2} = \omega I$? Conclude that this form of ILU factorization is in effect an SSOR preconditioning with a different relaxation factor ω for each equation.

11. Consider a general sparse matrix A (irregularly structured). We seek an approximate LU factorization of the form

    $$M = (D - E)D^{-1}(D - F),$$

    in which $-E$ and $-F$ are the strict lower and upper parts of A, respectively. It is assumed that A is such that

    $$a_{ii} > 0, \quad a_{ij}a_{ji} \geq 0 \quad \text{for} \quad i, j = 1, \ldots, n.$$

    a. By identifying the diagonal elements of A with those of M, derive an algorithm for generating the elements of the diagonal matrix D recursively.

    b. Establish that if $d_j > 0$ for $j < i$ then $d_i \leq a_{ii}$. Is it true in general that $d_j > 0$ for all j?

    c. Assume that for $i = 1, \ldots, j-1$ we have $d_i \geq \alpha > 0$. Show a sufficient condition under which $d_j \geq \alpha$. Are there cases in which this condition cannot be satisfied for any α?

    d. Assume now that all diagonal elements of A are equal to a constant; i.e., $a_{jj} = a$ for $j = 1, \ldots, n$. Define $\alpha \equiv \frac{a}{2}$ and let

    $$S_j \equiv \sum_{i=1}^{j-1} a_{ij}a_{ji}, \quad \sigma \equiv \max_{j=1,\ldots,n} S_j.$$

    Show a condition on σ under which $d_j \geq \alpha$, $j = 1, 2, \ldots, n$.

12. Show the second part of (10.81). [Hint: Exploit the formula $AB^T = \sum_{j=1}^n a_j b_j^T$, where a_j, b_j are the jth columns of A and B, respectively.]

13. Let a preconditioning matrix M be related to the original matrix A by $M = A + E$, in which E is a matrix of rank k.

    a. Assume that both A and M are SPD. How many steps at most are required for the preconditioned CG method to converge when M is used as a preconditioner?

    b. Answer the same question for the case when A and M are nonsymmetric and the full GMRES is used on the preconditioned system.

14. Formulate the problem for finding an approximate inverse M to a matrix A as a large $n^2 \times n^2$ linear system. What is the Frobenius norm in the space in which you formulated this problem?

15. The concept of *mask* is useful in the global iteration technique. For a sparsity pattern S, i.e., a set of pairs (i, j) and a matrix B, we define the product $C = B \odot S$ to be the matrix whose elements c_{ij} are zero if (i, j) does not belong to S and b_{ij} otherwise. This is called a mask operation since its effect is to ignore every value not in the pattern S. Consider a global minimization of the function $F_S(M) \equiv \|S \odot (I - AM)\|_F$.

    a. What does the result of Proposition 10.9 become for this new objective function?

    b. Formulate an algorithm based on a global masked iteration, in which the mask is fixed and equal to the pattern of A.

    c. Formulate an algorithm in which the mask is adapted at each outer step. What criteria would you use to select the mask?

16. Consider the global self-preconditioned MR iteration algorithm seen in Section 10.5.5. Define the acute angle between two matrices as

    $$\cos \angle(X, Y) \equiv \frac{\langle X, Y \rangle}{\|X\|_F \|Y\|_F}.$$

    a. Following what was done for the (standard) MR algorithm seen in Chapter 5, establish that the matrices $B_k = AM_k$ and $R_k = I - B_k$ produced by global MR without dropping are such that

    $$\|R_{k+1}\|_F \leq \|R_k\|_F \sin \angle(R_k, B_k R_k).$$

    b. Let now $M_0 = \alpha A^T$ so that B_k is symmetric for all k (see Section 10.5.5). Assume that, at a given step k, the matrix B_k is positive definite. Show that

    $$\cos \angle(R_k, B_k R_k) \geq \frac{\lambda_{min}(B_k)}{\lambda_{max}(B_k)},$$

    in which $\lambda_{min}(B_k)$ and $\lambda_{max}(B_k)$ are, respectively, the smallest and largest eigenvalues of B_k.

17. In the two-sided version of AINV preconditioners, the option of minimizing

    $$f(L, U) = \|I - LAU\|_F^2$$

    was mentioned, where L is unit lower triangular and U is upper triangular.

a. What is the gradient of $f(L, U)$?

b. Formulate an algorithm based on minimizing this function globally.

18. Consider the two-sided version of AINV preconditioners, in which a unit lower triangular L and an upper triangular U are sought so that $LAU \approx I$. One idea is to use an alternating procedure in which the first half-step computes a right AINV U to LA, which is restricted to be upper triangular, and the second half-step computes a left AINV L to AU, which is restricted to be lower triangular.

    a. Consider the first half-step. Since the candidate matrix U is restricted to be upper triangular, special care must be exercised when writing a column-oriented AINV algorithm. What are the differences with the standard MR approach described by Algorithm 10.12?

    b. Now consider seeking an upper triangular matrix U such that the matrix $(LA)U$ is close to the identity only in its upper triangular part. A similar approach is to be taken for the second half-step. Formulate an algorithm based on this approach.

19. Write all six variants of the preconditioned CG algorithm applied to the normal equations mentioned at the end of Section 10.8.1.

20. With the standard splitting $A = D - E - F$, in which D is the diagonal of A and $-E, -F$ its lower and upper triangular parts, respectively, we associate the factored factorization

$$(I + ED^{-1})A(I + D^{-1}F) = D + R. \quad (10.85)$$

    a. Determine R and show that it consists of second order terms, i.e., terms involving products of at least two matrices from the pair E, F.

    b. Now use the previous approximation for $D + R \equiv D_1 - E_1 - F_1$,

    $$(I + E_1 D_1^{-1})(D + R)(I + D_1^{-1} F_1) = D_1 + R_1.$$

    Show how the AINV factorization (10.85) can be improved using this new approximation. What is the order of the resulting approximation?

Notes and References

The idea of transforming a linear system into one that is easier to solve by iterations was known quite early on. In a 1937 paper, Cesari [71] proposed what is now known as polynomial preconditioning (see also [43, p. 156], where this is discussed). Other forms of preconditioning were also exploited in some earlier papers. For example, in [11] Axelsson discusses SSOR iteration, *accelerated* by either the CG or Chebyshev acceleration. Incomplete factorizations were also discussed quite early, for example, by Varga [291] and Buleev [68]. The breakthrough article by Meijerink and van der Vorst [207] established existence of the incomplete factorization for M-matrices and showed that preconditioning the CG by using an incomplete factorization can result in an extremely efficient combination. This article played an essential role in directing the attention of researchers and practitioners to a rather important topic and marked a turning point. Many of the early techniques were developed for regularly structured matrices. The generalization, using the definition of level of fill for high order ILU factorizations for unstructured matrices, was introduced by Watts [305] for petroleum engineering problems.

Recent research on iterative techniques has focused on preconditioning methods, while the importance of accelerators has diminished. Preconditioners are essential to the success of iterative methods in real-life applications. A general preconditioning approach based on modifying a given direct solver by including dropping was one of the first general-purpose approaches that was proposed [211, 220, 324, 137]. More economical alternatives, akin to ILU(p), were developed later [248, 97, 96, 313, 322, 244]. ILUT and ILUTP are relatively robust and efficient but they can nonetheless fail. Instances can also be encountered when a more accurate ILUT factorization leads to a larger number of steps to converge. One source of failure is the instability of the preconditioning operation. These phenomena of instability have been studied by Elman [116], who proposed a detailed analysis of ILU and MILU preconditioners for model problems. The theoretical analysis on ILUT stated as Theorem 10.8 is modeled after Theorem 1.14 in Axelsson and Barker [15] for ILU(0).

Some theory for block preconditioners is discussed in the book by O. Axelsson [14]. Different forms of block preconditioners were developed independently by Axelsson, Brinkkemper, and Il'in [16] and by Concus, Golub, and Meurant [89], initially for block matrices arising from PDEs in two dimensions. Later, some generalizations were proposed by Kolotilina and Yeremin [190]. Thus, the two-level implicit-explicit preconditioning introduced in [190] consists of using sparse inverse approximations to Δ_i^{-1} for obtaining Ω_i.

The rebirth of AINV preconditioners [158, 91, 190, 159, 34, 157, 33, 80, 78] has been spurred both by considerations related to parallel processing and the relative ineffectiveness of standard ILU preconditioners in dealing with highly indefinite matrices. Other preconditioners that are not covered here are those based on domain decomposition techniques. Some of these techniques will be reviewed in Chapter 14.

The primary motivation for ILUC is the overhead in ILUT due to the search for the leftmost pivot. The idea of exploiting condition number estimators in this context has been motivated by compelling results in Bollhöfer's work [44].

The effect of reordering on incomplete factorizations has been a subject of debate among researchers, but the problem is still not well understood. What experience shows is that some of the better reordering techniques used for sparse direct solution methods do not necessarily perform well for ILU [35, 64, 111, 112, 96, 97, 264]. As could be expected when fill-in is increased to high levels, the effect of reordering starts resembling that of direct solvers. A rule of thumb is that the RCM ordering does quite well on average. It appears that orderings that take into account the values of the matrix can perform better, but these may be expensive [87, 96, 97]. The use of nonsymmetric orderings as a means of enhancing robustness of ILU has been proposed in recent articles by Duff and Koster [109, 110]. The algorithms developed in this context are rather complex but lead to remarkable improvements, especially for matrices with very irregular patterns.

The saddle-point problem is a classic example of what can be achieved by a preconditioner developed by exploiting the physics versus a general-purpose preconditioner. An ILUT factorization for the saddle-point problem may work if a high level of fill is used. However, this usually results in poor performance. A better performance can be obtained by exploiting information about the original problem; see, for example, [152, 304, 263, 118, 119].

Notes and References

On another front, there is also some interest in methods that utilize normal equations in one way or another. Earlier, ideas revolved around shifting the matrix $B = A^T A$ before applying the IC(0) factorization, as was suggested by Kershaw [186] in 1978. Manteuffel [205] also made some suggestions on how to select a good α in the context of the Concus, Golub, and Widlund algorithm. Currently, new ways of exploiting the relationship with the QR (or LQ) factorization to define IC(0) more rigorously are being explored; see the work in [303]. Preconditioning normal equations remains a difficult problem.

Chapter 11
Parallel Implementations

Parallel computing has recently gained widespread acceptance as a means of handling very large computational tasks. Since iterative methods are appealing for large linear systems of equations, it is no surprise that they are the prime candidates for implementations on parallel architectures. There have been two traditional approaches for developing parallel iterative techniques thus far. The first extracts parallelism whenever possible from standard algorithms. The advantage of this viewpoint is that it is easier to understand in general since the underlying method has not changed from its sequential equivalent. The second approach is to develop alternative algorithms that have enhanced parallelism. This chapter will give an overview of implementations and will emphasize methods in the first category. The later chapters will consider alternative algorithms that have been developed specifically for parallel computing environments.

11.1 Introduction

Because of the increased importance of three-dimensional models and the high cost associated with sparse direct methods for solving these problems, iterative techniques play a major role in application areas. The main appeal of iterative methods is their low storage requirement. Another advantage is that they are far easier to implement on parallel computers than sparse direct methods because they only require a rather small set of computational kernels. Increasingly, direct solvers are being used in conjunction with iterative solvers to develop robust preconditioners.

The first considerations for high performance implementations of iterative methods involved implementations on vector computers. These efforts started in the mid-1970s, when the first vector computers appeared. Currently, there is a larger effort to develop new practical iterative methods that are not only efficient in a parallel environment, but also robust. Often, however, these two requirements seem to be in conflict.

This chapter begins with a short overview of the various ways in which parallelism has been exploited in the past and a description of the current architectural models for existing commercial parallel computers. Then the basic computations required in Krylov subspace methods will be discussed along with their implementations.

11.2 Forms of Parallelism

Parallelism has been exploited in a number of different forms since the first computers were built. The six major forms of parallelism are (1) multiple functional units, (2) pipelining, (3) vector processing, (4) multiple vector pipelining, (5) multiprocessing, and (6) distributed computing. Next is a brief description of each of these approaches.

11.2.1 Multiple Functional Units

This is one of the earliest forms of parallelism. It consists of multiplying the number of functional units, such as adders and multipliers. Thus, the control units and the registers are shared by the functional units. The detection of parallelism is done at compilation time with a *dependence analysis graph*, an example of which is shown in Figure 11.1.

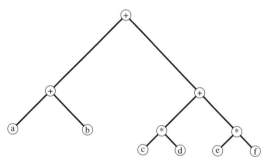

Figure 11.1. *Dependence analysis for arithmetic expression* $(a + b) + (c * d + d * e)$.

In the example of Figure 11.1, the two multiplications can be performed simultaneously, then the two additions in the middle are performed simultaneously. Finally, the addition at the root is performed.

11.2.2 Pipelining

The pipelining concept is essentially the same as that of an assembly line used in car manufacturing. Assume that an operation takes s stages to complete. Then the operands can be passed through the s stages instead of waiting for all stages to be completed for the first two operands.

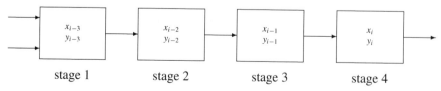

11.3. Types of Parallel Architectures

If each stage takes a time τ to complete, then an operation with n numbers will take the time $s\tau + (n-1)\tau = (n+s-1)\tau$. The speed-up would be the ratio of the time to complete the s stages in a nonpipelined unit versus, i.e., $s \times n \times \tau$, to the above obtained time:

$$S = \frac{ns}{n+s-1}.$$

For large n, this would be close to s.

11.2.3 Vector Processors

Vector computers appeared in the beginning of the 1970s with the CDC Star 100 and then the CRAY-1 and Cyber 205. These computers are equipped with vector pipelines, i.e., pipelined functional units, such as pipelined floating-point adders and multipliers. In addition, they incorporate vector instructions explicitly as part of their instruction sets. Typical vector instructions are, for example:

> VLOAD   To load a vector from memory to a vector register.
> VADD   To add the content of two vector registers.
> VMUL   To multiply the content of two vector registers.

Similarly to the case of multiple functional units for scalar machines, vector pipelines can be duplicated to take advantage of any fine-grain parallelism available in loops. For example, the Fujitsu and NEC computers tend to obtain a substantial portion of their performance in this fashion. There are many vector operations that can take advantage of *multiple vector pipelines*.

11.2.4 Multiprocessing and Distributed Computing

A multiprocessor system is a computer, or a set of several computers, consisting of several processing elements (PEs), each consisting of a CPU, a memory, an I/O subsystem, etc. These PEs are connected to one another with some communication medium, either a bus or some multistage network. There are numerous possible configurations, some of which will be covered in the next section.

Distributed computing is a more general form of multiprocessing, in which the processors are actually computers linked by some local area network. Currently, there are a number of libraries that offer communication mechanisms for exchanging messages between Unix-based systems. The best known of these are the parallel virtual machine (PVM) and the message passing interface (MPI). In heterogeneous networks of computers, the processors are separated by relatively large distances, which has a negative impact on the performance of distributed applications. In fact, this approach is cost effective only for large applications, in which a high volume of computation can be performed before more data are to be exchanged.

11.3 Types of Parallel Architectures

There are currently three leading architecture models. These are
- the shared memory model,
- single instruction multiple data (SIMD) or data parallel models, and
- the distributed memory message-passing models.

A brief overview of the characteristics of each of the three groups follows. Emphasis is on the possible effects these characteristics have on the implementations of iterative methods.

11.3.1 Shared Memory Computers

A shared memory computer has the processors connected to a large global memory with the same global view, meaning the address space is the same for all processors. One of the main benefits of shared memory models is that access to data depends very little on their location in memory. In a shared memory environment, transparent data access facilitates programming to a great extent. From the user's point of view, data are stored in a large global memory that is readily accessible to any processor. However, memory conflicts as well as the necessity to maintain data coherence can lead to degraded performance. In addition, shared memory computers cannot easily take advantage of data locality in problems that have an intrinsically local nature, as is the case with most discretized partial differential equations (PDEs). Some current machines have a physically distributed memory but they are logically shared; i.e., each processor has the same view of the global address space. There are two possible implementations of shared memory machines: (1) bus-based architectures and (2) switch-based architectures. These two model architectures are illustrated in Figure 11.2 and Figure 11.3, respectively. So far, shared memory computers have been implemented more often with buses than with switching networks.

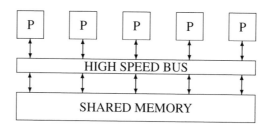

Figure 11.2. *A bus-based shared memory computer.*

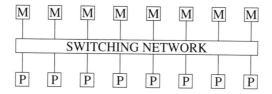

Figure 11.3. *A switch-based shared memory computer.*

Buses are the backbone for communication between the different units of most computers. Physically, a bus is nothing but a bundle of wires, made of either fiber or copper. These wires carry information consisting of data, control signals, and error correction bits. The speed of a bus, often measured in megabytes per second and called the *bandwidth* of the bus, is determined by the number of lines in the bus and the clock rate. Often, the limiting factor for parallel computers based on bus architectures is the bus bandwidth rather than the CPU speed.

11.3. Types of Parallel Architectures

The primary reason why bus-based multiprocessors are more common than switch-based ones is that the hardware involved in such implementations is simple. On the other hand, the difficulty with bus-based machines is that the number of processors that can be connected to the memory is small in general. Typically, the bus is time-shared, meaning slices of time are allocated to the different clients (processors, I/O processors, etc.) that request its use.

In a multiprocessor environment, the bus can easily be saturated. Several remedies are possible. The first, and most common, remedy is to attempt to reduce traffic by adding *local memories* or *caches* attached to each processor. Since a data item used by a given processor is likely to be reused by the same processor in the next instructions, storing the data item in local memory will help reduce traffic in general. However, this strategy causes some difficulties due to the requirement to maintain *data coherence*. If processor A reads some data from the shared memory and processor B modifies the same data in shared memory, immediately after, the result is two copies of the same data that have different values. A mechanism should be put in place to ensure that the most recent update of the data is always used. The additional overhead incurred by such memory coherence operations may well offset the savings involving memory traffic.

The main features here are the switching network and the fact that a global memory is shared by all processors through the switch. There can be p processors on one side connected to p memory units or banks on the other side. Alternative designs based on switches connect p processors to each other instead of p memory banks. The switching network can be a crossbar switch when the number of processors is small. A crossbar switch is analogous to a telephone switchboard and allows p inputs to be connected to m outputs without conflict. Since crossbar switches for large numbers of processors are typically expensive they are replaced with multistage networks. Signals travel across a small number of stages consisting of an array of elementary switches, e.g., 2×2 or 4×4 switches.

There have been two ways of exploiting multistage networks. In *circuit-switching* networks, the elementary switches are set up by sending electronic signals across all of the switches. The circuit is set up once in much the same way that telephone circuits are switched in a switchboard. Once the switch has been set up, communication between processors $P_1, \ldots, P_n$ is open to the memories

$$M_{\pi_1}, M_{\pi_2}, \ldots, M_{\pi_n},$$

in which π represents the desired permutation. This communication will remain functional for as long as it is not reset. Setting up the switch can be costly, but once it is done, communication can be quite fast. In *packet-switching* networks, a packet of data will be given an address token and the switching within the different stages will be determined based on this address token. The elementary switches have to provide for buffering capabilities, since messages may have to be queued at different stages.

11.3.2 Distributed Memory Architectures

The *distributed memory* model refers to the distributed memory *message-passing* architectures as well as to distributed memory SIMD computers. A typical distributed memory system consists of a large number of identical processors with their own memories, which are interconnected in a regular topology. Examples are depicted in Figure 11.4. In these diagrams, each processor unit can be viewed actually as a complete processor with its own memory, CPU, I/O subsystem, control unit, etc. These processors are linked to a number of "neighboring" processors, which in turn are linked to other neighboring processors, etc. In message-passing models there is no global synchronization of the parallel tasks. Instead, computations are *data driven* because a processor performs a given task only when the operands it requires become available. The programmer must program all the data exchanges explicitly between processors.

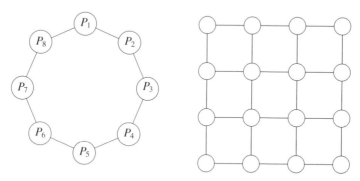

Figure 11.4. *An eight-processor ring (left) and a 4 × 4 multiprocessor mesh (right).*

In SIMD designs, a different approach is used. A host processor stores the program and each slave processor holds different data. The host then broadcasts instructions to processors, which execute them simultaneously. One advantage of this approach is that there is no need for large memories in each node to store large programs, since the instructions are broadcast one by one to all processors.

Distributed memory computers can exploit locality of data in order to keep communication costs to a minimum. Thus, a two-dimensional processor grid such as the one depicted in Figure 11.4 is perfectly suitable for solving discretized elliptic PDEs (e.g., by assigning each grid point to a corresponding processor) because some iterative methods for solving the resulting linear systems will require only interchange of data between adjacent grid points. A good general-purpose multiprocessor must have powerful *mapping capabilities* because it should be capable of easily emulating many of the common topologies, such as two- and three-dimensional grids or linear arrays, fast Fourier transform butterflies, finite element meshes, etc.

Three-dimensional configurations have also been popular. Hypercubes are highly concurrent multiprocessors based on the binary n-cube topology, which is well known for its rich interconnection capabilities. A parallel processor based on the n-cube topology, called a *hypercube* hereafter, consists of 2^n identical processors interconnected with n neighbors. A 3-cube can be represented as an ordinary cube in three dimensions, where the vertices are the $8 = 2^3$ nodes of the 3-cube; see Figure 11.5. More generally, one can construct an n-cube as follows: First, the 2^n nodes are labeled with the 2^n binary numbers from 0 to $2^n - 1$. Then a link between two nodes is drawn iff their binary numbers differ by one (and only one) bit.

An n-cube graph can be constructed recursively from lower dimensional cubes. More precisely, consider two identical $(n-1)$-cubes whose vertices are labeled likewise from 0 to 2^{n-1}. By joining every vertex of the first $(n-1)$-cube to the vertex of the second having

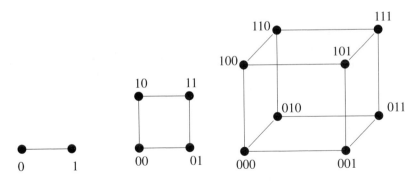

Figure 11.5. *The n-cubes of dimensions* $n = 1, 2, 3$.

the same number, one obtains an n-cube. Indeed, it suffices to renumber the nodes of the first cube as $0 \wedge a_i$ and those of the second as $1 \wedge a_i$, where a_i is a binary number representing the two similar nodes of the $(n-1)$-cubes and where $\wedge$ denotes the concatenation of binary numbers.

Distributed memory computers come in two different designs, namely, SIMD and multiple instruction multiple data (MIMD). Many of the early projects have adopted the SIMD organization. For example, the historical ILLIAC IV Project of the University of Illinois was a machine based on a mesh topology where all processors execute the same instructions.

SIMD distributed processors are sometimes called array processors because of the regular arrays that they constitute. In this category, systolic arrays can be classified as an example of distributed computing. Systolic arrays, which were popular in the 1980s, are organized in connected cells, which are programmed (possibly microcoded) to perform only one of a few operations. All the cells are synchronized and perform the same task. Systolic arrays are designed in very large scale integration (VLSI) technology and are meant to be used for special-purpose applications, primarily in signal processing.

In the last few years, parallel computing technologies have seen a healthy maturation. Currently, the architecture of choice is the distributed memory machine using message passing. There is no doubt that this is due to the availability of excellent communication software, such as MPI; see [156]. In addition, the topology is often hidden from the user, so there is no need to code communication on specific configurations such as hypercubes. Since this mode of computing has penetrated the applications areas and industrial applications, it is likely to remain for some time.

11.4 Types of Operations

Now consider two prototype Krylov subspace techniques, namely, the preconditioned conjugate gradient (PCG) method for the symmetric case and the preconditioned generalized minimal residual (GMRES) algorithm for the nonsymmetric case. It should be emphasized that all Krylov subspace techniques require the same basic operations.

Consider Algorithm 9.1. The first step when implementing this algorithm on a high performance computer is identifying the main operations that it requires. We distinguish five types of operations, which are (1) preconditioner setup, (2) matrix-by-vector multiplications, (3) vector updates, (4) dot products, and (5) preconditioning operations. In this list the potential bottlenecks are (1), setting up the preconditioner, and (5), solving linear systems with M, i.e., the preconditioning operation. Section 11.6 discusses the implementation of traditional preconditioners and Chapters 12 and 13 are devoted to preconditioners that are specialized to parallel environments. Next come the matrix-by-vector products, which deserve particular attention. The rest of the algorithm consists essentially of dot products and vector updates, which do not cause significant difficulties in parallel machines, although inner products can lead to some loss of efficiency on certain types of computers with large numbers of processors.

If we now consider the GMRES algorithm, the only new operation here with respect to the CG method is the orthogonalization of the vector Av_i against the previous v's. The usual way to accomplish this is via the modified Gram–Schmidt (MGS) process, which is basically a sequence of subprocesses of the following form:

- Compute $\alpha = (y, v)$.
- Compute $\hat{y} := y - \alpha v$.

This orthogonalizes a vector y against another vector v of norm one. Thus, the outer loop of the MGS is sequential, but the inner loop, i.e., each subprocess, can be parallelized by dividing the inner product and SAXPY operations among processors. Although this constitutes a perfectly acceptable approach for a small number of processors, the elementary subtasks may be too small to be efficient on a large number of processors. An alternative for this case is to use a standard Gram–Schmidt process with reorthogonalization. This replaces the previous sequential orthogonalization process with a matrix operation of the form $\hat{y} = y - VV^T y$; i.e., basic linear algebra subprogram (BLAS)-1 kernels are replaced with BLAS-2 kernels.

Recall that the next level of BLAS, i.e., level-3 BLAS, exploits blocking in dense matrix operations in order to obtain performance on machines with hierarchical memories. Unfortunately, level-3 BLAS kernels cannot be exploited here because, at every step, there is only one vector to orthogonalize against all previous ones. This may be remedied by using block Krylov methods.

Vector operations, such as linear combinations of vectors and dot products, are usually the simplest to implement on any computer. In shared memory computers, compilers are capable of recognizing these operations and invoking the appropriate machine instructions, possibly vector instructions. We consider now these operations in turn.

Vector Updates. Operations of the form

$$y(1:n) = y(1:n) + a * x(1:n),$$

where a is a scalar and y and x two vectors, are known as *vector updates* or SAXPY operations. They are typically straightforward to implement in all three machine models discussed earlier. For example, the above FORTRAN-90 code segment can be used on most shared memory (symmetric multiprocessing) computers and the compiler will translate it into the proper parallel version.

On distributed memory computers, some assumptions must be made about the way in which the vectors are distributed. The main assumption is that the vectors x and y are distributed in the same manner among the processors, meaning the indices of the components of any vector that are mapped to a given processor are the same. In this case, the vector update operation will be translated into p independent vector updates, requiring no communication.

Specifically, if *nloc* is the number of variables local to a given processor, this processor will simply execute a vector loop of the form

$$y(1:nloc) = y(1:nloc) + a * x(1:nloc)$$

and all processors will execute a similar operation simultaneously.

Dot Products. A number of operations use all the components of a given vector to compute a single floating-point result, which is then needed by all processors. These are termed *reduction operations* and the dot product is the prototype example. A distributed version of the dot product is needed to compute the inner product of two vectors x and y that are distributed the same way across the processors. In fact, to be more specific, this distributed dot product operation should compute the inner product $t = x^T y$ of these two vectors and then make the result t available in each processor. Typically, this result is needed to perform vector updates or other operations in each node. For a large number of processors, this sort of operation can be demanding in terms of communication costs. On the other hand, parallel computer designers have become aware of their importance and are starting to provide hardware and software support for performing *global reduction operations* efficiently. Reduction operations that can be useful include global sums, global max-min calculations, etc. A commonly adopted convention provides a single subroutine for all these operations and passes the type of operation to be performed (add, max, min, multiply, ...) as one of the arguments. With this in mind, a distributed dot product function can be programmed roughly as follows (using C syntax):

```
tloc = DDOT(nrow, x, incx, y, incy);
MPI_Allreduce(&t, &tsum, 1, MPI_DOUBLE, MPI_SUM, comm);
```

The function DDOT performs the usual BLAS-1 dot product of x and y with strides *incx* and *incy*, respectively. The MPI_Allreduce operation, which is called with MPI_SUM as the operation-type parameter, sums all the variables *tloc* from each processor and puts the resulting global sum in the variable *tsum* in each processor.

11.5 Matrix-by-Vector Products

Matrix-by-vector multiplications (sometimes called *Matvecs* for short) are relatively easy to implement efficiently on high performance computers. For a description of storage formats for sparse matrices, see Chapter 3. We will first discuss matrix-by-vector algorithms without consideration of sparsity. Then we will cover sparse matvec operations for a few different storage formats.

The computational kernels for performing sparse matrix operations such as matrix-by-vector products are intimately associated with the data structures used. However, there are a few general approaches that are common to different algorithms for matrix-by-vector products that can be described for dense matrices. Two popular ways of performing these operations are the inner product form and the SAXPY form. In the inner product form for computing $y = Ax$, the component y_i is obtained as a dot product of the ith row of i and the vector x. The SAXPY form computes y as a linear combination of the columns of A, specifically as the sum of $x_i A_{:,i}$ for $i = 1, \ldots, n$. A third option consists of performing the product by diagonals. This option bears no interest in the dense case, but it is at the basis of many important matrix-by-vector algorithms in the sparse case, as will be seen shortly.

11.5.1 The CSR and CSC Formats

Recall that the compressed sparse row (CSR) data structure seen in Chapter 3 consists of three arrays: a real array $A(1:nnz)$ to store the nonzero elements of the matrix row-wise, an integer array $JA(1:nnz)$ to store the column positions of the elements in the real array A, and, finally, a pointer array $IA(1:n+1)$, the ith entry of which points to the beginning of the ith row in the arrays A and JA. To perform the matrix-by-vector product $y = Ax$ in parallel using this format, note that each component of the resulting vector y can be computed independently as the dot product of the ith row of the matrix with the vector x.

ALGORITHM 11.1. CSR Format—Dot Product Form

1. Do $i = 1, n$
2. $\quad k1 = ia(i)$
3. $\quad k2 = ia(i+1) - 1$
4. $\quad y(i) = dotproduct(a(k1:k2), x(ja(k1:k2)))$
5. EndDo

Line 4 computes the dot product of the vector with components $a(k1), a(k1+1), \ldots, a(k2)$ with the vector with components $x(ja(k1)), x(ja(k1+1)), \ldots, x(ja(k2))$.

The fact that the outer loop can be performed in parallel can be exploited on any parallel platform. On some shared memory machines, the synchronization of this outer loop is inexpensive and the performance of the above program can be excellent. On distributed memory machines, the outer loop can be split into a number of steps to be executed on each processor. Thus, each processor will handle a few rows that are assigned to it. It is common to assign a certain number of rows (often contiguous) to each processor and to also assign the component of each of the vectors similarly. The part of the matrix that is needed is loaded in each processor initially. When performing a matrix-by-vector product, interprocessor communication will be necessary to get the needed components of the vector x that do not reside in a given processor. We will return to this important case in Section 11.5.5.

The indirect addressing involved in the second vector in the dot product is called a *gather* operation. The vector $(ja(k1:k2))$ is first gathered from memory into a vector of contiguous elements. The dot product is then carried out as a standard dot product operation between two dense vectors. This is illustrated in Figure 11.6.

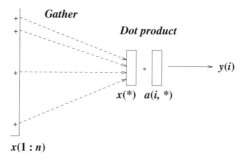

Figure 11.6. *Illustration of the row-oriented matrix-by-vector multiplication.*

11.5. Matrix-by-Vector Products

Now assume that the matrix is stored by columns (compressed sparse columns (CSC) format). The matrix-by-vector product can be performed by the following algorithm.

ALGORITHM 11.2. CSC Format—SAXPY Form

1. $y(1:n) = 0.0$
2. $Do\ i = 1, n$
3. $k1 = ia(i)$
4. $k2 = ia(i+1) - 1$
5. $y(ja(k1:k2)) = y(ja(k1:k2)) + x(j) * a(k1:k2)$
6. EndDo

The above code initializes y to zero and then adds the vectors $x(j) \times a(1:n, j)$ for $j = 1, \ldots, n$ to it. It can also be used to compute the product of the *transpose* of a matrix with a vector, when the matrix is stored (row-wise) in the CSR format. Normally, the vector $y(ja(k1:k2))$ is gathered and the SAXPY operation is performed in vector mode. Then the resulting vector is scattered back into the positions $ja(*)$ by what is called a *scatter* operation. This is illustrated in Figure 11.7.

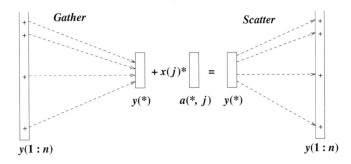

Figure 11.7. *Illustration of the column-oriented matrix-by-vector multiplication.*

A major difficulty with the above algorithm is that it is intrinsically sequential. First, the outer loop is not parallelizable as it is, but this may be remedied, as will be seen shortly. Second, the inner loop involves writing back results of the right-hand side into memory positions that are determined by the indirect address function ja. To be correct, $y(ja(1))$ must be copied first, followed by $y(ja(2))$, etc. However, if it is known that the mapping $ja(i)$ is one-to-one, then the order of the assignments no longer matters. Since compilers are not capable of deciding whether this is the case, a compiler directive from the user is necessary for the scatter to be invoked.

Going back to the outer loop, p subsums can be computed (independently) into p separate temporary vectors. Once all the p separate subsums are completed, then these p temporary vectors can be added to obtain the final result. Note that the final sum incurs some additional work but it is highly vectorizable and parallelizable.

11.5.2 Matvecs in the Diagonal Format

The *diagonal storage format* was one of the first data structures used in the context of high performance computing to take advantage of special sparse structures. Often, sparse matrices consist of a small number of diagonals, in which case the matrix-by-vector product can be performed by diagonals. There are again different variants of matvec algorithms for the diagonal format, related to different orderings of the loops in the basic FORTRAN program. Recall that the matrix is stored in a rectangular array $diag(1:n, 1:ndiag)$ and the offsets of these diagonals from the main diagonal may be stored in a small integer array $offset(1:ndiag)$. Consider a dot product variant first.

ALGORITHM 11.3. Diagonal Format—Dot Product Form

1. Do $i = 1, n$
2. $tmp = 0.0d0$
3.   Do $j = 1, ndiag$
4. $tmp = tmp + diag(i, j) * x(i + offset(j))$
5.   EndDo
6. $y(i) = tmp$
7. EndDo

In a second variant, the vector y is initialized to zero, and then x is multiplied by each of the diagonals and the separate results are added to y. The innermost loop in this computation is sometimes called a *triad* operation.

ALGORITHM 11.4. Matvec in Triad Form

1. $y = 0.0d0$
2. Do $j = 1, ndiag$
3. $joff = offset(j)$
4. $i1 = max(1, 1 - offset(j))$
5. $i2 = min(n, n - offset(j))$
6. $y(i1:i2) = y(i1:i2) + diag(i1:i2, j) * x(i1 + joff : i2 + joff)$
7. EndDo

Good speeds can be reached on vector machines for large enough matrices. A drawback with diagonal schemes is that they are not general. For general sparse matrices, we can either generalize the diagonal storage scheme or reorder the matrix in order to obtain a diagonal structure. The simplest generalization is the Ellpack-Itpack format.

11.5.3 The Ellpack-Itpack Format

The Ellpack-Itpack (or Ellpack) format is of interest only for matrices whose maximum number of nonzeros per row, *jmax*, is small. The nonzero entries are stored in a real array $ae(1:n, 1:jmax)$. Along with this is the integer array $jae(1:n, 1:jmax)$, which stores the column indices of each corresponding entry in *ae*. Similarly to the diagonal scheme, there are also two basic ways of implementing a matrix-by-vector product when using the Ellpack format. We begin with an analogue of Algorithm 11.3.

11.5. Matrix-by-Vector Products

ALGORITHM 11.5. Ellpack Format—Dot Product Form

1. $Do\ i = 1, n$
2. $\quad y_i = 0$
3. $\quad Do\ j = 1, ncol$
4. $\quad\quad y_i = y_i + ae(i, j) * x(jae(i, j))$
5. $\quad EndDo$
6. $\quad y(i) = y_i$
7. $EndDo$

If the number of nonzero elements per row varies substantially, many zero elements must be stored unnecessarily. Then the scheme becomes inefficient. As an extreme example, if all rows are very sparse except for one of them, which is full, then the arrays *ae, jae* must be full $n \times n$ arrays, containing mostly zeros. This is remedied by a variant of the format called the *jagged diagonal (JAD) format*.

11.5.4 The JAD Format

The JAD format can be viewed as a generalization of the Ellpack-Itpack format, which removes the assumption on the fixed-length rows. To build the JAD structure, start from the CSR data structure and sort the rows of the matrix by decreasing number of nonzero elements. To build the first JAD extract the first element from each row of the CSR data structure. The second JAD consists of the second elements of each row in the CSR data structure. The third, fourth, ..., JADs can then be extracted in the same fashion. The lengths of the successive JADs decrease. The number of JADs that can be extracted is equal to the number of nonzero elements of the first row of the permuted matrix, i.e., to the largest number of nonzero elements per row. To store this data structure, three arrays are needed: a real array *DJ* to store the values of the JADs, the associated array *JDIAG*, which stores the column positions of these values, and a pointer array *IDIAG*, which points to the beginning of each JAD in the *DJ, JDIAG* arrays.

Example 11.1. Consider the following matrix and its sorted version PA:

$$A = \begin{pmatrix} 1. & 0. & 2. & 0. & 0. \\ 3. & 4. & 0. & 5. & 0. \\ 0. & 6. & 7. & 0. & 8. \\ 0. & 0. & 9. & 10. & 0. \\ 0. & 0. & 0. & 11. & 12. \end{pmatrix} \rightarrow PA = \begin{pmatrix} 3. & 4. & 0. & 5. & 0. \\ 0. & 6. & 7. & 0. & 8. \\ 1. & 0. & 2. & 0. & 0. \\ 0. & 0. & 9. & 10. & 0. \\ 0. & 0. & 0. & 11. & 12. \end{pmatrix}.$$

The rows of PA have been obtained from those of A by sorting them by number of nonzero elements, from the largest to the smallest number. Then the JAD data structure for A is as follows:

DJ	3.	6.	1.	9.	11.	4.	7.	2.	10.	12.	5.	8.
JDIAG	1	2	1	3	4	2	3	3	4	5	4	5
IDIAG	1	6	11	13								

Thus, there are two JADs of full length (five) and one of length two.

A matrix-by-vector product with this storage scheme can be performed by the following code segment.

1. $Do\ j = 1, ndiag$
2. $\quad k1 = idiag(j)$
3. $\quad k2 = idiag(j+1) - 1$
4. $\quad len = idiag(j+1) - k1$
5. $\quad y(1:len) = y(1:len) + dj(k1:k2) * x(jdiag(k1:k2))$
6. $EndDo$

Since the rows of the matrix A have been permuted, the above code will compute PAx, a permutation of the vector Ax, rather than the desired Ax. It is possible to permute the result back to the original ordering after the execution of the above program. This operation can also be performed until the final solution has been computed, so that only two permutations of the solution vector are needed, one at the beginning and one at the end. For preconditioning operations, it may be necessary to perform a permutation before or within each call to the preconditioning subroutines. There are many possible variants of the JAD format. One variant, which does not require permuting the rows, is described in Exercise 8.

11.5.5 The Case of Distributed Sparse Matrices

Given a sparse linear system to be solved in a distributed memory environment, it is natural to map pairs of equations/unknowns to the same processor in a certain predetermined way. This mapping can be determined automatically by a graph partitioner or it can be assigned ad hoc from knowledge of the problem. Without any loss of generality, the matrix under consideration can be viewed as originating from the discretization of a PDE on a certain domain. This is illustrated in Figure 11.8. Assume that each subgraph (or subdomain, in the PDE literature) is assigned to a different processor, although this restriction can be relaxed; i.e., a processor can hold several subgraphs to increase parallelism.

A local data structure must be set up in each processor (or subdomain, or subgraph) to allow the basic operations, such as (global) matrix-by-vector products and preconditioning operations, to be performed efficiently. The only assumption to make regarding the mapping is that, if row number i is mapped into processor p, then so is the unknown i; i.e., the matrix is distributed row-wise across the processors according to the distribution of the variables. The graph is assumed to be undirected; i.e., the matrix has a symmetric pattern.

It is important to *preprocess the data* in order to facilitate the implementation of the communication tasks and to gain efficiency during the iterative process. The preprocessing requires setting up the following information *in each processor*.

1. A list of processors with which communication will take place. These are called *neighboring processors* although they may not be physically nearest neighbors.

2. A list of local nodes that are coupled with external nodes. These are the *local interface nodes*.

3. A local representation of the distributed matrix in each processor.

11.5. Matrix-by-Vector Products

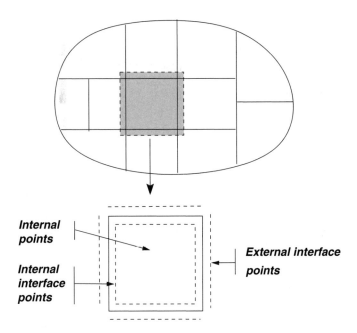

Figure 11.8. *Decomposition of physical domain or adjacency graph and the local data structure.*

To perform a matrix-by-vector product with the global matrix A, the matrix consisting of rows that are local to a given processor must be multiplied by some global vector v. Some components of this vector will be local and some components must be brought from external processors. These external variables correspond to interface points belonging to adjacent subdomains. When performing a matrix-by-vector product, neighboring processors must exchange values of their adjacent interface nodes.

Let A_{loc} be the local part of the matrix, i.e., the (rectangular) matrix consisting of all the rows that are mapped to *myproc*. Call A_{loc} the *diagonal block* of A located in A_{loc}, i.e., the submatrix of A_{loc} whose nonzero elements a_{ij} are such that j is a local variable. Similarly, call B_{ext} the *off-diagonal* block, i.e., the submatrix of A_{loc} whose nonzero elements a_{ij} are such that j is *not* a local variable (see Figure 11.9). To perform a matrix-by-vector product, start multiplying the diagonal block A_{loc} by the local variables. Then, multiply the external variables by the sparse matrix B_{ext}. Notice that, since the external interface points are not coupled with local internal points, only the rows $n_{int} + 1$ to n_{nloc} in the matrix B_{ext} will have nonzero elements.

Thus, the matrix-by-vector product can be separated into two such operations, one involving only the local variables and the other involving external variables. It is necessary to construct these two matrices and define a local numbering of the local variables in order to perform the two matrix-by-vector products efficiently each time.

To perform a global matrix-by-vector product, with the distributed data structure described above, each processor must perform the following operations. First, multiply the local variables by the matrix A_{loc}. Second, obtain the external variables from the neighboring processors in a certain order. Third, multiply these by the matrix B_{ext} and add the resulting vector to the one obtained from the first multiplication by A_{loc}. Note that the first and second steps can be done in parallel.

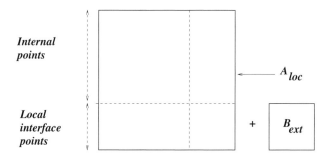

Figure 11.9. *The local matrices and data structure associated with each subdomain.*

With this decomposition, the global matrix-by-vector product can be implemented as indicated in Algorithm 11.6 below. In what follows, x_{loc} is a vector of variables that are local to a given processor. The components corresponding to the local interface points (ordered to be the last components in x_{loc} for convenience) are called x_{bnd}. The external interface points, listed in a certain order, constitute a vector called x_{ext}. The matrix A_{loc} is a sparse $nloc \times nloc$ matrix representing the restriction of A to the local variables x_{loc}. The matrix B_{ext} operates on the external variables x_{ext} to give the correction that must be added to the vector $A_{loc}x_{loc}$ in order to obtain the desired result $(Ax)_{loc}$.

ALGORITHM 11.6. Distributed Sparse Matrix Product Kernel

1. *Exchange interface data; i.e.,*
2. *Scatter x_{bnd} to neighbors*
3. *Gather x_{ext} from neighbors*
4. *Do local matvec:* $y = A_{loc}x_{loc}$
5. *Do external matvec:* $y = y + B_{ext}x_{ext}$

An important observation is that the matrix-by-vector products in lines 4 and 5 can use any convenient data structure that will improve efficiency by exploiting knowledge of the local architecture. An example of the implementation of this operation is illustrated next:

Call bdxchg(nloc, x, y, nproc, proc, ix, ipr, type, xlen, iout)
y(1 : nloc) = 0.0
Call amux1(nloc, x, y, aloc, jaloc, ialoc)
nrow = nloc − nbnd + 1
Call amux1(nrow, x, y(nbnd), aloc, jaloc, ialoc(nloc + 1))

The only routine requiring communication is *bdxchg*, whose purpose is to exchange interface values between nearest neighbor processors. The first call to *amux1* performs the operation $y := y + A_{loc}x_{loc}$, where y has been initialized to zero prior to the call. The second call to *amux1* performs $y := y + B_{ext}x_{ext}$. Notice that the data for the matrix B_{ext} are simply appended to those of A_{loc}, a standard technique used for storing a succession of sparse matrices. The B_{ext} matrix acts only on the subvector of x that starts at location $nbnd$ of x. The size of the B_{ext} matrix is $nrow = nloc - nbnd + 1$.

11.6 Standard Preconditioning Operations

Each preconditioned step requires the solution of a linear system of equations of the form $Mz = y$. This section only considers those traditional preconditioners, such as incomplete LU (ILU) or successive overrelaxation (SOR) or symmetric SOR (SSOR), in which the solution with M is the result of solving triangular systems. Since these are commonly used, it is important to explore ways to implement them efficiently in a parallel environment. It is also important to stress that the techniques to be described in this section are mostly useful on shared memory computers. Distributed memory computers utilize different strategies. We only consider lower triangular systems of the form

$$Lx = b. \tag{11.1}$$

Without loss of generality, it is assumed that L is unit lower triangular.

11.6.1 Parallelism in Forward Sweeps

Typically in solving a lower triangular system, the solution is overwritten onto the right-hand side on return. In other words, there is one array x for both the solution and the right-hand side. Therefore, the forward sweep for solving a lower triangular system with coefficients $al(i, j)$ and right-hand-side x is as follows.

ALGORITHM 11.7. Sparse Forward Elimination

1. $Do\ i = 2, n$
2.   For (all j such that $al(i, j)$ is nonzero), Do
3. $x(i) := x(i) - al(i, j) * x(j)$
4.   EndDo
5. EndDo

Assume that the matrix is stored row-wise in the general CSR format, except that the diagonal elements (ones) are not stored. Then the above algorithm translates into the following code segment:

1. $Do\ i = 2, n$
2. $Do\ j = ial(i), ial(i+1) - 1$
3. $x(i) = x(i) - al(j) * x(jal(j))$
4.   EndDo
5. EndDo

The outer loop corresponding to the variable i is sequential. The j loop is a sparse dot product of the ith row of L and the (dense) vector x. This dot product may be split among the processors and the partial results may be added at the end. However, the length of the vector involved in the dot product is typically short. So this approach is quite inefficient in general. We examine next a few alternative approaches. The regularly structured and irregularly structured cases are treated separately.

11.6.2 Level Scheduling: The Case of Five-Point Matrices

First, consider an example that consists of a five-point matrix associated with a 4 × 3 mesh, as represented in Figure 11.10. The lower triangular matrix associated with this mesh is represented on the left side of Figure 11.10. The stencil represented on the right side of Figure 11.10 establishes the data dependence between the unknowns in the lower triangular system solution when considered from the point of view of a grid of unknowns. It tells us that, in order to compute the unknown in position (i, j), only the two unknowns in positions $(i - 1, j)$ and $(i, j - 1)$ are needed. The unknown x_{11} does not depend on any other variable and can be computed first. Then the value of x_{11} can be used to get $x_{1,2}$ and $x_{2,1}$ simultaneously. Then these two values will in turn enable $x_{3,1}$, $x_{2,2}$, and $x_{1,3}$ to be obtained simultaneously, and so on. Thus, the computation can proceed in wave-fronts.

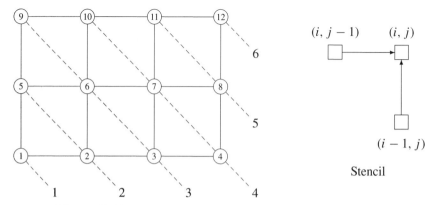

Figure 11.10. *Level scheduling for a 4 × 3 grid problem.*

The steps for this wave-front algorithm are shown with dashed lines in Figure 11.10. Observe that the maximum degree of parallelism (or vector length, in the case of vector processing) that can be reached is the minimum of n_x, n_y, the number of mesh points in the x and y directions, respectively, for two-dimensional problems.

For three-dimensional problems, the parallelism is of the order of the maximum size of the sets of domain points $x_{i,j,k}$, where $i + j + k = lev$, a constant level lev. It is important to note that there is little parallelism or vectorization at the beginning and at the end of the sweep. The degree of parallelism is equal to one initially, and then increases by one for each wave until it reaches its maximum, and then decreases back down to one at the end of the sweep. For example, for a 4 × 3 grid, the levels (sets of equations that can be solved in parallel) are {1}, {2, 5}, {3, 6, 9}, {4, 7, 10}, {8, 11}, and finally {12}. The first and last few steps may take a heavy toll on achievable speed-ups.

The idea of proceeding by *levels* or *wave-fronts* is a natural one for finite difference matrices on rectangles. Discussed next is the more general case of irregular matrices—a textbook example of scheduling, or *topological sorting*, it is well known in different forms to computer scientists.

11.6.3 Level Scheduling for Irregular Graphs

The simple scheme described above can be generalized for irregular grids. The objective of the technique, called *level scheduling*, is to group the unknowns in subsets so that they

11.6. Standard Preconditioning Operations

can be determined simultaneously. To explain the idea, consider again Algorithm 11.7 for solving a unit lower triangular system. The ith unknown can be determined once all the other ones that participate in equation i become available. In the ith step, all unknowns j such that $al(i, j) \neq 0$ must be known. To use graph terminology, these unknowns are *adjacent* to the unknown number i. Since L is lower triangular, the adjacency graph is a directed acyclic graph. The edge $j \to i$ in the graph simply indicates that x_j must be known before x_i can be determined. It is possible and quite easy to find a labeling of the nodes that satisfies the property that, if $label(j) < label(i)$, then task j must be executed before task i. This is called a topological sorting of the unknowns.

The first step computes x_1 and any other unknowns for which there are no predecessors in the graph, i.e., all those unknowns x_i for which the off-diagonal elements of row i are zero. These unknowns will constitute the elements of the first level. The next step computes in parallel all those unknowns that will have the nodes of the first level as their (only) predecessors in the graph. The following steps can be defined similarly: The unknowns that can be determined at step l are all those that have as predecessors equations that have been determined in steps $1, 2, \ldots, l - 1$. This leads naturally to the definition of a *depth* for each unknown. The *depth* of a vertex is defined by performing the following loop for $j = 1, 2, \ldots, n$, after initializing $depth(j)$ to zero for all j:

$$depth(i) = 1 + \max_{j}\{depth(j) \text{ for all } j \text{ such that } al(i, j) \neq 0\}.$$

By definition, a *level* of the graph is the set of nodes with the same depth. A data structure for the levels can be defined: A permutation $q(1 : n)$ defines the new ordering and $level(i), i = 1, \ldots, nlev + 1$ points to the beginning of the ith level in that array.

Once these level sets are found, there are two different ways to proceed. The permutation vector q can be used to permute the matrix according to the new order. In the 4×3 example mentioned in the previous subsection, this means renumbering the variables $\{1\}, \{2, 5\}, \{3, 6, 9\}, \ldots$, consecutively, i.e., as $\{1, 2, 3, \ldots\}$. The resulting matrix after the permutation is shown on the right side of Figure 11.11. An alternative is simply to keep the permutation array and use it to identify unknowns that correspond to a given level in the solution. Then the algorithm for solving the triangular systems can be written as follows, assuming that the matrix is stored in the usual row sparse matrix format.

ALGORITHM 11.8. Forward Elimination with Level Scheduling

1. $Do\ lev = 1, nlev$
2. $j1 = level(lev)$
3. $j2 = level(lev + 1) - 1$
4. $Do\ k = j1, j2$
5. $i = q(k)$
6. $Do\ j = ial(i), ial(i + 1) - 1$
7. $x(i) = x(i) - al(j) * x(jal(j))$
8. $EndDo$
9. $EndDo$
10. $EndDo$

An important observation here is that the outer loop, which corresponds to a level, performs an operation of the form

$$x := x - Bx,$$

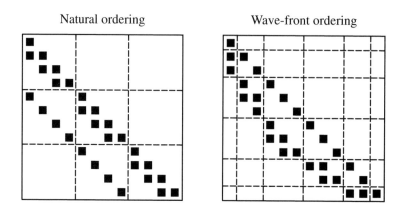

Figure 11.11. *Lower triangular matrix associated with mesh of Figure* 11.10.

where B is a submatrix consisting only of the rows of level lev, excluding the diagonal elements. This operation can in turn be optimized by using a proper data structure for these submatrices. For example, the JAD data structure can be used. The resulting performance can be quite good. On the other hand, implementation can be quite involved since two embedded data structures are required.

Example 11.2. Consider a finite element matrix obtained from the example shown in Figure 3.1. After an additional level of refinement, done in the same way as was described in Chapter 3, the resulting matrix, shown in the left part of Figure 11.12, is of size $n = 145$. In this case, eight levels are obtained. If the matrix is reordered by levels, the matrix shown on the right side of the figure results. The last level consists of only one element.

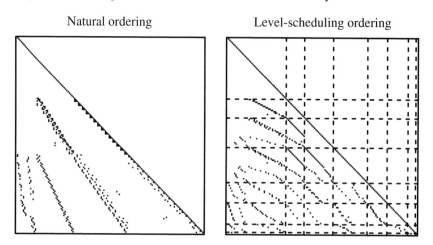

Figure 11.12. *Lower triangular matrix associated with a finite element matrix and its level-ordered version.*

Exercises

1. Give a short answer to each of the following questions:

   a. What is the main disadvantage of shared memory computers based on a bus architecture?

   b. What is the main factor in yielding the speed-up in pipelined processors?

   c. Related to the previous question: What is the main limitation of pipelined processors with regard to their potential for providing high speed-ups?

2. Show that the number of edges in a binary n-cube is $n2^{n-1}$.

3. Show that a binary 4-cube is identical to a *torus*, which is a 4×4 mesh with wrap-around connections. Are there hypercubes of any other dimensions that are equivalent topologically to toruses?

4. A Gray code of length $k = 2^n$ is a sequence $a_0, \ldots, a_{k-1}$ of n-bit binary numbers such that (i) any two successive numbers in the sequence differ by one and only one bit, (ii) all n-bit binary numbers are represented in the sequence, and (iii) a_0 and a_{k-1} differ by one bit.

   a. Find a Gray code sequence of length $k = 8$ and show the (closed) path defined by the sequence of nodes of a 3-cube, whose labels are the elements of the Gray code sequence. What type of paths does a Gray code define in a hypercube?

   b. To build a "binary-reflected" Gray code, start with the trivial Gray code sequence consisting of the two one-bit numbers 0 and 1. To build a two-bit Gray code, take the same sequence and insert a zero in front of each number, then take the sequence in *reverse order* and insert a one in front of each number. This gives $G_2 = \{00, 01, 11, 10\}$. The process is repeated until an n-bit sequence is generated. Show the binary-reflected Gray code sequences of lengths 2, 4, 8, and 16. Prove (by induction) that this process does indeed produce a valid Gray code sequence.

   c. Let an n-bit Gray code be given and consider the subsequence of all elements whose first bit is constant (e.g., zero). Is this an $(n-1)$-bit Gray code sequence? Generalize this to any of the n-bit positions. Generalize further to any set of $k < n$ bit positions.

   d. Use part (c) to find a strategy to map a $2^{n_1} \times 2^{n_2}$ mesh into an $(n_1 + n_2)$-cube.

5. Consider a ring of k processors characterized by the following communication performance characteristics. Each processor can communicate with its two neighbors *simultaneously*; i.e., it can send or receive a message while sending or receiving another message. The time for a message of length m to be transmitted between two nearest neighbors is of the form
$$\beta + m\tau.$$

   a. A message of length m is *broadcast* to all processors by sending it from P_1 to P_2 and then from P_2 to P_3, etc., until it reaches all destinations, i.e., until it reaches P_k. How much time does it take for the message to complete this process?

b. Now split the message into packets of equal size and pipeline the data transfer. Typically, each processor will receive packet number i from the previous processor, while sending packet $i - 1$, which it has already received, to the next processor. The packets will travel in a chain from P_1 to $P_2, \ldots,$ to P_k. In other words, each processor executes a program that is described roughly as follows:

```
Do i=1, Num_packets
   Receive packet number i from previous processor
   Send packet number i to next processor
EndDo
```

There are a few additional conditions. Assume that the number of packets is equal to $k - 1$. How much time does it take for all packets to reach all k processors? How does this compare with the simple method in (a)?

6. a. Write a short FORTRAN routine (or C function) that sets up the level number of each unknown of an upper triangular matrix. The input matrix is in CSR format and the output should be an array of length n containing the level number of each node.

   b. What data structure should be used to represent levels? Without writing the code, show how to determine this data structure from the output of your routine.

   c. Assuming the data structure of the levels has been determined, write a short FORTRAN routine (or C function) to solve an upper triangular system using the data structure resulting in part (b). Show clearly which loop should be executed in parallel.

7. In the JAD format described in Section 11.5.4, it is necessary to preprocess the matrix by sorting its rows by decreasing number of rows. What type of sorting should be used for this purpose?

8. In the JAD format described in Section 11.5.4, the matrix had to be preprocessed by sorting it by rows of decreasing number of elements.

   a. What is the main reason it is necessary to reorder the rows?

   b. Assume that the same process of extracting one element per row is used. At some point the extraction process will come to a stop and the remainder of the matrix can be put into a CSR data structure. Write down a good data structure to store the two pieces of data and a corresponding algorithm for matrix-by-vector products.

   c. This scheme is efficient in many situations but can lead to problems if the first row is very short. Suggest how to remedy the situation by padding with zero elements, as is done for the Ellpack format.

9. Many matrices that arise in PDE applications have a structure that consists of a few diagonals and a small number of nonzero elements scattered irregularly in the matrix. In such cases, it is advantageous to extract the diagonal part and put the rest in a general sparse (e.g., CSR) format. Write a pseudocode to extract the main diagonals and the sparse part. As input parameter, the number of diagonals desired must be specified.

Notes and References

General recommended reading on parallel computing are the books by Kumar et al. [193], Foster [131], and Wilkinson and Allen [315]. Trends in high performance architectures seem to come and go rapidly. In the 1980s, it seemed that the paradigm of shared memory computers with massive parallelism and coarse-grain parallelism was sure to win in the long run. Then, a decade ago, massive parallelism of the SIMD type dominated the scene for a while, with hypercube topologies at the forefront. Thereafter, computer vendors started mixing message-passing paradigms with *global address space*. Currently, it appears that distributed heterogeneous computing will dominate the high performance computing scene for some time to come. Another recent development is the advent of network computing or grid computing.

Until the advent of supercomputing in the mid-1970s, storage schemes for sparse matrices were chosen mostly for convenience, as performance was not an issue in general. The first paper showing the advantage of diagonal storage schemes in sparse matrix computations is probably [183]. The discovery by supercomputer manufacturers of the specificity of sparse matrix computations was accompanied by the painful realization that, without hardware support, vector computers could be inefficient. Indeed, the early vector machines (CRAY) did not have hardware instructions for gather and scatter operations but this was soon remedied in the second generation machines. For a detailed account of the beneficial impact of hardware for scatter and gather on vector machines, see [200].

Level scheduling is a textbook example of topological sorting in graph theory and was discussed from this viewpoint in, e.g., [8, 257, 317]. For the special case of finite difference matrices on rectangular domains, the idea was suggested by several authors independently [287, 288, 155, 251, 10]. In fact, the level-scheduling approach described in this chapter is a *greedy* approach and is unlikely to be optimal. It may be preferable to use a *backward scheduling* [7], which defines the levels from bottom up in the graph. Thus, the last level consists of the leaves of the graph, the previous level consists of their predecessors, etc. Instead of static scheduling, it is also possible to perform a dynamic scheduling whereby the order of the computation is determined at run time. The advantage over prescheduled triangular solutions is that it allows processors to always execute a task as soon as its predecessors have been completed, which reduces idle time. Some of the earlier references on implementations and tests with level scheduling are [30, 256, 165, 37, 7, 8, 293, 295].

Chapter 12
Parallel Preconditioners

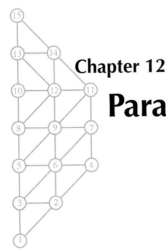

This chapter covers a few alternative methods for preconditioning a linear system. These methods are suitable when the desired goal is to maximize parallelism. The simplest approach is the diagonal (or Jacobi) preconditioning. Often, this preconditioner is not very useful, since the number of iterations of the resulting iteration tends to be much larger than the more standard variants, such as incomplete LU (ILU) or symmetric successive overrelaxation (SSOR). When developing parallel preconditioners, one should be aware that the benefits of increased parallelism are not outweighed by the increased number of computations. The main question to ask is whether or not it is possible to find preconditioning techniques that have a high degree of parallelism, as well as good intrinsic qualities.

12.1 Introduction

As seen in the previous chapter, a limited amount of parallelism can be extracted from the standard preconditioners, such as ILU and SSOR. Fortunately, a number of alternative techniques can be developed that are specifically targeted at parallel environments. These are preconditioning techniques that would normally not be used on a standard machine, but only for parallel computers. There are at least three such types of techniques discussed in this chapter. The simplest approach is to use a Jacobi or, even better, a block Jacobi approach. In the simplest case, a Jacobi preconditioner may consist of the diagonal or a block diagonal of A. To enhance performance, these preconditioners can themselves be accelerated by polynomial iterations, i.e., a second level of preconditioning called *polynomial preconditioning*.

A different strategy altogether is to enhance parallelism by using graph theory algorithms, such as graph-coloring techniques. These consist of coloring nodes such that two adjacent nodes have different colors. The gist of this approach is that all unknowns associated with the same color can be determined simultaneously in the forward and backward sweeps of the ILU preconditioning operation.

Finally, a third strategy uses generalizations of *partitioning* techniques, which can be put in the general framework of *domain decomposition* approaches. These will be covered in detail in Chapter 14.

Algorithms are emphasized rather than implementations. There are essentially two types of algorithms, namely, those that can be termed *coarse grain* and those that can be termed *fine grain*. In coarse-grain algorithms, the parallel tasks are relatively big and may, for example, involve the solution of small linear systems. In fine-grain parallelism, the subtasks can be elementary floating-point operations or consist of a few such operations. As always, the dividing line between the two classes of algorithms is somewhat blurred.

12.2 Block Jacobi Preconditioners

Overlapping block Jacobi preconditioning consists of a general block Jacobi approach, as described in Chapter 4, in which the sets S_i overlap. Thus, we define the index sets

$$S_i = \{j \mid l_i \leq j \leq r_i\},$$

with

$$l_1 = 1,$$
$$r_p = n,$$
$$r_i > l_{i+1}, \quad 1 \leq i \leq p-1,$$

where p is the number of blocks. Now we use the block Jacobi method with this particular partitioning, or employ the general framework of additive projection processes of Chapter 5, and use an additive projection method onto the sequence of subspaces

$$K_i = \text{span}\{V_i\}, \quad V_i = [e_{l_i}, e_{l_i+1}, \ldots, e_{r_i}].$$

Each of the blocks will give rise to a correction of the form

$$\xi_i^{(k+1)} = \xi_i^{(k)} + A_i^{-1} V_i^T (b - Ax^{(k)}). \tag{12.1}$$

One problem with the above formula is related to the overlapping portions of the x variables. The overlapping sections will receive two different corrections in general. According to the definition of *additive projection processes* seen in Chapter 5, the next iterate can be defined as

$$x_{k+1} = x_k + \sum_{i=1}^{p} V_i A_i^{-1} V_i^T r_k,$$

where $r_k = b - Ax_k$ is the residual vector at the previous iteration. Thus, the corrections for the overlapping regions are simply added together. It is also possible to weigh these contributions before adding them up. This is equivalent to redefining (12.1) as

$$\xi_i^{(k+1)} = \xi_i^{(k)} + D_i A_i^{-1} V_i^T (b - Ax_k),$$

in which D_i is a nonnegative diagonal matrix of weights. It is typical to weigh a nonoverlapping contribution by one and an overlapping contribution by $1/k$, where k is the number of times the unknown is represented in the partitioning.

12.3. Polynomial Preconditioners

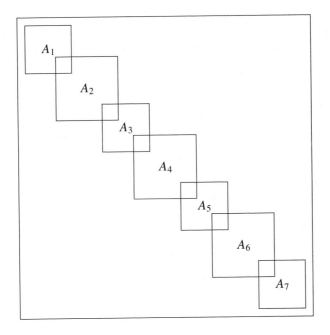

Figure 12.1. *The block Jacobi matrix with overlapping blocks.*

The block Jacobi iteration is often over- or underrelaxed, using a relaxation parameter ω. The iteration can be defined in the form

$$x_{k+1} = x_k + \sum_{i=1}^{p} \omega_i V_i A_i^{-1} V_i^T r_k.$$

Recall that the residual at step $k+1$ is then related to that at step k by

$$r_{k+1} = \left[I - \sum_{i=1}^{p} \omega_i A V_i \left(V_i^T A V_i \right)^{-1} V_i^T \right] r_k.$$

The solution of a sparse linear system is required at each projection step. These systems can be solved by direct methods if the sub-blocks are small enough. Otherwise, iterative methods may be used. The outer loop accelerator should then be a flexible variant, such as flexible generalized minimal residual (FGMRES), which can accommodate variations in the preconditioners.

12.3 Polynomial Preconditioners

In polynomial preconditioning the matrix M is defined by

$$M^{-1} = s(A),$$

where s is a polynomial, typically of low degree. Thus, the original system is replaced with the preconditioned system

$$s(A)Ax = s(A)b, \qquad (12.2)$$

which is then solved by a conjugate gradient (CG)-type technique. Note that $s(A)$ and A commute and, as a result, the preconditioned matrix is the same for right and left preconditioning. In addition, the matrix $s(A)$ or $As(A)$ does not need to be formed explicitly since $As(A)v$ can be computed for any vector v from a sequence of matrix-by-vector products.

Initially, this approach was motivated by the good performance of matrix-by-vector operations on vector computers for long vectors, e.g., the Cyber 205. However, the idea itself is an old one and was suggested by Stiefel [275] for eigenvalue calculations in the mid-1950s. Next, some of the popular choices for the polynomial s are described.

12.3.1 Neumann Polynomials

The simplest polynomial s that has been used is the polynomial of the Neumann series expansion

$$I + N + N^2 + \cdots + N^s,$$

in which

$$N = I - \omega A$$

and ω is a scaling parameter. The above series comes from expanding the inverse of ωA using the splitting

$$\omega A = I - (I - \omega A).$$

This approach can also be generalized by using a splitting of the form

$$\omega A = D - (D - \omega A),$$

where D can be the diagonal of A or, more appropriately, a block diagonal of A. Then

$$(\omega A)^{-1} = \left[D(I - (I - \omega D^{-1}A))\right]^{-1}$$
$$= \left[I - (I - \omega D^{-1}A)\right]^{-1} D^{-1}.$$

Thus, setting

$$N = I - \omega D^{-1} A$$

results in the approximate s-term expansion

$$(\omega A)^{-1} \approx M^{-1} \equiv [I + N + \cdots + N^s] D^{-1}. \qquad (12.3)$$

Since $D^{-1}A = \omega^{-1}[I - N]$, note that

$$M^{-1}A = [I + N + \cdots + N^s] D^{-1} A$$
$$= \frac{1}{\omega}[I + N + \cdots + N^s](I - N)$$
$$= \frac{1}{\omega}(I - N^{s+1}).$$

The matrix operation with the preconditioned matrix can be difficult numerically for large s. If the original matrix is symmetric positive definite (SPD), then $M^{-1}A$ is not symmetric, but it is self-adjoint with respect to the D inner product; see Exercise 1.

12.3.2 Chebyshev Polynomials

The polynomial s can be selected to be optimal in some sense, which leads to the use of Chebyshev polynomials. The criterion that is used makes the preconditioned matrix $s(A)A$ as close as possible to the identity matrix in some sense. For example, the spectrum of the preconditioned matrix can be made as close as possible to that of the identity. Denoting by $\sigma(A)$ the spectrum of A and by $\mathbb{P}_k$ the space of polynomials of degree not exceeding k, the following may be solved:

Find $s \in \mathbb{P}_k$ that minimizes
$$\max_{\lambda \in \sigma(A)} |1 - \lambda s(\lambda)|. \tag{12.4}$$

Unfortunately, this problem involves all the eigenvalues of A and is harder to solve than the original problem. Usually, problem (12.4) is replaced with the following problem:

Find $s \in \mathbb{P}_k$ that minimizes
$$\max_{\lambda \in E} |1 - \lambda s(\lambda)|, \tag{12.5}$$

which is obtained from replacing the set $\sigma(A)$ with some continuous set E that encloses it. Thus, a rough idea of the spectrum of the matrix A is needed. Consider first the particular case where A is SPD, in which case E can be taken to be an interval $[\alpha, \beta]$ containing the eigenvalues of A.

A variation of Theorem 6.25 is that, for any real scalar γ such that $\gamma \leq \alpha$, the minimum
$$\min_{p \in \mathbb{P}_k, p(\gamma)=1} \max_{t \in [\alpha, \beta]} |p(t)|$$
is reached for the shifted and scaled Chebyshev polynomial of the first kind
$$\hat{C}_k(t) \equiv \frac{C_k\left(1 + 2\frac{\alpha-t}{\beta-\alpha}\right)}{C_k\left(1 + 2\frac{\alpha-\gamma}{\beta-\alpha}\right)}.$$

Of interest is the case where $\gamma = 0$, which gives the polynomial
$$T_k(t) \equiv \frac{1}{\sigma_k} C_k\left(\frac{\beta + \alpha - 2t}{\beta - \alpha}\right), \quad \text{with} \quad \sigma_k \equiv C_k\left(\frac{\beta + \alpha}{\beta - \alpha}\right).$$

Denote the center and mid-width of the interval $[\alpha, \beta]$, respectively, by
$$\theta \equiv \frac{\beta + \alpha}{2}, \quad \delta \equiv \frac{\beta - \alpha}{2}.$$

Using these parameters instead of α, β, the above expressions then become
$$T_k(t) \equiv \frac{1}{\sigma_k} C_k\left(\frac{\theta - t}{\delta}\right), \quad \text{with} \quad \sigma_k \equiv C_k\left(\frac{\theta}{\delta}\right).$$

The three-term recurrence for the Chebyshev polynomials results in the following three-term recurrences:
$$\sigma_{k+1} = 2\frac{\theta}{\delta}\sigma_k - \sigma_{k-1}, \quad k = 1, 2 \ldots,$$

with
$$\sigma_1 = \frac{\theta}{\delta}, \quad \sigma_0 = 1,$$
and
$$T_{k+1}(t) \equiv \frac{1}{\sigma_{k+1}}\left[2\frac{\theta-t}{\delta}\sigma_k T_k(t) - \sigma_{k-1}T_{k-1}(t)\right]$$
$$= \frac{\sigma_k}{\sigma_{k+1}}\left[2\frac{\theta-t}{\delta}T_k(t) - \frac{\sigma_{k-1}}{\sigma_k}T_{k-1}(t)\right], \quad k \geq 1,$$
with
$$T_1(t) = 1 - \frac{t}{\theta}, \quad T_0(t) = 1.$$

Define
$$\rho_k \equiv \frac{\sigma_k}{\sigma_{k+1}}, \quad k = 1, 2, \ldots. \tag{12.6}$$

Note that the above recurrences can be put together as
$$\rho_k = \frac{1}{2\sigma_1 - \rho_{k-1}}, \tag{12.7}$$

$$T_{k+1}(t) = \rho_k\left[2\left(\sigma_1 - \frac{t}{\delta}\right)T_k(t) - \rho_{k-1}T_{k-1}(t)\right], \quad k \geq 1. \tag{12.8}$$

Observe that formulas (12.7)–(12.8) can be started at $k = 0$ provided we set $T_{-1} \equiv 0$ and $\rho_{-1} \equiv 0$, so that $\rho_0 = 1/(2\sigma_1)$.

The goal is to obtain an iteration that produces a residual vector of the form $r_{k+1} = T_{k+1}(A)r_0$, where T_k is the polynomial defined by the above recurrence. The difference between two successive residual vectors is given by
$$r_{k+1} - r_k = (T_{k+1}(A) - T_k(A))r_0.$$

The identity $1 = (2\sigma_1 - \rho_{k-1})\rho_k$ and the relations (12.8) yield
$$T_{k+1}(t) - T_k(t) = T_{k+1}(t) - (2\sigma_1 - \rho_{k-1})\rho_k T_k(t)$$
$$= \rho_k\left[-\frac{2t}{\delta}T_k(t) + \rho_{k-1}(T_k(t) - T_{k-1}(t))\right].$$

As a result,
$$\frac{T_{k+1}(t) - T_k(t)}{t} = \rho_k\left[\rho_{k-1}\frac{T_k(t) - T_{k-1}(t)}{t} - \frac{2}{\delta}T_k(t)\right]. \tag{12.9}$$

Define
$$d_k \equiv x_{k+1} - x_k$$
and note that $r_{k+1} - r_k = -Ad_k$. As a result, the relation (12.9) translates into the recurrence
$$d_k = \rho_k\left[\rho_{k-1}d_{k-1} + \frac{2}{\delta}r_k\right].$$

Finally, the following algorithm is obtained.

12.3. Polynomial Preconditioners

ALGORITHM 12.1. Chebyshev Acceleration

1. $r_0 = b - Ax_0; \sigma_1 = \theta/\delta$
2. $\rho_0 = 1/\sigma_1; d_0 = \frac{1}{\theta}r_0$
3. For $k = 0, \ldots$, until convergence, Do
4. $x_{k+1} = x_k + d_k$
5. $r_{k+1} = r_k - Ad_k$
6. $\rho_{k+1} = (2\sigma_1 - \rho_k)^{-1}$
7. $d_{k+1} = \rho_{k+1}\rho_k d_k + \frac{2\rho_{k+1}}{\delta}r_{k+1}$
8. EndDo

Note that the algorithm requires no inner products, which constitutes one of its attractions in a parallel computing environment. Lines 7 and 4 can also be recast into one single update of the form

$$x_{k+1} = x_k + \rho_k \left[\rho_{k-1}(x_k - x_{k-1}) + \frac{2}{\delta}(b - Ax_k) \right].$$

It can be shown that, when $\alpha = \lambda_1$ and $\beta = \lambda_N$, the resulting preconditioned matrix minimizes the condition number of the preconditioned matrices of the form $As(A)$ over all polynomials s of degree not exceeding $k - 1$. However, when used in conjunction with the CG method, it is observed that the polynomial that minimizes the total number of CG iterations *is far from being the one that minimizes the condition number.* If, instead of taking $\alpha = \lambda_1$ and $\beta = \lambda_N$, the interval $[\alpha, \beta]$ is chosen to be slightly inside the interval $[\lambda_1, \lambda_N]$, a much faster convergence might be achieved. The true optimal parameters, i.e., those that minimize the number of iterations of the polynomial preconditioned CG (PCG) method, are difficult to determine in practice.

There is a slight disadvantage to the approaches described above. The parameters α and β, which approximate the smallest and largest eigenvalues of A, are usually not available beforehand and must be obtained in some dynamic way. This may be a problem mainly because a software code based on Chebyshev acceleration could become quite complex.

To remedy this, one may ask whether the values provided by an application of Gershgorin's theorem can be used for α and β. Thus, in the symmetric case, the parameter α, which *estimates* the smallest eigenvalue of A, may be nonpositive even when A is a positive definite matrix. However, when $\alpha \leq 0$, the problem of minimizing (12.5) is not well defined, since it does not have a unique solution due to the nonstrict convexity of the uniform norm. An alternative uses the L_2-norm on $[\alpha, \beta]$ with respect to some weight function $w(\lambda)$. This *least-squares* polynomials approach is considered next.

12.3.3 Least-Squares Polynomials

Consider the inner product on the space $\mathbb{P}_k$:

$$\langle p, q \rangle = \int_\alpha^\beta p(\lambda)q(\lambda)w(\lambda)d\lambda, \tag{12.10}$$

where $w(\lambda)$ is some nonnegative weight function on (α, β). Denote by $\|p\|_w$ and call w-norm, the 2-norm induced by this inner product.

We seek the polynomial s_{k-1} that minimizes

$$\|1 - \lambda s(\lambda)\|_w \tag{12.11}$$

over all polynomials s of degree not exceeding $k - 1$. Call s_{k-1} the *least-squares iteration polynomial*, or simply the least-squares polynomial, and refer to $R_k(\lambda) \equiv 1 - \lambda s_{k-1}(\lambda)$ as the least-squares residual polynomial. A crucial observation is that the least-squares polynomial is well defined for arbitrary values of α and β. Computing the polynomial $s_{k-1}(\lambda)$ is not a difficult task when the weight function w is suitably chosen.

Computation of the Least-Squares Polynomials. There are three ways to compute the least-squares polynomials s_k defined in the previous section. The first approach is to use an explicit formula for R_k, known as the kernel polynomials formula:

$$R_k(\lambda) = \frac{\sum_{i=0}^{k} q_i(0) q_i(\lambda)}{\sum_{i=0}^{k} q_i(0)^2}, \tag{12.12}$$

in which the q_i's represent a sequence of polynomials orthogonal with respect to the weight function $w(\lambda)$. The second approach generates a three-term recurrence satisfied by the residual polynomials $R_k(\lambda)$. These polynomials are orthogonal with respect to the weight function $\lambda w(\lambda)$. From this three-term recurrence, we can proceed exactly as for the Chebyshev iteration to obtain a recurrence formula for the sequence of approximate solutions x_k. Finally, a third approach solves the normal equations associated with the minimization of (12.11); namely,

$$\langle 1 - \lambda s_{k-1}(\lambda), \lambda Q_j(\lambda) \rangle = 0, \quad j = 0, 1, 2, \ldots, k - 1,$$

where Q_j, $j = 1, \ldots, k - 1$, is any basis of the space P_{k-1} of polynomials of degree not exceeding $k - 1$.

These three approaches can all be useful in different situations. For example, the first approach can be useful for computing least-squares polynomials of low degree explicitly. For high degree polynomials, the last two approaches are preferable for their better numerical behavior. The second approach is restricted to the case where $\alpha \geq 0$, while the third is more general.

Since the degrees of the polynomial preconditioners are often low, e.g., not exceeding 5 or 10, we will give some details on the first formulation. Let $q_i(\lambda)$, $i = 0, 1, \ldots, n, \ldots$, be the *orthonormal* polynomials with respect to $w(\lambda)$. It is known that the least-squares residual polynomial $R_k(\lambda)$ of degree k is determined by the kernel polynomials formula (12.12). To obtain $s_{k-1}(\lambda)$, simply notice that

$$s_{k-1}(\lambda) = \frac{1 - R_k(\lambda)}{\lambda}$$

$$= \frac{\sum_{i=0}^{k} q_i(0) t_i(\lambda)}{\sum_{i=0}^{k} q_i(0)^2}, \quad \text{with} \tag{12.13}$$

$$t_i(\lambda) = \frac{q_i(0) - q_i(\lambda)}{\lambda}. \tag{12.14}$$

This allows s_{k-1} to be computed as a linear combination of the polynomials $t_i(\lambda)$. Thus, we can obtain the desired least-squares polynomials from the sequence of orthogonal polynomials q_i that satisfy a three-term recurrence of the form

$$\beta_{i+1} q_{i+1}(\lambda) = (\lambda - \alpha_i) q_i(\lambda) - \beta_i q_{i-1}(\lambda), \quad i = 1, 2, \ldots.$$

12.3. Polynomial Preconditioners

From this, the following recurrence for the t_i's can be derived:

$$\beta_{i+1}t_{i+1}(\lambda) = (\lambda - \alpha_i)t_i(\lambda) - \beta_i t_{i-1}(\lambda) + q_i(0), \quad i = 1, 2, \ldots.$$

The weight function w is chosen so that the three-term recurrence of the orthogonal polynomials q_i is known explicitly and/or is easy to generate. An interesting class of weight functions that satisfy this requirement is considered next.

Choice of the Weight Functions. This section assumes that $\alpha = 0$ and $\beta = 1$. Consider the Jacobi weights

$$w(\lambda) = \lambda^{\mu-1}(1-\lambda)^\nu, \text{ where } \mu > 0 \text{ and } \nu \geq -\frac{1}{2}. \tag{12.15}$$

For these weight functions, the recurrence relations are known explicitly for the polynomials that are orthogonal with respect to $w(\lambda)$, $\lambda w(\lambda)$, or $\lambda^2 w(\lambda)$. This allows the use of any of the three methods described in the previous section for computing $s_{k-1}(\lambda)$. Moreover, it has been shown [179] that the preconditioned matrix $As_k(A)$ is SPD when A is SPD, provided that $\mu - 1 \geq \nu \geq -\frac{1}{2}$.

The following explicit formula for $R_k(\lambda)$ can be derived easily from the explicit expression of the Jacobi polynomials and the fact that $\{R_k\}$ is orthogonal with respect to the weight $\lambda w(\lambda)$:

$$R_k(\lambda) = \sum_{j=0}^{k} \kappa_j^{(k)}(1-\lambda)^{k-j}(-\lambda)^j, \tag{12.16}$$

$$\kappa_j^{(k)} = \binom{k}{j} \prod_{i=0}^{j-1} \frac{k-i+\nu}{i+1+\mu}.$$

Using (12.13), the polynomial $s_{k-1}(\lambda) = (1 - R_k(\lambda))/\lambda$ can be derived easily by hand for small degrees; see Exercise 4.

Example 12.1. As an illustration, we list the least-squares polynomials s_k for $k = 1, \ldots, 8$ obtained for the Jacobi weights with $\mu = \frac{1}{2}$ and $\nu = -\frac{1}{2}$. The polynomials listed are for the interval $[0, 4]$, as this leads to integer coefficients. For a general interval $[0, \beta]$, the best polynomial of degree k is $s_k(4\lambda/\beta)$. Also, each polynomial s_k is rescaled by $(3 + 2k)/4$ to simplify the expressions. However, this scaling factor is unimportant if these polynomials are used for preconditioning.

	1	λ	λ^2	λ^3	λ^4	λ^5	λ^6	λ^7	λ^8
s_1	5	-1							
s_2	14	-7	1						
s_3	30	-27	9	-1					
s_4	55	-77	44	-11	1				
s_5	91	-182	156	-65	13	-1			
s_6	140	-378	450	-275	90	-15	1		
s_7	204	-714	1122	-935	442	-119	17	-1	
s_8	285	-1254	2508	-2717	1729	-665	152	-19	1

We selected $\mu = \frac{1}{2}$ and $\nu = -\frac{1}{2}$ only because these choices lead to a very simple recurrence for the polynomials q_i, which are the Chebyshev polynomials of the first kind.

Theoretical Considerations. An interesting theoretical question is whether the least-squares residual polynomial becomes small in some sense as its degree increases. Consider first the case $0 < \alpha < \beta$. Since the residual polynomial R_k minimizes the norm $\|R\|_w$ associated with the weight w over all polynomials R of degree not exceeding k such that $R(0) = 1$, the polynomial $(1 - (\lambda/\theta))^k$ with $\theta = (\alpha + \beta)/2$ satisfies

$$\|R_k\|_w \leq \left\|\left(1 - \frac{\lambda}{c}\right)^k\right\|_w \leq \left\|\left[\frac{b-a}{b+a}\right]^k\right\|_w = \kappa \left[\frac{\beta-\alpha}{\beta+\alpha}\right]^k,$$

where κ is the w-norm of the function equal to unity on the interval $[\alpha, \beta]$. The norm of R_k will tend to zero geometrically as k tends to infinity, provided that $\alpha > 0$.

Consider now the case $\alpha = 0$, $\beta = 1$ and the Jacobi weight (12.15). For this choice of the weight function, the least-squares residual polynomial is known to be $p_k(\lambda)/p_k(0)$, where p_k is the kth degree Jacobi polynomial associated with the weight function $w'(\lambda) = \lambda^\mu (1-\lambda)^\nu$. It can be shown that the 2-norm of such a residual polynomial with respect to this weight is given by

$$\|p_k/p_k(0)\|_{w'}^2 = \frac{\Gamma^2(\mu+1)\Gamma(k+\nu+1)}{(2k+\mu+\nu+1)\Gamma(k+\mu+\nu+1)} \frac{\Gamma(k+1)}{\Gamma(k+\mu+1)},$$

in which Γ is the gamma function. For the case $\mu = \frac{1}{2}$ and $\nu = -\frac{1}{2}$, this becomes

$$\|p_k/p_k(0)\|_{w'}^2 = \frac{[\Gamma(\frac{3}{2})]^2}{(2k+1)(k+\frac{1}{2})} = \frac{\pi}{2(2k+1)^2}.$$

Therefore, the w'-norm of the least-squares residual polynomial converges to zero as $1/k$ as the degree k increases (a much slower rate than when $\alpha > 0$). However, note that the condition $p(0) = 1$ implies that the polynomial must be large in some interval around the origin.

12.3.4 The Nonsymmetric Case

Given a set of approximate eigenvalues of a nonsymmetric matrix A, a simple region E can be constructed in the complex plane, e.g., a disk, an ellipse, or a polygon, which encloses the spectrum of the matrix A. There are several choices for E. The first idea uses an ellipse E that encloses an approximate convex hull of the spectrum. Consider an ellipse centered at θ and with focal distance δ. Then, as seen in Chapter 6, the shifted and scaled Chebyshev polynomials defined by

$$T_k(\lambda) = \frac{C_k\left(\frac{\theta-\lambda}{\delta}\right)}{C_k\left(\frac{\theta}{\delta}\right)}$$

are nearly optimal. The use of these polynomials leads again to an attractive three-term recurrence and to an algorithm similar to Algorithm 12.1. In fact, the recurrence is identical, except that the scalars involved can now be complex to accommodate cases where the ellipse has foci not necessarily located on the real axis. However, when A is real, then the symmetry of the foci with respect to the real axis can be exploited. The algorithm can still be written in real arithmetic.

An alternative to Chebyshev polynomials over ellipses employs a polygon H that contains $\sigma(A)$. Polygonal regions may better represent the shape of an arbitrary spectrum. The best polynomial for the infinity norm is not known explicitly but it may be computed by

12.3. Polynomial Preconditioners

an algorithm known in approximation theory as the Remez algorithm. It may be simpler to use an L_2-norm instead of the infinity norm, i.e., to solve (12.11), where w is some weight function defined on the boundary of the polygon H.

Now here is a sketch of an algorithm based on this approach. An L_2-norm associated with Chebyshev weights on the edges of the polygon is used. If the contour of H consists of k edges each with center θ_i and half-length δ_i, then the weight on each edge is defined by

$$w_i(\lambda) = \frac{2}{\pi} |\delta_i - (\lambda - \theta_i)^2|^{-1/2}, \quad i = 1, \ldots, k. \qquad (12.17)$$

Using the power basis to express the best polynomial is not viable. It is preferable to use the Chebyshev polynomials associated with the ellipse of smallest area containing H. With the above weights or any other Jacobi weights on the edges, there is a finite procedure *that does not require numerical integration* to compute the best polynomial. To do this, each of the polynomials of the basis (i.e., the Chebyshev polynomials associated with the ellipse of smallest area containing H) must be expressed as a linear combination of the Chebyshev polynomials associated with the different intervals $[\theta_i - \delta_i, \theta_i + \delta_i]$. This redundancy allows exact expressions for the integrals involved in computing the least-squares solution to (12.11).

Next, the main lines of a preconditioned GMRES algorithm are described based on least-squares polynomials. Eigenvalue estimates are obtained from a GMRES step at the beginning of the outer loop. This GMRES adaptive corrects the current solution and the eigenvalue estimates are used to update the current polygon H. Correcting the solution at this stage is particularly important since it often results in an improvement of a few orders of magnitude. This is because the polygon H may be inaccurate and the residual vector is dominated by components in one or two eigenvectors. The GMRES step will immediately annihilate those dominating components. In addition, the eigenvalues associated with these components will now be accurately represented by eigenvalues of the Hessenberg matrix.

ALGORITHM 12.2. Polynomial Preconditioned GMRES

1. *Start or Restart*
2. Compute current residual vector $r := b - Ax$
3. *Adaptive GMRES step*
4. Run m_1 steps of GMRES for solving $Ad = r$
5. Update x by $x := x + d$
6. Get eigenvalue estimates from the eigenvalues of the Hessenberg matrix
7. *Compute new polynomial*
8. Refine H from previous hull H and new eigenvalue estimates
9. Get new best polynomial s_k
10. *Polynomial iteration*
11. Compute the current residual vector $r = b - Ax$
12. Run m_2 steps of GMRES applied to $s_k(A)Ad = s_k(A)r$
13. Update x by $x := x + d$
14. Test for convergence
15. If solution converged then Stop; else go to 1

Example 12.2. Table 12.1 shows the results of applying GMRES(20) with polynomial preconditioning to the first four test problems described in Section 3.7.

Matrix	Iters	Kflops	Residual	Error
F2DA	56	2774	0.22E−05	0.51E−06
F3D	22	7203	0.18E−05	0.22E−05
ORS	78	4454	0.16E−05	0.32E−08
F2DB	100	4432	0.47E−05	0.19E−05

Table 12.1. *A test run of ILU(0)-GMRES accelerated with polynomial preconditioning.*

See Example 6.1 for the meaning of the column headers in the table. In fact, the system is preconditioned by ILU(0) before polynomial preconditioning is applied to it. Degree-10 polynomials (maximum) are used. The tolerance for stopping is 10^{-7}. Recall that *Iters* is the number of matrix-by-vector products rather than the number of GMRES iterations. Notice that, for most cases, the method does not compare well with the simpler ILU(0) example seen in Chapter 10. The notable exception is example F2DB, for which the method converges fairly fast in contrast with the simple ILU(0)-GMRES; see Example 10.2. An attempt to use the method for the fifth matrix in the test set, namely, the FIDAP matrix FID, failed because the matrix has eigenvalues on both sides of the imaginary axis and the code tested does not handle this situation.

It is interesting to follow the progress of the algorithm in the above examples. For the first example, the coordinates of the vertices of the upper part of the first polygon H are as follows:

$\Re e(c_i)$	$\Im m(c_i)$
0.06492	0.00000
0.17641	0.02035
0.29340	0.03545
0.62858	0.04977
1.18052	0.00000

This hull is computed from the 20 eigenvalues of the 20×20 Hessenberg matrix resulting from the first run of GMRES(20). In the ensuing GMRES loop, the outer iteration converges in three steps, each using a polynomial of degree 10; i.e., there is no further adaptation required. For the second problem, the method converges in the 20 first steps of GMRES, so polynomial acceleration was never invoked. For the third example, the initial convex hull found is the interval [0.06319, 1.67243] of the real line. The polynomial preconditioned GMRES then converges in five iterations. Finally, the initial convex hull found for the last example is as follows:

$\Re e(c_i)$	$\Im m(c_i)$
0.17131	0.00000
0.39337	0.10758
1.43826	0.00000

and the outer loop converges again without another adaptation step, this time in seven steps.

12.4 Multicoloring

The general idea of multicoloring, or graph coloring, has been used for a long time by numerical analysts. It was exploited, in particular, in the context of relaxation techniques both for understanding their theory and for deriving efficient algorithms. More recently, these techniques were found to be useful in improving parallelism in iterative solution techniques. This discussion begins with the two-color case, called *red-black* ordering.

12.4.1 Red-Black Ordering

The problem addressed by multicoloring is to determine a coloring of the nodes of the adjacency graph of a matrix such that any two adjacent nodes have different colors. For a two-dimensional finite difference grid (five-point operator), this can be achieved with two colors, typically referred to as red and black. This red-black coloring is illustrated in Figure 12.2 for a 6 × 4 mesh, where the black nodes are represented by filled circles.

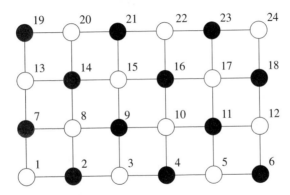

Figure 12.2. *Red-black coloring of a 6 × 4 grid with natural labeling of the nodes.*

Assume that the unknowns are labeled by first listing the red unknowns together, followed by the black ones. The new labeling of the unknowns is shown in Figure 12.3. Since the red nodes are not coupled with other red nodes and, similarly, the black nodes are not coupled with other black nodes, the system that results from this reordering has the structure

$$\begin{pmatrix} D_1 & F \\ E & D_2 \end{pmatrix} \begin{pmatrix} x_1 \\ x_2 \end{pmatrix} = \begin{pmatrix} b_1 \\ b_2 \end{pmatrix}, \tag{12.18}$$

in which D_1 and D_2 are diagonal matrices. The reordered matrix associated with this new labeling is shown in Figure 12.4.

Two issues will be explored regarding red-black ordering. The first is how to exploit this structure for solving linear systems. The second is how to generalize this approach for systems whose graphs are not necessarily two colorable.

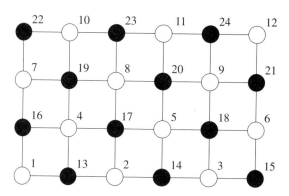

Figure 12.3. *Red-black coloring of a 6 × 4 grid with red-black labeling of the nodes.*

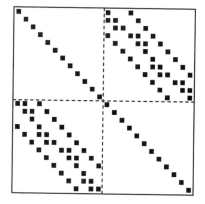

Figure 12.4. *Matrix associated with the red-black reordering of Figure 12.3.*

12.4.2 Solution of Red-Black Systems

The easiest way to exploit the red-black ordering is to use the standard SSOR or ILU(0) preconditioners for solving the block system (12.18), which is derived from the original system. The resulting preconditioning operations are highly parallel. For example, the linear system that arises from the forward solve in SSOR will have the form

$$\begin{pmatrix} D_1 & O \\ E & D_2 \end{pmatrix} \begin{pmatrix} x_1 \\ x_2 \end{pmatrix} = \begin{pmatrix} b_1 \\ b_2 \end{pmatrix}.$$

This system can be solved by performing the following sequence of operations:

1. Solve $D_1 x_1 = b_1$
2. Compute $\hat{b}_2 := b_2 - E x_1$
3. Solve $D_2 x_2 = \hat{b}_2$

This consists of two diagonal scalings (operations 1 and 3) and a sparse matrix-by-vector product. Therefore, the degree of parallelism is at least $n/2$ if an atomic task is considered

12.4. Multicoloring

to be any arithmetic operation. The situation is identical to the ILU(0) preconditioning. However, since the matrix has been reordered before ILU(0) is applied to it, the resulting LU factors are not related in any simple way to those associated with the original matrix. In fact, a simple look at the structure of the ILU factors reveals that many more elements are dropped with the red-black ordering than with the natural ordering. The result is that the number of iterations to achieve convergence can be much higher with red-black ordering than with the natural ordering.

A second method that has been used in connection with the red-black ordering solves the reduced system involving only the black unknowns. Eliminating the red unknowns from (12.18) results in the reduced system

$$(D_2 - ED_1^{-1}F)x_2 = b_2 - ED_1^{-1}b_1.$$

Note that this new system is again a sparse linear system with about half as many unknowns. In addition, it has been observed that, for easy problems, the reduced system can often be solved efficiently with only diagonal preconditioning. The computation of the reduced system is a highly parallel and inexpensive process. Note that it is not necessary to form the reduced system. This strategy is more often employed when D_1 is not diagonal, such as in domain decomposition methods, but it can also have some uses in other situations. For example, applying the matrix to a given vector x can be performed using nearest neighbor communication, which can be more efficient than the standard approach of multiplying the vector by the Schur complement matrix $D_2 - ED_1^{-1}F$. In addition, this can save storage, which may be more critical in some cases.

12.4.3 Multicoloring for General Sparse Matrices

Chapter 3 discussed a general greedy approach for multicoloring a graph. Given a general sparse matrix A, this inexpensive technique allows us to reorder it into a block form where the diagonal blocks are diagonal matrices. The number of blocks is the number of colors. For example, for six colors, a matrix would result with the structure shown in Figure 12.5, where the D_i's are diagonal and E, F are general sparse. This structure is obviously a generalization of the red-black ordering.

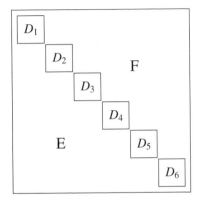

Figure 12.5. *A six-color ordering of a general sparse matrix.*

Just as for the red-black ordering, ILU(0), SOR, or SSOR preconditioning can be used on this reordered system. The parallelism of SOR/SSOR is now of order n/p, where p is the number of colors. A loss in efficiency may occur since the number of iterations is likely to increase.

A Gauss–Seidel sweep will essentially consist of p scalings and $p - 1$ matrix-by-vector products, where p is the number of colors. Specifically, assume that the matrix is stored in the well-known Ellpack-Itpack format and that the block structure of the permuted matrix is defined by a pointer array $iptr$. The index $iptr(j)$ is the index of the first row in the jth block. Thus, the pair $A(n1 : n2, *)$, $JA(n1 : n2, *)$ represents the sparse matrix consisting of the rows $n1$ to $n2$ in the Ellpack-Itpack format. The main diagonal of A is assumed to be stored separately in inverted form in a one-dimensional array $diag$. One single step of the multicolor SOR iteration will then take the following form.

ALGORITHM 12.3. Multicolor SOR Sweep in the Ellpack Format

1. $\text{Do } col = 1, ncol$
2. $\quad n1 = iptr(col)$
3. $\quad n2 = iptr(col + 1) - 1$
4. $\quad y(n1 : n2) = rhs(n1 : n2)$
5. $\quad \text{Do } j = 1, ndiag$
6. $\quad\quad \text{Do } i = n1, n2$
7. $\quad\quad\quad y(i) = y(i) - a(i, j) * y(ja(i, j))$
8. $\quad\quad \text{EndDo}$
9. $\quad \text{EndDo}$
10. $\quad y(n1 : n2) = diag(n1 : n2) * y(n1 : n2)$
11. EndDo

In the above algorithm, $ncol$ is the number of colors. The integers $n1$ and $n2$ set in lines 2 and 3 represent the beginning and the end of block col. In line 10, $y(n1 : n2)$ is multiplied by the diagonal D^{-1}, which is kept in inverted form in the array $diag$. The outer loop, i.e., the loop starting in line 1, is sequential. The loop starting in line 6 is vectorizable/parallelizable. There is additional parallelism, which can be extracted in the combination of the two loops starting in lines 5 and 6.

12.5 Multi-Elimination Incomplete LU

The discussion in this section begins with the Gaussian elimination algorithm for a general sparse linear system. Parallelism in sparse Gaussian elimination can be obtained by finding unknowns that are independent at a given stage of the elimination, i.e., unknowns that do not depend on each other according to the binary relation defined by the graph of the matrix. A set of independent unknowns of a linear system is called an independent set. Thus, independent set orderings (ISOs) can be viewed as permutations to put the original matrix in the form

$$\begin{pmatrix} D & E \\ F & C \end{pmatrix}, \tag{12.19}$$

12.5. Multi-Elimination Incomplete LU

in which D is diagonal, but C can be arbitrary. This amounts to a less restrictive form of multicoloring, in which a set of vertices in the adjacency graph is found so that no equation in the set involves unknowns from the same set. A few algorithms for finding ISOs of a general sparse graph were discussed in Chapter 3.

The rows associated with an independent set can be used as pivots simultaneously. When such rows are eliminated, a smaller linear system results, which is again sparse. Then we can find an independent set for this reduced system and repeat the process of reduction. The resulting second reduced system is called the second level reduced system.

The process can be repeated recursively a few times. As the level of the reduction increases, the reduced systems gradually lose their sparsity. A direct solution method would continue the reduction until the reduced system is small enough or dense enough to switch to a dense Gaussian elimination to solve it. This process is illustrated in Figure 12.6. There exist a number of sparse direct solution techniques based on this approach.

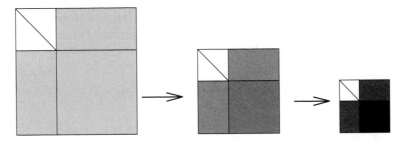

Figure 12.6. *Illustration of two levels of multi-elimination for sparse linear systems.*

After a brief review of the direct solution method based on ISOs, we will explain how to exploit this approach for deriving ILU factorizations by incorporating drop tolerance strategies.

12.5.1 Multi-Elimination

We start with a discussion of an *exact* reduction step. Let A_j be the matrix obtained at the jth step of the reduction, $j = 0, \ldots, nlev$, with $A_0 = A$. Assume that an ISO is applied to A_j and that the matrix is permuted accordingly as follows:

$$P_j A_j P_j^T = \begin{pmatrix} D_j & F_j \\ E_j & C_j \end{pmatrix}, \qquad (12.20)$$

where D_j is a diagonal matrix. Now eliminate the unknowns of the independent set to get the next reduced matrix,

$$A_{j+1} = C_j - E_j D_j^{-1} F_j. \qquad (12.21)$$

This results, implicitly, in a block LU factorization

$$P_j A_j P_j^T = \begin{pmatrix} D_j & F_j \\ E_j & C_j \end{pmatrix} = \begin{pmatrix} I & O \\ E_j D_j^{-1} & I \end{pmatrix} \times \begin{pmatrix} D_j & F_j \\ O & A_{j+1} \end{pmatrix},$$

with A_{j+1} defined above. Thus, in order to solve a system with the matrix A_j, both a forward and a backward substitution need to be performed with the block matrices on the right-hand side of the above system. The backward solution involves solving a system with the matrix A_{j+1}.

This block factorization approach can be used recursively until a system results that is small enough to be solved with a standard method. The transformations used in the elimination process, i.e., the matrices $E_j D_j^{-1}$ and F_j, must be saved. The permutation matrices P_j can also be saved. Alternatively, the matrices involved in the factorization at each new reordering step can be permuted explicitly.

12.5.2 ILUM

The successive reduction steps described above will give rise to matrices that become more and more dense due to the fill-ins introduced by the elimination process. In iterative methods, a common cure for this is to neglect some of the fill-ins introduced by using a simple dropping strategy as the reduced systems are formed. For example, any fill-in element introduced is dropped whenever its size is less than a given tolerance times the 2-norm of the original row. Thus, an *approximate* version of the successive reduction steps can be used to provide an approximate solution $M^{-1}v$ to $A^{-1}v$ for any given v. This can be used to precondition the original linear system. Conceptually, the modification leading to an *incomplete* factorization replaces (12.21) with

$$A_{j+1} = (C_j - E_j D_j^{-1} F_j) - R_j, \tag{12.22}$$

in which R_j is the matrix of the elements that are dropped in this reduction step. Globally, the algorithm can be viewed as a form of incomplete block LU with permutations.

Thus, there is a succession of block ILU factorizations of the form

$$P_j A_j P_j^T = \begin{pmatrix} D_j & F_j \\ E_j & C_j \end{pmatrix}$$

$$= \begin{pmatrix} I & O \\ E_j D_j^{-1} & I \end{pmatrix} \times \begin{pmatrix} D_j & F_j \\ O & A_{j+1} \end{pmatrix} + \begin{pmatrix} O & O \\ O & R_j \end{pmatrix},$$

with A_{j+1} defined by (12.22). An ISO for the new matrix A_{j+1} will then be found and this matrix reduced again in the same manner. It is not necessary to save the successive A_j-matrices, but only the last one that is generated. We need also to save the sequence of sparse matrices

$$B_{j+1} = \begin{pmatrix} D_j & F_j \\ E_j D_j^{-1} & O \end{pmatrix}, \tag{12.23}$$

which contain the transformation needed at level j of the reduction. The successive permutation matrices P_j can be discarded if they are applied to the previous B_i-matrices as soon as these permutation matrices are known. Then only the global permutation is needed, which is the product of all these successive permutations.

An illustration of the matrices obtained after three reduction steps is shown in Figure 12.7. The original matrix is a five-point matrix associated with a 15×15 grid and is therefore of size $N = 225$. Here, the successive matrices B_i (with permutations applied) are shown together with the last A_j-matrix, which occupies the location of the O block in (12.23).

12.5. Multi-Elimination Incomplete LU

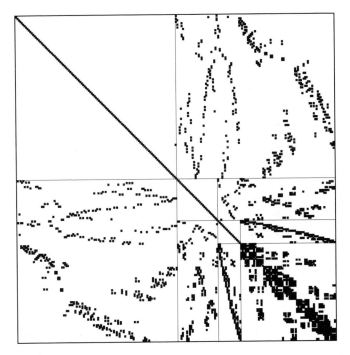

Figure 12.7. *Illustration of the processed matrices obtained from three steps of ISO and reductions.*

We refer to this incomplete factorization as ILU with multi-elimination (ILUM). The preprocessing phase consists of a succession of *nlev* applications of the following three steps: (1) finding the ISO, (2) permuting the matrix, and (3) reducing the matrix.

ALGORITHM 12.4. ILUM: Preprocessing Phase

1. Set $A_0 = A$
2. For $j = 0, 1, \ldots, nlev - 1$, Do
3.   Find an ISO permutation P_j for A_j
4.   Apply P_j to A_j to permute it into the form (12.20)
5.   Apply P_j to $B_1, \ldots, B_j$
6.   Apply P_j to $P_0, \ldots, P_{j-1}$
7.   Compute the matrices A_{j+1} and B_{j+1} defined by (12.22) and (12.23)
8. EndDo

In the backward and forward solution phases, the last reduced system must be solved but not necessarily with high accuracy. For example, we can solve it according to the level of tolerance allowed in the dropping strategy during the preprocessing phase. Observe that, if the linear system is solved inaccurately, only an accelerator that allows variations in the preconditioning should be used. Such algorithms have been discussed in Chapter 9. Alternatively, we can use a fixed number of multicolor SOR or SSOR steps or a fixed polynomial iteration. The implementation of the ILUM preconditioner corresponding to this strategy is rather complicated and involves several parameters.

In order to describe the forward and backward solution, we introduce some notation. We start by applying the *global permutation*, i.e., the product

$$P_{nlev-1}, P_{nlev-2}, \ldots, P_0,$$

to the right-hand side. We overwrite the result on the current solution vector, an N-vector called x_0. Now partition this vector into

$$x_0 = \begin{pmatrix} y_0 \\ x_1 \end{pmatrix}$$

according to the partitioning (12.20). The forward step consists of transforming the second component of the right-hand side as

$$x_1 := x_1 - E_0 D_0^{-1} y_0.$$

Now x_1 is partitioned in the same manner as x_0 and the forward elimination is continued in the same way. Thus, at each step, each x_j is partitioned as

$$x_j = \begin{pmatrix} y_j \\ x_{j+1} \end{pmatrix}.$$

A forward elimination step defines the new x_{j+1} using the old x_{j+1} and y_j for $j = 0, \ldots, nlev - 1$, while a backward step defines y_j using the old y_j and x_{j+1} for $j = nlev - 1, \ldots, 0$. Algorithm 12.5 describes the general structure of the forward and backward solution sweeps. Because the global permutation was applied at the beginning, the successive permutations need not be applied. However, the final result obtained must be permuted back into the original ordering.

ALGORITHM 12.5. ILUM: Forward and Backward Solutions

1. Apply global permutation to right-hand side b and copy into x_0
2. For $j = 0, 1, \ldots, nlev - 1$, Do [Forward sweep]
3. $\quad x_{j+1} := x_{j+1} - E_j D_j^{-1} y_j$
4. EndDo
5. Solve with a relative tolerance ϵ
6. $\quad A_{nlev} x_{nlev} := x_{nlev}$
7. For $j = nlev - 1, \ldots, 1, 0$, Do [Backward sweep]
8. $\quad y_j := D_j^{-1}(y_j - F_j x_{j+1})$
9. EndDo
10. Permute the resulting solution vector back to the original ordering to obtain the solution x

Computer implementations of ILUM can be rather tedious. The implementation issues are similar to those of parallel direct solution methods for sparse linear systems.

12.6 Distributed Incomplete LU and Symmetric Successive Overrelaxation

This section describes parallel variants of the block SOR (BSOR) and ILU(0) preconditioners that are suitable for distributed memory environments. Chapter 11 briefly discussed *distributed sparse matrices*. A distributed matrix is a matrix whose entries are located in the

12.6. Distributed Incomplete LU and Symmetric Successive Overrelaxation

memories of different processors in a multiprocessor system. These types of data structures are very convenient for distributed memory computers and it is useful to discuss implementations of preconditioners that are specifically developed for them. Refer to Section 11.5.5 for the terminology used here. In particular, the term *subdomain* is used in the very general sense of subgraph. For both ILU and SOR, multicoloring or level scheduling can be used at the macro level to extract parallelism. Here, macro level means the level of parallelism corresponding to the processors, blocks, or subdomains.

In the ILU(0) factorization, the LU factors have the same nonzero patterns as the original matrix A, so that the references of the entries belonging to the external subdomains in the ILU(0) factorization are identical to those of the matrix-by-vector product operation with the matrix A. This is not the case for the more accurate ILU(p) factorization, with $p > 0$. If an attempt is made to implement a wave-front ILU preconditioner on a distributed memory computer, a difficulty arises because the natural ordering for the original sparse problem may put an unnecessary limit on the amount of parallelism available. Instead, a two-level ordering is used. First, define a *global* ordering, which is a wave-front ordering for the subdomains. This is based on the graph that describes the coupling between the subdomains: Two subdomains are coupled iff they contain at least a pair of coupled unknowns, one from each subdomain. Then, within each subdomain, define a local ordering.

To describe the possible parallel implementations of these ILU(0) preconditioners, it is sufficient to consider a local view of the distributed sparse matrix, illustrated in Figure 12.8. The problem is partitioned into p subdomains or subgraphs using some graph-partitioning technique. This results in a mapping of the matrix into processors, where it is assumed that the ith equation (row) and the ith unknown are mapped to the same processor. We distinguish between *interior* points and *interface* points. The interior points are those nodes that are not coupled with nodes belonging to other processors. Interface nodes are those local nodes that are coupled with at least one node belonging to another processor. Thus, processor number 10 in the figure holds a certain number of rows that are local rows.

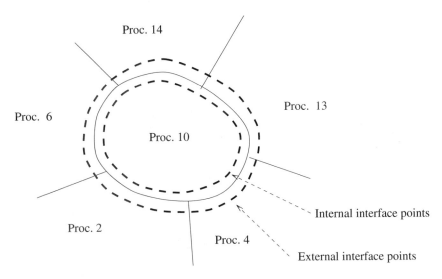

Figure 12.8. *A local view of the distributed ILU(0).*

Consider the rows associated with the interior nodes. The unknowns associated with these nodes are not coupled with variables from other processors. As a result, the rows associated with these nodes can be eliminated independently in the ILU(0) process. The rows associated with the nodes on the interface of the subdomain will require more attention. Recall that an ILU(0) factorization is determined entirely by the order in which the rows are processed.

The interior nodes can be eliminated first. Once this is done, the interface rows can be eliminated *in a certain order*. There are two natural choices for this order. The first would be to impose a global order based on the labels of the processors. Thus, in the illustration, the interface rows belonging to processors 2, 4, and 6 are processed before those in processor 10. The interface rows in processor 10 must in turn be processed before those in processors 13 and 14.

The local order, i.e., the order in which we process the interface rows in the same processor (e.g., processor 10), may not be as important. This global order based on processing element (PE) number defines a natural priority graph and parallelism can be exploited easily in a data-driven implementation.

It is somewhat unnatural to base the ordering just on the processor labeling. Observe that a proper order can also be defined for performing the elimination by *replacing the PE numbers with any labels, provided that any two neighboring processors have a different label*. The most natural way to do this is by performing a multicoloring of the subdomains and using the colors in exactly the same way as before to define an order of the tasks. The algorithms will be written in this general form, i.e., with a label associated with each processor. Thus, the simplest valid labels are the PE numbers, which lead to the PE label-based order. In the following, we define Lab_j as the label of processor number j.

ALGORITHM 12.6. Distributed ILU(0) Factorization

1. *In each processor P_i, $i = 1, \ldots, p$, Do*
2.   *Perform the ILU(0) factorization for interior local rows*
3.   *Receive the factored rows from the adjacent processors j with*
                   $Lab_j < Lab_i$
4.   *Perform the ILU(0) factorization for the interface rows with*
                   pivots received from the external processors in line 3
5.   *Perform the ILU(0) factorization for the boundary nodes with*
                   pivots from the interior rows completed in line 2
6.   *Send the completed interface rows to adjacent processors j with*
                   $Lab_j > Lab_i$
7. *EndDo*

Line 2 of the above algorithm can be performed in parallel because it does not depend on data from other subdomains. Once this distributed ILU(0) factorization is completed, the preconditioned Krylov subspace algorithm will require a forward and backward sweep at each step. The distributed forward/backward solution based on this factorization can be implemented as follows.

ALGORITHM 12.7. Distributed Forward and Backward Sweep

1. In each processor P_i, $i = 1, \ldots, p$, Do
2.   *Forward solve*
3.   Perform the forward solve for the interior nodes
4.   Receive the updated values from the adjacent processors j with $Lab_j < Lab_i$
5.   Perform the forward solve for the interface nodes
6.   Send the updated values of boundary nodes to the adjacent processors j with $Lab_j > Lab_i$
7.   *Backward solve*
8.   Receive the updated values from the adjacent processors j with $Lab_j > Lab_i$
9.   Perform the backward solve for the boundary nodes
10.   Send the updated values of boundary nodes to the adjacent processors j with $Lab_j < Lab_i$
11.   Perform the backward solve for the interior nodes
12. EndDo

As in the ILU(0) factorization, the interior nodes do not depend on the nodes from the external processors and can be computed in parallel in lines 3 and 11. In the forward solve, the solution of the interior nodes is followed by an exchange of data and the solution on the interface. The backward solve works in reverse in that the boundary nodes are first computed, then they are sent to adjacent processors. Finally, interior nodes are updated.

12.7 Other Techniques

This section gives a brief account of other parallel preconditioning techniques that are sometimes used. Chapter 14 also examines another important class of methods, which were briefly mentioned before, namely, the class of domain decomposition methods.

12.7.1 AINVs

Another class of preconditioners that require only matrix-by-vector products is the class of approximate inverse (AINV) preconditioners. Discussed in Chapter 10, these can be used in many different ways. Besides being simple to implement, both their preprocessing phase and iteration phase allow a large degree of parallelism. Their disadvantage is similar to polynomial preconditioners; namely, the number of steps required for convergence may be large, possibly substantially larger than with the standard techniques. On the positive side, they are fairly robust techniques that can work well where standard methods may fail.

12.7.2 EBE Techniques

A somewhat specialized set of techniques is the class of element-by-element (EBE) preconditioners, which are geared toward finite element problems and are motivated by the desire to avoid assembling finite element matrices. Many finite element codes keep the data related to the linear system in unassembled form. The element matrices associated with each element are stored and never added together. This is convenient when using direct methods, since there are techniques, known as frontal methods, that allow Gaussian elimination to be performed by using a few elements at a time.

It was seen in Chapter 2 that the global stiffness matrix A is the sum of matrices $A^{[e]}$ associated with each element; i.e.,

$$A = \sum_{e=1}^{Nel} A^{[e]}.$$

Here the matrix $A^{[e]}$ is an $n \times n$ matrix defined as

$$A^{[e]} = P_e A_{K_e} P_e^T,$$

in which A_{K_e} is the element matrix and P_e is a Boolean connectivity matrix that maps the coordinates of the small A_{K_e}-matrix into those of the full matrix A. Chapter 2 showed how matrix-by-vector products can be performed in unassembled form. To perform this product in parallel, note that the only potential obstacle to performing the matrix-by-vector product in parallel, i.e., across all elements, is in the last phase, i.e., when the contributions are summed to the resulting vector y. In order to add the contributions $A^{[e]}x$ in parallel, group elements that do not have nodes in common. Referring to (2.46), the contributions

$$y_e = A_{K_e}(P_e^T x)$$

can all be computed in parallel and do not depend on one another. The operations

$$y := y + P_e y_e$$

can be processed in parallel for any group of elements that do not share any vertices. This grouping can be found by performing a multicoloring of the elements. Any two elements with a node in common receive different colors. Using this idea, good performance can be achieved on vector computers.

EBE preconditioners are based on similar principles and many different variants have been developed. They are defined by first normalizing each of the element matrices. In what follows, assume that A is an SPD matrix. Typically, a diagonal, or block diagonal, scaling is first applied to A to obtain a scaled matrix $\tilde{A}$:

$$\tilde{A} = D^{-1/2} A D^{-1/2}. \tag{12.24}$$

This results in each matrix $A^{[e]}$ and element matrix A_{K_e} being transformed similarly:

$$\begin{aligned}
\tilde{A}^{[e]} &= D^{-1/2} A^{[e]} D^{-1/2} \\
&= D^{-1/2} P_e A_{K_e} D^{-1/2} \\
&= P_e (P_e^T D^{-1/2} P_e) A^{[e]} (P_e D^{-1/2} P_e^T) \\
&\equiv P_e \tilde{A}_{K_e} P_e^T.
\end{aligned}$$

The second step in defining an EBE preconditioner is to *regularize* each of these transformed matrices. Indeed, each of the matrices $A^{[e]}$ is of rank p_e at most, where p_e is the size of

12.7. Other Techniques

the element matrix A_{K_e}, i.e., the number of nodes that constitute the eth element. In the so-called Winget regularization, the diagonal of each $A^{[e]}$ is forced to be the identity matrix. In other words, the regularized matrix is defined as

$$\bar{A}^{[e]} = I + \tilde{A}^{[e]} - \text{diag}(\tilde{A}^{[e]}). \tag{12.25}$$

These matrices are positive definite; see Exercise 8.

The third and final step in defining an EBE preconditioner is to choose the factorization itself. In the EBE Cholesky factorization, the Cholesky (or Crout) factorization of each regularized matrix $\bar{A}^{[e]}$ is performed:

$$\bar{A}^{[e]} = L_e D_e L_e^T. \tag{12.26}$$

The preconditioner from it is defined as

$$M = \prod_{e=1}^{nel} L_e \times \prod_{e=1}^{nel} D_e \times \prod_{e=nel}^{1} L_e^T. \tag{12.27}$$

Note that, to ensure symmetry, the last product is in reverse order of the first one. The factorization (12.26) consists of a factorization of the small $p_e \times p_e$ matrix $\bar{A}_{K_e}$. Performing the preconditioning operations will therefore consist of a sequence of small $p_e \times p_e$ backward or forward solves. The gather and scatter matrices P_e defined in Chapter 2 must also be applied for each element. These solves are applied to the right-hand side in sequence. In addition, the same multicoloring idea as for the matrix-by-vector product can be exploited to perform these sweeps in parallel.

One of the drawbacks of the EBE Cholesky preconditioner is that an additional set of element matrices must be stored. That is because the factorizations (12.26) must be stored for each element. In EBE/SSOR, this is avoided. Instead of factoring each $\bar{A}^{[e]}$, the usual splitting of each $\bar{A}^{[e]}$ is exploited. Assuming the Winget regularization, we have

$$\bar{A}^{[e]} = I - E_e - E_e^T, \tag{12.28}$$

in which $-E_e$ is the strict lower part of $\bar{A}^{[e]}$. By analogy with the SSOR preconditioner, the EBE/SSOR preconditioner is defined by

$$M = \prod_{e=1}^{nel} (I - \omega E_e) \times \prod_{e=1}^{nel} D_e \times \prod_{e=nel}^{1} (I - \omega E_e^T). \tag{12.29}$$

12.7.3 Parallel Row Projection Preconditioners

One of the attractions of row projection methods seen in Chapter 8 is their high degree of parallelism. In Cimmino's method, the scalars δ_i as well as the new residual vector can be computed in parallel. In the Gauss–Seidel normal error (resp. Gauss–Seidel normal residual), it is also possible to group the unknowns in such a way that any pair of rows (resp. columns) have disjointed nonzero patterns. Updates of components in the same group can then be performed in parallel. This approach essentially requires finding a multicolor ordering for the matrix $B = AA^T$ (resp. $B = A^T A$).

It is necessary to first identify a partition of the set $\{1, 2, \ldots, N\}$ into subsets $S_1, \ldots, S_k$ such that the rows (resp. columns) whose indices belong to the same set S_i are *structurally orthogonal* to each other, i.e., have no nonzero elements in the same column locations. When implementing a BSOR scheme where the blocking is identical to that defined by the partition, all of the unknowns belonging to the same set S_j can be updated in parallel. To be more specific, the rows are reordered by scanning those in S_1 followed by those in S_2, etc. Denote by A_i the matrix consisting of the rows belonging to the ith block. We assume that all rows of the same set are orthogonal to each other and that they have been normalized so that their 2-norm is unity. Then a block Gauss–Seidel sweep, which generalizes Algorithm 8.1, follows.

ALGORITHM 12.8. Forward Block Normal Error Gauss–Seidel Sweep

1. Select an initial x_0
2. For $i = 1, 2, \ldots, k$, Do
3. $d_i = b_i - A_i x$
4. $x := x + A_i^T d_i$
5. EndDo

Here, x_i and b_i are subvectors corresponding to the blocking and d_i is a vector of length the size of the block, which replaces the scalar δ_i of Algorithm 8.1. There is parallelism in each of the lines 3 and 4.

The question that arises is how to find good partitions S_i. In simple cases, such as block tridiagonal matrices, this can easily be done; see Exercise 7. For general sparse matrices, a multicoloring algorithm on the graph of AA^T (resp. $A^T A$) can be employed. However, these matrices are never stored explicitly. Their rows can be generated, used, and then discarded.

Exercises

1. Let A be an SPD matrix and consider $N = I - D^{-1}A$, where D is a block diagonal of A.
    a. Show that D is an SPD matrix. Denote by $(\cdot, \cdot)_D$ the associated inner product.
    b. Show that N is self-adjoint with respect to $(\cdot, \cdot)_D$.
    c. Show that N^k is self-adjoint with respect to $(\cdot, \cdot)_D$ for any integer k.
    d. Show that the Neumann series expansion preconditioner defined by the right-hand side of (12.3) leads to a preconditioned matrix that is self-adjoint with respect to the D inner product.
    e. Describe an implementation of the PCG algorithm using this preconditioner.

2. The development of the Chebyshev iteration algorithm seen in Section 12.3.2 can be exploited to derive yet another formulation of the conjugate algorithm from the Lanczos algorithm. Observe that the recurrence relation (12.8) is not restricted to scaled Chebyshev polynomials.

a. The scaled Lanczos polynomials, i.e., the polynomials $p_k(t)/p_k(0)$, in which $p_k(t)$ is the polynomial such that $v_{k+1} = p_k(A)v_1$ in the Lanczos algorithm, satisfy a relation of the form (12.8). What are the coefficients ρ_k and δ in this case?

b. Proceed in the same manner as in Section 12.3.2 to derive a version of the CG algorithm.

3. Show that ρ_k as defined by (12.7) has a limit ρ. What is this limit? Assume that Algorithm 12.1 is to be executed with the ρ_k's all replaced by this limit ρ. Will the method converge? What is the asymptotic rate of convergence of this modified method?

4. Derive the least-squares polynomials for $\alpha = -\frac{1}{2}, \beta = \frac{1}{2}$ for the interval $[0, 1]$ for $k = 1, 2, 3, 4$. Check that these results agree with those of the table shown at the end of Section 12.3.3.

5. Consider the mesh shown below. Assume that the objective is to solve the Poisson equation with Dirichlet boundary conditions.

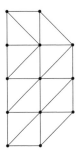

a. Consider the resulting matrix obtained (before boundary conditions are applied) from ordering the nodes from bottom up, and left to right (thus, the bottom-left vertex is labeled 1 and the top-right vertex is labeled 13). What is the bandwidth of the linear system? How many memory locations would be needed to store the matrix in skyline format? (Assume that the matrix is nonsymmetric so both upper and lower triangular parts must be stored.)

b. Is it possible to find a two-color ordering of the mesh points? If so, show the ordering, or otherwise prove that it is not possible.

c. Find an independent set of size five. Show the pattern of the matrix associated with this ISO.

d. Find a multicolor ordering of the mesh by using the greedy multicolor algorithm. Can you find a better coloring (i.e., a coloring with fewer colors)? If so, show the coloring (use letters to represent the colors).

6. A linear system $Ax = b$, where A is a five-point matrix, is reordered using red-black ordering as

$$\begin{pmatrix} D_1 & F \\ E & D_2 \end{pmatrix} \begin{pmatrix} x \\ y \end{pmatrix} = \begin{pmatrix} f \\ g \end{pmatrix}.$$

a. Write the block Gauss–Seidel iteration associated with the above partitioned system (where the blocking in block Gauss–Seidel is the same as the above blocking).

b. Express the y iterates independently of the x iterates; i.e., find an iteration that involves only y iterates. What type of iteration is the resulting scheme?

7. Consider a tridiagonal matrix $T = \text{tridiag}(a_i, b_i, c_i)$. Find a grouping of the rows such that rows in each group are *structurally* orthogonal, i.e., orthogonal regardless of the values of the entry. Find a set of three groups at most. How can this be generalized to block tridiagonal matrices such as those arising from 2-D and 3-D centered difference matrices?

8. Why are the Winget regularized matrices $\bar{A}^{[e]}$ defined by (12.25) positive definite when the matrix $\tilde{A}$ is obtained from A by a *diagonal* scaling from A?

Notes and References

When vector processing appeared in the middle to late 1970s, a number of efforts were made to change algorithms, or implementations of standard methods, to exploit the new architectures. One of the first ideas in this context was to perform matrix-by-vector products by diagonals [183]. Matrix-by-vector products using this format can yield excellent performance. Hence the idea came of using polynomial preconditioning.

Polynomial preconditioning was exploited, independently of supercomputing, as early as 1937 in a paper by Cesari [71], and then in a 1952 paper by Lanczos [195]. The same idea was later applied for eigenvalue problems by Stiefel, who employed least-squares polynomials [275], and Rutishauser [236], who combined the QD algorithm with Chebyshev acceleration. Dubois et al. [105] suggested using polynomial preconditioning, specifically, the Neumann series expansion, for solving SPD linear systems on vector computers. Johnson et al. [179] later extended the idea by exploiting Chebyshev polynomials and other orthogonal polynomials. It was observed in [179] that least-squares polynomials tend to perform better than those based on the uniform norm, in that they lead to a better overall clustering of the spectrum. Moreover, as was already observed by Rutishauser [236], in the symmetric case there is no need for accurate eigenvalue estimates: it suffices to use the simple bounds that are provided by Gershgorin's theorem. In [240] it was also observed that in some cases the least-squares polynomial approach, which requires less information than the Chebyshev approach, tends to perform better.

The use of least-squares polynomials over polygons was first advocated by Smolarski and Saylor [270] and later by Saad [241]. The application to the indefinite case was examined in detail in [239]. Still in the context of using polygons instead of ellipses, yet another attractive possibility proposed by Fischer and Reichel [129] avoids the problem of best approximation altogether. The polygon can be conformally transformed into a circle and the theory of Faber polynomials yields a simple way of deriving good polynomials from exploiting specific points on the circle.

Although only approaches based on the formulations (12.5) and (12.11) have been discussed in this book, there are other lesser known possibilities based on minimizing $\|1/\lambda - s(\lambda)\|_\infty$. There has been very little work on polynomial preconditioning or Krylov subspace methods for highly non-normal matrices; see, however, the recent analysis in [284]. Another important point is that polynomial preconditioning can be combined with a subsidiary relaxation-type preconditioning such as SSOR [2, 216]. Finally, polynomial preconditionings can be useful in some special situations such as that of complex linear systems arising from the Helmholtz equation [132].

Multicoloring has been known for a long time in the numerical analysis literature and was used in particular for understanding the theory of relaxation techniques [321, 292] as well as for deriving efficient alternative formulations of some relaxation algorithms [292, 151]. With the advent of parallel processing, it became an essential ingredient in parallelizing iterative algorithms; see, for example, [4, 2, 117, 218, 217, 227]. In [98] and [247] it was observed that k-step SOR preconditioning was very competitive relative to the standard ILU preconditioners. Combined with multicolor ordering, multiple-step SOR can perform quite well on vector computers. Multicoloring is also useful in finite element methods, where elements instead of nodes are colored [31, 296]. In EBE techniques, multicoloring is used when forming the residual, i.e., when multiplying an unassembled matrix by a vector [173, 126, 261]. The contributions of the elements of the same color can all be evaluated and applied simultaneously to the resulting vector.

ISOs have been used in the context of parallel direct solution techniques for sparse matrices [95, 198, 199], and multifrontal techniques [107] can be viewed as a particular case. The gist of all these techniques is that it is possible to reorder the system in groups of equations that can be solved simultaneously. A parallel direct sparse solver based on performing several successive levels of ISOs and reduction was suggested in [198] and in a more general form in [94].

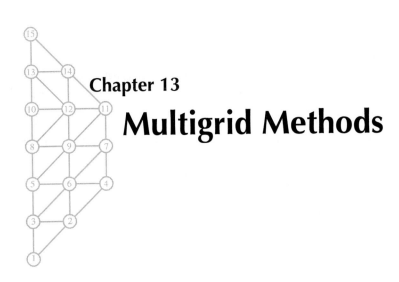

Chapter 13
Multigrid Methods

The convergence of preconditioned Krylov subspace methods for solving systems arising from discretized partial differential equations (PDEs) tends to slow down considerably as these systems become larger. This deterioration in the convergence rate, compounded with the increased operation count per step due to the sheer problem size, results in a severe loss of efficiency. In contrast, the class of methods to be described in this chapter is capable of achieving convergence rates that are, in theory, independent of the mesh size. One significant difference with the preconditioned Krylov subspace approach is that multigrid (MG) methods were initially designed specifically for the solution of discretized elliptic PDEs. The method was later extended in different ways to handle other PDE problems, including nonlinear ones, as well as problems not modeled by PDEs. Because these methods exploit more information on the problem than do standard preconditioned Krylov subspace methods, their performance can be vastly superior. On the other hand, they may require implementations that are specific to the physical problem at hand, in contrast with preconditioned Krylov subspace methods, which attempt to be general purpose.

13.1 Introduction

MG techniques exploit discretizations with different mesh sizes of a given problem to obtain optimal convergence from relaxation techniques. At the foundation of these techniques is the basic and powerful principle of divide and conquer. Though most relaxation-type iterative processes, such as Gauss–Seidel, may converge slowly for typical problems, it can be noticed that the components of the errors (or residuals) in the directions of the eigenvectors of the iteration matrix corresponding to the large eigenvalues are damped very rapidly. These eigenvectors are known as the oscillatory or high frequency modes. The other components, associated with low frequency or smooth modes, are difficult to damp with standard relaxation. This causes the observed slowdown of all basic iterative methods. However, many of these modes (say half) are mapped naturally into high frequency modes on a coarser mesh, hence the idea of moving to a coarser mesh to eliminate the corresponding error components. The process can obviously be repeated with the help of recursion, using a hierarchy of meshes.

The methods described in this chapter will differ in one essential way from those seen so far. They will require taking a special look at the original physical problem and in particular at the modes associated with different meshes. The availability of a hierarchy of meshes and the corresponding linear problems opens up possibilities that were not available with the methods seen so far, which only have access to the coefficient matrix and the right-hand side. There are, however, generalizations of MG methods, termed algebraic MG (AMG), which attempt to reproduce the outstanding performance enjoyed by multigrid in the regularly structured elliptic case. This is done by extending in a purely algebraic manner the fundamental principles just described to general sparse linear systems.

This chapter will begin with a description of the model problems and the spectra of the associated matrices. This is required for understanding the motivation and theory behind MG.

13.2 Matrices and Spectra of Model Problems

Consider first the one-dimensional model problem seen in Chapter 2:

$$-u''(x) = f(x) \text{ for } x \in (0, 1), \tag{13.1}$$
$$u(0) = u(1) = 0. \tag{13.2}$$

The interval $[0, 1]$ is discretized with centered difference approximations, using the equally spaced $n + 2$ points

$$x_i = i \times h, \ i = 0, \ldots, n + 1,$$

where $h = 1/(n + 1)$. A common notation is to call the original (continuous) domain Ω and its discrete version Ω_h. So $\Omega = (0, 1)$ and $\Omega_h = \{x_i\}_{i=0,\ldots,n+1}$. The discretization results in the system

$$Ax = b, \tag{13.3}$$

where

$$A = \begin{pmatrix} 2 & -1 & & & \\ -1 & 2 & -1 & & \\ & \ddots & \ddots & \ddots & \\ & & -1 & 2 & -1 \\ & & & -1 & 2 \end{pmatrix}, \quad b = h^2 \begin{pmatrix} f(x_0) \\ f(x_1) \\ \vdots \\ f(x_{n-2}) \\ f(x_{n-1}) \end{pmatrix}. \tag{13.4}$$

The above system is of size $n \times n$.

Next, the eigenvalues and eigenvectors of the matrix A will be determined. The following trigonometric relation will be useful:

$$\sin((j + 1)\theta) + \sin((j - 1)\theta) = 2 \sin(j\theta) \cos \theta. \tag{13.5}$$

13.2. Matrices and Spectra of Model Problems

Consider the vector u whose components are $\sin\theta, \sin 2\theta, \ldots, \sin n\theta$. Using the relation (13.5) we find that

$$(A - 2(1 - \cos\theta)I)u = \sin((n+1)\theta)e_n,$$

where e_n is the nth column of the identity. The right-hand side in the above relation is equal to zero for the following values of θ:

$$\theta_k = \frac{k\pi}{n+1}, \qquad (13.6)$$

for any integer value k. Therefore, the eigenvalues of A are

$$\lambda_k = 2(1 - \cos\theta_k) = 4\sin^2\frac{\theta_k}{2}, \quad k = 1, \ldots, n, \qquad (13.7)$$

and the associated eigenvectors are given by

$$w_k = \begin{pmatrix} \sin\theta_k \\ \sin(2\theta_k) \\ \vdots \\ \sin(n\theta_k) \end{pmatrix}. \qquad (13.8)$$

The ith component of w_k can be rewritten in the form

$$\sin\frac{ik\pi}{n+1} = \sin(k\pi x_i)$$

and represents the value of the function $\sin(k\pi x)$ at the discretization point x_i. This component of the eigenvector may therefore be written

$$w_k(x_i) = \sin(k\pi x_i). \qquad (13.9)$$

Note that these eigenfunctions satisfy the boundary conditions $w_k(x_0) = w_k(x_{n+1}) = 0$. These eigenfunctions are illustrated in Figure 13.1 for the case $n = 7$.

Now consider the two-dimensional Poisson equation

$$-\left(\frac{\partial^2 u}{\partial x^2} + \frac{\partial^2 u}{\partial y^2}\right) = f \quad \text{in } \Omega, \qquad (13.10)$$

$$u = 0 \quad \text{on } \Gamma, \qquad (13.11)$$

where Ω is the rectangle $(0, l_1) \times (0, l_2)$ and Γ its boundary. Both intervals can be discretized uniformly by taking $n+2$ points in the x direction and $m+2$ points in the y direction:

$$x_i = i \times h_1, i = 0, \ldots, n+1; \quad y_j = j \times h_2, j = 0, \ldots, m+1,$$

where

$$h_1 = \frac{l_1}{n+1}, \quad h_2 = \frac{l_2}{m+1}.$$

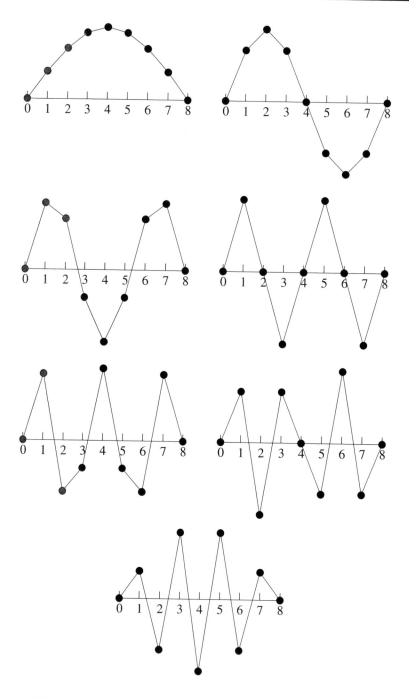

Figure 13.1. *The seven eigenfunctions of the discretized one-dimensional Laplacian when $n = 7$.*

13.2. Matrices and Spectra of Model Problems

For simplicity we now assume that $h_1 = h_2$. In this case the linear system has the form (13.3), where A has the form

$$A = \begin{pmatrix} B & -I & & & \\ -I & B & -I & & \\ & \ddots & \ddots & \ddots & \\ & & -I & B & -I \\ & & & -I & B \end{pmatrix}, \quad \text{with} \quad B = \begin{pmatrix} 4 & -1 & & & \\ -1 & 4 & -1 & & \\ & \ddots & \ddots & \ddots & \\ & & -1 & 4 & -1 \\ & & & -1 & 4 \end{pmatrix}.$$

The right-hand side is again the discrete version of the function f scaled by h^2. The above matrix can be represented in a succinct way using tensor product notation. Given an $m \times p$ matrix X and an $n \times q$ matrix Y, the matrix

$$X \otimes Y$$

can be viewed as a block matrix that has in (block) location (i, j) the matrix $x_{ij}Y$. In other words, $X \otimes Y$ is of size $(nm) \times (pq)$ and is obtained by expanding each entry x_{ij} of X into the block $x_{ij}Y$. This definition is also valid for vectors by considering them as matrices ($p = q = 1$).

With this notation, the matrix A given above can be written as

$$A = I \otimes T_x + T_y \otimes I, \tag{13.12}$$

in which T_x and T_y are tridiagonal matrices of the same form as the matrix A in (13.4) and of dimension n and m, respectively. Often, the right-hand side of (13.12) is called the *tensor sum* of T_x and T_y and is denoted by $T_x \oplus T_y$. A few simple properties are easy to show (see Exercise 1) for tensor products and tensor sums. One that is important for determining the spectrum of A is

$$(T_x \oplus T_y)(v \otimes w) = v \otimes (T_x w) + (T_y v) \otimes w. \tag{13.13}$$

In particular, if w_k is an eigenvector of T_x associated with σ_k and v_l is an eigenvector of T_l associated with μ_l, it is clear that

$$(T_x \oplus T_y)(v_l \otimes w_k) = v_l \otimes (T_x w_k) + (T_y v_k) \otimes w_k = (\sigma_k + \mu_l) v_l \otimes w_k.$$

So $\lambda_{kl} = \sigma_k + \mu_l$ is an eigenvalue of A for each pair of eigenvalues $\sigma_k \in \Lambda(T_x)$ and $\mu_k \in \Lambda(T_y)$. The associated eigenvector is $v_l \otimes w_k$. These eigenvalues and associated eigenvectors are best labeled with two indices:

$$\begin{aligned} \lambda_{kl} &= 2\left(1 - \cos\frac{k\pi}{n+1}\right) + 2\left(1 - \cos\frac{l\pi}{m+1}\right) \\ &= 4\left(\sin^2\frac{k\pi}{2(n+1)} + \sin^2\frac{l\pi}{2(m+1)}\right). \end{aligned} \tag{13.14}$$

Their associated eigenvectors $z_{k,l}$ are

$$z_{k,l} = v_l \otimes w_k$$

and they are best expressed by their values at the points (x_i, y_j) on the grid:

$$z_{k,l}(x_i, y_j) = \sin(k\pi x_i) \sin(l\pi y_j).$$

When all the sums $\sigma_k + \mu_l$ are distinct, this gives all the eigenvalues and eigenvectors of A. Otherwise, we must show that the multiple eigenvalues correspond to independent eigenvectors. In fact it can be shown that the system

$$\{v_l \otimes w_k\}_{k=1,\ldots,n,\ l=1,\ldots,m}$$

is an orthonormal system if the systems of both the v_l's and the w_k's are orthonormal.

13.2.1 The Richardson Iteration

MG can be easily motivated by taking an in-depth look at simple iterative schemes such as the Richardson iteration and the Jacobi method. Note that these two methods are essentially identical for the model problems under consideration, because the diagonal of the matrix is a multiple of the identity matrix. The Richardson iteration is considered here for the one-dimensional case, using a fixed parameter ω. In the next section, the weighted Jacobi iteration is fully analyzed with an emphasis on studying the effect of varying the parameter ω.

The Richardson iteration takes the form

$$u_{j+1} = u_j + \omega(b - Au_j) = (I - \omega A)u_j + \omega b.$$

Thus, the iteration matrix is

$$M_\omega = I - \omega A. \tag{13.15}$$

Recall from Example 4.1 in Chapter 4 that convergence takes place for $0 < \omega < 2/\rho(A)$. In realistic situations, the optimal ω given by (4.33) is difficult to use. Instead, an upper bound $\rho(A) \leq \gamma$ is often available from, e.g., Gershgorin's theorem, and we can simply take $\omega = 1/\gamma$. This yields a converging iteration since $1/\gamma \leq 1/\rho(A) < 2/\rho(A)$.

By the relation (13.15), the eigenvalues of the iteration matrix are $1 - \omega \lambda_k$, where λ_k is given by (13.7). The eigenvectors are the same as those of A. If u_* is the exact solution then, as was seen in Chapter 4, the error vector $d_j \equiv u_* - u_j$ obeys the relation

$$d_j = M_\omega^j d_0. \tag{13.16}$$

It is useful to expand the error vector d_0 in the eigenbasis of M_ω as

$$d_0 = \sum_{k=1}^{n} \xi_k w_k.$$

From (13.16) and (13.15) this implies that, at step j,

$$d_j = \sum_{k=1}^{n} \left(1 - \frac{\lambda_k}{\gamma}\right)^j \xi_k w_k.$$

Each component is reduced by $(1 - \lambda_k/\gamma)^j$. The slowest converging component corresponds to the smallest eigenvalue λ_1, which could yield a very slow convergence rate when $|\lambda_1/\gamma| \ll 1$.

13.2. Matrices and Spectra of Model Problems

For the model problem seen above, in the one-dimensional case, Gershgorin's theorem yields $\gamma = 4$ and the corresponding reduction coefficient is

$$1 - \sin^2 \frac{\pi}{2(n+1)} \approx 1 - (\pi h/2)^2 = 1 - O(h^2).$$

As a result, convergence can be quite slow for fine meshes, i.e., when h is small. However, the basic observation on which MG methods are founded is that convergence is not similar for all components. Half of the error components actually see a very good decrease. This is the case for the *high frequency* components, that is, all those components corresponding to $k > n/2$. This part of the error is often referred to as the *oscillatory part*, for obvious reasons. The reduction factors for these components are

$$\eta_k = 1 - \sin^2 \frac{k\pi}{2(n+1)} = \cos^2 \frac{k\pi}{2(n+1)} \leq \frac{1}{2}.$$

These coefficients are illustrated in Figure 13.2. Two important observations can be made. The first is that the oscillatory components, i.e., those corresponding to $\theta_{n/2+1}, \ldots, \theta_n$, undergo excellent reduction, better than $1/2$, at each step of the iteration. It is also important to note that this factor is independent of the step size h. The next observation will give a hint as to what might be done to reduce the other components. In order to see this we now introduce, for the first time, a coarse-grid problem. Assume that n is odd and consider the problem issuing from discretizing the original PDE (13.1) on a mesh Ω_{2h} with the mesh size $2h$. The superscripts h and $2h$ will now be used to distinguish between quantities related to each grid. The grid points on the coarser mesh are $x_i^{2h} = i * (2h)$. The second observation is based on the simple fact that $x_i^{2h} = x_{2i}^h$, from which it follows that, for $k \leq n/2$,

$$w_k^h(x_{2i}^h) = \sin(k\pi x_{2i}^h) = \sin(k\pi x_i^{2h}) = w_k^{2h}(x_i^{2h}).$$

In other words, taking a smooth mode on the fine grid (w_k^h with $k \leq n/2$) and canonically injecting it into the coarse grid, i.e., defining its values on the coarse points to be the same

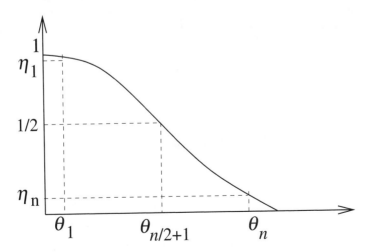

Figure 13.2. *Reduction coefficients for the Richardson method applied to the one-dimensional model problem.*

as those on the fine points, yields the kth mode on the coarse grid. This is illustrated in Figure 13.3 for $k = 2$ and grids of nine points ($n = 7$) and five points ($n = 3$).

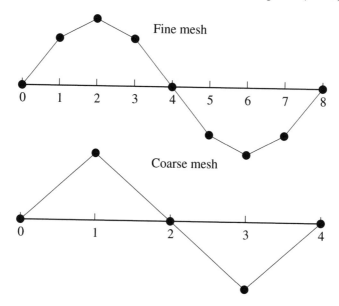

Figure 13.3. *The mode w_2 on a fine grid ($n = 7$) and a coarse grid ($n = 3$).*

Some of the modes that were smooth on the fine grid become oscillatory. For example, when n is odd, the mode $w^h_{(n+1)/2}$ becomes precisely the highest mode on Ω_{2h}. At the same time the oscillatory modes on the fine mesh are no longer represented on the coarse mesh. The iteration fails to make progress on the fine grid when the only components left are those associated with the smooth modes. MG strategies do not attempt to eliminate these components on the fine grid. Instead, they first move down to a coarser grid, where smooth modes are translated into oscillatory ones. Practically, this requires going back and forth between different grids. The necessary grid transfer operations will be discussed in detail later.

13.2.2 Weighted Jacobi Iteration

In this section a weighted Jacobi iteration is considered and analyzed for both one- and two-dimensional model problems. The standard Jacobi iteration is of the form

$$u_{j+1} = D^{-1}(E + F)u_j + D^{-1}f.$$

The weighted version of this iteration uses a parameter ω and combines the above iterate with the current u_j:

$$\begin{aligned} u_{j+1} &= \omega\left(D^{-1}(E + F)u_j + D^{-1}f\right) + (1 - \omega)u_j \\ &= \left[(1 - \omega)I + \omega D^{-1}(E + F)\right]u_j + \omega D^{-1}f \quad (13.17) \\ &\equiv J_\omega u_j + f_\omega. \quad (13.18) \end{aligned}$$

13.2. Matrices and Spectra of Model Problems

Using the relation $E + F = D - A$ it follows that

$$J_\omega = I - \omega D^{-1} A. \tag{13.19}$$

In particular note that, when A is symmetric positive definite (SPD), the weighted Jacobi iteration will converge when $0 < \omega < 2/\rho(D^{-1}A)$. In the case of our one-dimensional model problem the diagonal is $D = 2I$, so the following expression for J_ω is obtained:

$$J_\omega = (1 - \omega)I + \frac{\omega}{2}(2I - A) = I - \frac{\omega}{2}A. \tag{13.20}$$

For the two-dimensional case, a similar result can be obtained in which the denominator 2 is replaced with 4. The eigenvalues of the iteration matrix follow immediately from the expression (13.7):

$$\mu_k(\omega) = 1 - \omega\left(1 - \cos\frac{k\pi}{n+1}\right) = 1 - 2\omega\left(\sin^2\frac{k\pi}{2(n+1)}\right). \tag{13.21}$$

In the two-dimensional case, these become

$$\mu_{k,l}(\omega) = 1 - \omega\left(\sin^2\frac{k\pi}{2(n+1)} + \sin^2\frac{l\pi}{2(m+1)}\right).$$

Consider the one-dimensional case first. The sine terms $\sin^2(k\pi/2(n+1))$ lie between $1 - s^2$ and s^2, in which $s = \sin(\pi/2(n+1))$. Therefore, the eigenvalues are bounded as follows:

$$(1 - 2\omega) + 2\omega s^2 \le \mu_k(\omega) \le 1 - 2\omega s^2. \tag{13.22}$$

The spectral radius of J_ω is

$$\rho(J_\omega) = \max\{|(1 - 2\omega) + 2\omega s^2|, |1 - 2\omega s^2|\}.$$

When ω is less than 0 or greater than 1, it can be shown that $\rho(J_\omega) > 1$ for h small enough (see Exercise 1). When ω is between 0 and 1, then the spectral radius is simply $1 - 2\omega s^2 \approx 1 - \omega\pi^2 h^2/2$.

It is interesting to note that the best ω in the interval $[0, 1]$ is $\omega = 1$, so no acceleration of the original Jacobi iteration can be achieved. On the other hand, if the weighted Jacobi iteration is regarded as a smoother, the situation is different. For those modes associated with $k > n/2$, the term $\sin^2\theta_k$ is not less than $1/2$, so

$$1 - 2\omega < (1 - 2\omega) + 2\omega s^2 \le \mu_k(\omega) \le 1 - \frac{1}{2}\omega. \tag{13.23}$$

For example, when $\omega = 1/2$, then all reduction coefficients for the oscillatory modes will be between 0 and 3/4, thus guaranteeing again a reduction of h. For $\omega = 2/3$ the eigenvalues are between $-1/3$ and $1/3$, leading to a smoothing factor of 1/3. This is the best that can be achieved *independently of h*.

For two-dimensional problems, the situation is qualitatively the same. The bound (13.22) becomes

$$(1 - 2\omega) + \omega(s_x^2 + s_y^2) \le \mu_{k,l}(\omega) \le 1 - \omega(s_x^2 + s_y^2), \tag{13.24}$$

in which s_x is the same as s and $s_y = \sin(\pi/(2(m+1)))$. The conclusion for the spectral radius and rate of convergence of the iteration is similar, in that $\rho(J_\omega) \approx 1 - O(h^2)$ and the best ω is one. In addition, the high frequency modes are damped with coefficients that satisfy

$$1 - 2\omega < (1 - 2\omega) + \omega(s_x^2 + s_y^2) \le \mu_{k,l}(\omega) \le 1 - \frac{1}{2}\omega. \tag{13.25}$$

As before, $\omega = 1/2$ yields a smoothing factor of 3/4 and $\omega = 3/5$ yields a smoothing factor of 4/5. Here the best that can be done is to take $\omega = 4/5$.

13.2.3 Gauss–Seidel Iteration

In practice, Gauss–Seidel and red-black Gauss–Seidel relaxation are more common smoothers than the Jacobi or the Richardson iterations. Also, successive overrelaxation (SOR) (with $\omega \ne 1$) is rarely used as it is known that overrelaxation adds no benefit in general. Gauss–Seidel and SOR schemes are somewhat more difficult to analyze.

Consider the iteration matrix

$$G = (D - E)^{-1} F \tag{13.26}$$

in the one-dimensional case. The eigenvalues and eigenvectors of G satisfy the relation

$$[F - \lambda(D - E)]u = 0,$$

the jth row of which is

$$\xi_{j+1} - 2\lambda \xi_j + \lambda \xi_{j-1} = 0, \tag{13.27}$$

where ξ_j is the j component of the vector u. The boundary conditions $\xi_0 = \xi_{n+1} = 0$ should be added to the above equations. Note that, because $\xi_{n+1} = 0$, (13.27) is valid when $j = n$ (despite the fact that entry $(n, n+1)$ of F is not defined). This is a difference equation, similar to (2.21) encountered in Chapter 2, and it can be solved similarly by seeking a general solution in the form $\xi_j = r^j$. Substituting in (13.27), r must satisfy the quadratic equation

$$r^2 - 2\lambda r + \lambda = 0,$$

whose roots are

$$r_1 = \lambda + \sqrt{\lambda^2 - \lambda}, \quad r_2 = \lambda - \sqrt{\lambda^2 - \lambda}.$$

This leads to the general solution $\xi_j = \alpha r_1^j + \beta r_2^j$. The first boundary condition, $\xi_0 = 0$, implies that $\beta = -\alpha$. The boundary condition $\xi_{n+1} = 0$ yields the following equation in λ:

$$\left(\lambda + \sqrt{\lambda^2 - \lambda}\right)^{n+1} - \left(\lambda - \sqrt{\lambda^2 - \lambda}\right)^{n+1} = 0 \quad \rightarrow \quad \left(\frac{(\lambda + \sqrt{\lambda^2 - \lambda})^2}{\lambda}\right)^{n+1} = 1,$$

in which it is assumed that $\lambda \ne 0$. With the change of variables $\lambda \equiv \cos^2 \theta$, this becomes $(\cos\theta \pm i \sin\theta)^{2(n+1)} = 1$, where the sign $\pm$ is positive when $\cos\theta$ and $\sin\theta$ are of the same sign and negative otherwise. Hence,

$$\pm 2(n+1)\theta = \pm 2k\pi \quad \rightarrow \quad \theta = \theta_k \equiv \frac{k\pi}{n+1}, \quad k = 1, \ldots, n. \tag{13.28}$$

13.2. Matrices and Spectra of Model Problems

Therefore, the eigenvalues are of the form $\lambda_k = \cos^2 \theta_k$, where θ_k was defined above; i.e.,

$$\lambda_k = \cos^2 \frac{k\pi}{n+1}.$$

In fact this result could have been obtained in a simpler way. According to Theorem 4.16 seen in Chapter 4, when $\omega = 1$, the eigenvalues of an SOR iteration matrix are the squares of those of the corresponding Jacobi iteration matrix with the same ω, which, according to (13.21) (left side), are $\mu_k = \cos[k\pi/(n+1)]$.

Some care must be exercised when computing the eigenvectors. The jth component of the eigenvector is given by $\xi_j = r_1^j - r_2^j$. Proceeding as before, we have

$$r_1^j = \left(\cos^2 \theta_k + \sqrt{\cos^4 \theta_k - \cos^2 \theta_k}\right)^j = (\cos \theta_k)^j (\cos \theta_k \pm i \sin \theta_k)^j,$$

where the $\pm$ sign was defined before. Similarly, $r_2^j = (\cos \theta_k)^j (\cos \theta_k \mp i \sin \theta_k)^j$, where $\mp$ is the opposite sign from $\pm$. Therefore,

$$\xi_j = (\cos \theta_k)^j \left[(\cos \theta_k \mp i \sin \theta_k)^j - (\cos \theta_k \pm i \sin \theta_k)^j\right] = 2i \, (\cos \theta_k)^j \left[\pm \sin(j\theta_k)\right].$$

Since θ_k is defined by (13.28), $\sin \theta_k$ is nonnegative, and therefore the $\pm$ sign is simply the sign of $\cos \theta_k$. In addition the constant factor $2i$ can be deleted since eigenvectors are defined up to a scalar constant. Therefore, we can set

$$u_k = \left[|\cos \theta_k|^j \sin(j\theta_k)\right]_{j=1,\ldots,n}. \tag{13.29}$$

The above formula would yield an incorrect answer (a zero vector) for the situation when $\lambda_k = 0$. This special case can be handled by going back to (13.27), which yields the vector e_1 as an eigenvector. In addition, it is left to show that indeed the above set of vectors constitutes a basis. This is the subject of Exercise 2.

The smallest eigenvalues are those for which k is close to $n/2$, as is illustrated in Figure 13.5. Components in the directions of the corresponding eigenvectors, i.e., those associated with the eigenvalues of G in the middle of the spectrum, are damped rapidly by the process. The others are harder to eliminate. This is in contrast with the situation for the Jacobi iteration, where the modes corresponding to the largest eigenvalues are damped first.

Interestingly, the eigenvectors corresponding to the middle eigenvalues are not the most oscillatory in the proper sense of the word. Specifically, a look at the eigenfunctions of G illustrated in Figure 13.4 reveals that the modes with high oscillations are those corresponding to eigenvalues with the larger values of k, which are not damped rapidly. The figure shows the eigenfunctions of G for a 13-point discretization of the one-dimensional Laplacian. It omits the case $k = 6$, which corresponds to the special case of a zero eigenvalue mentioned above.

The eigenvectors of the Gauss–Seidel iteration matrix are not related in a simple way to those of the original matrix A. As it turns out, the low frequency eigenfunctions of the original matrix A are damped slowly while the high frequency modes are damped rapidly, just as is the case for the Jacobi and Richardson iterations. This can be readily verified experimentally; see Exercise 3.

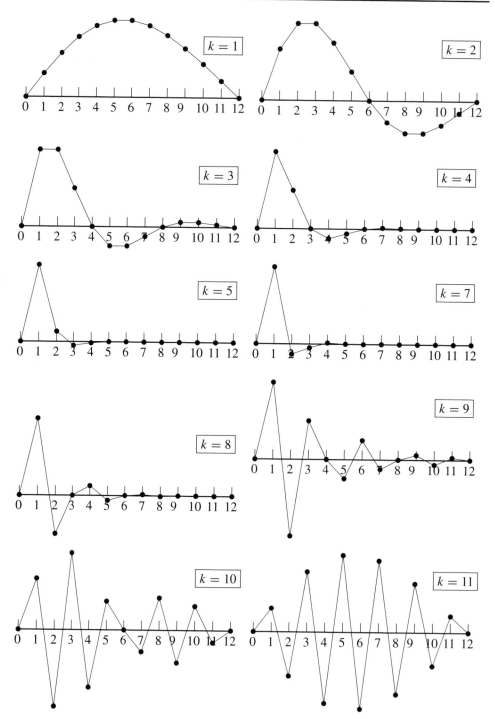

Figure 13.4. *The eigenfunctions of a 13-point one-dimensional mesh (n = 11). The case k = 6 is omitted.*

13.3. Intergrid Operations

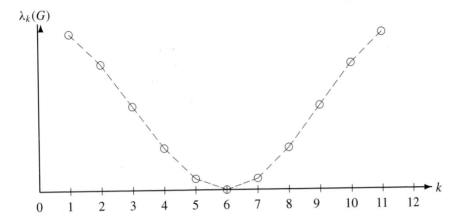

Figure 13.5. *Eigenvalues of the Gauss–Seidel iteration for a 13-point one-dimensional mesh ($n = 11$).*

13.3 Intergrid Operations

MG algorithms require going back and forth between several grid problems related to the solution of the same original equations. It is sufficient to present these grid transfer operations for the simple case of two meshes, Ω_h (fine) and Ω_H (coarse), and to only consider the situation when $H = 2h$. In particular, the problem size of the fine-mesh problem will be about 2^d times as large as that of the coarse-mesh problem, where d is the space dimension. In the previous section, the one-dimensional case was already considered and the subscript h corresponding to the mesh problem under consideration was introduced.

13.3.1 Prolongation

A prolongation operation takes a vector from Ω_H and defines the analogue vector in Ω_h. A common notation in use is

$$I_H^h : \Omega_H \longrightarrow \Omega_h.$$

The simplest way to define a prolongation operator is through linear interpolation, which is now presented for the one-dimensional case. The generic situation is that of $n + 2$ points, $x_0, x_1, \ldots, x_{n+1}$, where x_0 and x_{n+1} are boundary points. The number of internal points n is assumed to be odd, so that halving the size of each subinterval amounts to introducing the middle points. Given a vector $(v_i^{2h})_{i=0,\ldots,(n+1)/2}$, the vector $v^h = I_{2h}^h v^{2h}$ of Ω_h is defined as follows:

$$\begin{cases} v_{2j}^h &= v_j^{2h} \\ v_{2j+1}^h &= (v_j^{2h} + v_{j+1}^{2h})/2 \end{cases} \quad \text{for} \quad j = 0, \ldots, \frac{n+1}{2}.$$

In matrix form, the above-defined prolongation can be written as

$$v^h = \frac{1}{2} \begin{bmatrix} 1 & & & & \\ 2 & & & & \\ 1 & 1 & & & \\ & 2 & & & \\ & 1 & 1 & & \\ & & \vdots & & \\ & & & 1 & 1 \\ & & & & 2 \\ & & & & 1 \end{bmatrix} v^{2h}. \tag{13.30}$$

In two dimensions, the linear interpolation can be defined in a straightforward manner from the one-dimensional case. Thinking in terms of a matrix v_{ij} representing the coordinates of a function v at the points x_i, y_j, it is possible to define the interpolation in two dimensions in two stages. In the following, $I^h_{x,2h}$ denotes the interpolation in the x direction only and $I^h_{y,2h}$ the interpolation for the y variable only. First, interpolate all values in the x direction only:

$$v^{h,x} = I^h_{x,2h} v, \quad \text{where} \quad \begin{cases} v^{h,x}_{2i,:} = v^{2h}_{i,:} \\ v^{h,x}_{2i+1,:} = (v^{2h}_{i,:} + v^{2h}_{i+1,:})/2 \end{cases} \quad \text{for} \quad i = 0, \ldots, \frac{m+1}{2}.$$

Then interpolate this semi-interpolated result with respect to the y variable:

$$v^h = I^h_{y,2h} v^{h,x}, \quad \text{where} \quad \begin{cases} v^h_{:,2j} = v^{x,2h}_{:,j} \\ v^h_{:,2j+1} = (v^{x,2h}_{:,j} + v^{x,2h}_{:,j+1})/2 \end{cases} \quad \text{for} \quad j = 0, \ldots, \frac{n+1}{2}.$$

This gives the following formulas for the two-dimensional interpolation of an element v^H in Ω_H into the corresponding element $v^h = I^h_H$ in Ω^h:

$$\begin{cases} v^h_{2i,2j} = v^{2h}_{ij} \\ v^h_{2i+1,2j} = (v^{2h}_{ij} + v^{2h}_{i+1,j})/2 \\ v^h_{2i,2j+1} = (v^{2h}_{ij} + v^{2h}_{i,j+1})/2 \\ v^h_{2i+1,2j+1} = (v^{2h}_{ij} + v^{2h}_{i+1,j} + v^{2h}_{i,j+1} + v^{2h}_{i+1,j+1})/4 \end{cases} \quad \text{for} \quad \begin{cases} i = 0, \ldots, \frac{n+1}{2}, \\ j = 0, \ldots, \frac{m+1}{2}. \end{cases}$$

From the above derivation, it is useful to observe that the two-dimensional interpolation can be expressed as the tensor product of the two one-dimensional interpolations; i.e.,

$$I^h_{2h} = I^h_{y,2h} \otimes I^h_{x,2h}. \tag{13.31}$$

This is the subject of Exercise 4.

It is common to represent the prolongation operators using a variation of the stencil notation employed in Chapter 4 and Chapter 10. The stencil now operates on a grid to give values on a different grid. The one-dimensional stencil is denoted by

$$p = \left]\frac{1}{2} \quad 1 \quad \frac{1}{2}\right[.$$

13.3. Intergrid Operations

The open bracket notation only means that the stencil must be interpreted as a fan-out rather than fan-in operation, as in the cases we have seen in earlier chapters. In other words, it is a column instead of a row operation, as can also be understood by a look at the matrix in (13.30). Each stencil is associated with a coarse grid point. The results of the stencil operation are the values $v_i^H/2$, v_i^H, $v_i^H/2$ contributed to the three fine mesh points x_{2i-1}^h, x_{2i}^h, and x_{2i+1}^h by the value v_i^H. Another, possibly clearer, interpretation is that the function with value one at the coarse grid point x_i^{2h}, and zero elsewhere, will be interpolated to a function in the fine mesh with the values $0.5, 1, 0.5$ at the points $x_{2i-1}^h, x_{2i}^h, x_{2i+1}^h$, respectively, and zero elsewhere. Under this interpretation, the stencil for the two-dimensional linear interpolation is

$$\frac{1}{4} \begin{bmatrix} 1 & 2 & 1 \\ 2 & 4 & 2 \\ 1 & 2 & 1 \end{bmatrix} [.$$

It is also interesting to note that the two-dimensional stencil can be viewed as a tensor product of the one-dimensional stencil p and its transpose p^T. The stencil p^T acts on the vertical coordinates in exactly the same way as p acts on the horizontal coordinates.

Example 13.1. This example illustrates the use of the tensor product notation to determine the two-dimensional stencil. The stencil can also be understood in terms of the action of the interpolation operation on a unit vector. Using the stencil notation, this unit vector is of the form $e_i \otimes e_j$ and we have (see Exercise 1)

$$I_{2h}^h(e_i \otimes e_j) = (I_{y,2h}^h \otimes I_{x,2h}^h)(e_i \otimes e_j) = (I_{y,2h}^h e_i) \otimes (I_{x,2h}^h e_j).$$

When written in coordinate (or matrix) form, this is a vector that corresponds to the outer product pp^T, with $p^T \equiv [\frac{1}{2} \ 1 \ \frac{1}{2}]$, centered at the point with coordinates x_i, y_j.

13.3.2 Restriction

The restriction operation is the reverse of prolongation. Given a function v^h on the fine mesh, a corresponding function in Ω_H must be defined from v^h. In the earlier analysis one such operation was encountered. It was simply based on defining the function v^{2h} from the function v^h as follows:

$$v_i^{2h} = v_{2i}^h. \tag{13.32}$$

Because this is simply a canonical injection from Ω_h to Ω_{2h}, it is termed the *injection operator*. This injection has an obvious two-dimensional analogue: $v_{i,j}^{2h} = v_{2i,2j}^h$.

A more common restriction operator, called full weighting (FW), defines v^{2h} as follows in the one-dimensional case:

$$v_j^{2h} = \frac{1}{4}\left(v_{2j-1}^h + 2v_{2j}^h + v_{2j+1}^h\right). \tag{13.33}$$

This averages the neighboring values using the weights $0.25, 0.5, 0.25$. An important property can be seen by considering the matrix associated with this definition of I_h^{2h}:

$$I_h^{2h} = \frac{1}{4} \begin{bmatrix} 1 & 2 & 1 & & & & & & \\ & & 1 & 2 & 1 & & & & \\ & & & & 1 & 2 & 1 & & \\ & & & \cdots & \cdots & \cdots & & & \\ & & & & & & 1 & 2 & 1 \end{bmatrix}. \tag{13.34}$$

Apart from a scaling factor, this matrix is the transpose of the interpolation operator seen earlier. Specifically,
$$I_{2h}^{h} = 2\, (I_{h}^{2h})^{T}. \tag{13.35}$$
The stencil for the above operator is
$$\frac{1}{4}[1 \quad 2 \quad 1],$$
where the closed brackets are now used to indicate the standard fan-in (row) operation.

In the two-dimensional case, the stencil for the FW averaging is given by
$$\frac{1}{16}\begin{bmatrix} 1 & 2 & 1 \\ 2 & 4 & 2 \\ 1 & 2 & 1 \end{bmatrix}.$$

This takes for $u_{2i,2j}^{h}$ the result of a weighted average of the nine points $u_{i+q,j+p}^{H}$ with $|p|, |q| \leq 1$, with the associated weights $2^{-|p|-|q|-2}$. Note that, because the FW stencil is a scaled row (fan-in) version of the linear interpolation stencil, the matrix associated with the operator I_{h}^{2h} is essentially a transpose of the prolongation (interpolation) operator:
$$I_{2h}^{h} = 4(I_{h}^{2h})^{T}. \tag{13.36}$$

The statements (13.35) and (13.36) can be summarized by
$$I_{H}^{h} = 2^{d}(I_{h}^{H})^{T}, \tag{13.37}$$
where d is the space dimension.

The following relation can be shown:
$$I_{h}^{2h} = I_{y,h}^{2h} \otimes I_{x,h}^{2h}, \tag{13.38}$$
which is analogous to (13.31) (see Exercise 5).

13.4 Standard Multigrid Techniques

One of the most natural ways to exploit a hierarchy of grids when solving PDEs is to obtain an initial guess from interpolating a solution computed on a coarser grid. The process can be recursively repeated until a given grid is reached. This interpolation from a coarser grid can be followed by a few steps of a smoothing iteration. This is known as nested iteration. General MG cycles are intrinsically recursive processes that use essentially two main ingredients. The first is a hierarchy of grid problems along with restrictions and prolongations to move between grids. The second is a smoother, i.e., any scheme that has the smoothing property of damping quickly the high frequency components of the error. A few such schemes, such as the Richardson and weighted Jacobi iterations, have been seen in earlier sections. Other smoothers that are often used in practice are the Gauss–Seidel and red-black Gauss–Seidel iterations seen in Chapter 4.

13.4.1 Coarse Problems and Smoothers

At the highest level (finest grid) a mesh size of h is used and the resulting problem to solve is of the form

$$A_h u^h = f^h.$$

One of the requirements of MG techniques is that a system similar to the one above be solved at the coarser levels. It is natural to define this problem at the next level, where a mesh of size, say, H, is used as simply the system arising from discretizing the same problem on the coarser mesh Ω_H. In other cases, it may be more useful to define the linear system by a *Galerkin projection*, where the coarse-grid problem is defined by

$$A_H = I_h^H A_h I_H^h, \qquad f^H = I_h^H f^h. \tag{13.39}$$

This formulation is more common in finite element methods. It also has some advantages from a theoretical point of view.

Example 13.2. Consider the model problem in one dimension and the situation when A_H is defined from the Galerkin projection, i.e., via formula (13.39), where the prolongation and restriction operators are related by (13.35) (one dimension) or (13.36) (two dimensions). In one dimension, A_H can be easily defined for the model problem when FW is used. Indeed,

$$\begin{aligned} A_H e_j^H &= I_h^H A_h I_H^h e_j^H \\ &= I_h^H A_h \left[\frac{1}{2} e_{2j-1}^h + e_{2j}^h + \frac{1}{2} e_{2j+1}^h \right] \\ &= I_h^H \left[-\frac{1}{2} e_{2j-2}^h + e_{2j}^h - \frac{1}{2} e_{2j+2}^h \right] \\ &= -e_{j-1}^H + 2 e_j^H - e_{j+1}^H. \end{aligned}$$

This defines the jth column of A_H, which has a 2 in the diagonal, -1 in the super- and subdiagonals, and zero elsewhere. This means that the operator A_H defined by the Galerkin property is identical to the operator that would be defined from a coarse discretization. This property is not true in two dimensions when FW is used; see Exercise 6.

The notation

$$u_\nu^h = \texttt{smooth}^\nu(A_h, u_0^h, f_h)$$

means that u_ν^h is the result of ν smoothing steps for solving the above system, starting with the initial guess u_0^h. Smoothing iterations are of the form

$$u_{j+1}^h = S_h u_j^h + g^h, \tag{13.40}$$

where S_h is the iteration matrix associated with one smoothing step. As was seen in earlier chapters, the above iteration can always be rewritten in the *preconditioning* form

$$u_{j+1}^h = u_j^h + B_h(f^h - A_h u_j^h), \tag{13.41}$$

where
$$S_h \equiv I - B_h A_h, \qquad B_h \equiv (I - S_h)A_h^{-1}, \qquad g^h \equiv B_h f^h. \qquad (13.42)$$

The error d_ν^h and residual r_ν^h resulting from ν smoothing steps satisfy
$$d_\nu^h = (S_h)^\nu d_0^h = (I - B_h A_h)^\nu d_0^h, \qquad r_h^\nu = (I - A_h B_h)^\nu r_0^h.$$

It will be useful later to make use of the following observation. When $f^h = 0$, then g^h is also zero and, as a result, one step of the iteration (13.40) will provide the result of one product with the operator S_h.

Example 13.3. For example, setting $f \equiv 0$ in (13.17) yields the Jacobi iteration matrix
$$B = (I - D^{-1}(E + F))A^{-1} = D^{-1}(D - E - F)A^{-1} = D^{-1}.$$

In a similar way one finds that, for the Gauss–Seidel iteration, $B = (D - E)^{-1}F$, and, for the Richardson iteration, $B = \omega I$.

Nested iteration was mentioned earlier as a means of obtaining good initial guesses from coarser meshes in a recursive way. The algorithm, presented here to illustrate the notation just introduced, is described below. The assumption is that there are $p + 1$ grids, with mesh sizes $h, 2h, \ldots, 2^p n \equiv h_0$.

ALGORITHM 13.1. Nested Iteration

1. Set $h := h_0$. Given an initial guess u_0^h, set $u^h = \texttt{smooth}^{\nu_p}(A_h, u_0^h, f^h)$
2. For $l = p - 1, \ldots, 0$, Do
3. $\quad u^{h/2} = I_h^{h/2} u^h$
4. $\quad h := h/2$
5. $\quad u^h := \texttt{smooth}^{\nu_l}(A_h, u^h, f^h)$
6. EndDo

In practice, nested iteration is not much used in this form. However, it provides the foundation for one of the most effective MG algorithms, namely, the full MG (FMG), which will be described in a later section.

13.4.2 Two-Grid Cycles

When a smoother is applied to a linear system at a fine level, the residual
$$r^h = f^h - Au^h$$

obtained at the end of the smoothing step will typically still be large. However, it will have small components in the space associated with the high frequency modes. If these components are removed by solving the above system (exactly) at the lower level, then a better approximation should result. Two-grid methods are rarely practical because the coarse-mesh problem may still be too large to be solved exactly. However, they are useful from a theoretical point of view. In the following algorithm $H = 2h$.

13.4. Standard Multigrid Techniques

ALGORITHM 13.2. Two-Grid Cycle

1. *Presmooth:* $u^h := \text{smooth}^{\nu_1}(A_h, u_0^h, f^h)$
2. *Get residual:* $r^h = f^h - A_h u^h$
3. *Coarsen:* $r^H = I_h^H r^h$
4. *Solve:* $A_H \delta^H = r^H$
5. *Correct:* $u^h := u^h + I_H^h \delta^H$
6. *Postsmooth:* $u^h := \text{smooth}^{\nu_2}(A_h, u^h, f^h)$

It is clear that the result of one iteration of the above algorithm corresponds to some iteration process of the form

$$u_{new}^h = M_h u_0^h + g_{M_h}.$$

In order to determine the operator M_h we exploit the observation made above that taking $f^h = 0$ provides the product $M_h u_0^h$. When $f^h = 0$, then, in line 1 of the algorithm, u^h becomes $S_h^{\nu_1} u_0^h$. In line 3, we have $r^H = I_h^H(f_h - A_h S_h^{\nu_1}) = I_h^H(-A_h S_h^{\nu_1})$. Following this process, the vector u^h resulting from one cycle of the algorithm becomes

$$u_{new}^h = S_h^{\nu_2}[S_h^{\nu_1} u_0^h + I_H^h A_H^{-1} I_h^H(-A_h S_h^{\nu_1} u_0^h)].$$

Therefore, the two-grid iteration operator is given by

$$M_H^h = S_h^{\nu_2}[I - I_H^h A_H^{-1} I_h^H A_h] S_h^{\nu_1}.$$

The matrix inside the brackets,

$$T_h^H = I - I_H^h A_H^{-1} I_h^H A_h, \qquad (13.43)$$

acts as another iteration by itself known as the *coarse-grid correction,* which can be viewed as a particular case of the two-grid operator with no smoothing, i.e., with $\nu_1 = \nu_2 = 0$. Note that the B preconditioning matrix associated with this iteration is, according to (13.42), $B_h = I_H^h A_H^{-1} I_h^H$.

An important property of the coarse grid correction operator is discussed in the following lemma. It is assumed that A_h is SPD.

Lemma 13.1. *When the coarse-grid matrix is defined via (13.39), then the coarse grid correction operator (13.43) is a projector that is orthogonal with respect to the A_h inner product. In addition, the range of T_h^H is A_h-orthogonal to the range of I_h^H.*

Proof. It suffices to show that $I - T_h^H = I_H^h A_H^{-1} I_h^H A_h$ is a projector:

$$(I_H^h A_H^{-1} I_h^H A_h) \times (I_H^h A_H^{-1} I_h^H A_h) = I_H^h A_H^{-1} \underbrace{(I_h^H A_h I_H^h)}_{A_H} A_H^{-1} I_h^H A_h = I_H^h A_H^{-1} I_h^H A_h.$$

That $I_H^h A_H^{-1} I_h^H A_h$ is an A-orthogonal projector follows from its self-adjointness with respect to the A_h inner product (see Chapter 1):

$$(T_h^H x, y)_{A_h} = (A_h I_H^h A_H^{-1} I_h^H A_h x, y) = (x, A_h I_H^h A_H^{-1} I_h^H A_h y) = (x, T_h^H y)_{A_h}.$$

Finally, the statement that $\text{Ran}(T_h^H)$ is orthogonal to $\text{Ran}(I_h^H)$ is equivalent to stating that, for all x of the form $x = T_h^H y$, we have $I_h^H A_h x = 0$, which is readily verified. □

13.4.3 V-Cycles and W-Cycles

Anyone familiar with recursivity will immediately think of the following practical version of the two-grid iteration: apply the two-grid cycle recursively until a coarse enough level is reached and then solve exactly (typically using a direct solver). This gives the algorithm described below, called the V-cycle MG. In the algorithm, H stands for $2h$ and h_0 for the coarsest mesh size.

ALGORITHM 13.3. u^h = V-Cycle(A_h, u_0^h, f^h)

1. Presmooth: $u^h := \text{smooth}^{\nu_1}(A_h, u_0^h, f^h)$
2. Get residual: $r^h = f^h - A_h u^h$
3. Coarsen: $r^H = I_h^H r^h$
4. If ($H == h_0$)
5.   Solve: $A_H \delta^H = r^H$
6. Else
7.   Recursion: $\delta^H = \text{V-cycle}(A_H, 0, r^H)$
8. EndIf
9. Correct: $u^h := u^h + I_H^h \delta^H$
10. Postsmooth: $u^h := \text{smooth}^{\nu_2}(A_h, u^h, f^h)$
11. Return u^h

Consider the cost of one cycle, i.e., one iteration of the above algorithm. A few simple assumptions are needed along with new notation. The number of nonzero elements of A_h is denoted by nnz_h. It is assumed that $nnz_h \leq \alpha n_h$, where α does not depend on h. The cost of each smoothing step is equal to nnz_h, while the cost of the transfer operations (interpolation and restriction) is of the form βn_h, where again β does not depend on h. The cost at the level where the grid size is h is given by

$$C(n_h) = (\alpha(\nu_1 + \nu_2) + 2\beta)n_h + C(n_{2h}).$$

Noting that $n_{2h} = n_h/2$ in the one-dimensional case, this gives the recurrence relation

$$C(n) = \eta n + C(n/2), \qquad (13.44)$$

in which $\eta = (\alpha(\nu_1 + \nu_2) + 2\beta)$. The solution of this recurrence relation yields $C(n) \leq 2\eta n$. For two-dimensional problems, $n_h = 4n_{2h}$, and in this case the cost becomes not greater than $(4/3)\eta n$.

We now introduce the general MG cycle that generalizes the V-cycle seen above. Once more, the implementation of the MG cycle is of a recursive nature.

ALGORITHM 13.4. u^h = MG(A_h, u_0^h, f^h, ν_1, ν_2, γ)

1. Presmooth: $u^h := \text{smooth}^{\nu_1}(A_h, u_0^h, f^h)$
2. Get residual: $r^h = f^h - A_h u^h$
3. Coarsen: $r^H = I_h^H r^h$
4. If ($H == h_0$)
5.   Solve: $A_H \delta^H = r^H$
6. Else
7.   Recursion: $\delta^H = MG^\gamma(A_H, 0, r^H, \nu_1, \nu_2, \gamma)$
8. EndIf
9. Correct: $u^h := u^h + I_H^h \delta^H$
10. Postsmooth: $u^h := \text{smooth}^{\nu_2}(A_h, u^h, f^h)$
11. Return u^h

13.4. Standard Multigrid Techniques

Notice now that there is a new parameter, γ, which determines how many times MG is iterated in line 7. Each of the MG iterations in line 7 takes the form

$$\delta^H_{new} = MG(A_H, \delta^H, r_H, \nu_1, \nu_2, \gamma), \qquad (13.45)$$

which is iterated γ times. The initial guess for the iteration is $\delta_H = 0$, which the second argument to the MG call in line 7 shows. The case $\gamma = 1$ yields the V-cycle MG. The case $\gamma = 2$ is known as the *W-cycle* MG. The resulting intergrid up and down moves can be complex, as is illustrated by the diagrams in Figure 13.6. The case $\gamma = 3$ is rarely used.

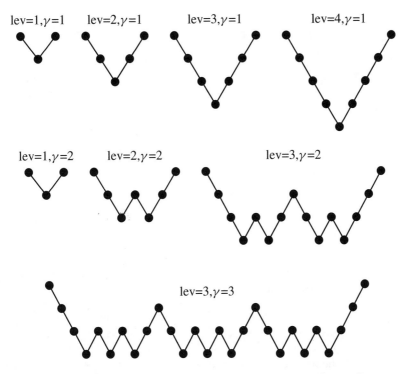

Figure 13.6. *Representations of various V-cycles and W-cycles.*

Now consider the cost of the general MG cycle. The only significant difference with the V-cycle algorithm is that the recursive call to MG is iterated γ times instead of only once for the V-cycle. Therefore, the cost formula (13.44) becomes

$$C(n) = \eta n + \gamma C(n/2) \quad \text{(1-D case)}, \qquad C(n) = \eta n + \gamma C(n/4) \quad \text{(2-D case)}. \qquad (13.46)$$

It can easily be shown that the cost of each loop is still linear when $\gamma < 2$ in 1-D and $\gamma < 4$ in 2-D; see Exercise 12. In other cases, the cost per cycle increases to $O(n \log_2 n)$.

Example 13.4. This example illustrates the convergence behavior of the V-cycle MG for solving a model Poisson problem with Dirichlet boundary conditions in two-dimensional space. The problem considered is of the form

$$-\Delta u = 13 \sin(2\pi x) \times \sin(3\pi y) \qquad (13.47)$$

and has the exact solution $u(x, y) = \sin(2\pi x) \times \sin(3\pi y)$. The Poisson equation is set on a square grid and discretized using $n_x = n_y = 33$ points, including the two boundary points in each direction. This leads to a linear system of dimension $N = 31^2 = 961$. The V-cycle MG was tested with three smoothers: (1) the weighted Jacobi relaxation with $\omega = 2/3$, (2) Gauss–Seidel relaxation, and (3) red-black Gauss–Seidel relaxation. Various values of ν_1 and ν_2, the number of pre- and postsmoothing steps, respectively, were used. Table 13.1 shows the convergence factors ρ as estimated from the expression

$$\rho = \exp\left(\frac{1}{k} \log \frac{\|r_k\|_2}{\|r_0\|_2}\right)$$

for each of the smoothers. Here k is the total number of smoothing steps taken. The convergence was stopped as soon as the 2-norm of the residual was reduced by a factor of $tol = 10^{-8}$.

(ν_1, ν_2)	smoother	ρ	(ν_1, ν_2)	smoother	ρ
(0, 1)	w-Jac	0.570674	(1, 1)	w-Jac	0.387701
(0, 1)	GS	0.308054	(1, 1)	GS	0.148234
(0, 1)	RB-GS	0.170635	(1, 1)	RB-GS	0.087510
(0, 2)	w-Jac	0.358478	(1, 2)	w-Jac	0.240107
(0, 2)	GS	0.138477	(1, 2)	GS	0.107802
(0, 2)	RB-GS	0.122895	(1, 2)	RB-GS	0.069331
(0, 3)	w-Jac	0.213354	(1, 3)	w-Jac	0.155938
(0, 3)	GS	0.105081	(1, 3)	GS	0.083473
(0, 3)	RB-GS	0.095490	(1, 3)	RB-GS	0.055480

Table 13.1. *Tests with V-cycle MG for a model Poisson equation using three smoothers and various numbers of presmoothing steps (ν_1) and postsmoothing steps (ν_2).*

The overall winner is clearly the red-black Gauss–Seidel smoother. It is remarkable that, even with a number of total smoothing steps $\nu_1 + \nu_2$ as small as two, a reduction factor of less than 0.1 is achieved with red-black Gauss–Seidel. Also, it is worth pointing out that, when $\nu_1 + \nu_2$ is constant, the red-black Gauss–Seidel smoother tends to perform better when ν_1 and ν_2 are more or less balanced (compare the case $(\nu_1, \nu_2) = (0, 2)$ versus $(\nu_1, \nu_2) = (1, 1)$, for example). In the asymptotic regime (or very large k), the two ratios should be identical in theory.

It is important to determine the iteration operator corresponding to the application of one MG loop. We start with the two-grid operator seen earlier, which is

$$M_H^h = S_h^{\nu_2}[I - I_H^h A_H^{-1} I_h^H A_h] S_h^{\nu_1}.$$

The only difference between this operator and the MG operator sought is that the inverse of A_H is replaced with an application of γ steps of MG on the grid Ω_H. Each of these steps is of the form (13.45). However, the above formula uses the inverse of A_H, so it is necessary to replace A_H with the corresponding B form (preconditioned form) of the MG operator, which, according to (13.42), is given by

$$(I - M_H) A_H^{-1}.$$

13.4. Standard Multigrid Techniques

Therefore,

$$
\begin{aligned}
M_h &= S_h^{\nu_2}[I - I_H^h(I - M_H)A_H^{-1}I_h^H A_h]S_h^{\nu_1} \\
&= S_h^{\nu_2}[I - I_H^h A_H^{-1}I_h^H A_h + I_H^h M_H A_H^{-1}I_h^H A_h]S_h^{\nu_1} \\
&= M_H^h + S_h^{\nu_2}I_H^h M_H A_H^{-1}I_h^H A_h S_h^{\nu_1},
\end{aligned}
$$

showing that the MG operator M_h can be viewed as a perturbation of the two-grid operator M_h^H.

13.4.4 FMG

FMG, sometimes also referred to as nested iteration, takes a slightly different approach from the MG algorithms seen in the previous section. FMG can be viewed as an improvement of nested iteration, seen earlier, whereby the smoothing step in line 5 is replaced with an MG cycle. The difference in viewpoint is that it seeks to find an approximation to the solution with only one sweep through the levels, going from bottom to top. The error of the resulting approximation is guaranteed, under certain conditions, to be of the order of the discretization. In practice, no more accuracy than this should ever be required. The algorithm is described below.

ALGORITHM 13.5. FMG

1. Set $h := h_0$. Solve $A_h u^h = f^h$
2. For $l = 1, \ldots, p$, Do
3. $\quad u^{h/2} = \hat{I}_h^{h/2} u^h$
4. $\quad h := h/2$
5. $\quad u^h := MG^\mu(A_h, u^h, f^h, \nu_1, \nu_2, \gamma)$
6. EndDo

Notice that the interpolation operator in line 3 is denoted with a hat. This is in order to distinguish it from the interpolation operator used in the MG loop, which is sometimes different, typically of a lower order. The tasks and transfer operations of FMG are illustrated in Figure 13.7. The MG iteration requires the standard parameters ν_1, ν_2, γ, in addition to the other choices of smoothers and interpolation operators.

In the following, u^h represents the exact (discrete) solution of the problem on the grid Ω_h and $\tilde{u}^h$ will be the approximation resulting from the FMG cycle on the grid Ω_h. Thus, $\tilde{u}^h$ is the result of line 5 in Algorithm 13.5. One of the main assumptions made in analyzing FMG is that the exact solution of the discrete linear system $A_h u^h = f^h$ is close, within the discretization accuracy, to the exact solution of the PDE problem:

$$\|u - u^h\| \leq ch^\kappa. \tag{13.48}$$

The left-hand side represents the norm of the difference between the exact solution u^h of the discrete problem and the solution of the continuous problem sampled at the grid points of Ω_h. Any norm on Ω_h can be used and the choice of the norm will be reexamined shortly. Using an argument based on the triangle inequality, a particular consequence of

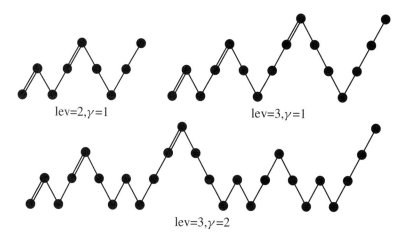

Figure 13.7. *Representation of various FMG cycles (with $\mu = 1$). The doubled lines correspond to the FMG interpolation.*

the above assumption is that u^h and $\hat{I}_H^h u^H$ should also be close since they are close to the same (continuous) function u. Specifically, the assumption (13.48) is replaced with

$$\|u^h - \hat{I}_H^h u^H\| \leq c_1 h^\kappa. \tag{13.49}$$

A bound of this type can be shown by making a more direct assumption on the interpolation operator; see Exercise 13. The next important assumption to make is that the MG iteration operator is uniformly bounded:

$$\|M_h\| \leq \xi < 1. \tag{13.50}$$

Finally, the interpolation $\hat{I}_H^h$ must also be bounded, a condition that is convenient to state as follows:

$$\|\hat{I}_H^h\| \leq c_2 2^{-\kappa}. \tag{13.51}$$

Theorem 13.2. *Assume that* (13.49), (13.50), *and* (13.51) *are satisfied and that μ is sufficiently large that*

$$c_2 \xi^\mu < 1. \tag{13.52}$$

Then the FMG iteration produces approximations $\tilde{u}_h$, which at each level satisfy

$$\|u^h - \tilde{u}^h\| \leq c_3 c_1 h^\kappa, \tag{13.53}$$

with

$$c_3 = \xi^\mu / (1 - c_2 \xi^\mu). \tag{13.54}$$

Proof. The proof is by induction. At the lowest level, (13.53) is clearly satisfied because the solution $\tilde{u}^h$ is exact at the coarsest level and so the error is zero. Consider now the problem associated with the mesh size h and assume that the theorem is valid for the mesh size H. The error vector is given by

$$u^h - \tilde{u}^h = (M_h)^\mu (u^h - u_0^h). \tag{13.55}$$

13.4. Standard Multigrid Techniques

The initial guess is defined by $u_0^h = \hat{I}_H^h \tilde{u}^H$. Therefore,

$$\begin{aligned}
\|u^h - u_0^h\| &= \|u^h - \hat{I}_H^h u^H + \hat{I}_H^h (u^H - \tilde{u}^H)\| \\
&\leq \|u^h - \hat{I}_H^h u^H\| + \|\hat{I}_H^h (u^H - \tilde{u}^H)\| \\
&\leq c_1 h^\kappa + \|\hat{I}_H^h\| c_1 c_3 H^\kappa \quad \text{(by (13.49) and induction hypothesis)} \\
&\leq h^\kappa (c_1 + 2^{-\kappa} c_2 H^\kappa c_1 c_3) \quad \text{(by (13.51))} \\
&\leq h^\kappa c_1 (1 + c_2 c_3).
\end{aligned}$$

Combining the above with (13.55) and (13.50) yields

$$\|u^h - \tilde{u}^h\| \leq \xi^\mu h^\kappa c_1 (1 + c_2 c_3).$$

From the relation (13.54), we get $\xi^\mu = c_3/(1 + c_2 c_3)$, which shows the result (13.53) for the next level and completes the induction proof. □

In practice it is accepted that taking $\mu = 1$ is generally sufficient to satisfy the assumptions of the theorem. For example, if $\|\hat{I}_H^h\| \leq 1$ and $\kappa = 1$, then $c_2 = 4$. In this case, with $\mu = 1$, the result of the theorem will be valid provided $\xi < 0.25$, which is easily achieved by a simple V-cycle using Gauss–Seidel smoothers.

Example 13.5. This example illustrates the behavior of the FMG cycle when solving the same model Poisson problem as in Example 13.4. As before, the Poisson equation is set on a square grid and discretized with centered differences. The problem is solved using the mesh sizes $n_x = n_y = 9, 17, 33, 65$, and 129 points (including the two boundary points) in each direction. Thus, for example, the last problem leads to a linear system of dimension $N = 127^2 = 16129$.

Figure 13.8 shows in log scale the 2-norm of the actual error achieved for three FMG schemes as a function of $\log(n_x - 1)$. It also shows the 2-norm of the discretization error. Note that, when $n_x = 9$, all methods show the same error as the discretization error because the system is solved exactly at the coarsest level, i.e., when $n_x = 9$. The first FMG scheme uses a weighted Jacobi iteration with the weight $\omega = 2/3$ and $(\nu_1, \nu_2) = (1, 0)$. As can be seen the error achieved becomes too large relative to the discretization error when the number of levels increases. On the other hand, the other two schemes, red-black Gauss–Seidel with $(\nu_1, \nu_2) = (4, 0)$ and Gauss–Seidel with $(\nu_1, \nu_2) = (2, 0)$, perform well. It is remarkable that the error achieved by red-black Gauss–Seidel is actually slightly smaller than the discretization error itself.

The result of the above theorem is valid in any norm. However, it is important to note that the type of bound obtained will depend on the norm used.

Example 13.6. It is useful to illustrate the basic discretization error bound (13.48) for the one-dimensional model problem. As before, we abuse the notation slightly by denoting by u the vector in Ω_h whose values at the grid points are the values of the (continuous) solution **u** of the differential equation (13.1)–(13.2). Now the discrete L_2-norm on Ω_h, denoted by $\|v\|_h$, will be used. In this particular case, this norm is also equal to $h^{1/2}\|v\|_2$, the Euclidean

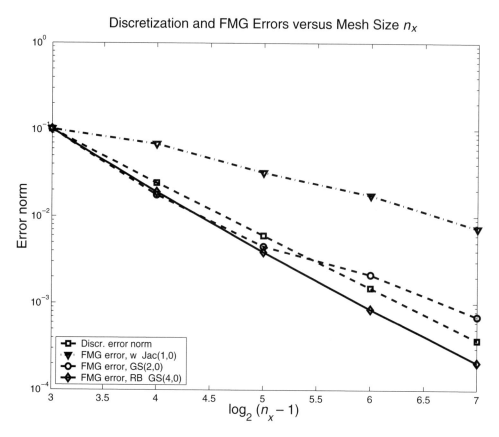

Figure 13.8. *FMG error norms with various smoothers versus the discretization error as a function of the mesh size.*

norm scaled by $\sqrt{h}$. Then we note that

$$\|u^h - u\|_h = \|(A_h)^{-1} A_h(u^h - u)\|_h = \|(A_h)^{-1}[f^h - A_h u]\|_h$$
$$\leq \|(A_h)^{-1}\|_h \, \|f^h - A_h u\|_h. \tag{13.56}$$

Assuming that the continuous **u** is in C^4 (four times differentiable with continuous fourth derivative), (2.12) from Chapter 2 gives

$$(f - A_h u)_i = f_i + u''(x_i) + \frac{h^2}{12} u^{(4)}(\xi_i) = \frac{h^2}{12} u^{(4)}(\xi_i),$$

where ξ_i is in the interval $(x_i - h, x_i + h)$. Since $\mathbf{u} \in C^4(\Omega)$, we have $|u^{(4)}(\xi_i)| \leq K$, where K, is the maximum of $u^{(4)}$ over Ω and, therefore,

$$\left\| \left(u^{(4)}(\xi_i)\right)_{i=1,\ldots,n} \right\|_h \leq h^{1/2} \| (K)_{i=1,\ldots,n} \|_2 \leq K.$$

This provides the bound $\|f^h - A_h u\|_h \leq K h^2 / 12$ for the second term in (13.56).

The norm $\|(A_h)^{-1}\|_h$ in (13.56) can be computed by noting that

$$\|(A_h)^{-1}\|_h = \|(A_h)^{-1}\|_2 = 1/\lambda_{min}(A_h).$$

According to (13.7),

$$\lambda_{min}(A_h) = \frac{4}{h^2}\sin^2(\pi h/2) = \pi^2 \frac{\sin^2(\pi h/2)}{(\pi h/2)^2}.$$

It can be shown that when, for example, $x < 1$, then $1 \geq \sin(x)/x \geq 1 - x^2/6$. Therefore, when $\pi h < 2$, we have

$$\frac{1}{\pi^2} \leq \|(A_h)^{-1}\|_h \leq \frac{1}{\pi^2 \left(1 - \frac{1}{6}\left(\frac{\pi h}{2}\right)^2\right)}.$$

Putting these results together yields the inequality

$$\|u^h - u\|_h \leq \frac{K}{12\pi^2 \left(1 - \frac{1}{6}\left(\frac{\pi h}{2}\right)^2\right)} h^2.$$

Exercise 16 considers an extension of this argument to two-dimensional problems. Exercise 17 explores what happens if other norms are used.

13.5 Analysis of the Two-Grid Cycle

The two-grid correction cycle is at the basis of most of the more complex MG cycles. For example, it was seen that the general MG cycle can be viewed as a perturbation of a two-grid correction cycle. Similarly, practical FMG schemes use a V-cycle iteration for their inner loop. This section will take an in-depth look at the convergence of the two-grid cycle for the case when the coarse-grid problem is defined by the Galerkin approximation (13.39). This case is important in particular because it is at the basis of all the AMG techniques that will be covered in the next section.

13.5.1 Two Important Subspaces

Consider the two-grid correction operator T_h^H defined by (13.43). As was seen in Section 13.4.2 (see Lemma 13.1), this is an A_h-orthogonal projector onto the subspace Ω_h. It is of the form $I - Q_h$, where

$$Q_h = I_H^h A_H^{-1} I_h^H A_h.$$

Clearly, Q_h is also an A_h-orthogonal projector (since $I - Q_h$ is: see Chapter 1), and we have

$$\Omega_h = \text{Ran}(Q_h) \oplus \text{Ker}(Q_h) \equiv \text{Ran}(Q_h) \oplus \text{Ran}(I - Q_h). \quad (13.57)$$

The definition of Q_h implies that

$$\text{Ran}(Q_h) \subset \text{Ran}(I_H^h).$$

As it turns out, the inclusion also holds in the other direction, which means that the two subspaces are the same. To show this, take a vector z in the range of I_H^h, so $z = I_H^h y$ for a certain $y \in \Omega_h$. Remembering that $A_H = I_h^H A_h I_H^h$, we obtain

$$Q_h z = I_H^h A_H^{-1} I_h^H A_h \, I_H^h y = I_H^h y = z,$$

which shows that z belongs to $\mathrm{Ran}(Q_h)$. Hence,

$$\mathrm{Ran}(Q_h) = \mathrm{Ran}(I_H^h).$$

This says that Q_h is the A_h-orthogonal projector onto the space $\mathrm{Ran}(I_H^h)$, while T_h^H is the A_h-orthogonal projector onto the orthogonal complement. This orthogonal complement, which is the range of $I - Q_h$, is also the null space of Q_h according to the fundamental relation (1.57) of Chapter 1. Finally, the null space of Q_h is identical to the null space of the restriction operator I_h^H. It is clear that $\mathrm{Ker}(I_h^H) \subset \mathrm{Ker}(Q_h)$. The reverse inclusion is not as clear and may be derived from the fundamental relation (1.17) seen in Chapter 1. This relation implies that

$$\Omega_h = \mathrm{Ran}(I_H^h) \oplus \mathrm{Ker}\left((I_H^h)^T\right) = \mathrm{Ran}(Q_h) \oplus \mathrm{Ker}\left((I_H^h)^T\right).$$

However, by (13.37), $\mathrm{Ker}\left((I_H^h)^T\right) = \mathrm{Ker}(I_h^H)$. Comparing this with the decomposition (13.57), it follows that

$$\mathrm{Ker}(Q_h) = \mathrm{Ker}(I_h^H).$$

In summary, if we set

$$\mathcal{S}_h \equiv \mathrm{Ran}(Q_h), \qquad \mathcal{T}_h \equiv \mathrm{Ran}(T_h^H), \tag{13.58}$$

then the following relations can be stated:

$$\Omega_h = \mathcal{S}_h \oplus \mathcal{T}_h, \tag{13.59}$$
$$\mathcal{S}_h = \mathrm{Ran}(Q_h) = \mathrm{Ker}(T_h) = \mathrm{Ran}(I_H^h), \tag{13.60}$$
$$\mathcal{T}_h = \mathrm{Ker}(Q_h) = \mathrm{Ran}(T_h) = \mathrm{Ker}(I_h^H). \tag{13.61}$$

These two subspaces are fundamental when analyzing MG methods. Intuitively, it can be guessed that the null space of T_h^H is somewhat close to the space of smooth modes. This is because it is constructed so that its desired action on a smooth component is to annihilate it. On the other hand, it should leave an oscillatory component more or less unchanged. If s is a smooth mode and t an oscillatory one, then this translates into the rough statements

$$T_h^H s \approx 0, \qquad T_h^H t \approx t.$$

Clearly, opposite relations are true with Q_h, namely, $Q_h t \approx 0$ and $Q_h s \approx s$.

Example 13.7. Consider the case when the prolongation operator I_h^H corresponds to the case of FW in the one-dimensional case. Consider the effect of this operator on any eigenmode w_k^h, which has components $\sin(j\theta_k)$ for $j = 1, \ldots, n$, where $\theta_k = k\pi/(n+1)$. Then,

13.5. Analysis of the Two-Grid Cycle

according to (13.33),

$$
\begin{aligned}
(I_h^H w_k^h)_j &= \frac{1}{4}[\sin((2j-1)\theta_k) + 2\sin(2j\theta_k) + \sin((2j+1)\theta_k)] \\
&= \frac{1}{4}[2\sin(2j\theta_k)\cos\theta_k + 2\sin(2j\theta_k)] \\
&= \frac{1}{2}(1+\cos\theta_k)\sin(2j\theta_k) \\
&= \cos^2\left(\frac{\theta_k}{2}\right)\sin(2j\theta_k).
\end{aligned}
$$

Consider a mode w_k where k is large, i.e., close to n. Then $\theta_k \approx \pi$. In this case, the restriction operator will transform this mode into a constant times the same mode on the coarser grid. The multiplicative constant, which is $\cos^2(\theta_k/2)$, is close to zero in this situation, indicating that $I_h^H w_k \approx 0$, i.e., that w_k is near the null space of I_h^H. Oscillatory modes are close to being in the null space of I_h^H or, equivalently, the range of T_h^H.

When k is small, i.e., for smooth modes, the constant $\cos^2(\theta_k/2)$ is close to one. In this situation the interpolation produces the equivalent smooth mode in Ω_H without damping it.

13.5.2 Convergence Analysis

When analyzing convergence for the Galerkin case, the A_h-norm is often used. In addition, 2-norms weighted by $D^{1/2}$, or $D^{-1/2}$, where D is the diagonal of A, are convenient. For example, we will use the notation

$$\|x\|_D = (Dx,x)^{1/2} \equiv \|D^{1/2}x\|_2.$$

The following norm also plays a significant role:

$$\|e\|_{A_h D^{-1} A_h} = (D^{-1} A_h e, A_h e)^{1/2} \equiv \|A_h e\|_{D^{-1}}.$$

To avoid burdening the notation unnecessarily we simply use $\|A_h e\|_{D^{-1}}$ to denote this particular norm of e. It can be shown that standard two-grid cycles satisfy an inequality of the form

$$\|S_h e^h\|_{A_h}^2 \leq \|e^h\|_{A_h}^2 - \alpha \|A e^h\|_{D^{-1}}^2 \quad \forall \ e^h \in \Omega_h, \tag{13.62}$$

independently of h. This is referred to as the *smoothing property*.

In addition to the above requirement, which characterizes the smoother, another assumption will be made that characterizes the discretization. This assumption is referred to as the *approximation property* and can be stated as follows:

$$\min_{u_H \in \Omega_H} \|e^h - I_H^h e^H\|_D^2 \leq \beta \|e^h\|_{A_h}^2, \tag{13.63}$$

where β does not depend on h. In the following theorem, it is assumed that A is SPD and that the restriction and prolongation operators are linked by a relation of the form (13.37), with I_H^h being of full rank.

Theorem 13.3. *Assume that inequalities (13.62) and (13.63) are satisfied for a certain smoother, where $\alpha > 0$ and $\beta > 0$. Then $\alpha \leq \beta$, the two-level iteration converges, and the norm of its operator is bounded as follows:*

$$\|S_h T_h^H\|_{A_h} \leq \sqrt{1 - \frac{\alpha}{\beta}}. \qquad (13.64)$$

Proof. It was seen in the previous section that $\text{Ran}(T_h^H) = \mathcal{T}_h$ is A_h-orthogonal to $\text{Ran}(I_H^h) = \mathcal{S}_h$. As a result, $(e^h, I_H^h e^H)_{A_h} = 0$ for any $e^h \in \text{Ran}(T_h^H)$ and so

$$\|e^h\|_{A_h}^2 = (A_h e^h, e^h - I_H^h e^H) \quad \forall\, e^h \in \text{Ran}(T_h^H).$$

For any $e^h \in \text{Ran}(T_h^H)$, the Cauchy–Schwarz inequality gives

$$\begin{aligned}
\|e^h\|_{A_h}^2 &= (D^{-1/2} A_h e^h, D^{1/2}(e^h - I_H^h e^H)) \\
&\leq \|D^{-1/2} A_h e^h\|_2 \, \|D^{1/2}(e^h - I_H^h e^H)\|_2 \\
&= \|A_h e^h\|_{D^{-1}} \, \|e^h - I_H^h e^H\|_D.
\end{aligned}$$

By (13.63), this implies that $\|e^h\|_{A_h} \leq \sqrt{\beta} \|A_h e^h\|_{D^{-1}}$ for any $e^h \in \text{Ran}(T_H^h)$ or, equivalently, $\|T_h^H e^h\|_{A_h}^2 \leq \beta \|A_h T_h^H e^h\|_{D^{-1}}^2$ for any e^h in Ω_h. The proof is completed by exploiting the smoothing property, i.e., inequality (13.62):

$$\begin{aligned}
0 \leq \|S_h T_h^H e^h\|_{A_h}^2 &\leq \|T_h^H e^h\|_{A_h}^2 - \alpha \|A_h T_h^H e^h\|_{D^{-1}}^2 \\
&\leq \|T_h^H e^h\|_{A_h}^2 - \frac{\alpha}{\beta} \|T_h^H e^h\|_{A_h}^2 \\
&= \left(1 - \frac{\alpha}{\beta}\right) \|T_h^H e^h\|_{A_h}^2 \\
&\leq \left(1 - \frac{\alpha}{\beta}\right) \|e^h\|_{A_h}^2.
\end{aligned}$$

The fact that T_h^H is an A_h-orthogonal projector was used to show the last step. $\square$

Example 13.8. As an example, we will explore the smoothing property (13.62) in the case of the weighted Jacobi iteration. The index h is now dropped for clarity. From (13.19) the smoothing operator in this case is

$$S(\omega) \equiv I - \omega D^{-1} A.$$

When A is SPD, then the weighted Jacobi iteration will converge for $0 < \omega < 2/\rho(D^{-1}A)$. For any vector e we have

$$\begin{aligned}
\|S(\omega) e\|_A^2 &= (A(I - \omega D^{-1} A)e, (I - \omega D^{-1} A e)) \\
&= (Ae, e) - 2\omega(AD^{-1} Ae, e) + \omega^2 (AD^{-1} Ae, D^{-1} Ae) \\
&= (Ae, e) - 2\omega(D^{-1/2} Ae, D^{-1/2} Ae) + \omega^2 ((D^{-1/2} A D^{-1/2}) D^{-1/2} Ae, D^{-1/2} Ae) \\
&= (Ae, e) - ([\omega(2I - \omega D^{-1/2} A D^{-1/2})] D^{-1/2} Ae, D^{-1/2} Ae) \\
&\leq \|e\|_A^2 - \lambda_{min}[\omega(2I - \omega D^{-1/2} A D^{-1/2})] \|Ae\|_{D^{-1}}^2. \qquad (13.65)
\end{aligned}$$

Let $\gamma = \rho(D^{-1/2}AD^{-1/2}) = \rho(D^{-1}A)$. Then the above restriction on ω implies that $2 - \omega\gamma > 0$ and the matrix in the brackets in (13.65) is positive definite with minimum eigenvalue $\omega(2 - \omega\gamma)$. Then it suffices to take

$$\alpha = \omega(2 - \omega\gamma)$$

to satisfy the requirement (13.62). Note that (13.62) is also valid with α replaced by any positive number that does not exceed the above value, but the resulting inequality will be less sharp. Exercise 15 explores the same question when the Richardson iteration is used instead of weighted Jacobi.

13.6 Algebraic Multigrid

Throughout the previous sections of this chapter, it was seen that MG methods depend in a fundamental way on the availability of an underlying mesh. In addition to this, the performance of MG deteriorates for problems with anisotropic or discontinuous coefficients. It is also difficult to define MG on domains that are not rectangular, especially in three dimensions. Given the success of these techniques, it is clear that it is important to consider alternatives that use similar principles that do not face the same disadvantages. AMG methods have been defined to fill this gap. The main strategy used in AMG is to exploit the Galerkin approach (13.39), in which the interpolation and prolongation operators are defined in an algebraic way, i.e., only from the knowledge of the matrix.

In this section the matrix A is assumed to be positive definite. Since meshes are no longer available, the notation must be changed, or interpreted differently, to reflect levels rather than grid sizes. Here h is no longer a mesh size but an index to a certain level and H is used to index a coarser level. The mesh Ω_h is now replaced with a subspace X_h of R^n at a certain level and X_H denotes the subspace of the coarse problem. Since there are no meshes, one might wonder how the coarse problems can be defined.

In AMG, the coarse problem is typically defined using the Galerkin approach, which we restate here:

$$A_H = I_h^H A_h I_H^h, \qquad f^H = I_h^H f^h, \qquad (13.66)$$

where I_h^H is the restriction operator and I_H^h is the prolongation operator, both defined algebraically. The prolongation and restriction operators are now related by transposition:

$$I_h^H = (I_H^h)^T. \qquad (13.67)$$

A minimum assumption made on the prolongation operator is that it is of full rank.

It can therefore be said that only two ingredients are required to generalize the MG framework:

1. a way to define the *coarse* subspace X_H from a fine subspace X_h,

2. a way to define the interpolation operator I_h^H from X_h to X_H.

In other words, all that is required is a scheme for *coarsening* a fine space along with an interpolation operator that would map a coarse node into a fine one.

In order to understand the motivations for the choices made in AMG when defining the above two components, it is necessary to extend the notion of smooth and oscillatory modes. This is examined in the next section.

Note that Lemma 13.1 is valid and it implies that T_h^H is a projector, which is orthogonal when the A_h inner product is used. The corresponding relations (13.59)–(13.61) also hold. Therefore, Theorem 13.3 is also valid and is a fundamental tool used in the analysis of AMG.

13.6.1 Smoothness in AMG

By analogy with MG, an error is decomposed into smooth and oscillatory components. However, these concepts are now defined with respect to the ability or inability of the smoother to reduce these modes. Specifically, an error is smooth when its convergence with respect to the smoother is slow. The common way to state this is to say that, for a smooth error s,

$$\|S_h s\|_A \approx \|s\|_A.$$

Note the use of the energy norm, which simplifies the analysis. If the smoother satisfies the smoothing property (13.62), then this means that, for a smooth error s, we will have

$$\|As\|_{D^{-1}} \ll \|s\|_{A_h}.$$

Expanding the norms and using the Cauchy–Schwarz inequality gives

$$\|s\|_{A_h}^2 = (D^{-1/2} A_h s, D^{1/2} s)$$
$$\leq \|D^{-1/2} A_h s\|_2 \, \|D^{1/2} s\|_2$$
$$= \|A_h s\|_{D^{-1}} \, \|s\|_D.$$

Since $\|As\|_{D^{-1}} \ll \|s\|_{A_h}$, this means that $\|s\|_{A_h} \ll \|s\|_D$ or

$$(As, s) \ll (Ds, s). \tag{13.68}$$

It simplifies the analysis to set $v = D^{1/2} s$. Then

$$(D^{-1/2} A D^{-1/2} v, v) \ll (v, v).$$

The matrix $\hat{A} \equiv D^{-1/2} A D^{-1/2}$ is a scaled version of A in which the diagonal entries are transformed into ones. The above requirement states that the Rayleigh quotient of $D^{1/2} s$ is small. This is in good agreement with standard MG since a small Rayleigh quotient implies that the vector v is a linear combination of the eigenvectors of A with smallest eigenvalues. In particular, $(As, s) \approx 0$ also implies that $As \approx 0$; i.e.,

$$a_{ii} s_i \approx -\sum_{j \neq i} a_{ij} s_j. \tag{13.69}$$

It is also interesting to see how to interpret smoothness in terms of the matrix coefficients. A common argument held in AMG methods exploits the following expansion of (As, s):

$$(As, s) = \sum_{i,j} a_{ij} s_i s_j$$
$$= \frac{1}{2} \sum_{i,j} -a_{ij} \left((s_j - s_i)^2 - s_i^2 - s_j^2 \right)$$
$$= \frac{1}{2} \sum_{i,j} -a_{ij} (s_j - s_i)^2 + \sum_i \left(\sum_j a_{ij} \right) s_i^2.$$

13.6. Algebraic Multigrid

The condition (13.68) can be rewritten as $(As, s) = \epsilon(Ds, s)$, in which $0 < \epsilon \ll 1$. For the special case when the row sums of the matrix are zero and the off-diagonal elements are negative, this gives

$$\frac{1}{2}\sum_{i,j} |a_{ij}|(s_j - s_i)^2 = \epsilon \sum_i a_{ii} s_i^2 \quad \rightarrow \quad \sum_i a_{ii} s_i^2 \left[\sum_{j \neq i} \frac{|a_{ij}|}{a_{ii}} \left(\frac{s_i - s_j}{s_i}\right)^2 - 2\epsilon \right] = 0.$$

A weighted sum, with nonnegative weights, of the bracketed terms must vanish. It cannot be rigorously argued that the first bracketed term must be of the order 2ϵ, but one can say that *on average* this will be true; i.e.,

$$\sum_{j \neq i} \frac{|a_{ij}|}{a_{ii}} \left(\frac{s_i - s_j}{s_i}\right)^2 \ll 1. \tag{13.70}$$

For the above relation to hold, $|s_i - s_j|/|s_i|$ must be small when $|a_{ji}/a_{ii}|$ is large. In other words, the components of s vary slowly in the direction of the strong connections. This observation is used in AMG when defining interpolation operators and for coarsening.

13.6.2 Interpolation in AMG

The argument given at the end of the previous section is at the basis of many AMG techniques. Consider a coarse node i and its adjacent nodes j, i.e., those indices such that $a_{ij} \neq 0$. The argument following (13.70) makes it possible to distinguish between weak couplings, when $|a_{ij}/a_{ii}|$ is smaller than a certain threshold σ, and strong couplings, when it is larger. Therefore, there are three types of nodes among the nearest neighbors of a fine node i. First there is a set of coarse nodes, denoted by C_i. Then among the fine nodes we have a set F_i^s of nodes that are strongly connected with i and a set F_i^w of nodes that are weakly connected with i. An illustration is shown in Figure 13.9. The smaller filled circles represent the fine nodes and the thin dashed lines represent the weak connections. The thick dash-dot lines represent the strong connections.

According to the argument given above, a good criterion for finding an interpolation formula is to use the relation (13.69), which heuristically characterizes a smooth error. This is because interpolation should average out, i.e., eliminate, highly oscillatory elements in X_h and produce a function that is smooth in the coarser space. Then we rewrite (13.69) as

$$a_{ii} s_i \approx -\sum_{j \in C_i} a_{ij} s_j - \sum_{j \in F_i^s} a_{ij} s_j - \sum_{j \in F_i^w} a_{ij} s_j. \tag{13.71}$$

Consider eliminating the weak connections first. Instead of just removing them from the picture, it is natural to lump their action and add the result into the diagonal term, in a manner similar to the compensation strategy used in incomplete LU (ILU). This gives

$$\left(a_{ii} + \sum_{j \in F_i^w} a_{ij}\right) s_i \approx -\sum_{j \in C_i} a_{ij} s_j - \sum_{j \in F_i^s} a_{ij} s_j. \tag{13.72}$$

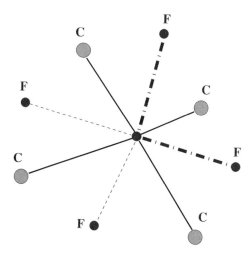

Figure 13.9. *Example of nodes adjacent to a fine node i (center). Fine mesh nodes are labeled with F, coarse nodes with C.*

The end result should be a formula in which the right-hand side depends only on coarse points. Therefore, there remains to express each of the terms of the second sum on the right-hand side of the above equation in terms of values at coarse points. Consider the term s_j for $j \in F_i^s$. At node j, the following expression can be written that is similar to (13.69):

$$a_{jj}s_j \approx -\sum_{l \in C_j} a_{jl}s_l - \sum_{l \in F_j^s} a_{jl}s_l - \sum_{l \in F_j^w} a_{jl}s_l.$$

If the aim is to invoke only those nodes in C_i, then a rough approximation is to remove all other nodes from the formula, so the first sum is replaced by a sum over all $k \in C_i$ (in effect l will belong to $C_i \cap C_j$), and write

$$a_{jj}s_j \approx -\sum_{l \in C_i} a_{jl}s_l.$$

However, this would not be a consistent formula in the sense that it would lead to incorrect approximations for constant functions. To remedy this, a_{jj} should be changed to the opposite of the sum of the coefficients a_{jl}. This gives

$$\left(-\sum_{l \in C_i} a_{jl}\right) s_j \approx -\sum_{l \in C_i} a_{jl}s_l \quad \to \quad s_j \approx \sum_{l \in C_i} \frac{a_{jl}}{\delta_j} s_l, \quad \text{with} \quad \delta_j \equiv \sum_{l \in C_i} a_{jl}.$$

Substituting this into (13.72) yields

$$\left(a_{ii} + \sum_{j \in F_i^w} a_{ij}\right) s_i \approx -\sum_{j \in C_i} a_{ij}s_j - \sum_{j \in F_i^s} a_{ij} \sum_{l \in C_i} \frac{a_{jl}}{\delta_j} s_l. \qquad (13.73)$$

13.6. Algebraic Multigrid

This is the desired formula since it expresses the new fine value s_i in terms of coarse values s_j and s_l for j, l in C_i. A little manipulation will help put it in a *matrix form*, in which s_i is expressed as a combination of the s_j's for $j \in C_i$:

$$s_i = \sum_{j \in C_i} w_{ij} s_j, \quad \text{with} \quad w_{ij} \equiv -\frac{a_{ij} + \sum_{k \in F_i^s} \frac{a_{ik} a_{kj}}{\delta_k}}{a_{ii} + \sum_{k \in F_i^w} a_{ik}}. \tag{13.74}$$

Once the weights are determined, the resulting interpolation formula generalizes the formulas seen for standard MG:

$$(I_H^h x)_i = \begin{cases} x_i & \text{if } i \in X_H, \\ \sum_{j \in C_i} w_{ij} x_j & \text{otherwise.} \end{cases}$$

Example 13.9. Consider the situation depicted in Figure 13.10, which can correspond to a nine-point discretization of some convection diffusion equation on a regular grid. The coarse and fine nodes can be obtained by a red-black coloring of the corresponding five-point graph. For example, black nodes can be the coarse nodes and red nodes the fine nodes.

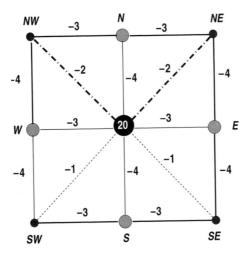

Figure 13.10. *Darker filled circles represent the fine nodes. Thick dash-dot lines represent the strong connections. Values on the edges are the a_{ij}'s. The value 20 at the center (fine) point is a_{ii}.*

In this case, (13.73) yields

$$s_i = \frac{1}{18}\left[4s_S + 4s_N + 3s_W + 3s_E + 2\frac{3s_N + 4s_W}{7} + 2\frac{3s_N + 4s_E}{7}\right]$$

$$= \frac{1}{18}\left[4s_S + \left(4 + \frac{12}{7}\right)s_N + \left(3 + \frac{8}{7}\right)s_W + \left(3 + \frac{8}{7}\right)s_E\right].$$

Notice that, as is expected from an interpolation formula, the weights are all nonnegative and they add up to one.

13.6.3 Defining Coarse Spaces in AMG

Coarsening, i.e., the mechanism by which the coarse subspace X_H is defined from X_h, can be achieved in several heuristic ways. One of the simplest methods, mentioned in the above example, uses the ideas of multicoloring, or independent set orderings (ISOs) seen in Chapter 3. These techniques do not utilize information about strong and weak connections seen in the previous section. A detailed description of these techniques is beyond the scope of this book. However, some of the guiding principles used to defined coarsening heuristics are formulated below.

- When defining the coarse problem, it is important to ensure that it will provide a good representation of smooth functions. In addition, interpolation of smooth functions should be accurate.

- The number of points is much smaller than on the finer problem.

- Strong couplings should not be lost in coarsening. For example, if i is strongly coupled with j, then j must be either a C node or an F node that is strongly coupled with a C node.

- The process should reach a balance between the size of X_H and the accuracy of the interpolation/restriction functions.

13.6.4 AMG via Multilevel ILU

It was stated in the introduction to this section that the main ingredients needed for defining an AMG method are a coarsening scheme and an interpolation operator. A number of techniques have been recently developed that attempt to use the framework of incomplete factorizations to define AMG preconditioners. Let us consider coarsening first. Coarsening can be achieved by using variations of ISOs, which were covered in Chapter 3. Often the independent set is called the fine set and the complement is the coarse set, though this naming is now somewhat arbitrary.

Recall that ISOs transform the original linear system into a system of the form

$$\begin{pmatrix} B & F \\ E & C \end{pmatrix} \begin{pmatrix} x \\ y \end{pmatrix} = \begin{pmatrix} f \\ g \end{pmatrix}, \qquad (13.75)$$

in which the B block is a diagonal matrix. A block LU factorization will help establish the link with AMG-type methods:

$$\begin{pmatrix} B & F \\ E & C \end{pmatrix} = \begin{pmatrix} I & 0 \\ EB^{-1} & I \end{pmatrix} \begin{pmatrix} B & F \\ 0 & S \end{pmatrix},$$

where S is the Schur complement

$$S = C - EB^{-1}F.$$

The above factorization, using independent sets, was the foundation for ILU factorization with multi-elimination (ILUM) seen in Chapter 12. [Since the Schur complement matrix S is sparse, the above procedure can be repeated recursively for a few levels.] Clearly,

13.6. Algebraic Multigrid

dropping is applied each time to prune the Schur complement S, which becomes denser as the number of levels increases. In this section we consider this factorization again from the angle of AMG and will define block generalizations.

Factorizations that generalize the one shown above are now considered in which B is not necessarily diagonal. Such generalizations use the concept of *block* or *group*, independent sets, which generalize standard independent sets. A group independent set is a collection of subsets of unknowns such that there is no coupling between unknowns of any two different groups. Unknowns within the same group may be coupled. An illustration is shown in Figure 13.11.

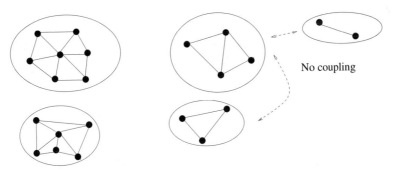

Figure 13.11. *Group (or block) independent sets.*

If the unknowns are permuted such that those associated with the group independent set are listed first, followed by the other unknowns, the original coefficient system will take the form (13.75), where now the matrix B is no longer diagonal but block diagonal. An illustration of two such matrices is given in Figure 13.12. Consider now an LU factorization (exact or incomplete) of B:

$$B = LU + R.$$

Then the matrix A can be factored as follows:

$$\begin{pmatrix} B & F \\ E & C \end{pmatrix} \approx \begin{pmatrix} L & 0 \\ EU^{-1} & I \end{pmatrix} \begin{pmatrix} I & 0 \\ 0 & S \end{pmatrix} \begin{pmatrix} U & L^{-1}F \\ 0 & I \end{pmatrix}. \tag{13.76}$$

The above factorization, which is of the form $A = \mathcal{L}\mathcal{D}\mathcal{U}$, gives rise to an analogue of a two-grid cycle. Solving with the $\mathcal{L}$-matrix would take a vector with components u, y in the fine and coarse spaces, respectively, to produce the vector $y_H = y - EU^{-1}u$ in the coarse space. The Schur complement system can now be solved in some unspecified manner. Once this is done, we need to back solve with the $\mathcal{U}$-matrix. This takes a vector from the coarse space and produces the u variable from the fine space: $u := u - L^{-1}Fy$.

ALGORITHM 13.6. Two-Level Block Solve

1. $f := L^{-1}f$
2. $g := g - EU^{-1}f_1$
3. Solve $Sy = g$
4. $f := f - L^{-1}Fy$
5. $x = U^{-1}f$

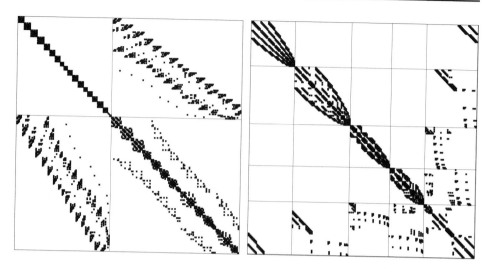

Figure 13.12. *Group independent set reorderings of a nine-point matrix. Left: Small groups (fine grain). Right: Large groups (coarse grain).*

The above solution steps are reminiscent of the two-grid cycle algorithm (Algorithm 13.2). The interpolation and restriction operations are replaced with those in lines 2 and 4, respectively.

A few strategies have recently been developed based on this parallel between a recursive ILU factorization and AMG. One such technique is the algebraic recursive multilevel solver (ARMS) [252]. In ARMS, the block factorization (13.76) is repeated recursively on the Schur complement S, which is kept sparse by dropping small elements. At the lth level, we would write

$$\begin{pmatrix} B_l & F_l \\ E_l & C_l \end{pmatrix} \approx \begin{pmatrix} L_l & 0 \\ E_l U_l^{-1} & I \end{pmatrix} \times \begin{pmatrix} I & 0 \\ 0 & A_{l+1} \end{pmatrix} \times \begin{pmatrix} U_l & L_l^{-1} F_l \\ 0 & I \end{pmatrix}, \qquad (13.77)$$

where $L_l U_l \approx B_l$ and $A_{l+1} \approx C_l - (E_l U_l^{-1})(L_l^{-1} F_l)$.

In a nutshell the ARMS procedure consists of essentially three steps: first, obtain a group independent set and reorder the matrix in the form (13.75); second, obtain an ILU factorization $B_l \approx L_l U_l$ for B_l; and third, obtain approximations to the matrices $L_l^{-1} F_l$, $E_l U_l^{-1}$, and A_{l+1}, and use these to compute an approximation to the Schur complement A_{l+1}. The process is repeated recursively on the matrix A_{l+1} until a selected number of levels is reached. At the last level, a simple ILU with threshold (ILUT) factorization, possibly with pivoting, or an approximate inverse method can be applied.

Each of the A_i's is sparse but will become denser as the number of levels increases, so small elements are dropped in the block factorization to maintain sparsity. The matrices $G_l \equiv E_l U_l^{-1}$ and $W_l \equiv L_l^{-1} F_l$ are only computed in order to obtain the Schur complement

$$A_{l+1} \approx C_l - G_l W_l. \qquad (13.78)$$

Once A_{l+1} is available, W_l and G_l are discarded to save storage. Subsequent operations with $L_l^{-1} F_l$ and $E_l U_l^{-1}$ are performed using U_l, L_l and the blocks E_l and F_l. It is important

13.7. Multigrid versus Krylov Methods

to distinguish between possible variants. To conform to the Galerkin approach, we may elect not to drop terms once A_{l+1} is obtained from (13.78). In this case (13.78) is not an approximation but an exact equality.

There are also many possible variations in the solution phase, which can be viewed as a recursive version of Algorithm 13.6. Line 3 of the algorithm now reads

3. Solve $A_{l+1} y_l = g_l$,

which essentially means *solve in some unspecified way*. At the lth level, this recursive solution step, which we call RSolve for reference, would be replaced by a sequence of statements like

3.0 If $lev = last$
3.1   Solve $A_{l+1} y_l = g_l$
3.2 Else
3.3   RSolve(A_{l+1}, g_l)
3.4 EndIf

Iterative processes can be used in line 3.3. The preconditioner for this iterative process can, for example, be defined using the next, $(l + 2)$th level (without iterating at each level). This is the simplest approach. It is also possible to use an iterative procedure at each level preconditioned (recursively) with the ARMS preconditioner below that level. This variation leads to a procedure similar to the MG cycle if the number of steps γ is specified. Finally, the local B_l block can be used to precondition the system at the lth level.

13.7 Multigrid versus Krylov Methods

The main differences between preconditioned Krylov subspace methods and the MG approach may now have become clearer to the reader. In brief, Krylov methods take a matrix A and a right-hand side b and try to produce a solution, using no other information. The term *black box* is often used for those methods that require minimal input from the user, a good example being sparse direct solvers. Preconditioned Krylov subspace methods attempt to duplicate this attribute of direct solvers, but they are not black box solvers since they require parameters and do not always succeed in solving the problem.

The approach taken by MG methods is to tackle the original problem, e.g., the PDE, directly instead. By doing so, it is possible to exploit properties that are not always readily available from the data A, b. For example, what makes MG work in the Poisson equation case is the strong relation between eigenfunctions of the iteration matrix M and the mesh. It is this strong relation that makes it possible to take advantage of coarser meshes and to exploit a divide-and-conquer principle based on the spectral decomposition of M. AMG methods try to recover similar relationships directly from A, but this is not always easy.

The answer to the question of which method to use cannot be a simple one because it is related to two other important and subjective considerations. The first is the cost of the coding effort. Whether or not one is willing to spend a substantial amount of time coding and testing is now a factor. The second is how important it is to develop an *optimal* code for the problem at hand. If the goal is to solve a single linear system, then a direct solver (assuming enough memory is available) or a preconditioned Krylov solver (in case memory is an issue) may be best. Here, optimality is a secondary consideration. On the other extreme, the best possible performance may be required from the solver if it is meant to be part of a large simulation code that may take, say, days on a high performance computer to complete one run. In this case, it may be worth the time and cost to build the best solver possible, because this cost will be amortized over the lifetime of the simulation code. Here, multilevel techniques can constitute a significant part of the solution scheme. A wide grey zone lies between these two extremes wherein Krylov subspace methods are often invoked.

It may be important to comment on another practical consideration, which is that most industrial solvers are not monolithic schemes based on one single approach. Rather, they comprise building blocks that include tools extracted from various methodologies: direct sparse techniques, multilevel methods, and ILU-type preconditioners, as well as strategies that exploit the specificity of the problem. For example, a solver could utilize the knowledge of the problem to reduce the system by eliminating part of the unknowns, then invoke an AMG or MG scheme to solve the resulting reduced system in cases when it is known to be Poisson-like and an ILU-Krylov approach combined with some reordering schemes (from sparse direct solvers) in other cases. Good iterative solvers must rely on a battery of techniques if they are to be robust and efficient at the same time. To ensure robustness, industrial codes may include an option to resort to direct solvers for those, hopefully rare, instances when the main iterative scheme fails.

Exercises

1. The following notation will be used. Given a vector z of size $n \cdot m$, denote by

$$Z = [z]_{n,m}$$

the matrix of dimension $n \times m$ with entries $Z_{ij} = z_{(j-1)*n+i}$. When there is no ambiguity the subscripts n, m are omitted. In other words n consecutive entries of z will form the columns of Z. The opposite operation, which consists of stacking the consecutive columns of a matrix Z into a vector z, is denoted by

$$z = Z_|.$$

   a. Let $u \in \mathbb{R}^m$, $v \in \mathbb{R}^n$. What is the matrix $Z = [z]_{n,m}$ when $z = u \otimes v$?

   b. Show that

   $$(I \otimes A)z = (A \cdot [z])_| \quad \text{and} \quad (A \otimes I)z = ([z] \cdot A^T)_|.$$

   c. Show, more generally, that

   $$(A \otimes B)z = (B \cdot [z] \cdot A^T)_|.$$

   d. What becomes of the above relation when $z = u \otimes v$? Find an eigenvector of $A \otimes B$ based on this.

   e. Show that $(A \otimes B)^T = (A^T \otimes B^T)$.

Exercises

2. Establish that the eigenvectors of the Gauss–Seidel operator given by (13.29) are indeed a set of n linearly independent vectors. [Hint: notice that the eigenvalues other than for $k = (n + 1)/2$ are all double, so it suffices to show that the two eigenvectors defined by the formula are independent.] What happens if the absolute values are removed from the expression (13.29)?

3. Consider the Gauss–Seidel iteration as a smoother for the one-dimensional model problem when $n = 11$ (spectrum illustrated in Figure 13.4). For each eigenvector u_i of the original matrix A, compute the norm reduction $\|Gu_i\|/\|u_i\|$, where G is the Gauss–Seidel iteration matrix. Plot these ratios against i in a way that is similar to Figure 13.4. What can you conclude? Repeat for the powers G^2, G^4, and G^8. What can you conclude (see the statement made at the end of Section 13.2.3)?

4. Show relation (13.31). Consider, as an example, a 7×5 grid and illustrate for this case the semi-interpolation operators $I^h_{x,2h}$ and $I^h_{y,2h}$. Then derive the relation by seeing how the two-dimensional interpolation operator was defined in Section 13.3.1.

5. Show relation (13.38). Consider first, as an example, a 7×5 grid and illustrate for this case the semirestriction operators $I^{2h}_{x,h}$ and $I^{2h}_{y,h}$. Then use (13.31) (see previous exercise) and part (e) of Exercise 1.

6. What is the matrix $I^H_h A_h I^h_H$ for the two-dimensional model problem when FW is used? [Hint: Use the tensor product notation and the results of Exercises 5 and 4.]

7. Consider the matrix J_ω given by (13.20). Show that it is possible to find $\omega > 1$ such that $\rho(J_\omega) > 1$. Repeat for $\omega < 0$.

8. Derive the FW formula by applying the trapezoidal rule to approximate the numerator and denominator in the following approximation:

$$u(x) \approx \frac{\int_{x-h}^{x+h} u(t)\,dt}{\int_{x-h}^{x+h} 1 \cdot dt}.$$

9. Derive the B form (or preconditioning form; see (13.41)) of the weighted Jacobi iteration.

10. Do the following experiment using an interactive package such as MATLAB (or code in FORTRAN or C). Consider the linear system $Ax = 0$, where A arises from the discretization of $-u''$ on $[0, 1]$ using 64 internal points. Take u_0 to be the average of the two modes $u_{n/4} = u_{16}$ and $u_{3n/4} = u_{48}$. Plot the initial error. [Hint: the error is just u_0.] Then plot the error after two steps of the Richardson process, then after five steps of the Richardson process. Plot also the components of the final error after the five Richardson steps, with respect to the eigenbasis. Now obtain the residual on the grid Ω_{2h} and plot the residual obtained after two and five Richardson steps on the coarse-grid problem. Show also the components of the error in the eigenbasis of the original problem (on the fine mesh). Finally, interpolate to the fine grid and repeat the process again, doing two and then five steps of the Richardson process.

11. Repeat Exercise 10 using the Gauss–Seidel iteration instead of the Richardson iteration.

12. Consider the cost of the general MG algorithm as given by the recurrence formula (13.46). Solve the recurrence equation (in terms of η and γ) for the one-dimensional case. You may assume that $n = 2^k + 1$ and that the maximum number of levels is used so that the cost of the last system to solve is zero. For the two-dimensional case, you may assume that $n = m = 2^k + 1$. Under which condition is the cost $O(n \log n)$, where n is the size of the finest grid under consideration? What would be the situation for three-dimensional problems?

13. It was stated in Section 13.4.4 that condition (13.48) implies condition (13.49) provided an assumption is made on the interpolation $\hat{I}_H^h$. Prove a rigorous bound of the type (13.49) (i.e., find c_1) by assuming the conditions (13.48), (13.51), and

$$\|u - \hat{I}_H^h u\| \le c_4 h^\kappa,$$

in which as before u represents the discretization of the solution of the continuous problem (i.e., the continuous solution sampled at the grid points of Ω_h or Ω_H).

14. Justify the definition of the norm $\|v\|_h$ in Example 13.6 by considering that the integral

$$\int_0^1 \mathbf{v}(t)^2 \, dt$$

is approximated by the trapezoidal rule. It is assumed that $v(x_0) = v(x_{n+1}) = 0$ and the composite trapezoidal rule uses all points $x_0, \ldots, x_{n+1}$.

15. Find the constant α in the smoothing property (13.62) for the case of the Richardson iteration when A is SPD. [Hint: The Richardson iteration is like a Jacobi iteration where the diagonal is replaced with the identity.]

16. Extend the argument of Example 13.6 to the two-dimensional case. Start with the case of the square $(0, 1)^2$, which uses the same discretization in each direction. Then consider the more general situation. Define the norm $\|v\|_h$ from the discrete L_2-norm (see also Exercise 15).

17. Establish a bound of the type shown in Example 13.6 using the 2-norm instead of the discrete L_2-norm. What if the A_h-norm is used?

18. The energy norm can be used to establish a result similar to that of Theorem 13.2, leading to a slightly simpler argument. It is now assumed that (13.49) is satisfied with respect to the A_h-norm, i.e., that

$$\|u^h - \hat{I}_H^h u^H\|_{A_h} \le c_1 h^\kappa.$$

a. Show that for any vector v in Ω_H we have

$$\|I_H^h v\|_{A_h} = \|v\|_{A_H}.$$

b. Let u_0^h be the initial guess at grid Ω_h in FMG and assume that the error achieved by the system at level $H = 2h$ satisfies $\|u^H - \tilde{u}^H\|_{A_H} \le c_1 c_3 H^\kappa$, in which c_3 is to be determined. Follow the argument of the proof of Theorem 13.2 and use the relation established in (a) to show that

$$\|u^h - u_0^h\|_{A_h} \le \|u^h - \hat{I}_H^h u^H\|_{A_h} + \|u^H - \tilde{u}^H\|_{A_H} \le c_1 h^\kappa + c_1 c_3 H^\kappa.$$

c. Show a result analogous to that of Theorem 13.2 that uses the A_h-norm; i.e., find c_3 such that $\|u^h - \tilde{u}^h\|_{A_h} \leq c_1 c_3 h^\kappa$ on each grid.

19. Starting from relation (13.73), establish (13.74).

Notes and References

The material presented in this chapter is based on several sources. Foremost among these are the references [65, 206, 163, 285, 300]. A highly recommended reference is the *Multigrid Tutorial* by Briggs, Henson, and McCormick [65], for its excellent introduction to the subject. This tutorial includes enough theory to understand how MG methods work. More detailed volumes include the books by McCormick et al. [206], Hackbusch [162, 163], and Wesseling [310], and the more recent book by Trottenberg et al. [285].

Early work on MG methods dates back to the 1960s and includes the papers by Brakhage [46], Fedorenko [124, 125], Bakhvalov [23], and Kronsjö and Dahlquist [192]. However, MG methods saw much of their modern development in the 1970s and early 1980s, essentially under the pioneering work of Brandt [54, 55, 56]. Brandt played a key role in promoting the use of MG methods by establishing their overwhelming superiority over existing techniques for elliptic PDEs and by introducing many new concepts that are now widely used in MG literature. AMG methods were later developed to attempt to obtain similar performance. These methods were introduced in [58] and analyzed in a number of papers; see, e.g., [57, 234].

Closely related to the MG approach is the aggregation-disaggregation technique, which is popular in Markov chain modeling. A recommended book for these methods and others used in the context of Markov chain modeling is [274].

Today MG methods are still among the most efficient techniques available for solving elliptic PDEs on regularly structured problems. Their algebraic variants do not seem to have proven as effective and the search for the elusive black box iterative solver is still underway, with research on multilevel methods in general and AMG in particular still quite active. With computer power constantly improving, problems are becoming larger and more complex, which makes mesh independent convergence look ever more attractive.

The paper [252] describes a scalar version of ARMS and the report [202] describes a parallel implementation. The related method named MLILU described in [28] also exploits the connection between ILU and AMG. The parallel version of ARMS (called pARMS) is available from the author's web site: www.cs.umn.edu/~saad.

Resources for MG are available at www.mgnet.org, which provides bibliographical references, software, and a newsletter. In particular, Examples 13.4 and 13.5 have been run with the MGLAB MATLAB codes (contributed by James Bordner and Faisal Saied) available from this site. A parallel code named HYPRE, which is available from the Lawrence Livermore National Lab, includes implementations of AMG.

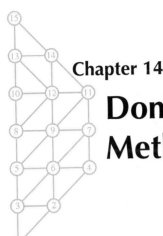

Chapter 14
Domain Decomposition Methods

As multiprocessing technology steadily gains ground, new classes of numerical methods that can take better advantage of parallelism are emerging. Among these techniques, domain decomposition methods are undoubtedly the best known and perhaps the most promising for certain types of problems. These methods combine ideas from partial differential equations (PDEs), linear algebra, and mathematical analysis, and techniques from graph theory. This chapter is devoted to *decomposition* methods, which are based on the general concepts of graph partitionings.

14.1 Introduction

Domain decomposition methods refer to a collection of techniques that revolve around the principle of divide and conquer. Such methods have been primarily developed for solving PDEs over regions in two or three dimensions. However, similar principles have been exploited in other contexts of science and engineering. In fact, one of the earliest practical uses for domain decomposition approaches was in structural engineering, a discipline that is not dominated by PDEs. Although this chapter considers these techniques from a purely linear algebra viewpoint, the basic concepts, as well as the terminology, are introduced from a model PDE.

Consider the problem of solving the Laplace equation on an L-shaped domain Ω partitioned as shown in Figure 14.1. Domain decomposition or substructuring methods attempt to solve the problem on the entire domain

$$\Omega = \bigcup_{i=1}^{s} \Omega_i$$

from problem solutions on the subdomains Ω_i. There are several reasons why such techniques can be advantageous. In the case of the above picture, one obvious reason is that the subproblems are much simpler because of their rectangular geometry. For example, fast

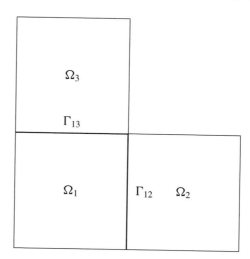

Figure 14.1. *An L-shaped domain subdivided into three subdomains.*

Poisson solvers (FPS) can be used on each subdomain in this case. A second reason is that the physical problem can sometimes be split naturally into a small number of subregions where the modeling equations are different (e.g., Euler's equations in one region and Navier–Stokes in another).

Substructuring can also be used to develop *out-of-core* solution techniques. As already mentioned, such techniques were often used in the past to analyze very large mechanical structures. The original structure is partitioned into s pieces, each of which is small enough to fit into memory. Then a form of block Gaussian elimination is used to solve the global linear system from a sequence of solutions using s subsystems. More recent interest in domain decomposition techniques has been motivated by parallel processing.

14.1.1 Notation

In order to review the issues and techniques in use and to introduce some notation, assume that the following problem is to be solved:

$$\Delta u = f \text{ in } \Omega,$$
$$u = u_\Gamma \text{ on } \Gamma = \partial\Omega.$$

Domain decomposition methods are all implicitly or explicitly based on different ways of handling the unknown at the interfaces. From the PDE point of view, if the value of the solution is known at the interfaces between the different regions, these values could be used in Dirichlet-type boundary conditions and we will obtain s uncoupled Poisson equations. We can then solve these equations to obtain the value of the solution at the interior points. If the whole domain is discretized by either finite element or finite difference techniques, then this is easily translated into the resulting linear system.

14.1. Introduction

Now some terminology and notation will be introduced for use throughout this chapter. Assume that the problem associated with the domain shown in Figure 14.1 is discretized with centered differences. We can label the nodes by subdomain, as shown in Figure 14.3. Note that the interface nodes are labeled last. As a result, the matrix associated with this problem will have the structure shown in Figure 14.4. For a general partitioning into s subdomains, the linear system associated with the problem has the following structure:

$$\begin{pmatrix} B_1 & & & & E_1 \\ & B_2 & & & E_2 \\ & & \ddots & & \vdots \\ & & & B_s & E_s \\ F_1 & F_2 & \cdots & F_s & C \end{pmatrix} \begin{pmatrix} x_1 \\ x_2 \\ \vdots \\ x_s \\ y \end{pmatrix} = \begin{pmatrix} f_1 \\ f_2 \\ \vdots \\ f_s \\ g \end{pmatrix}, \qquad (14.1)$$

where each x_i represents the subvector of unknowns that are interior to subdomain Ω_i and y represents the vector of all interface unknowns. It is useful to express the above system in the simpler form

$$A \begin{pmatrix} x \\ y \end{pmatrix} = \begin{pmatrix} f \\ g \end{pmatrix}, \quad \text{with} \quad A = \begin{pmatrix} B & E \\ F & C \end{pmatrix}. \qquad (14.2)$$

Thus, E represents the subdomain-to-interface coupling seen from the subdomains, while F represents the interface-to-subdomain coupling seen from the interface nodes.

14.1.2 Types of Partitionings

When partitioning a problem, it is common to use graph representations. Since the subproblems obtained from a given partitioning will eventually be mapped into distinct processors, there are some restrictions regarding the type of partitioning needed. For example, in element-by-element (EBE) finite element techniques, it may be desirable to map elements into processors instead of vertices. In this case, the restriction means no element should be split between two subdomains; i.e., all information related to a given element is mapped to the same processor. These partitionings are termed *element based*. A somewhat less restrictive class of partitionings are the edge-based partitionings, which do not allow edges to be split between two subdomains. These may be useful for finite volume techniques where computations are expressed in terms of fluxes across edges in two dimensions. Finally, vertex-based partitionings work by dividing the original vertex set into subsets of vertices and have no restrictions on the edges; i.e., they allow edges or elements to straddle between subdomains. See Figure 14.2.

14.1.3 Types of Techniques

The interface values can be obtained by employing a form of block Gaussian elimination that may be too expensive for large problems. In some simple cases, using fast Fourier transforms (FFTs), it is possible to explicitly obtain the solution of the problem on the interfaces inexpensively.

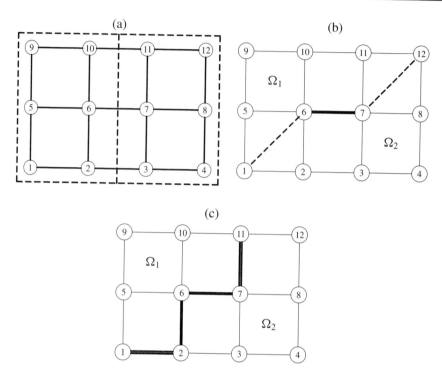

Figure 14.2. (a) *Vertex-based,* (b) *edge-based, and* (c) *element-based partitioning of a* 4 × 3 *mesh into two subregions.*

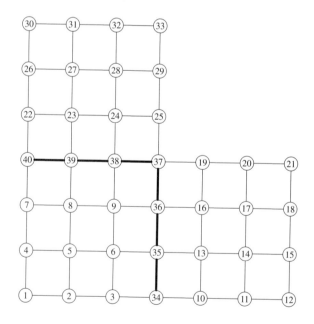

Figure 14.3. *Discretization of problem shown in Figure* 14.1.

14.1. Introduction

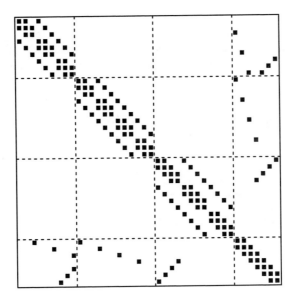

Figure 14.4. *Matrix associated with the finite difference mesh of Figure* 14.3.

Other methods alternate between the subdomains, solving a new problem each time, with boundary conditions updated from the most recent subdomain solutions. These methods are called *Schwarz alternating procedures* (SAPs), after the Swiss mathematician who used the idea to prove the existence of a solution of the Dirichlet problem on irregular regions.

The subdomains may be allowed to *overlap*. This means that the Ω_i's are such that

$$\Omega = \bigcup_{i=1,s} \Omega_i, \quad \Omega_i \cap \Omega_j \neq \emptyset.$$

For a discretized problem, it is typical to quantify the extent of overlapping by the number of mesh lines that are common to the two subdomains. In the particular case of Figure 14.3, the overlap is of order one.

The various domain decomposition techniques are distinguished by four features:

1. *Type of partitioning.* For example, should partitioning occur along edges, along vertices, or by elements? Is the union of the subdomains equal to the original domain or a superset of it (fictitious domain methods)?

2. *Overlap.* Should subdomains overlap or not, and by how much?

3. *Processing of interface values.* For example, is the Schur complement approach used? Should there be successive updates to the interface values?

4. *Subdomain solution.* Should the subdomain problems be solved exactly or approximately by an iterative method?

The methods to be discussed in this chapter will be classified in four distinct groups. First, direct methods and the substructuring approach are useful for introducing some definitions and for providing practical insight. Second, among the simplest and oldest techniques are the SAPs. Then there are methods based on preconditioning the Schur complement system. The last category groups all the methods based on solving the linear system with the matrix A, by using a preconditioning derived from domain decomposition concepts.

14.2 Direct Solution and the Schur Complement

One of the first divide-and-conquer ideas used in structural analysis exploited the partitioning (14.1) in a direct solution framework. This approach, which is covered in this section, introduces the Schur complement and explains some of its properties.

14.2.1 Block Gaussian Elimination

Consider the linear system written in the form (14.2), in which B is assumed to be nonsingular. From the first equation the unknown x can be expressed as

$$x = B^{-1}(f - Ey). \tag{14.3}$$

Upon substituting this into the second equation, the following *reduced system* is obtained:

$$(C - FB^{-1}E)y = g - FB^{-1}f. \tag{14.4}$$

The matrix

$$S = C - FB^{-1}E \tag{14.5}$$

is called the *Schur complement* matrix associated with the y variable. If this matrix can be formed and the linear system (14.4) can be solved, all the interface variables y will become available. Once these variables are known, the remaining unknowns can be computed, via (14.3). Because of the particular structure of B, observe that any linear system solution with it decouples in s separate systems. The parallelism in this situation arises from this natural decoupling.

A solution method based on this approach involves four steps:

1. Obtain the right-hand side of the reduced system (14.4).

2. Form the Schur complement matrix (14.5).

3. Solve the reduced system (14.4).

4. Back substitute using (14.3) to obtain the other unknowns.

One linear system solution with the matrix B can be saved by reformulating the algorithm in a more elegant form. Define

$$E' = B^{-1}E \quad \text{and} \quad f' = B^{-1}f.$$

The matrix E' and the vector f' are needed in steps 1 and 2. Then rewrite step 4 as

$$x = B^{-1}f - B^{-1}Ey = f' - E'y,$$

which gives the following algorithm.

14.2. Direct Solution and the Schur Complement

ALGORITHM 14.1. Block Gaussian Elimination

1. Solve $BE' = E$ and $Bf' = f$ for E' and f', respectively
2. Compute $g' = g - Ff'$
3. Compute $S = C - FE'$
4. Solve $Sy = g'$
5. Compute $x = f' - E'y$

In a practical implementation, all the B_i-matrices are factored and then the systems $B_i E_i' = E_i$ and $B_i f_i' = f_i$ are solved. In general, many columns in E_i will be zero. These zero columns correspond to interfaces that are not adjacent to subdomain i. Therefore, any efficient code based on the above algorithm should start by identifying the nonzero columns.

14.2.2 Properties of the Schur Complement

Now the connections between the Schur complement and standard Gaussian elimination will be explored and a few simple properties will be established. Start with the block LU factorization of A:

$$\begin{pmatrix} B & E \\ F & C \end{pmatrix} = \begin{pmatrix} I & O \\ FB^{-1} & I \end{pmatrix} \begin{pmatrix} B & E \\ O & S \end{pmatrix}, \qquad (14.6)$$

which is readily verified. The Schur complement can therefore be regarded as the $(2, 2)$ block in the U part of the block LU factorization of A. From the above relation, note that, if A is nonsingular, then so is S. Taking the inverse of A with the help of the above equality yields

$$\begin{pmatrix} B & E \\ F & C \end{pmatrix}^{-1} = \begin{pmatrix} B^{-1} & -B^{-1}ES^{-1} \\ O & S^{-1} \end{pmatrix} \begin{pmatrix} I & O \\ -FB^{-1} & I \end{pmatrix}$$

$$= \begin{pmatrix} B^{-1} + B^{-1}ES^{-1}FB^{-1} & -B^{-1}ES^{-1} \\ -S^{-1}FB^{-1} & S^{-1} \end{pmatrix}. \qquad (14.7)$$

Observe that S^{-1} is the $(2, 2)$ block in the block inverse of A. In particular, if the original matrix A is symmetric positive definite (SPD), then so is A^{-1}. As a result, S is also SPD in this case.

Although simple to prove, the above properties are nonetheless important. They are summarized in the following proposition.

Proposition 14.1. *Let A be a nonsingular matrix partitioned as in* (14.2) *such that the submatrix B is nonsingular and let R_y be the restriction operator onto the interface variables, i.e., the linear operator defined by*

$$R_y \begin{pmatrix} x \\ y \end{pmatrix} = y.$$

Then the following properties are true.

1. *The Schur complement matrix S is nonsingular.*
2. *If A is SPD, then so is S.*
3. *For any y, $S^{-1}y = R_y A^{-1} \begin{pmatrix} 0 \\ y \end{pmatrix}$.*

The first property indicates that a method that uses the above block Gaussian elimination algorithm is feasible since S is nonsingular. A consequence of the second property is that, when A is positive definite, an algorithm such as the conjugate gradient (CG) algorithm can be used to solve the reduced system (14.4). Finally, the third property establishes a relation that may allow preconditioners for S to be defined based on solution techniques with the matrix A.

14.2.3 Schur Complement for Vertex-Based Partitionings

The partitioning used in Figure 14.3 is edge based, meaning that a given edge in the graph does not straddle two subdomains or that, if two vertices are coupled, then they cannot belong to the two distinct subdomains. From the graph theory point of view, this is perhaps less common than vertex-based partitionings, in which a vertex is not shared by two partitions (except when domains overlap). A vertex-based partitioning is illustrated in Figure 14.5.

We will call interface edges all edges that link vertices that do not belong to the same subdomain. In the case of overlapping, this needs clarification. An overlapping edge or vertex belongs to the same subdomain. Interface edges are only those that link a vertex to another vertex that is not in the same subdomain already, whether in the overlapping portion or elsewhere. Interface vertices are those vertices in a given subdomain that are adjacent to an interface edge. For the example of the figure, the interface vertices for subdomain one (bottom-left subsquare) are the vertices labeled 10 to 16. The matrix shown at the bottom of Figure 14.5 differs from the one of Figure 14.4 because here the interface nodes are not relabeled as the last in the global labeling, as was done in Figure 14.3. Instead, the interface nodes are labeled as the last nodes in each subdomain. The number of interface nodes is about twice that of the edge-based partitioning.

Consider the Schur complement system obtained with this new labeling. It can be written similarly to the edge-based case using a reordering in which all interface variables are listed last. The matrix associated with the domain partitioning of the variables will have a natural s block structure, where s is the number of subdomains. For example, when $s = 3$ (as is the case in the above illustration), the matrix has the block structure defined by the solid lines in the figure; i.e.,

$$A = \begin{pmatrix} A_1 & A_{12} & A_{13} \\ A_{21} & A_2 & A_{23} \\ A_{31} & A_{32} & A_3 \end{pmatrix}. \tag{14.8}$$

In each subdomain, the variables are of the form

$$z_i = \begin{pmatrix} x_i \\ y_i \end{pmatrix},$$

where x_i denotes interior nodes and y_i denotes the interface nodes associated with subdomain i. Each matrix A_i will be called the local matrix.

The structure of A_i is as follows:

$$A_i = \begin{pmatrix} B_i & E_i \\ F_i & C_i \end{pmatrix}, \tag{14.9}$$

14.2. Direct Solution and the Schur Complement

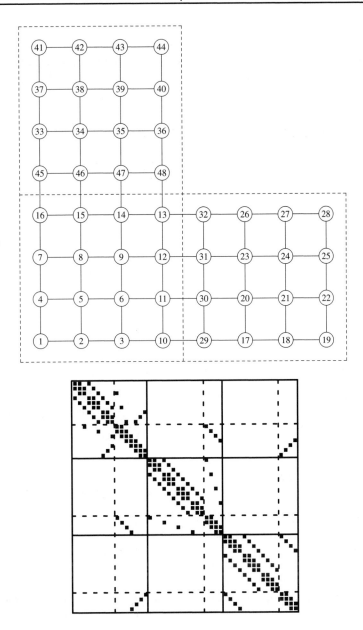

Figure 14.5. *Discretization of problem shown in Figure* 14.1 *and associated matrix.*

in which, as before, B_i represents the matrix associated with the internal nodes of subdomain i and E_i and F_i represent the couplings to/from local interface nodes. The matrix C_i is the local part of the interface matrix C defined before and represents the coupling between local interface points. A careful look at the matrix in Figure 14.5 reveals an additional structure for the blocks A_{ij}, $j \neq i$. Partitioning A_{ij} according to the variables x_i, y_i on the one hand (rows), and x_j, y_j on the other, reveals that it comprises only one nonzero block. Indeed, there is no coupling between x_i and x_j, between x_i and y_j, or between y_i and x_j. Therefore, the submatrix A_{ij} has the following structure:

$$A_{ij} = \begin{pmatrix} 0 & 0 \\ 0 & E_{ij} \end{pmatrix}. \qquad (14.10)$$

In addition, most of the E_{ij}-matrices are zero since only those indices j of the subdomains that have couplings with subdomain i will yield a nonzero E_{ij}.

Now write the part of the linear system that is local to subdomain i as

$$\begin{array}{rcl} B_i x_i + E_i y_i & = & f_i, \\ F_i x_i + C_i y_i + \sum_{j \in N_i} E_{ij} y_j & = & g_i. \end{array} \qquad (14.11)$$

The term $E_{ij} y_j$ is the contribution to the equation from the neighboring subdomain number j and N_i is the set of subdomains adjacent to subdomain i. Assuming that B_i is nonsingular, the variable x_i can be eliminated from this system by extracting from the first equation $x_i = B_i^{-1}(f_i - E_i y_i)$, which yields, upon substitution in the second equation,

$$S_i y_i + \sum_{j \in N_i} E_{ij} y_j = g_i - F_i B_i^{-1} f_i, \quad i = 1, \ldots, s, \qquad (14.12)$$

in which S_i is the *local* Schur complement

$$S_i = C_i - F_i B_i^{-1} E_i. \qquad (14.13)$$

When written for all subdomains i, the equations (14.12) yield a system of equations that involves only the interface points y_j, $j = 1, 2, \ldots, s$, and that has a natural block structure associated with these vector variables:

$$S = \begin{pmatrix} S_1 & E_{12} & E_{13} & \cdots & E_{1s} \\ E_{21} & S_2 & E_{23} & \cdots & E_{2s} \\ \vdots & & \ddots & & \vdots \\ \vdots & & & \ddots & \vdots \\ E_{s1} & E_{s2} & E_{s3} & \cdots & S_s \end{pmatrix}. \qquad (14.14)$$

The diagonal blocks in this system, namely, the matrices S_i, are dense in general, but the off-diagonal blocks E_{ij} are sparse and most of them are zero. Specifically, $E_{ij} \neq 0$ only if subdomains i and j have at least one equation that couples them.

A structure of the global Schur complement S has been unraveled, which has the following important implication: *For vertex-based partitionings, the Schur complement matrix can be assembled from local Schur complement matrices (the S_i's) and interface-to-interface information (the E_{ij}'s).* The term *assembled* was used on purpose because a similar idea will be exploited for finite element partitionings.

14.2.4 Schur Complement for Finite Element Partitionings

In finite element partitionings, the original discrete set Ω is subdivided into s subsets Ω_i, each consisting of a distinct set of elements. Given a finite element discretization of the

14.2. Direct Solution and the Schur Complement

domain Ω, a finite dimensional space V_h of functions over Ω is defined, e.g., functions that are piecewise linear and continuous on Ω and that vanish on the boundary Γ of Ω. Consider now the Dirichlet problem on Ω and recall that its weak formulation on the finite element discretization can be stated as follows (see Section 2.3):

$$\text{Find} \quad u \in V_h \quad \text{such that} \quad a(u, v) = (f, v) \quad \forall \ v \in V_h,$$

where the bilinear form $a(\cdot, \cdot)$ is defined by

$$a(u, v) = \int_\Omega \nabla u \cdot \nabla v \ dx = \int_\Omega \left(\frac{\partial u}{\partial x_1} \frac{\partial v}{\partial x_1} + \frac{\partial u}{\partial x_2} \frac{\partial u}{\partial x_2} \right) dx.$$

It is interesting to observe that, since the set of the elements of the different Ω_i's are disjoint, $a(\cdot, \cdot)$ can be decomposed as

$$a(u, v) = \sum_{i=1}^{s} a_i(u, v),$$

where

$$a_i(u, v) = \int_{\Omega_i} \nabla u \cdot \nabla v \ dx.$$

In fact, this is a generalization of the technique used to assemble the stiffness matrix from element matrices, which corresponds to the extreme case where each Ω_i consists of exactly one element.

If the unknowns are ordered again by subdomains and the interface nodes are placed last, as was done in Section 14.1, immediately the system shows the same structure:

$$\begin{pmatrix} B_1 & & & & E_1 \\ & B_2 & & & E_2 \\ & & \ddots & & \vdots \\ & & & B_s & E_s \\ F_1 & F_2 & \cdots & F_s & C \end{pmatrix} \begin{pmatrix} x_1 \\ x_2 \\ \vdots \\ x_s \\ y \end{pmatrix} = \begin{pmatrix} f_1 \\ f_2 \\ \vdots \\ f_s \\ g \end{pmatrix}, \tag{14.15}$$

where each B_i represents the coupling between interior nodes and E_i and F_i represent the coupling between the interface nodes and the nodes interior to Ω_i. Note that each of these matrices has been assembled from element matrices and can therefore be obtained from contributions over all subdomains Ω_j that contain any node of Ω_i.

In particular, assume that the assembly is considered only with respect to Ω_i. Then the assembled matrix will have the structure

$$A_i = \begin{pmatrix} B_i & E_i \\ F_i & C_i \end{pmatrix},$$

where C_i contains only contributions from local elements, i.e., elements that are in Ω_i. Clearly, C is the sum of the C_i's:

$$C = \sum_{i=1}^{s} C_i.$$

The Schur complement associated with the interface variables is such that

$$S = C - FB^{-1}E$$
$$= C - \sum_{i=1}^{s} F_i B_i^{-1} E_i$$
$$= \sum_{i=1}^{s} C_i - \sum_{i=1}^{s} F_i B_i^{-1} E_i$$
$$= \sum_{i=1}^{s} \left[C_i - F_i B_i^{-1} E_i \right].$$

Therefore, if S_i denotes the *local* Schur complement

$$S_i = C_i - F_i B_i^{-1} E_i,$$

then the above proves that

$$S = \sum_{i=1}^{s} S_i, \tag{14.16}$$

showing again that the Schur complement can be obtained easily from smaller Schur complement matrices.

Another important observation is that the stiffness matrix A_k, defined above by restricting the assembly to Ω_k, solves a Neumann–Dirichlet problem on Ω_k. Indeed, consider the problem

$$\begin{pmatrix} B_k & E_k \\ F_k & C_k \end{pmatrix} \begin{pmatrix} x_k \\ y_k \end{pmatrix} = \begin{pmatrix} f_k \\ g_k \end{pmatrix}. \tag{14.17}$$

The elements of the submatrix C_k are the terms $a_k(\phi_i, \phi_j)$ where ϕ_i, ϕ_j are the basis functions associated with nodes belonging to the interface Γ_k. As was stated above, the matrix C is the sum of these submatrices. Consider the problem of solving the Poisson equation on Ω_k with boundary conditions defined as follows: On Γ_{k0}, the part of the boundary that belongs to Γ_k, use the original boundary conditions; on the interfaces Γ_{kj} with other subdomains, use a Neumann boundary condition. According to (2.47) seen in Section 2.3, the jth equation will be of the form

$$\int_{\Omega_k} \nabla u \cdot \nabla \phi_j \, dx = \int_{\Omega_k} f \phi_j dx + \int_{\Gamma_k} \phi_j \frac{\partial u}{\partial \vec{n}} \, ds. \tag{14.18}$$

This gives rise to a system of the form (14.17), in which the g_k part of the right-hand side incorporates the Neumann data related to the second integral on the right-hand side of (14.18).

It is interesting to note that, if a problem were to be solved with all Dirichlet conditions, i.e., if the Neumann conditions at the interfaces were replaced by Dirichlet conditions, the resulting matrix problem would be of the form

$$\begin{pmatrix} B_k & E_k \\ 0 & I \end{pmatrix} \begin{pmatrix} x_k \\ y_k \end{pmatrix} = \begin{pmatrix} f_k \\ g_k \end{pmatrix}, \tag{14.19}$$

where g_k represents precisely the Dirichlet data. Indeed, according to what was seen in Section 2.3, Dirichlet conditions are handled simply by replacing equations associated with boundary points with identity equations.

14.2.5 Schur Complement for the Model Problem

An explicit expression for the Schur complement can be found in the simple case of a rectangular region partitioned into two subdomains, as illustrated in Figure 14.6. The figure shows a vertex-based partitioning, but what follows is also valid for edge-based partitionings, since we will only compute the local Schur complements S_1, S_2, from which the global Schur complement is constituted. For an edge-based partitioning, the Schur complement S is the sum of the local Schur complements S_1 and S_2. For a vertex-based partitioning, S is of the form (14.14), with $s = 2$.

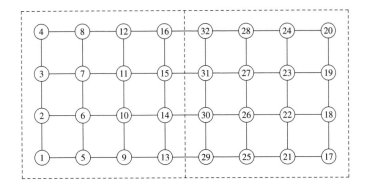

Figure 14.6. *A two-domain partitioning of the model problem on a rectangular domain.*

To determine S_1, start by writing the discretized matrix for the model problem in the subdomain Ω_1:

$$A = \begin{pmatrix} B & -I & & & \\ -I & B & -I & & \\ & \ddots & \ddots & \ddots & \\ & & -I & B & -I \\ & & & -I & B \end{pmatrix}, \quad \text{with} \quad B = \begin{pmatrix} 4 & -1 & & & \\ -1 & 4 & -1 & & \\ & \ddots & \ddots & \ddots & \\ & & -1 & 4 & -1 \\ & & & -1 & 4 \end{pmatrix}.$$

Assume that the size of each block (i.e., the number of points in the vertical direction in Ω_1) is m, while the number of blocks (i.e., the number of points in the horizontal direction in Ω_1) is n. Also, the points are ordered naturally, with the last vertical line forming the interface. Then A is factored in the following block LU decomposition:

$$A = \begin{pmatrix} I & & & & \\ -T_1^{-1} & I & & & \\ & \ddots & \ddots & & \\ & & -T_{j-1}^{-1} & I & \\ & & & -T_{n-1}^{-1} & I \end{pmatrix} \begin{pmatrix} T_1 & -I & & & \\ & T_2 & -I & & \\ & & \ddots & \ddots & \\ & & & T_{n-1} & -I \\ & & & & T_n \end{pmatrix}.$$

The matrices T_i satisfy the recurrence

$$T_1 = B, \qquad T_{k+1} = B - T_k^{-1}, \quad k = 1, \ldots, n-1. \qquad (14.20)$$

It can easily be shown that the above recurrence does not break down, i.e., that each inverse does indeed exist. Also, the matrix T_n is the desired local Schur complement S_1.

Each T_k is a rational function of B; i.e., $T_k = f_k(B)$, where f_k is a rational function defined by

$$f_{k+1}(\mu) = \mu - \frac{1}{f_k(\mu)}.$$

With each eigenvalue μ of B is associated an eigenvalue λ_k. This sequence of eigenvalues satisfies the recurrence

$$\lambda_{k+1} = \mu - \frac{1}{\lambda_k}.$$

To calculate f_n it is sufficient to calculate λ_k in terms of μ. The above difference equation differs from the ones we have encountered in other chapters in that it is nonlinear. It can be solved by defining the auxiliary unknown

$$\eta_k = \prod_{j=0}^{k} \lambda_j.$$

By definition $\lambda_0 = 1$, $\lambda_1 = \mu$ so that $\eta_0 = 1$, $\eta_1 = \mu$. The sequence η_k satisfies the recurrence

$$\eta_{k+1} = \mu \eta_k - \eta_{k-1},$$

which is now a linear difference equation. The characteristic roots of the equation are $(\mu \pm \sqrt{\mu^2 - 4})/2$. Let ρ denote the largest root and note that the other root is equal to $1/\rho$. The general solution of the difference equation is therefore

$$\eta_k = \alpha \rho^k + \beta \rho^{-k} = \alpha \left[\frac{\mu + \sqrt{\mu^2 - 4}}{2}\right]^k + \beta \left[\frac{\mu - \sqrt{\mu^2 - 4}}{2}\right]^k.$$

The condition at $k = 0$ yields $\alpha + \beta = 1$. Then, writing the condition $\eta_1 = \mu$ yields

$$\alpha = \frac{\sqrt{\mu^2 - 4} + \mu}{2\sqrt{\mu^2 - 4}}, \qquad \beta = \frac{\sqrt{\mu^2 - 4} - \mu}{2\sqrt{\mu^2 - 4}}.$$

Therefore,

$$\eta_k = \frac{1}{\sqrt{\mu^2 - 4}} \left[\rho^{k+1} - \rho^{-k-1}\right].$$

The sequence λ_k is η_k/η_{k-1}, which yields

$$\lambda_k = \rho \frac{1 - \rho^{-2(k+1)}}{1 - \rho^{-2k}}.$$

This gives the desired expression for f_n and T_n. Specifically, if we define

$$\hat{S} = \frac{B + \sqrt{B^2 - 4I}}{2},$$

then

$$S_1 = T_n = \hat{S} X, \quad \text{where} \quad X = \left(I - \hat{B}^{-2(k+1)}\right)\left(I - \hat{B}^{-2k}\right)^{-1}.$$

Despite the apparent nonsymmetry of the above expression, it is worth noting that the operator thus defined is SPD. In addition, the factor X is usually very close to the identity matrix because the powers B^{-2j} decay exponentially to zero (the eigenvalues of B are all larger than 2).

The result can be stated in terms of the one-dimensional finite difference operator T instead of B because $B = T + 2I$. Either way, the final expression is a rather complex one, since it involves a square root, even when $\hat{S}$, the approximation to S_1, is used. It is possible, however, to use FFTs or sine transforms to perform a solve with the matrix $\hat{S}$. This is because, if the spectral decomposition of B is written as $B = Q \Lambda Q^T$, then $S_1 = Q f_n(\Lambda) Q^T$, and the products with Q and Q^T can be performed with FFT; see Section 2.2.6.

14.3 Schwarz Alternating Procedures

The original alternating procedure described by Schwarz in 1870 consisted of three parts: alternating between two overlapping domains, solving the Dirichlet problem on one domain at each iteration, and taking boundary conditions based on the most recent solution obtained from the other domain. This procedure is called the multiplicative Schwarz procedure. In matrix terms, this is very reminiscent of the block Gauss–Seidel iteration with overlap defined with the help of projectors, as seen in Chapter 5. The analogue of the block Jacobi procedure is known as the additive Schwarz procedure.

14.3.1 Multiplicative Schwarz Procedure

In the following, assume that each subdomain Ω_i extends into its neighboring subdomains by one level, which will be used as a boundary for Ω_i. The boundary of subdomain Ω_i that is included in subdomain j is denoted by Γ_{ij}. This is illustrated in Figure 14.7 for the L-shaped domain example. A more specific illustration is in Figure 14.5, where, for example, $\Gamma_{12} = \{29, 30, 31, 32\}$ and $\Gamma_{31} = \{13, 14, 15, 16\}$. Call Γ_i the boundary of Ω_i consisting of its original boundary (which consists of the Γ_{i0} pieces in the figure) and the Γ_{ij}'s and denote by u_{ji} the restriction of the solution u to the boundary Γ_{ji}. Then SAP can be described as follows.

ALGORITHM 14.2. SAP

1. *Choose an initial guess u to the solution*
2. *Until convergence, Do*
3.   *For $i = 1, \ldots, s$, Do*
4.   *Solve $\Delta u = f$ in Ω_i with $u = u_{ij}$ in Γ_{ij}*
5.   *Update u values on Γ_{ji} $\forall j$*
6.   *EndDo*
7. *EndDo*

The algorithm sweeps through the s subdomains and solves the original equation in each of them by using boundary conditions that are updated from the most recent values of u. Since each of the subproblems is likely to be solved by some iterative method, we can take advantage of a good initial guess. It is natural to take as the initial guess for a given

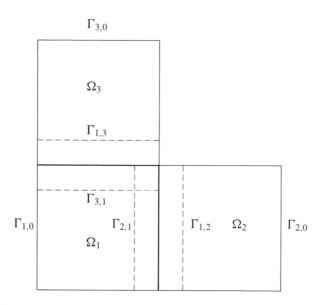

Figure 14.7. *An L-shaped domain subdivided into three overlapping subdomains.*

subproblem the most recent approximation. Going back to the expression (14.11) of the local problems, observe that each of the solutions in line 4 of the algorithm will be translated into an update of the form

$$u_i := u_i + \delta_i,$$

where the correction δ_i solves the system

$$A_i \delta_i = r_i.$$

Here r_i is the local part of the most recent global residual vector $b - Ax$ and the above system represents the system associated with the problem in line 4 of the algorithm when a nonzero initial guess is used in some iterative procedure. The matrix A_i has the block structure (14.9). Writing

$$u_i = \begin{pmatrix} x_i \\ y_i \end{pmatrix}, \quad \delta_i = \begin{pmatrix} \delta_{x,i} \\ \delta_{y,i} \end{pmatrix}, \quad r_i = \begin{pmatrix} r_{x,i} \\ r_{y,i} \end{pmatrix},$$

the correction to the current solution step in the algorithm leads to

$$\begin{pmatrix} x_i \\ y_i \end{pmatrix} := \begin{pmatrix} x_i \\ y_i \end{pmatrix} + \begin{pmatrix} B_i & E_i \\ F_i & C_i \end{pmatrix}^{-1} \begin{pmatrix} r_{x,i} \\ r_{y,i} \end{pmatrix}. \tag{14.21}$$

After this step is taken, normally a residual vector r has to be computed again to get the components associated with domain $i + 1$ and to proceed with a similar step for the next subdomain. However, only those residual components that have been affected by the change of the solution need to be updated. Specifically, employing the same notation used in (14.11),

14.3. Schwarz Alternating Procedures

we can simply update the residual $r_{y,j}$ for each subdomain j for which $i \in N_j$ as

$$r_{y,j} := r_{y,j} - E_{ji}\delta_{y,i}.$$

This amounts implicitly to performing line 5 of the above algorithm. Note that, since the matrix pattern is assumed to be symmetric, then the set of all indices j such that $i \in N_j$, i.e., $N_i^* = \{j \mid i \in N_i\}$, is identical to N_i. Now the loop starting in line 3 of Algorithm 14.2 and called *domain sweep* can be restated as follows.

ALGORITHM 14.3. Multiplicative Schwarz Sweep—Matrix Form

1. For $i = 1, \ldots, s$, Do
2.   Solve $A_i \delta_i = r_i$
3.   Compute $x_i := x_i + \delta_{x,i}$, $y_i := y_i + \delta_{y,i}$ and set $r_i := 0$
4.   For each $j \in N_i$, Compute $r_{y,j} := r_{y,j} - E_{ji}\delta_{y,i}$
5. EndDo

Considering only the y iterates, the above iteration would resemble a form of Gauss–Seidel procedure on the Schur complement matrix (14.14). In fact, it is mathematically equivalent, provided a consistent initial guess is taken. This is stated in the next result, established by Chan and Goovaerts [73].

Theorem 14.2. *Let the guess $\binom{x_i^{(0)}}{y_i^{(0)}}$ for the Schwarz procedure in each subdomain be chosen such that*

$$x_i^{(0)} = B_i^{-1}[f_i - E_i y_i^{(0)}]. \tag{14.22}$$

Then the y iterates produced by Algorithm 14.3 are identical to those of a Gauss–Seidel sweep applied to the Schur complement system (14.12).

Proof. We start by showing that, with the choice (14.22), the y components of the initial residuals produced by the algorithm are identical to those of the Schur complement system (14.12). Refer to Section 14.2.3 and the relation (14.10), which defines the E_{ij}'s from the block structure (14.8) of the global matrix. Observe that $A_{ij} u_j = \binom{0}{E_{ij} y_j}$ and note from (14.11) that for the global system the y components of the initial residual vectors are

$$r_{y,i}^{(0)} = g_i - F_i x_i^{(0)} - C_i y_i^{(0)} - \sum_{j \in N_i} E_{ij} y_j^{(0)}$$

$$= g_i - F_i B_i^{-1}[f_i - E_i y_i^{(0)}] - C_i y_i^{(0)} - \sum_{j \in N_i} E_{ij} y_j^{(0)}$$

$$= g_i - F_i B_i^{-1} f_i - S_i y_i^{(0)} - \sum_{j \in N_i} E_{ij} y_j^{(0)}.$$

This is precisely the expression of the residual vector associated with the Schur complement system (14.12) with the initial guess $y_i^{(0)}$.

Now observe that the initial guess has been selected so that $r_{x,i}^{(0)} = 0$ for all i. Because only the y components of the residual vector are modified, according to line 4 of Algorithm

14.3, this property remains valid throughout the iterative process. By the updating equation (14.21) and the relation (14.7), we have

$$y_i := y_i + S_i^{-1} r_{y,i},$$

which is precisely a Gauss–Seidel step associated with the system (14.14). Note that the update of the residual vector in the algorithm results in the same update for the y components as in the Gauss–Seidel iteration for (14.14). □

It is interesting to interpret Algorithm 14.2, or rather its discrete version, in terms of projectors. For this we follow the model of the overlapping block Jacobi technique seen in Chapter 12. Let $\mathcal{S}_i$ be an index set

$$\mathcal{S}_i = \{j_1, j_2, \ldots, j_{n_i}\},$$

where the indices j_k are those associated with the n_i mesh points of the interior of the discrete subdomain Ω_i. Note that, as before, the $\mathcal{S}_i$'s form a collection of index sets such that

$$\bigcup_{i=1,\ldots,s} \mathcal{S}_i = \{1, \ldots, n\}$$

and are not necessarily disjoint. Let R_i be a *restriction operator* from Ω to Ω_i. By definition, $R_i x$ belongs to Ω_i and keeps only those components of an arbitrary vector x that are in Ω_i. It is represented by an $n_i \times n$ matrix of zeros and ones. The matrices R_i associated with the partitioning of Figure 14.4 are represented in the three diagrams of Figure 14.8, where each square represents a nonzero element (equal to one) and every other element is a zero. These matrices depend on the ordering chosen for the local problem. Here, boundary nodes are labeled last, for simplicity. Observe that each row of each R_i has exactly one nonzero element (equal to one). Boundary points such as the nodes 36 and 37 are represented several times in the matrices R_1, R_2, and R_3 because of the overlapping of the boundary points. Thus, node 36 is represented in matrices R_1 and R_2, while 37 is represented in all three matrices. From the linear algebra point of view, the restriction operator R_i is an $n_i \times n$ matrix formed by the transposes of columns e_j of the $n \times n$ identity matrix, where j belongs to the index set $\mathcal{S}_i$. The transpose R_i^T of this matrix is a *prolongation operator*, which takes a variable from Ω_i and *extends* it to the equivalent variable in Ω. The matrix

$$A_i = R_i A R_i^T$$

of dimension $n_i \times n_i$ defines a restriction of A to Ω_i. Now a problem associated with A_i can be solved that would update the unknowns in the domain Ω_i. With this notation, the multiplicative Schwarz procedure can be described as follows:

1. *For* $i = 1, \ldots, s$, *Do*
2. $x := x + R_i^T A_i^{-1} R_i (b - Ax)$
3. *EndDo*

We change notation and rewrite line 2 as

$$x_{new} = x + R_i^T A_i^{-1} R_i (b - Ax). \qquad (14.23)$$

If the errors $d = x_* - x$ are considered, where x_* is the exact solution, then notice that $b - Ax = A(x_* - x)$ and, at each iteration, the following equation relates the new error

14.3. Schwarz Alternating Procedures

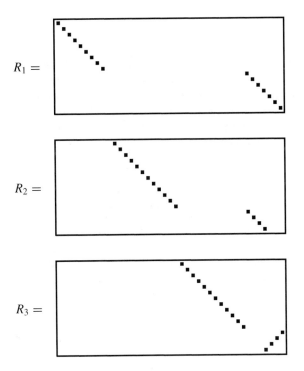

Figure 14.8. *Patterns of the three matrices R_i associated with the partitioning of Figure 14.4.*

d_{new} and the previous error d:

$$d_{new} = d - R_i^T A_i^{-1} R_i A d.$$

Starting from a given x_0 whose error vector is $d_0 = x_* - x$, each subiteration produces an error vector that satisfies the relation

$$d_i = d_{i-1} - R_i^T A_i^{-1} R_i A d_{i-1}$$

for $i = 1, \ldots, s$. As a result,

$$d_i = (I - P_i)d_{i-1},$$

in which

$$P_i = R_i^T A_i^{-1} R_i A. \tag{14.24}$$

Observe that the operator $P_i \equiv R_i^T A_i^{-1} R_i A$ is a projector since

$$(R_i^T A_i^{-1} R_i A)^2 = R_i^T A_i^{-1} (R_i A R_i^T) A_i^{-1} R_i A = R_i^T A_i^{-1} R_i A.$$

Thus, one sweep produces an error that satisfies the relation

$$d_s = (I - P_s)(I - P_{s-1}) \cdots (I - P_1)d_0. \tag{14.25}$$

In the following, we use the notation

$$Q_s \equiv (I - P_s)(I - P_{s-1}) \cdots (I - P_1). \tag{14.26}$$

14.3.2 Multiplicative Schwarz Preconditioning

Because of the equivalence of the multiplicative Schwarz procedure and a block Gauss–Seidel iteration, it is possible to recast one multiplicative Schwarz sweep in the form of a global fixed-point iteration of the form $x_{new} = Gx + f$. Recall that this is a fixed-point iteration for solving the *preconditioned* system $M^{-1}Ax = M^{-1}b$, where the preconditioning matrix M and the matrix G are related by $G = I - M^{-1}A$. To interpret the operation associated with M^{-1}, it is helpful to identify the result of the error vector produced by this iteration with that of (14.25), which is $x_{new} - x_* = Q_s(x - x_*)$. This comparison yields

$$x_{new} = Q_s x + (I - Q_s)x_*$$

and, therefore,

$$G = Q_s, \quad f = (I - Q_s)x_*.$$

Hence, the preconditioned matrix is $M^{-1}A = I - Q_s$. This result is restated as follows.

Proposition 14.3. *The multiplicative Schwarz procedure is equivalent to a fixed-point iteration for the preconditioned problem*

$$M^{-1}Ax = M^{-1}b,$$

in which

$$M^{-1}A = I - Q_s, \tag{14.27}$$
$$M^{-1}b = (I - Q_s)x_* = (I - Q_s)A^{-1}b. \tag{14.28}$$

The transformed right-hand side in the proposition is not known explicitly since it is expressed in terms of the exact solution. However, a procedure can be found to compute it. In other words, it is possible to operate with M^{-1} without invoking A^{-1}. Note that $M^{-1} = (I - Q_s)A^{-1}$. As the next lemma indicates, M^{-1}, as well as $M^{-1}A$, can be computed recursively.

Lemma 14.4. *Define the matrices*

$$Z_i = I - Q_i, \tag{14.29}$$
$$M_i = Z_i A^{-1}, \tag{14.30}$$
$$T_i = P_i A^{-1} = R_i^T A_i^{-1} R_i \tag{14.31}$$

for $i = 1, \ldots, s$. Then $M^{-1} = M_s$, $M^{-1}A = Z_s$, and the matrices Z_i and M_i satisfy the recurrence relations

$$Z_1 = P_1,$$
$$Z_i = Z_{i-1} + P_i(I - Z_{i-1}), \quad i = 2, \ldots, s, \tag{14.32}$$

and

$$M_1 = T_1,$$
$$M_i = M_{i-1} + T_i(I - AM_{i-1}), \quad i = 2, \ldots, s. \tag{14.33}$$

14.3. Schwarz Alternating Procedures

Proof. It is clear by the definitions (14.29) and (14.30) that $M_s = M^{-1}$ and that $M_1 = T_1$, $Z_1 = P_1$. For the cases $i > 1$, by definition of Q_i and Q_{i-1},

$$Z_i = I - (I - P_i)(I - Z_{i-1}) = P_i + Z_{i-1} - P_i Z_{i-1}, \quad (14.34)$$

which gives the relation (14.32). Multiplying (14.34) to the right by A^{-1} yields

$$M_i = T_i + M_{i-1} - P_i M_{i-1}.$$

Rewriting the term P_i as $T_i A$ above yields the desired formula (14.33). □

Note that (14.32) yields immediately the important relation

$$Z_i = \sum_{j=1}^{i} P_j Q_{j-1}. \quad (14.35)$$

If the relation (14.33) is multiplied to the right by a vector v and if the vector $M_i v$ is denoted by z_i, then the following recurrence results:

$$z_i = z_{i-1} + T_i(v - A z_{i-1}).$$

Since $z_s = (I - Q_s)A^{-1}v = M^{-1}v$, the end result is that $M^{-1}v$ can be computed for an arbitrary vector v by the following procedure.

ALGORITHM 14.4. Multiplicative Schwarz Preconditioner

1. Input: v; Output: $z = M^{-1}v$
2. $z := T_1 v$
3. For $i = 2, \ldots, s$, Do
4. $z := z + T_i(v - Az)$
5. EndDo

By a similar argument, a procedure can be found to compute vectors of the form $z = M^{-1}Av$. In this case, the following algorithm results.

ALGORITHM 14.5. Multiplicative Schwarz Preconditioned Operator

1. Input: v, Output: $z = M^{-1}Av$
2. $z := P_1 v$
3. For $i = 2, \ldots, s$, Do
4. $z := z + P_i(v - z)$
5. EndDo

In summary, the multiplicative Schwarz procedure is equivalent to solving the preconditioned system

$$(I - Q_s)x = g, \quad (14.36)$$

where the operation $z = (I - Q_s)v$ can be computed from Algorithm 14.5 and $g = M^{-1}b$ can be computed from Algorithm 14.4. Now the above procedures can be used within an accelerator such as generalized minimal residual (GMRES). First, to obtain the right-hand side g of the preconditioned system (14.36), Algorithm 14.4 must be applied to the original right-hand side b. Then GMRES can be applied to (14.36), in which the preconditioned operations $I - Q_s$ are performed by Algorithm 14.5.

Another important aspect of the multiplicative Schwarz procedure is that multicoloring can be exploited in the same way as it is done traditionally for block successive overrelaxation (SOR). Finally, note that symmetry is lost in the preconditioned system but it can be recovered by following the sweep $1, 2, \ldots, s$ by a sweep in the other direction, namely, $s-1, s-2, \ldots, 1$. This yields a form of the block symmetric SOR (SSOR) algorithm.

14.3.3 Additive Schwarz Procedure

The additive Schwarz procedure is similar to a block Jacobi iteration and consists of updating all the new (block) components from the same residual. Thus, it differs from the multiplicative procedure only because the components in each subdomain are not updated until a whole cycle of updates through all domains are completed. The basic additive Schwarz iteration is therefore as follows:

1. For $i = 1, \ldots, s$, Do
2.   Compute $\delta_i = R_i^T A_i^{-1} R_i (b - Ax)$
3. EndDo
4. $x_{new} = x + \sum_{i=1}^{s} \delta_i$

The new approximation (obtained after a cycle of the s substeps in the above algorithm is applied) is

$$x_{new} = x + \sum_{i=1}^{s} R_i^T A_i^{-1} R_i (b - Ax).$$

Each instance of the loop redefines different components of the new approximation and there is no data dependency between the subproblems involved in the loop.

The preconditioning matrix is rather simple to obtain for the additive Schwarz procedure. Using the matrix notation defined in the previous section, notice that the new iterate satisfies the relation

$$x_{new} = x + \sum_{i=1}^{s} T_i (b - Ax) = \left(I - \sum_{i=1}^{s} P_i \right) x + \sum_{i=1}^{s} T_i b.$$

Thus, using the same analogy as in the previous section, this iteration corresponds to a fixed-point iteration $x_{new} = Gx + f$, with

$$G = I - \sum_{i=1}^{s} P_i, \quad f = \sum_{i=1}^{s} T_i b.$$

With the relation $G = I - M^{-1}A$ between G and the preconditioning matrix M, the result is that

$$M^{-1}A = \sum_{i=1}^{s} P_i$$

14.3. Schwarz Alternating Procedures

and
$$M^{-1} = \sum_{i=1}^{s} P_i A^{-1} = \sum_{i=1}^{s} T_i.$$

Now the procedure for applying the preconditioned operator M^{-1} becomes clear.

ALGORITHM 14.6. Additive Schwarz Preconditioner

1. Input: v; Output: $z = M^{-1} v$
2. For $i = 1, \ldots, s$, Do
3.   Compute $z_i := T_i v$
4. EndDo
5. Compute $z := z_1 + z_2 + \cdots + z_s$

Note that the do loop can be performed in parallel. Line 5 sums up the vectors z_i in each domain to obtain a global vector z. In the nonoverlapping case, this step is parallel and consists of just forming these different components, since the addition is trivial. In the presence of overlap, the situation is similar except that the overlapping components are added up from the different results obtained in each subdomain.

The procedure for computing $M^{-1} A v$ is identical to the one above except that T_i in line 3 is replaced by P_i.

14.3.4 Convergence

Throughout this section, it is assumed that A is SPD. The projectors P_i defined by (14.24) play an important role in the convergence theory of both additive and multiplicative Schwarz. A crucial observation here is that these projectors are orthogonal with respect to the A inner product. Indeed, it is sufficient to show that P_i is self-adjoint with respect to the A inner product:

$$(P_i x, y)_A = (A R_i^T A_i^{-1} R_i A x, y) = (Ax, R_i^T A_i^{-1} R_i A y) = (x, P_i y)_A.$$

Consider the operator
$$A_J = \sum_{i=1}^{s} P_i. \tag{14.37}$$

Since each P_j is self-adjoint with respect to the A inner product, i.e., A self-adjoint, their sum A_J is also A self-adjoint. Therefore, it has real eigenvalues. An immediate consequence of the fact that the P_i's are projectors is stated in the following theorem.

Theorem 14.5. *The largest eigenvalue of A_J is such that*
$$\lambda_{max}(A_J) \leq s,$$

where s is the number of subdomains.

Proof. For any matrix norm, $\lambda_{max}(A_J) \leq \|A_J\|$. In particular, if the A-norm is used, we have
$$\lambda_{max}(A_J) \leq \sum_{i=1}^{s} \|P_i\|_A.$$

Each of the A-norms of P_i is equal to one since P_i is an A-orthogonal projector. This proves the desired result. $\square$

This result can be improved substantially by observing that the projectors can be grouped in sets that have disjoint ranges. Graph-coloring techniques seen in Chapter 3 can be used to obtain such colorings of the subdomains. Assume that c sets of indices $\Theta_i, i = 1, \ldots, c$, are such that all the subdomains Ω_j for $j \in \Theta_i$ have no intersection with one another. Then

$$P_{\Theta_i} = \sum_{j \in \Theta_i} P_j \qquad (14.38)$$

is again an orthogonal projector.

This shows that the result of the previous theorem can be improved trivially into the following.

Theorem 14.6. *Suppose that the subdomains can be colored in such a way that two subdomains with the same color have no common nodes. Then the largest eigenvalue of A_J is such that*

$$\lambda_{max}(A_J) \leq c,$$

where c is the number of colors.

In order to estimate the lowest eigenvalue of the preconditioned matrix, an assumption must be made regarding the decomposition of an arbitrary vector x into components of Ω_i.

Assumption 1. *There exists a constant K_0 such that the inequality*

$$\sum_{i=1}^{s}(Au_i, u_i) \leq K_0(Au, u)$$

is satisfied by the representation of $u \in \Omega$ as the sum

$$u = \sum_{i=1}^{s} u_i, \quad u_i \in \Omega_i.$$

The following theorem has been proved by several authors in slightly different forms and contexts.

Theorem 14.7. *If Assumption 1 holds, then*

$$\lambda_{min}(A_J) \geq \frac{1}{K_0}.$$

Proof. Unless otherwise stated, all summations in this proof are from 1 to s. Start with an arbitrary u decomposed as $u = \sum u_i$ and write

$$(u, u)_A = \sum (u_i, u)_A = \sum (P_i u_i, u)_A = \sum (u_i, P_i u)_A.$$

The last equality is due to the fact that P_i is an A-orthogonal projector onto Ω_i and it is therefore self-adjoint. Now, using the Cauchy–Schwarz inequality, we get

$$(u, u)_A = \sum (u_i, P_i u)_A \leq \left(\sum (u_i, u_i)_A\right)^{1/2} \left(\sum (P_i u, P_i u)_A\right)^{1/2}.$$

14.3. Schwarz Alternating Procedures

By Assumption 1, this leads to

$$\|u\|_A^2 \leq K_0^{1/2} \|u\|_A \left(\sum (P_i u, P_i u)_A\right)^{1/2},$$

which, after squaring, yields

$$\|u\|_A^2 \leq K_0 \sum (P_i u, P_i u)_A.$$

Finally, observe that, since each P_i is an A-orthogonal projector, we have

$$\sum (P_i u, P_i u)_A = \sum (P_i u, u)_A = \left(\sum P_i u, u\right)_A.$$

Therefore, for any u, the inequality

$$(A_J u, u)_A \geq \frac{1}{K_0} (u, u)_A$$

holds, which yields the desired upper bound by the min-max theorem. □

Note that the proof uses the following form of the Cauchy–Schwarz inequality:

$$\sum_{i=1}^{p} (x_i, y_i) \leq \left(\sum_{i=1}^{p} (x_i, x_i)\right)^{1/2} \left(\sum_{i=1}^{p} (y_i, y_i)\right)^{1/2}.$$

See Exercise 1 for a proof of this variation.

We now turn to the analysis of the multiplicative Schwarz procedure. We start by recalling that the error after each outer iteration (sweep) is given by

$$d = Q_s d_0.$$

We wish to find an upper bound for $\|Q_s\|_A$. First note that (14.32) in Lemma 14.4 results in

$$Q_i = Q_{i-1} - P_i Q_{i-1},$$

from which we get, using the A-orthogonality of P_i,

$$\|Q_i v\|_A^2 = \|Q_{i-1} v\|_A^2 - \|P_i Q_{i-1} v\|_A^2.$$

The above equality is valid for $i = 1$, provided $Q_0 \equiv I$. Summing these equalities from $i = 1$ to s gives the result

$$\|Q_s v\|_A^2 = \|v\|_A^2 - \sum_{i=1}^{s} \|P_i Q_{i-1} v\|_A^2. \tag{14.39}$$

This indicates that the A-norm of the error will not increase at each substep of the sweep.

Now a second assumption must be made to prove the next lemma.

Assumption 2. *For any subset S of $\{1, 2, \ldots, s\}^2$ and $u_i, v_j \in \Omega$, the following inequality holds:*

$$\sum_{(i,j) \in S} (P_i u_i, P_j v_j)_A \leq K_1 \left(\sum_{i=1}^{s} \|P_i u_i\|_A^2\right)^{1/2} \left(\sum_{j=1}^{s} \|P_j v_j\|_A^2\right)^{1/2}. \tag{14.40}$$

Lemma 14.8. *If Assumptions 1 and 2 are satisfied, then the following is true:*

$$\sum_{i=1}^{s} \|P_i v\|_A^2 \leq (1+K_1)^2 \sum_{i=1}^{s} \|P_i Q_{i-1} v\|_A^2. \tag{14.41}$$

Proof. Begin with the relation that follows from the fact that P_i is an A-orthogonal projector:

$$(P_i v, P_i v)_A = (P_i v, P_i Q_{i-1} v)_A + (P_i v, (I - Q_{i-1}) v)_A,$$

which yields, with the help of (14.35),

$$\sum_{i=1}^{s} \|P_i v\|_A^2 = \sum_{i=1}^{s} (P_i v, P_i Q_{i-1} v)_A + \sum_{i=1}^{s} \sum_{j=1}^{i-1} (P_i v, P_j Q_{j-1} v)_A. \tag{14.42}$$

For the first term on the right-hand side, use the Cauchy–Schwarz inequality to obtain

$$\sum_{i=1}^{s} (P_i v, P_i Q_{i-1} v)_A \leq \left(\sum_{i=1}^{s} \|P_i v\|_A^2 \right)^{1/2} \left(\sum_{i=1}^{s} \|P_i Q_{i-1} v\|_A^2 \right)^{1/2}.$$

For the second term on the right-hand side of (14.42), use the assumption (14.40) to get

$$\sum_{i=1}^{s} \sum_{j=1}^{i-1} (P_i v, P_j Q_{j-1} v)_A \leq K_1 \left(\sum_{i=1}^{s} \|P_i v\|_A^2 \right)^{1/2} \left(\sum_{j=1}^{s} \|P_j Q_{j-1} v\|_A^2 \right)^{1/2}.$$

Adding these two inequalities, squaring the result, and using (14.42) leads to the inequality (14.41). □

From (14.39), it can be deduced that, if Assumption 2 holds, then

$$\|Q_s v\|_A^2 \leq \|v\|_A^2 - \frac{1}{(1+K_1)^2} \sum_{i=1}^{s} \|P_i v\|_A^2. \tag{14.43}$$

Assumption 1 can now be exploited to derive a lower bound on $\sum_{i=1}^{s} \|P_i v\|_A^2$. This will yield the following theorem.

Theorem 14.9. *Assume that Assumptions 1 and 2 hold. Then*

$$\|Q_s\|_A \leq \left[1 - \frac{1}{K_0 (1+K_1)^2} \right]^{1/2}. \tag{14.44}$$

Proof. Using the notation of Section 14.3.3, the relation $\|P_i v\|_A^2 = (P_i v, v)_A$ yields

$$\sum_{i=1}^{s} \|P_i v\|_A^2 = \left(\sum_{i=1}^{s} P_i v, v \right)_A = (A_J v, v)_A.$$

14.4. Schur Complement Approaches

According to Theorem 14.7, $\lambda_{min}(A_J) \geq \frac{1}{K_0}$, which implies $(A_J v, v)_A \geq (v,v)_A/K_0$. Thus,

$$\sum_{i=1}^{s} \|P_i v\|_A^2 \geq \frac{(v,v)_A}{K_0},$$

which upon substitution into (14.43) gives the inequality

$$\frac{\|Q_s v\|_A^2}{\|v\|_A^2} \leq 1 - \frac{1}{K_0(1+K_1)^2}.$$

The result follows by taking the maximum over all vectors v. □

This result provides information on the speed of convergence of the multiplicative Schwarz procedure by making two key assumptions. These assumptions are not verifiable from linear algebra arguments alone. In other words, given a linear system, it is unlikely that one can establish that these assumptions are satisfied. However, they are satisfied for equations originating from finite element discretization of elliptic PDEs. For details, refer to Dryja and Widlund [102, 103, 104] and Xu [319].

14.4 Schur Complement Approaches

Schur complement methods are based on solving the reduced system (14.4) by some preconditioned Krylov subspace method. Procedures of this type involve three steps.

1. Get the right-hand side $g' = g - FB^{-1}f$
2. Solve the reduced system $Sy = g'$ via an iterative method
3. Back substitute; i.e., compute x via (14.3)

The different methods relate to the way in which line 2 is performed. First, observe that the matrix S *need not be formed explicitly* in order to solve the reduced system by an iterative method. For example, if a Krylov subspace method without preconditioning is used, then the only operations that are required with the matrix S are matrix-by-vector operations $w = Sv$. Such operations can be performed as follows.

1. Compute $v' = Ev$
2. Solve $Bz = v'$
3. Compute $w = Cv - Fz$

The above procedure involves only matrix-by-vector multiplications and one linear system solution with B. Recall that a linear system involving B translates into s independent linear systems. Also note that the linear systems with B must be solved exactly, either by a direct solution technique or by an iterative technique with a high level of accuracy.

While matrix-by-vector multiplications with S cause little difficulty, it is much harder to precondition the matrix S, since this full matrix is often not available explicitly. There have been a number of methods, derived mostly using arguments from PDEs, to precondition the Schur complement. Here we consider only those preconditioners that are derived from a linear algebra viewpoint.

14.4.1 Induced Preconditioners

One of the easiest ways to derive an approximation to S is to exploit Proposition 14.1 and the intimate relation between the Schur complement and Gaussian elimination. This proposition tells us that a preconditioning operator M to S can be defined from the (approximate) solution obtained with A. To precondition a given vector v, i.e., to compute $w = M^{-1}v$, where M is the desired preconditioner to S, first solve the system

$$A \begin{pmatrix} x \\ y \end{pmatrix} = \begin{pmatrix} 0 \\ v \end{pmatrix}, \qquad (14.45)$$

then take $w = y$. Use any approximate solution technique to solve the above system. Let M_A be any preconditioner for A. Using the notation defined earlier, let R_y represent the restriction operator on the interface variables, as defined in Proposition 14.1. Then the preconditioning operation for S that is induced from M_A is defined by

$$M_S^{-1} v = R_y M_A^{-1} \begin{pmatrix} 0 \\ v \end{pmatrix} = R_y M_A^{-1} R_y^T v.$$

Observe that, when M_A is an exact preconditioner, i.e., when $M_A = A$, then, according to Proposition 14.1, M_S is also an exact preconditioner; i.e., $M_S = S$. This induced preconditioner can be expressed as

$$M_S = \left(R_y M_A^{-1} R_y^T \right)^{-1}. \qquad (14.46)$$

It may be argued that this uses a preconditioner related to the original problem to be solved in the first place. However, even though the preconditioning on S may be defined from a preconditioning of A, the linear system is being solved for the interface variables. That is typically much smaller than the original linear system. For example, GMRES can be used with a much larger dimension of the Krylov subspace, since the Arnoldi vectors to keep in memory are much smaller. Also note that, from a PDEs viewpoint, systems of the form (14.45) correspond to the Laplace equation, the solutions of which are *harmonic* functions. There are fast techniques that provide the solution of such equations inexpensively.

In the case where M_A is an incomplete LU (ILU) factorization of A, M_S can be expressed in an explicit form in terms of the entries of the factors of M_A. This defines a preconditioner to S that is induced canonically from an ILU factorization of A. Assume that the preconditioner M_A is in a factored form $M_A = L_A U_A$, where

$$L_A = \begin{pmatrix} L_B & 0 \\ F U_B^{-1} & L_S \end{pmatrix}, \quad U_A = \begin{pmatrix} U_B & L_B^{-1} E \\ 0 & U_S \end{pmatrix}.$$

Then the inverse of M_A will have the following structure:

$$M_A^{-1} = U_A^{-1} L_A^{-1}$$

$$= \begin{pmatrix} \star & \star \\ 0 & U_S^{-1} \end{pmatrix} \begin{pmatrix} \star & 0 \\ \star & L_S^{-1} \end{pmatrix}$$

$$= \begin{pmatrix} \star & \star \\ \star & U_S^{-1} L_S^{-1} \end{pmatrix},$$

14.4. Schur Complement Approaches

where a star denotes a matrix whose actual expression is unimportant. Recall that, by definition,
$$R_y = (0 \quad I),$$
where this partitioning conforms to the above ones. This means that
$$R_y M_A^{-1} R_y^T = U_S^{-1} L_S^{-1}$$
and, therefore, according to (14.46), $M_S = L_S U_S$. This result is stated in the following proposition.

Proposition 14.10. *Let $M_A = L_A U_A$ be an ILU preconditioner for A. Then the preconditioner M_S for S induced by M_A, as defined by (14.46), is given by*
$$M_S = L_S U_S, \quad \text{with} \quad L_S = R_y L_A R_y^T, \quad U_S = R_y U_A R_y^T.$$

In other words, the proposition states that the L and U factors for M_S are the $(2,2)$ blocks of the L and U factors of the ILU factorization of A. An important consequence of the above idea is that the parallel Gaussian elimination can be exploited for deriving an ILU preconditioner for S by using a general-purpose ILU factorization. In fact, the L and U factors of M_A have the following structure:

$$A = L_A U_A - R, \quad \text{with}$$

$$L_A = \begin{pmatrix} L_1 & & & & \\ & L_2 & & & \\ & & \ddots & & \\ & & & L_s & \\ F_1 U_1^{-1} & F_2 U_2^{-1} & \cdots & F_s U_s^{-1} & L \end{pmatrix},$$

$$U_A = \begin{pmatrix} U_1 & & & & L_1^{-1} E_1 \\ & U_2 & & & L_2^{-1} E_2 \\ & & \ddots & & \vdots \\ & & & U_s & L_s^{-1} E_s \\ & & & & U \end{pmatrix}.$$

Each L_i, U_i pair is an ILU factorization of the local B_i-matrix. These ILU factorizations can be computed independently. Similarly, the matrices $L_i^{-1} E_i$ and $F_i U_i^{-1}$ can also be computed independently once the LU factors are obtained. Then each of the matrices
$$\tilde{S}_i = C_i - F_i U_i^{-1} L_i^{-1} E_i,$$
which are the approximate local Schur complements, is obtained. Note that, since an ILU factorization is being performed, some drop strategy is applied to the elements in $\tilde{S}_i$. Let T_i be the matrix obtained after this is done:
$$T_i = \tilde{S}_i - R_i.$$

Then a final stage would be to compute the ILU factorization of the matrix (14.14), where each S_i is replaced by T_i.

14.4.2 Probing

To derive preconditioners for the Schur complement, another general-purpose technique exploits ideas used in approximating sparse Jacobians when solving nonlinear equations. In general, S is a dense matrix. However, it can be observed, and there are physical justifications for model problems, that its entries decay away from the main diagonal. Assume that S is nearly tridiagonal; i.e., neglect all diagonals apart from the main diagonal and the two codiagonals, and write the corresponding tridiagonal approximation to S as

$$T = \begin{pmatrix} a_1 & b_2 & & & & \\ c_2 & a_2 & b_3 & & & \\ & \ddots & \ddots & \ddots & & \\ & & c_{m-1} & a_{m-1} & b_m \\ & & & c_m & a_m \end{pmatrix}.$$

Then it is easy to recover T by applying it to three well-chosen vectors. Consider the three vectors

$$w_1 = (1, 0, 0, 1, 0, 0, 1, 0, 0, \ldots)^T,$$
$$w_2 = (0, 1, 0, 0, 1, 0, 0, 1, 0, \ldots)^T,$$
$$w_3 = (0, 0, 1, 0, 0, 1, 0, 0, 1, \ldots)^T.$$

Then we have

$$Tw_1 = (a_1, c_2, b_4, a_4, c_5, \ldots, b_{3i+1}, a_{3i+1}, c_{3i+2}, \ldots)^T,$$
$$Tw_2 = (b_2, a_2, c_3, b_5, a_5, c_6, \ldots, b_{3i+2}, a_{3i+2}, c_{3i+3}, \ldots)^T,$$
$$Tw_3 = (b_3, a_3, c_4, b_6, a_6, c_7, \ldots, b_{3i}, a_{3i}, c_{3i+1}, \ldots)^T.$$

This shows that all the coefficients of the matrix T are indeed represented in the above three vectors. The first vector contains the nonzero elements of the columns $1, 4, 7, \ldots, 3i + 1, \ldots$, in succession, written as a long vector. Similarly, Tw_2 contains the columns $2, 5, 8, \ldots$, and Tw_3 contains the columns $3, 6, 9, \ldots$. We can easily compute $Sw_i, i = 1, 3$, and obtain a resulting approximation T that can be used as a preconditioner to S. The idea can be extended to compute any banded approximation to S. For details and analysis see [74].

14.4.3 Preconditioning Vertex-Based Schur Complements

We now discuss some issues related to the preconditioning of a linear system with the matrix coefficient of (14.14) associated with a vertex-based partitioning. As was mentioned before, this structure is helpful in the direct solution context because it allows the Schur complement to be formed by local pieces. Since ILU factorizations will utilize the same structure, this can be exploited as well.

Note that multicolor SOR or SSOR can also be exploited and that graph coloring can be used to color the interface values y_i in such a way that no two adjacent interface variables will have the same color. In fact, this can be achieved by coloring the domains. In the course

14.5. Full Matrix Methods

of a multicolor block SOR iteration, a linear system must be solved with the diagonal blocks S_i. For this purpose, it is helpful to interpret the Schur complement. Call P the canonical injection matrix from the local interface points to the local nodes. If n_i points are local and if m_i is the number of local interface points, then P is an $n_i \times m_i$ matrix whose columns are the last m_i columns of the $n_i \times n_i$ identity matrix. Then it is easy to see that

$$S_i = (P^T A_{loc,i}^{-1} P)^{-1}. \tag{14.47}$$

If $A_{loc,i} = LU$ is the LU factorization of $A_{loc,i}$, then it can be verified that

$$S_i^{-1} = P^T U^{-1} L^{-1} P = P^T U^{-1} P P^T L^{-1} P, \tag{14.48}$$

which indicates that, in order to operate with $P^T L^{-1} P$, the last $m_i \times m_i$ principal submatrix of L must be used. The same is true for $P^T U^{-1} P$, which requires only a back solve with the last $m_i \times m_i$ principal submatrix of U. Therefore, only the LU factorization of $A_{loc,i}$ is needed to solve a system with the matrix S_i. Interestingly, approximate solution methods associated with incomplete factorizations of $A_{loc,i}$ can be exploited.

14.5 Full Matrix Methods

We call any technique that iterates on the original system (14.2) a *full matrix method*. In the same way that preconditioners were derived from the LU factorization of A for the Schur complement, preconditioners for A can be derived from approximating interface values.

Before starting with preconditioning techniques, we establish a few simple relations between iterations involving A and S.

Proposition 14.11. *Let*

$$L_A = \begin{pmatrix} I & O \\ FB^{-1} & I \end{pmatrix}, \quad U_A = \begin{pmatrix} B & E \\ O & I \end{pmatrix} \tag{14.49}$$

and assume that a Krylov subspace method is applied to the original system (14.1) with left preconditioning L_A and right preconditioning U_A and with an initial guess of the form

$$\begin{pmatrix} x_0 \\ y_0 \end{pmatrix} = \begin{pmatrix} B^{-1}(f - E y_0) \\ y_0 \end{pmatrix}. \tag{14.50}$$

Then this preconditioned Krylov iteration will produce iterates of the form

$$\begin{pmatrix} x_m \\ y_m \end{pmatrix} = \begin{pmatrix} B^{-1}(f - E y_m) \\ y_m \end{pmatrix}, \tag{14.51}$$

in which the sequence y_m is the result of the same Krylov subspace method applied without preconditioning to the reduced linear system $Sy = g'$ with $g' = g - FB^{-1}f$ starting with the vector y_0.

Proof. The proof is a consequence of the factorization

$$\begin{pmatrix} B & E \\ F & C \end{pmatrix} = \begin{pmatrix} I & O \\ FB^{-1} & I \end{pmatrix} \begin{pmatrix} I & O \\ O & S \end{pmatrix} \begin{pmatrix} B & E \\ O & I \end{pmatrix}. \tag{14.52}$$

Applying an iterative method (e.g., GMRES) on the original system, preconditioned from the left by L_A and from the right by U_A, is equivalent to applying this iterative method to

$$L_A^{-1} A U_A^{-1} = \begin{pmatrix} I & O \\ O & S \end{pmatrix} \equiv A'. \tag{14.53}$$

The initial residual for the preconditioned system is

$$L_A^{-1} \begin{pmatrix} f \\ g \end{pmatrix} - (L_A^{-1} A U_A^{-1}) U_A \begin{pmatrix} x_0 \\ y_0 \end{pmatrix}$$

$$= \begin{pmatrix} I & O \\ -FB^{-1} & I \end{pmatrix} \left(\begin{pmatrix} f \\ g \end{pmatrix} - \begin{pmatrix} f \\ FB^{-1}(f - Ey_0) + Cy_0 \end{pmatrix} \right)$$

$$= \begin{pmatrix} 0 \\ g' - Sy_0 \end{pmatrix} \equiv \begin{pmatrix} 0 \\ r_0 \end{pmatrix}.$$

As a result, the Krylov vectors obtained from the preconditioned linear system associated with the matrix A' have the form

$$\begin{pmatrix} 0 \\ r_0 \end{pmatrix}, \begin{pmatrix} 0 \\ Sr_0 \end{pmatrix}, \ldots, \begin{pmatrix} 0 \\ S^{m-1} r_0 \end{pmatrix} \tag{14.54}$$

and the associated approximate solution will be of the form

$$\begin{pmatrix} x_m \\ y_m \end{pmatrix} = \begin{pmatrix} x_0 \\ y_0 \end{pmatrix} + \begin{pmatrix} B^{-1} & -B^{-1} E \\ O & I \end{pmatrix} \begin{pmatrix} 0 \\ \sum_{i=0}^{m-1} \alpha_i S^i r_0 \end{pmatrix}$$

$$= \begin{pmatrix} B^{-1}(f - Ey_0) - B^{-1} E(y_m - y_0) \\ y_m \end{pmatrix}$$

$$= \begin{pmatrix} B^{-1}(f - Ey_m) \\ y_m \end{pmatrix}.$$

Finally, the scalars α_i that express the approximate solution in the Krylov basis are obtained implicitly via inner products of vectors among the vector sequence (14.54). These inner products are identical to those of the sequence $r_0, Sr_0, \ldots, S^{m-1} r_0$. Therefore, these coefficients will achieve the same result as the same Krylov method applied to the reduced system $Sy = g'$ if the initial guess gives the residual guess r_0. □

A version of this proposition should allow S to be preconditioned. The following result is an immediate extension that achieves this goal.

Proposition 14.12. *Let* $S = L_S U_S - R$ *be an approximate factorization of S and define*

$$L_A = \begin{pmatrix} I & O \\ FB^{-1} & L_S \end{pmatrix}, \quad U_A = \begin{pmatrix} B & E \\ O & U_S \end{pmatrix}. \tag{14.55}$$

Assume that a Krylov subspace method is applied to the original system (14.1) with left preconditioning L_A and right preconditioning U_A and with an initial guess of the form

$$\begin{pmatrix} x_0 \\ y_0 \end{pmatrix} = \begin{pmatrix} B^{-1}(f - Ey_0) \\ y_0 \end{pmatrix}. \tag{14.56}$$

Then this preconditioned Krylov iteration will produce iterates of the form

$$\begin{pmatrix} x_m \\ y_m \end{pmatrix} = \begin{pmatrix} B^{-1}(f - Ey_m) \\ y_m \end{pmatrix}. \tag{14.57}$$

Moreover, the sequence y_m is the result of the same Krylov subspace method applied to the reduced linear system $Sy = g - FB^{-1}f$, left preconditioned with L_S, right preconditioned with U_S, and starting with the vector y_0.

Proof. The proof starts with the equality

$$\begin{pmatrix} B & E \\ F & C \end{pmatrix} = \begin{pmatrix} I & O \\ FB^{-1} & L_S \end{pmatrix} \begin{pmatrix} I & O \\ O & L_S^{-1}SU_S^{-1} \end{pmatrix} \begin{pmatrix} B & E \\ O & U_S \end{pmatrix}. \tag{14.58}$$

The rest of the proof is similar to that of the previous result and is omitted. □

Also, there are two other versions in which S is allowed to be preconditioned from the left or from the right. Thus, if M_S is a certain preconditioner for S, use the following factorizations:

$$\begin{pmatrix} B & E \\ F & C \end{pmatrix} = \begin{pmatrix} I & O \\ FB^{-1} & M_S \end{pmatrix} \begin{pmatrix} I & O \\ O & M_S^{-1}S \end{pmatrix} \begin{pmatrix} B & E \\ O & I \end{pmatrix} \tag{14.59}$$

$$= \begin{pmatrix} I & O \\ FB^{-1} & I \end{pmatrix} \begin{pmatrix} I & O \\ O & SM_S^{-1} \end{pmatrix} \begin{pmatrix} B & E \\ O & M_S \end{pmatrix}, \tag{14.60}$$

to derive the appropriate left or right preconditioners. Observe that, when the preconditioner M_S to S is exact, i.e., when $M = S$, then the block preconditioner L_A, U_A to A induced from M_S is also exact.

Although the previous results indicate that a preconditioned Schur complement iteration is mathematically equivalent to a certain preconditioned full matrix method, there are some practical benefits in iterating with the nonreduced system. The main benefit involves the requirement in the Schur complement techniques to compute Sx exactly at each Krylov subspace iteration. Indeed, the matrix S represents the coefficient matrix of the linear system and inaccuracies in the matrix-by-vector operation may result in loss of convergence. In the full matrix techniques, the operation Sx is never needed explicitly. In addition, this opens up the possibility of preconditioning the original matrix with approximate solves with the matrix B in the preconditioning operations L_A and U_A.

14.6 Graph Partitioning

The very first task that a programmer faces when solving a problem on a parallel computer, be it a dense or a sparse linear system, is to decide how to subdivide and map the data into the processors. Distributed memory computers allow mapping the data in an arbitrary fashion, but this added flexibility puts the burden on the user to find good mappings. When implementing domain decomposition–type ideas on a parallel computer, efficient techniques must be available for partitioning an arbitrary graph. This section gives an overview of the issues and covers a few techniques.

14.6.1 Basic Definitions

Consider a general sparse linear system whose adjacency graph is $G = (V, E)$. Graph-partitioning algorithms aim at subdividing the original linear system into smaller sets of equations, which will be assigned to different processors for their parallel solution. This translates into partitioning the graph into p subgraphs, with the underlying goal to achieve a good load balance of the work among the processors as well as to ensure that the ratio of communication to computation is small for the given task. We begin with a general definition.

Definition 14.13. *We call any set $V_1, V_2, \ldots, V_s$ of subsets of the vertex set V whose union is equal to V a map of V:*

$$V_i \subseteq V, \quad \bigcup_{i=1,s} V_i = V.$$

When all the V_i subsets are disjoint, the map is called a proper partition; otherwise we refer to it as an overlapping partition. Figure 14.9 shows an example of a 4×3 mesh being mapped into four processors.

The most general way to describe a node-to-processor mapping is by setting up a list for each processor, containing all the nodes that are mapped to that processor. Three distinct classes of algorithms have been developed for partitioning graphs. An overview of each of these three approaches is given next.

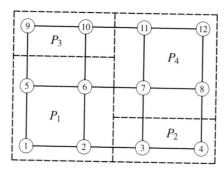

Figure 14.9. *Mapping of a simple 4×3 mesh to four processors.*

14.6.2 Geometric Approach

The geometric approach works on the physical mesh and requires the coordinates of the mesh points. In the simplest case, for a two-dimensional rectangular grid, stripes in the horizontal and vertical directions can be defined to get square subregions with roughly the same number of points. Other techniques utilize notions of moment of inertia to divide the region recursively into two roughly equal-sized subregions.

Next is a brief description of a technique based on work by Miller, Teng, Thurston, and Vavasis [210]. This technique finds good separators for a mesh using projections into a higher space. Given a mesh in $\mathbb{R}^d$, the method projects the mesh points into a unit sphere centered at the origin in $\mathbb{R}^{d+1}$. Stereographic projection is used: A line is drawn from a

14.6. Graph Partitioning

given point p in the plane to the North Pole $(0, \ldots, 0, 1)$ and the stereographic projection of p is the point where this line intersects the sphere. In the next step, a *centerpoint* of the projected points is found. A centerpoint c of a discrete set S is defined as a point where every hyperplane passing through c will divide S approximately evenly.

Once the centerpoint is found, the points of the sphere are rotated so that the centerpoint is aligned with the North Pole, i.e., so that coordinates of c are transformed into $(0, \ldots, 0, r)$. The points are further transformed by dilating them so that the centerpoint becomes the origin. Through all these transformations, the point c remains a centerpoint. Therefore, if any hyperplane is taken that passes through the centerpoint, which is now the origin, it should cut the sphere into two roughly equal-sized subsets. Any hyperplane passing through the origin will intersect the sphere along a large circle C. Transforming this circle back into the original space will give a desired separator. Notice that there is an infinity of circles to choose from.

One of the main ingredients in the above technique is a heuristic for finding centerpoints in $\mathbb{R}^d$ space (actually, $\mathbb{R}^{d+1}$ in the technique). The heuristic that is used repeatedly replaces randomly chosen sets of $d+2$ points with their centerpoints, which are easy to find in this case. There are a number of interesting results that analyze the quality of geometric graph partitionings based on separators. With some minimal assumptions on the meshes, it is possible to show that there exist *good* separators. In addition, the technique discussed above constructs such separators. We start with two definitions.

Definition 14.14. *A k-ply neighborhood system in $\mathbb{R}^d$ is a set of n closed disks D_i, $i = 1, \ldots, n$, in $\mathbb{R}^d$ such that no point in $\mathbb{R}^d$ is (strictly) interior to more than k disks.*

Definition 14.15. *Let $\alpha \geq 1$ and let $D_1, \ldots, D_n$ be a k-ply neighborhood system in $\mathbb{R}^d$. The (α, k)-overlap graph for the neighborhood system is the graph with vertex set $V = \{1, 2, \ldots, n\}$ and edge set the subset of $V \times V$ defined by*

$$\{(i, j) : (D_i \cap (\alpha \cdot D_j) \neq \emptyset) \text{ and } (D_j \cap (\alpha \cdot D_i) \neq \emptyset)\}.$$

A mesh in $\mathbb{R}^d$ is associated with an overlap graph by assigning the coordinate of the center c_i of disk i to each node i of the graph. Overlap graphs model computational meshes in d dimensions. Indeed, every mesh with bounded *aspect ratio* elements (ratio of largest to smallest edge length of each element) is contained in an overlap graph. In addition, any planar graph is an overlap graph. The main result regarding separators of overlap graphs is the following theorem [210].

Theorem 14.16. *Let G be an n-vertex (α, k)-overlap graph in d dimensions. Then the vertices of G can be partitioned into three sets A, B, and C such that*

1. *no edge joins A and B,*

2. *A and B each have at most $n(d+1)/(d+2)$ vertices, and*

3. *C has only $O(\alpha \, k^{1/d} n^{(d-1)/d})$ vertices.*

Thus, for $d = 2$, the theorem states that it is possible to partition the graph into two subgraphs A and B, with a separator C, such that the number of nodes for each of A and B

does not exceed $\frac{3}{4}n$ vertices in the worst case and such that the separator has a number of nodes of the order $O(\alpha\, k^{1/2} n^{1/2})$.

14.6.3 Spectral Techniques

Spectral bisection refers to a technique that exploits some known properties of the eigenvectors of the *Laplacian of a graph*. Given an adjacency graph $G = (V, E)$, we associate with it a Laplacian matrix L that is a sparse matrix having the same adjacency graph G and defined as follows:

$$l_{ij} = \begin{cases} -1 & \text{if } (v_i, v_j) \in E \text{ and } i \neq j, \\ \deg(i) & \text{if } i = j, \\ 0 & \text{otherwise.} \end{cases}$$

These matrices have some interesting fundamental properties. When the graph is undirected, L is symmetric. It can also be shown to be *negative semidefinite* (see Exercise 10). Zero is an eigenvalue and it is the smallest one. An eigenvector associated with this eigenvalue is any constant vector and it bears little interest. The second smallest eigenvector, called the *Fiedler vector*, has the useful property that the signs of its components divide the domain into roughly two equal subdomains.

The recursive spectral bisection (RSB) algorithm consists of sorting the components of the Fiedler vector and assigning the first half of the sorted vertices to the first subdomain and the second half to the second subdomain. The two subdomains are then partitioned into two recursively, until a desirable number of domains is reached.

ALGORITHM 14.7. RSB

1. *Compute the Fiedler vector f of the graph G*
2. *Sort the components of f, e.g., increasingly*
3. *Assign the first $\lfloor n/2 \rfloor$ nodes to V_1 and the rest to V_2*
4. *Apply RSB recursively to V_1, V_2 until the desired number of partitions is reached*

The main theoretical property that is exploited here is that the differences between the components of the Fiedler vector represent some sort of distance between the corresponding nodes. Thus, if these components were sorted, they would be effectively grouping the associated node by preserving nearness. Another interesting fact is that the algorithm will also tend to minimize the number n_c of *edge cuts*, i.e., the number of edges (v_i, v_j) such that $v_i \in V_1$ and $v_j \in V_2$. Assume that V_1 and V_2 are of equal size and define a *partition vector* p whose ith component is $+1$ if $v_i \in V_1$ and -1 if $v_i \in V_2$. By the assumptions, the sum of all the p_i's is zero. Then notice that

$$(Lp, p) = 4n_c, \quad (p, e) = 0.$$

Ideally, the objective function (Lp, p) should be minimized subject to the constraint $(p, e) = 0$. Here p is a vector of signs. If, instead, the objective function $(Lx, x)/(x, x)$ were minimized for x real, subject to $(x, e) = 0$, the solution would be the Fiedler vector, since e

is the eigenvector associated with the eigenvalue zero. The Fiedler vector can be computed by the Lanczos algorithm or any other method efficient for large sparse matrices. RSB gives excellent partitionings. On the other hand, it is rather unattractive because it requires computing an eigenvector.

14.6.4 Graph Theory Techniques

A number of other techniques exist that, like spectral techniques, are also based on the adjacency graph only. The simplest idea is one that is borrowed from the technique of *nested dissection* (ND) in the context of direct sparse solution methods; see Sections 3.6.2 and 3.3.3. An initial node is given that constitutes the level zero. Then the method recursively traverses the kth level ($k \geq 1$), which consists of the neighbors of all the elements that constitute level $k-1$. A simple idea for partitioning the graph into two traverses enough levels to visit about half of all the nodes. The visited nodes will be assigned to one subdomain and the others will constitute the second subdomain. The process can then be repeated recursively on each of the subdomains.

A key ingredient for this technique to be successful is to determine a good initial node from which to start the traversal. Often, a heuristic is used for this purpose. Recall that $d(x, y)$ is the distance between vertices x and y in the graph, i.e., the length of the shortest path between x and y.

If the diameter of a graph is defined as

$$\delta(G) = \max\{d(x, y) \mid x \in V, y \in V\},$$

then, ideally, one of two nodes in a pair (x, y) that achieves the diameter can be used as a starting node. These *peripheral nodes* are expensive to determine. Instead, a *pseudo-peripheral* node, as defined through the following procedure, is often employed [144].

ALGORITHM 14.8. Pseudo-Peripheral Node

---
1. Select an initial node x. Set $\delta = 0$
2. Do a level-set traversal from x
3. Select a node y in the last level set with minimum degree
4. If $d(x, y) > \delta$ then
5.   Set $x := y$ and $\delta := d(x, y)$
6.   Go to 2
7.   Else Stop: x is a pseudo-peripheral node
8. EndIf

The distance $d(x, y)$ in line 5 is the number of levels in the level-set traversal needed in line 2. The algorithm traverses the graph from a node of the last level in the previous traversal, until the number of levels stabilizes. It is easy to see that the algorithm does indeed stop after a finite number of steps, typically small.

A first heuristic approach based on level-set traversals is the recursive dissection procedure mentioned above and described next.

ALGORITHM 14.9. Recursive Graph Bisection (RGB)

1. Set $G_* := G$, $S := \{G\}$, $n_{dom} := 1$
2. While $n_{dom} < s$, Do
3.   Select in S the subgraph G_* with largest size
4.   Find a pseudo-peripheral node p in G_*
5.   Do a level set traversal from p. Let $lev :=$ number of levels
6.   Let G_1 be the subgraph of G_* consisting of the first $lev/2$ levels and G_2 the subgraph containing the rest of G_*
7.   Remove G_* from S and add G_1 and G_2 to it
8. $n_{dom} := n_{dom} + 1$
9. EndWhile

The cost of this algorithm is rather small. Each traversal of a graph $G = (V, E)$ costs around $|E|$, where $|E|$ is the number of edges (assuming that $|V| = O(|E|)$). Since there are s traversals of graphs whose size decreases by 2 at each step, it is clear that the cost is $O(|E|)$, the order of edges in the original graph. As can be expected, the results of such an algorithm are not always good. Typically, two qualities that are measured are the sizes of the domains and the number of edge cuts.

Ideally, the domains should be equal. In addition, since the values at the interface points should be exchanged with those of neighboring processors, their total number, as determined by the number of edge cuts, should be as small as possible. The first measure can be easily controlled in an RGB algorithm—for example, by using variants in which the number of nodes is forced to be exactly half that of the original subdomain. The second measure is more difficult to control.

As an example, the top part of Figure 14.10 shows the result of the RGB algorithm on a sample finite element mesh. This is a vertex-based partitioning. The dashed lines represent the edge cuts.

An approach that is competitive with the one described above is that of *double striping*. This method uses two parameters p_1, p_2 such that $p_1 p_2 = s$. The original graph is first partitioned into p_1 large partitions, using one-way partitioning, then each of these partitions is subdivided into p_2 partitions similarly. One-way partitioning into p subgraphs consists of performing a level-set traversal from a pseudo-peripheral node and assigning each set of roughly n/p consecutive nodes in the traversal to a different subgraph. The result of this approach with $p_1 = p_2 = 4$ is shown in Figure 14.10 on the same graph as before.

As can be observed, the subregions obtained by both methods have elongated and twisted shapes. This has the effect of giving a larger number of edge cuts. There are a number of heuristic ways to remedy this. One strategy is based on the fact that a level-set traversal from k nodes can be defined instead of from only one node. These k nodes are called the *centers* or *sites*. Each subdomain will expand from one of these k centers and the expansion will stop when it is no longer possible to acquire another point that is not already assigned. The boundary of each domain that is formed this way will tend to be more *circular*. To smooth the boundaries of an initial partition, find some center point of each domain and perform a level-set expansion from the set of points. The process can be repeated a few times.

14.6. Graph Partitioning

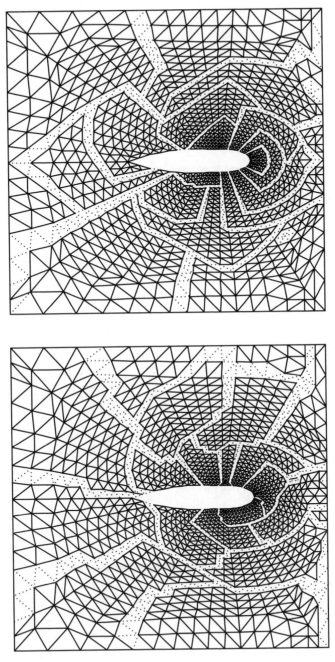

Figure 14.10. *The RGB algorithm (top) and the double-striping algorithm (bottom) for partitioning a graph into 16 subgraphs.*

ALGORITHM 14.10. Multinode Level-Set Expansion Algorithm

1. Find a partition $S = \{G_1, G_2, \ldots, G_s\}$
2. For $iter = 1, \ldots, nouter$, Do
3.   For $k = 1, \ldots, s$, Do
4.   Find a center c_k of G_k. Set $label(c_k) = k$
5.   EndDo
6.   Do a level-set traversal from $\{c_1, c_2, \ldots, c_s\}$. Label each child in the traversal with the same label as its parent
7.   For $k = 1, \ldots, s$, set $G_k :=$ subgraph of all nodes having label k
8. EndDo

For this method, a total number of edge cuts equal to 548 and a rather small standard deviation of 0.5 are obtained for the example seen earlier. Still to be decided is how to select the center nodes mentioned in line 4 of the algorithm. Once more, the pseudo-peripheral algorithm will be helpful. Find a pseudo-peripheral node, then do a traversal from it until about one-half of the nodes have been traversed. Then, traverse the latest level set (typically a line or a very narrow graph) and take the middle point as the center.

A typical number of outer steps, *nouter*, to be used in line 2, is less than five. This heuristic works well in spite of its simplicity. For example, if this is applied to the graph obtained from the RGB algorithm, with $nouter = 3$, the partition shown in Figure 14.11 is obtained. With this technique, the resulting total number of edge cuts is equal to 441 and the standard deviation is 7.04. As is somewhat expected, the number of edge cuts has decreased dramatically, while the standard deviation of the various sizes has increased.

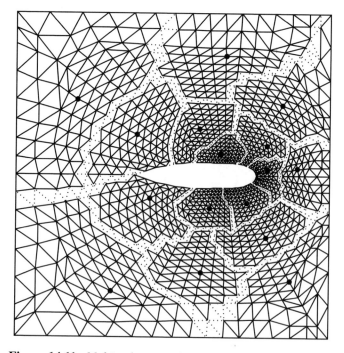

Figure 14.11. *Multinode expansion starting with the partition obtained in Figure* 14.10.

Exercises

1. In the proof of Theorem 14.7, the following form of the Cauchy–Schwarz inequality was used:
$$\sum_{i=1}^{p}(x_i, y_i) \le \left(\sum_{i=1}^{p}(x_i, x_i)\right)^{1/2}\left(\sum_{i=1}^{p}(y_i, y_i)\right)^{1/2}.$$
   a. Prove that this result is a consequence of the standard Cauchy–Schwarz inequality.
   b. Extend the result to the A inner product.
   c. Assume that the x_i's and y_i's are the columns of two $n \times p$ matrices X and Y. Rewrite the result in terms of these matrices.

2. Using Lemma 14.4, write explicitly the vector $M^{-1}b$ for the multiplicative Schwarz procedure in terms of the matrix A and the R_i's when $s = 2$, and then when $s = 3$.

3. Justify Algorithm 14.5; i.e., show that it does indeed compute the vector $M^{-1}Av$ for an input vector v, where M is the multiplicative Schwarz preconditioner. Then find a similar algorithm that computes $AM^{-1}v$ (right preconditioning).

4. a. Show that, in the multiplicative Schwarz procedure, the residual vectors $r_i = b - Ax_i$ obtained at each step satisfy the recurrence
$$r_i = r_{i-1} - AR_i^T A_i^{-1} R_i r_{i-1}$$
for $i = 1, \ldots, s$.
   b. Consider the operator $Q_i \equiv AR_i^T A_i^{-1} R_i$. Show that Q_i is a projector.
   c. Is Q_i an orthogonal projector with respect to the A inner product? With respect to which inner product is it orthogonal?

5. The analysis of the additive Schwarz procedure assumes that A_i^{-1} is *exact*, i.e., that linear systems $A_i x = b$ are solved exactly each time A_i^{-1} is applied. Assume that A_i^{-1} is replaced with some approximation Θ_i^{-1}.
   a. Is P_i still a projector?
   b. Show that, if Θ_i is SPD, then so is P_i.
   c. Now make the assumption that $\lambda_{max}(P_i) \le \omega_*$. What becomes of the result of Theorem 14.5?

6. In EBE methods, the extreme cases of the additive or the multiplicative Schwarz procedure are considered, in which the subdomain partition corresponds to taking Ω_i to be an element. The advantage here is that the matrices do not have to be assembled. Instead, they are kept in unassembled form (see Chapter 2). Assume that Poisson's equation is being solved.
   a. What are the matrices A_i?
   b. Are they SPD?
   c. Write down the EBE preconditioning corresponding to the multiplicative Schwarz procedure, its multicolor version, and the additive Schwarz procedure.

7. Theorem 14.2 was stated only for the multiplicative version of the Schwarz procedure. There is a similar result for the additive Schwarz procedure. State this result and prove it.

8. Show that the matrix defined by (14.38) is indeed a projector. Is it possible to formulate Schwarz procedures in terms of projection processes, as seen in Chapter 5?

9. It was stated at the end of the proof of Theorem 14.7 that, if
$$(A_J u, u)_A \geq \frac{1}{C}(u, u)_A$$
for any nonzero u, then $\lambda_{min}(A_J) \geq \frac{1}{C}$.

   a. Prove this result without invoking the min-max theorem.
   b. Prove a version of the min-max theorem with the A inner product; i.e., prove that the min-max theorem is valid for any inner product for which A is self-adjoint.

10. Consider the Laplacian of a graph as defined in Section 14.6. Show that
$$(Lx, x) = \sum_{(i,j) \in E} (x_i - x_j)^2.$$

11. Consider a rectangular finite difference mesh, with mesh size $\Delta x = h$ in the x direction and $\Delta y = h$ closest to the y direction.

   a. With each mesh point $p = (x_i, y_j)$, associate the closed disk D_{ij} of radius h centered at p_i. What is the smallest k such that the family $\{D_{ij}\}$ is a k-ply system?
   b. Answer the same question for the case where the radius is reduced to $h/2$. What is the overlap graph (and associated mesh) for any α such that
$$\frac{1}{2} < \alpha < \frac{\sqrt{2}}{2}?$$
   What about when $\alpha = 2$?

12. Determine the cost of a level-set expansion algorithm starting from p distinct centers.

13. Write recursive versions of the recursive graph bisection algorithm and recursive spectral bisection algorithm. [Hint: Recall that a recursive program unit is a subprogram or function, say *foo*, that calls itself, so *foo* is allowed to make a subroutine call to *foo* within its body.]

   a. Give a pseudocode for the recursive graph bisection algorithm that processes the subgraphs in any order.
   b. Give a pseudocode for the RGB algorithm case when the larger subgraph is to be processed before the smaller one in any dissection. Is this second version equivalent to Algorithm 14.9?

14. Write a FORTRAN-90 subroutine (or C function) to implement the recursive graph partitioning algorithm.

Notes and References

To start with, the original paper by Schwarz is the reference [260], but an earlier note appeared in 1870. In recent years, research on domain decomposition techniques has been very active and productive. This rebirth of an old technique has been in large part motivated by parallel processing. However, the first practical use of domain decomposition ideas has been in applications to very large structures [229, 41] and elasticity problems [233, 282, 268, 76, 40].

The book by Smith, Bjørstad, and Gropp [267] gives a thorough survey of domain decomposition methods. Two other monographs, one by P. LeTallec [197], and the other by C. Farhat and J. X. Roux [123], describe the use of domain decomposition approaches specifically for solving problems in structural mechanics. Survey papers include those by Keyes and Gropp [188] and by Chan and Mathew [75]. The volume [189] discusses the various uses of *domain-based* parallelism in computational sciences and engineering.

The bulk of recent work on domain decomposition methods has been geared toward a PDEs viewpoint. Often, there appears to be a dichotomy between this viewpoint and that of applied domain decomposition, in that the good methods from a theoretical point of view are hard to implement in practice. The additive Schwarz procedure with overlapping represents a compromise between good intrinsic properties and ease of implementation. For example, Venkatakrishnan concludes in [294] that, although the use of global coarse meshes may accelerate convergence of local, domain-based ILU preconditioners, it does not necessarily reduce the overall time to solve a practical aerodynamics problem.

Much is known about the convergence of the Schwarz procedure; refer in particular to the work by Widlund and co-authors [42, 102, 103, 104, 70]. The convergence results of Section 14.3.4 have been adapted from Xu [319] as well as Hackbusch [163]. The result on the equivalence between Schwarz and Schur complement iterations stated in Theorem 14.2 seems to have been originally proved by Chan and Goovaerts [73]; see also the more recent article by Wilders and Brakkee [314].

The results on the equivalence between the full matrix techniques and the Schur matrix techniques seen in Section 14.5 have been adapted from results by S. E. Eisenstat, reported in [188]. These connections are rather interesting and useful in practice since they provide some flexibility on ways to implement a method. A number of preconditioners have also been derived using similar connections in the PDE framework [48, 47, 49, 50, 51].

Research on graph partitioning has slowed in recent years, no doubt due to the appearance of Metis, a well-designed and efficient graph-partitioning code [184]. Variations of the RSB algorithm [228] seem to give the best results in terms of overall quality of the subgraphs. However, the algorithm is rather expensive, and the less costly multilevel techniques, such as the ones in the codes Metis [184] and Chaco [166], are usually preferred. The description of the geometric partitioning techniques in Section 14.6.2 is based on the papers [145] and [210]. Earlier approaches were developed in [81, 82, 83].

Bibliography

[1] J. ABAFFY AND E. SPEDICATO, *ABS Projection Algorithms*, Halsted Press, New York, 1989.

[2] L. M. ADAMS, *Iterative algorithms for large sparse linear systems on parallel computers*, PhD thesis, Applied Mathematics, University of Virginia, Charlottesville, VA, 1982. Also NASA Contractor Report 166027.

[3] L. M. ADAMS AND H. JORDAN, *Is SOR color-blind?*, SIAM Journal on Scientific and Statistical Computing, 7 (1986), pp. 490–506.

[4] L. M. ADAMS AND J. ORTEGA, *A multi-color SOR method for parallel computers*, in Proceedings of the 1982 International Conference on Pararallel Processing, 1982, pp. 53–56.

[5] J. I. ALIAGA, D. L. BOLEY, R. W. FREUND, AND V. HERNÁNDEZ, *A Lanczos-type algorithm for multiple starting vectors*, Tech. Rep. Numerical Analysis Manuscript No 95-11, AT&T Bell Laboratories, Murray Hill, NJ, 1995.

[6] F. L. ALVARADO, *Manipulation and visualization of sparse matrices*, ORSA Journal on Computing, 2 (1990), pp. 186–206.

[7] E. C. ANDERSON, *Parallel implementation of preconditioned conjugate gradient methods for solving sparse systems of linear equations*, Tech. Rep. 805, CSRD, University of Illinois, Urbana, IL, 1988. MS Thesis.

[8] E. C. ANDERSON AND Y. SAAD, *Solving sparse triangular systems on parallel computers*, International Journal of High Speed Computing, 1 (1989), pp. 73–96.

[9] W. E. ARNOLDI, *The principle of minimized iteration in the solution of the matrix eigenvalue problem*, Quarterly of Applied Mathematics, 9 (1951), pp. 17–29.

[10] C. C. ASHCRAFT AND R. G. GRIMES, *On vectorizing incomplete factorization and SSOR preconditioners*, SIAM Journal on Scientific and Statistical Computing, 9 (1988), pp. 122–151.

[11] O. AXELSSON, *A generalized SSOR method*, BIT, 12 (1972), pp. 443–467.

[12] O. AXELSSON, *Conjugate gradient-type methods for unsymmetric and inconsistent systems of linear equations*, Linear Algebra and Its Applications, 29 (1980), pp. 1–16.

[13] O. AXELSSON, *A generalized conjugate gradient, least squares method*, Numerische Mathematik, 51 (1987), pp. 209–227.

[14] O. AXELSSON, *Iterative Solution Methods*, Cambridge University Press, New York, 1994.

[15] O. AXELSSON AND V. A. BARKER, *Finite Element Solution of Boundary Value Problems*, Academic Press, Orlando, FL, 1984.

[16] O. AXELSSON, S. BRINKKEMPER, AND V. P. IL'IN, *On some versions of incomplete block-matrix factorization iterative methods*, Linear Algebra and Its Applications, 58 (1984), pp. 3–15.

[17] O. AXELSSON AND M. NEYTCHEVA, *Algebraic multilevel iteration method for Stieltjes matrices*, Numerical Linear Algebra with Applications, 1 (1994), pp. 213–236.

[18] O. AXELSSON AND B. POLMAN, *A robust preconditioner based on algebraic substructuring and two-level grids*, in Robust Multigrid Methods, Proceedings, Kiel, January 1988, W. Hackbusch, ed., Notes on Numerical Fluid Mechanics, Volume 23, Vieweg, Braunschweig, 1988, pp. 1–26.

[19] O. AXELSSON AND P. VASSILEVSKI, *Algebraic multilevel preconditioning methods. I*, Numerische Mathematik, 56 (1989), pp. 157–177.

[20] O. AXELSSON AND P. VASSILEVSKI, *A survey of multilevel preconditioned iterative methods*, BIT, 29 (1989), pp. 769–793.

[21] O. AXELSSON AND P. VASSILEVSKI, *Algebraic multilevel preconditioning methods. II*, SIAM Journal on Numerical Analysis, 27 (1990), pp. 1569–1590.

[22] O. AXELSSON AND P. S. VASSILEVSKI, *A black box generalized conjugate gradient solver with inner iterations and variable-step preconditioning*, SIAM Journal on Matrix Analysis and Applications, 12 (1991), pp. 625–644.

[23] N. S. BAKHVALOV, *On the convergence of a relaxation method with natural constraints on the elliptic operator*, U.S.S.R. Computational Mathematics and Mathematical Physics, 6 (1966), pp. 101–135.

[24] S. BALAY, W. D. GROPP, L. C. MCINNES, AND B. F. SMITH, *PETSc 2.0 users manual*, Tech. Rep. ANL-95/11 - Revision 2.0.24, Argonne National Laboratory, 1999.

[25] R. BANK, T. DUPONT, AND H. YSERENTANT, *The hierarchical basis multigrid method*, Numerische Mathematik, 52 (1988), pp. 427–458.

[26] R. BANK AND J. XU, *The hierarchical basis multigrid method and incomplete LU decomposition*, Tech. Rep., Department of Mathematics, University of California at San Diego, 1994.

[27] R. E. BANK AND T. F. CHAN, *An analysis of the composite step biconjugate gradient method*, Numerische Mathematik, 66 (1993), pp. 259–319.

[28] R. E. BANK AND C. WAGNER, *Multilevel ILU decomposition*, Numerische Mathematik, 82 (1999), pp. 543–576.

[29] T. BARTH AND T. MANTEUFFEL, *Variable metric conjugate gradient methods*, in Advances in Numerical Methods for Large Sparse Sets of Linear Equations, Number 10, Matrix Analysis and Parallel Computing, PCG 94, Keio University, Yokohama, Japan, 1994, pp. 165–188.

[30] D. BAXTER, J. SALTZ, M. H. SCHULTZ, S. C. EISENSTAT, AND K. CROWLEY, *An experimental study of methods for parallel preconditioned Krylov methods*, in Proceedings of the 1988 Hypercube Multiprocessors Conference, Pasadena, CA, January 1988, pp. 1698–1711.

[31] M. BENANTAR AND J. E. FLAHERTY, *A six color procedure for the parallel solution of Elliptic systems using the finite quadtree structures*, in Proceedings of the Fourth SIAM Conference on Parallel Processing for Scientific Computing, J. Dongarra, P. Messina, D. C. Sorenson, and R. G. Voigt, eds., SIAM, Philadelphia, 1990, pp. 230–236.

[32] M. BENZI, J. C. HAWS, AND M. TŮMA, *Preconditioning highly indefinite and nonsymmetric matrices*, SIAM Journal on Scientific Computing, 22 (2000), pp. 1333–1353.

[33] M. BENZI, J. MARÍN, AND M. TŮMA, *A two-level parallel preconditioner based on sparse approximate inverses*, in Iterative Methods in Scientific Computation, II, D. R. Kincaid and A. C. Elster, eds., IMACS, New Brunswick, NJ, 1999.

[34] M. BENZI, C. D. MEYER, AND M. TŮMA, *A sparse approximate inverse preconditioner for the conjugate gradient method*, SIAM Journal on Scientific Computing, 17 (1996), pp. 1135–1149.

[35] M. BENZI, D. B. SZYLD, AND A. VAN DUIN, *Orderings for incomplete factorization preconditioning of nonsymmetric problems*, SIAM Journal on Scientific Computing, 20 (1999), pp. 1652–1670.

[36] M. BENZI AND M. TŮMA, *A sparse approximate inverse preconditioner for nonsymmetric linear systems*, SIAM Journal on Scientific Computing, 19 (1998), pp. 968–994.

[37] H. BERRYMAN, J. SALTZ, W. GROPP, AND R. MIRCHANDANEY, *Krylov methods preconditioned with incompletely factored matrices on the CM-2*, Journal of Parallel and Distributed Computing, 8 (1990), pp. 186–190.

[38] G. BIRKHOFF, R. S. VARGA, AND D. M. YOUNG, *Alternating direction implicit methods*, in Advances in Computers, F. Alt and M. Rubinov, eds., Academic Press, New York, 1962, pp. 189–273.

[39] A. BJÖRCK AND T. ELFVING, *Accelerated projection methods for computing pseudo-inverse solutions of systems of linear equations*, BIT, 19 (1979), pp. 145–163.

[40] P. E. BJØRSTAD AND A. HVIDSTEN, *Iterative methods for substructured elasticity problems in structural analysis*, in Domain Decomposition Methods for Partial Differential Equations, R. Glowinski, G. H. Golub, G. A. Meurant, and J. Périaux, eds., SIAM, Philadelphia, 1988, pp. 301–312.

[41] P. E. BJØRSTAD AND O. B. WIDLUND, *Solving elliptic problems on regions partitioned into substructures*, in Elliptic Problem Solvers II, G. Birkhoff and A. Schoenstadt, eds., Academic Press, New York, 1984, pp. 245–256.

[42] P. E. BJØRSTAD AND O. B. WIDLUND, *Iterative methods for the solution of elliptic problems on regions partitioned into substructures*, SIAM Journal on Numerical Analysis, 23 (1986), pp. 1097–1120.

[43] E. BODEWIG, *Matrix Calculus*, North-Holland, Amsterdam, 1956.

[44] M. BOLLHÖFER, *A robust ILU with pivoting based on monitoring the growth of the inverse factors*, Linear Algebra and Its Applications, 338 (2001), pp. 201–213.

[45] E. BOTTA, A. PLOEG, AND F. WUBS, *Nested grids ILU-decomposition (NGILU)*, Journal of Computational and Applied Mathematics, 66 (1996), pp. 515–526.

[46] H. BRAKHAGE, *Über die numerische Behandlung von Integralgleichungen nach der Quadratureformelmethode*, Numerische Mathematik, 2 (1960), pp. 183–196.

[47] J. H. BRAMBLE, J. E. PASCIAK, AND A. H. SCHATZ, *The construction of preconditioners for elliptic problems by substructuring*, I, Mathematics of Computation, 47 (1986), pp. 103–134.

[48] J. H. BRAMBLE, J. E. PASCIAK, AND A. H. SCHATZ, *An iterative method for elliptic problems on regions partitioned into substructures*, Mathematics of Computation, 46 (1986), pp. 361–369.

[49] J. H. BRAMBLE, J. E. PASCIAK, AND A. H. SCHATZ, *The construction of preconditioners for elliptic problems by substructuring*, II, Mathematics of Computation, 49 (1987), pp. 1–16.

[50] J. H. BRAMBLE, J. E. PASCIAK, AND A. H. SCHATZ, *The construction of preconditioners for elliptic problems by substructuring*, III, Mathematics of Computation, 51 (1988), pp. 415–430.

[51] J. H. BRAMBLE, J. E. PASCIAK, AND A. H. SCHATZ, *The construction of preconditioners for elliptic problems by substructuring*, IV, Mathematics of Computation, 53 (1989), pp. 1–24.

[52] R. BRAMLEY AND A. SAMEH, *Row projection methods for large nonsymmetric linear systems*, SIAM Journal on Scientific and Statistical Computing, 13 (1992), pp. 168–193.

[53] R. BRAMLEY AND A. SAMEH, *A robust parallel solver for block tridiagonal systems*, in Proceedings of the International Conference on Supercomputing, ACM, July 1988, pp. 39–54.

[54] A. BRANDT, *Multi-level adaptive technique (MLAT) for fast numerical solutions to boundary problems*, in Proceedings of the 3rd International Conference on Numerical Methods in Fluid Mechanics, Paris, 1972, H. Cabannes and R. Temam, eds., Springer-Verlag, Berlin, 1973, pp. 82–89.

[55] A. BRANDT, *Multi-level adaptive solutions to boundary value problems*, Mathematics of Computation, 31 (1977), pp. 333–390.

[56] A. BRANDT, *A guide to multigrid development*, in Multigrid Methods, W. Hackbusch and U. Trottenberg, eds., Springer Verlag, Berlin, 1982, pp. 220–312.

[57] A. BRANDT, *Algebraic multigrid theory: The symmetric case*, Applied Mathematics and Computation, 19 (1986), pp. 23–56.

[58] A. BRANDT, S. F. MCCORMICK, AND J. RUGE, *Algebraic multigrid (AMG) for sparse matrix equations*, in Sparsity and Its Applications, D. J. Evans, ed., Cambridge University Press, Cambridge, 1984, pp. 257–284.

[59] C. BREZINSKI, *Padé Type Approximation and General Orthogonal Polynomials*, Birkhäuser-Verlag, Basel, Boston, Stuttgart, 1980.

[60] C. BREZINSKI AND M. REDIVO ZAGLIA, *Extrapolation Methods: Theory and Practice*, North-Holland, Amsterdam, 1991.

[61] C. BREZINSKI AND M. REDIVO ZAGLIA, *Hybrid procedures for solving systems of linear equations*, Numerische Mathematik, 67 (1994), pp. 1–19.

[62] C. BREZINSKI, M. REDIVO ZAGLIA, AND H. SADOK, *Avoiding breakdown and near-breakdown in Lanczos-type algorithms*, Numerical Algorithms, 1 (1991), pp. 261–284.

[63] C. BREZINSKI, M. REDIVO ZAGLIA, AND H. SADOK, *A breakdown-free Lanczos-type algorithm for solving linear systems*, Numerische Mathematik, 63 (1992), pp. 29–38.

[64] R. BRIDSON AND W.-P. TANG, *Ordering, anisotropy, and factored sparse approximate inverses*, SIAM Journal on Scientific Computing, 21 (1999), pp. 867–882.

[65] W. L. BRIGGS, V. E. HENSON, AND S. F. MCCORMICK, *A Multigrid Tutorial*, SIAM, Philadelphia, 2000. Second edition.

[66] P. N. BROWN, *A theoretical comparison of the Arnoldi and GMRES algorithms*, SIAM Journal on Scientific and Statistical Computing, 12 (1991), pp. 58–78.

[67] P. N. BROWN AND A. C. HINDMARSH, *Matrix-free methods for stiff systems of ODEs*, SIAM Journal on Numerical Analysis, 23 (1986), pp. 610–638.

[68] N. I. BULEEV, *A numerical method for the solution of two-dimensional and three-dimensional equations of diffusion*, Matematicheskij Sbornik, 51 (1960), pp. 227–238 (in Russian).

[69] O. BUNEMAN, *A compact non-iterative Poisson solver*, Tech. Rep. 294, Stanford University, Stanford, CA, 1969.

[70] X.-C. CAI AND O. B. WIDLUND, *Multiplicative Schwarz algorithms for some non-symmetric and indefinite problems*, SIAM Journal on Numerical Analysis, 30 (1993), pp. 936–952.

[71] L. CESARI, *Sulla risoluzione dei sistemi di equazioni lineari per approssimazioni successive*, Atti della Accademia Nazionale dei Lincei Rendiconti Classe di Scienze Fisiche, Matematiche e Naturali Serie. 6a, 25 (1937), pp. 422–428.

[72] T. F. CHAN, E. GALLOPOULOS, V. SIMONCINI, T. SZETO, AND C. H. TONG, *A quasi-minimal residual variant of the Bi-CGSTAB algorithm for nonsymmetric systems*, SIAM Journal on Scientific Computing, 15 (1994), pp. 338–347.

[73] T. F. CHAN AND D. GOOVAERTS, *On the relationship between overlapping and nonoverlapping domain decomposition methods*, SIAM Journal on Matrix Analysis and Applications, 13 (1992), pp. 663–670.

[74] T. F. CHAN AND T. P. MATHEW, *The interface probing technique in domain decomposition*, SIAM Journal on Matrix Analysis and Applications, 13 (1992), pp. 212–238.

[75] T. F. CHAN AND T. P. MATHEW, *Domain decomposition algorithms*, Acta Numerica, (1994), pp. 61–143.

[76] H.-C. CHEN AND A. H. SAMEH, *A matrix decomposition method for orthotropic elasticity problems*, SIAM Journal on Matrix Analysis and Applications, 10 (1989), pp. 39–64.

[77] C. C. CHENEY, *Introduction to Approximation Theory*, McGraw-Hill, New York, 1966.

[78] E. CHOW AND Y. SAAD, *Approximate inverse techniques for block-partitioned matrices*, SIAM Journal on Scientific Computing, 18 (1997), pp. 1657–1675.

[79] E. CHOW AND Y. SAAD, *ILUS: An incomplete LU factorization for matrices in sparse skyline format*, International Journal for Numerical Methods in Fluids, 25 (1997), pp. 739–748.

[80] E. CHOW AND Y. SAAD, *Approximate inverse preconditioners via sparse-sparse iterations*, SIAM Journal on Scientific Computing, 19 (1998), pp. 995–1023.

[81] N. CHRISOCHOIDES, G. FOX, AND J. THOMPSON, *MENUS-PGG mapping environment for unstructured and structured numerical parallel grid generation*, in Proceedings of the Seventh International Conference on Domain Decomposition Methods in Scientific and Engineering Computing, AMS, Providence, RI, 1993, pp. 381–386.

[82] N. CHRISOCHOIDES, C. E. HOUSTIS, E. N. HOUSTIS, P. N. PAPACHIOU, S. K. KORTESIS, AND J. RICE, *DOMAIN DECOMPOSER: A software tool for mapping PDE computations to parallel architectures*, in Domain Decomposition Methods for Partial Differential Equations, Roland Glowinski et al., eds., SIAM Philadelphia, 1991, pp. 341–357.

[83] N. CHRISOCHOIDES, E. HOUSTIS, AND J. RICE, *Mapping algorithms and software environment for data parallel PDE iterative solvers*, Journal of Parallel and Distributed Computing, 21 (1994), pp. 75–95.

[84] P. G. CIARLET, *The Finite Element Method for Elliptic Problems*, North-Holland, Amsterdam, 1978.

[85] G. CIMMINO, *Calcolo approssimato per le soluzioni dei sistemi di equazioni lineari*, La Ricerca Scientifica, II, 9 (1938), pp. 326–333.

[86] A. CLAYTON, *Further results on polynomials having least maximum modulus over an ellipse in the complex plane*, Tech. Rep. AEEW-7348, UKAEA, Harewell-UK, 1963.

[87] S. CLIFT AND W. TANG, *Weighted graph based ordering techniques for preconditioned conjugate gradient methods*, BIT, 35 (1995), pp. 30–47.

[88] P. CONCUS AND G. H. GOLUB, *A generalized conjugate gradient method for nonsymmetric systems of linear equations*, in Computing Methods in Applied Sciences and Engineering, R. Glowinski and J. L. Lions, eds., Springer-Verlag, New York, 1976, pp. 56–65.

[89] P. CONCUS, G. H. GOLUB, AND G. MEURANT, *Block preconditioning for the conjugate gradient method*, SIAM Journal on Scientific and Statistical Computing, 6 (1985), pp. 220–252.

[90] T. H. CORMEN, C. E. LEISERSON, AND R. L. RIVEST, *Introduction to Algorithms*, McGraw-Hill, New York, 1990.

[91] J. D. F. COSGROVE, J. C. DIAZ, AND A. GRIEWANK, *Approximate inverse preconditioning for sparse linear systems*, International Journal of Computational Mathematics, 44 (1992), pp. 91–110.

[92] J. CULLUM AND A. GREENBAUM, *Relations between Galerkin and norm-minimizing iterative methods for solving linear systems*, SIAM Journal on Matrix Analysis and Applications, 17 (1996), pp. 223–247.

[93] B. N. DATTA, *Numerical Linear Algebra and Applications*, Brooks/Cole Publishing, Pacific Grove, CA, 1995.

[94] P. J. DAVIS, *Interpolation and Approximation*, Blaisdell, Waltham, MA, 1963.

[95] T. A. DAVIS, *A parallel algorithm for sparse unsymmetric LU factorizations*, PhD thesis, University of Illinois at Urbana Champaign, Urbana, IL, 1989.

[96] E. F. D'AZEVEDO, P. A. FORSYTH, AND W.-P. TANG, *Ordering methods for preconditioned conjugate gradient methods applied to unstructured grid problems*, SIAM Journal on Matrix Analysis and Applications, 13 (1992), pp. 944–961.

[97] E. F. D'AZEVEDO, P. A. FORSYTH, AND W.-P. TANG, *Towards a cost effective ILU preconditioner with high level fill*, BIT, 31 (1992), pp. 442–463.

[98] M. A. DELONG AND J. M. ORTEGA, *SOR as a preconditioner*, Applied Numerical Mathematics, 18 (1995), pp. 431–440.

[99] J. W. DEMMEL, *Applied Numerical Linear Algebra*, SIAM, Philadelphia, 1997.

[100] P. DEUFLHARD, R. W. FREUND, AND A. WALTER, *Fast secant methods for the iterative solution of large nonsymmetric linear systems*, IMPACT of Computing in Science and Engineering, 2 (1990), pp. 244–276.

[101] J. J. DONGARRA, I. S. DUFF, D. SORENSEN, AND H. A. VAN DER VORST, *Solving Linear Systems on Vector and Shared Memory Computers*, SIAM, Philadelphia, 1991.

[102] M. DRYJA AND O. B. WIDLUND, *Some domain decomposition algorithms for elliptic problems*, in Iterative Methods for Large Linear Systems, L. Hayes and D. Kincaid, eds., Academic Press, New York, 1989, pp. 273–291.

[103] M. DRYJA AND O. B. WIDLUND, *Towards a unified theory of domain decomposition algorithms for elliptic problems*, in Proceedings of the Third International Symposium on Domain Decomposition Methods for Partial Differential Equations, Houston, TX, March, 1989, T. Chan, R. Glowinski, J. Périaux, and O. Widlund, eds., SIAM, Philadelphia, 1990, pp. 3–21.

[104] M. DRYJA AND O. B. WIDLUND, *Additive Schwarz methods for elliptic finite element problems in three dimensions*, in Proceedings of the Fifth International Symposium on Domain Decomposition Methods for Partial Differential Equations, D. E. Keyes, T. F. Chan, G. A. Meurant, J. S. Scroggs, and R. G. Voigt, eds., SIAM, Philadelphia, 1992, pp. 3–18.

[105] P. F. DUBOIS, A. GREENBAUM, AND G. H. RODRIGUE, *Approximating the inverse of a matrix for use on iterative algorithms on vector processors*, Computing, 22 (1979), pp. 257–268.

[106] I. S. DUFF, *A survey of sparse matrix research*, in Proceedings of the IEEE, 65, Prentice-Hall, New York, 1977, pp. 500–535.

[107] I. S. DUFF, A. M. ERISMAN, AND J. K. REID, *Direct Methods for Sparse Matrices*, Clarendon Press, Oxford, U.K., 1986.

[108] I. S. DUFF, R. G. GRIMES, AND J. G. LEWIS, *Sparse matrix test problems*, ACM Transactions on Mathematical Software, 15 (1989), pp. 1–14.

[109] I. S. DUFF AND J. KOSTER, *The design and use of algorithms for permuting large entries to the diagonal of sparse matrices*, SIAM Journal on Matrix Analysis and Applications, 20 (1999), pp. 889–901.

[110] I. S. DUFF AND J. KOSTER, *On algorithms for permuting large entries to the diagonal of a sparse matrix*, SIAM Journal on Matrix Analysis and Applications, 22 (2001), pp. 973–996.

[111] I. S. DUFF AND G. A. MEURANT, *The effect of ordering on preconditioned conjugate gradients*, BIT, 29 (1989), pp. 635–657.

[112] L. C. DUTTO, *The effect of reordering on the preconditioned GMRES algorithm for solving the compressible Navier-Stokes equations*, International Journal for Numerical Methods in Engineering, 36 (1993), pp. 457–497.

[113] T. EIROLA AND O. NEVANLINNA, *Accelerating with rank-one updates*, Linear Algebra and Its Applications, 121 (1989), pp. 511–520.

[114] S. EISENSTAT, *Efficient implementation of a class of preconditioned conjugate gradient methods*, SIAM Journal on Scientific and Statistical Computing, 2 (1981), pp. 1–4.

[115] S. C. EISENSTAT, M. H. SCHULTZ, AND A. H. SHERMAN, *Algorithms and data structures for sparse symmetric Gaussian elimination*, SIAM Journal on Scientific Computing, 2 (1981), pp. 225–237.

[116] H. C. ELMAN, *A stability analysis of incomplete LU factorizations*, Mathematics of Computation, 47 (1986), pp. 191–217.

[117] H. C. ELMAN AND E. AGRON, *Ordering techniques for the preconditioned conjugate gradient method on parallel computers*, Computer Physics Communications, 53 (1989), pp. 253–269.

[118] H. C. ELMAN AND G. H. GOLUB, *Inexact and preconditioned Uzawa algorithms for saddle point problems*, SIAM Journal on Numerical Analysis, 31 (1994), pp. 1645–1661.

[119] H. C. ELMAN AND D. J. SILVESTER, *Fast nonsymmetric iterations and preconditioning for Navier-Stokes equations*, SIAM Journal on Scientific Computing, 17 (1996), pp. 33–46.

[120] M. ENGLEMAN, *FIDAP manuals*, Tech. Rep. Volumes 1, 2, and 3, Fluid Dynamics International, Evanston, IL, 1986.

[121] V. FABER AND T. MANTEUFFEL, *Necessary and sufficient conditions for the existence of a conjugate gradient method*, SIAM Journal on Numerical Analysis, 21 (1984), pp. 352–362.

[122] K. FAN, *Note on M-matrices*, Quarterly Journal of Mathematics, Oxford series (2), 11 (1960), pp. 43–49.

[123] C. FARHAT AND J. X. ROUX, *Implicit parallel processing in structural mechanics*, Computational Mechanics Advances, 2 (1994), pp. 1–124.

[124] R. P. FEDORENKO, *A relaxation method for solving elliptic difference equations*, U.S.S.R. Computational Mathematics and Mathematical Physics, 1 (1962), pp. 1092–1096.

[125] R. P. FEDORENKO, *The speed of convergence of one iterative process*, U.S.S.R. Computational Mathematics and Mathematical Physics, 4 (1964), pp. 227–235.

[126] R. M. FERENCZ, *Element-by-element preconditioning techniques for large scale vectorized finite element analysis in nonlinear solid and structural mechanics*, PhD thesis, Department of Applied Mathematics, Stanford University, Stanford, CA, 1989.

[127] B. FISCHER AND R. W. FREUND, *On the constrained Chebyshev approximation problem on ellipses*, Journal of Approximation Theory, 62 (1990), pp. 297–315.

[128] B. FISCHER AND R. W. FREUND, *Chebyshev polynomials are not always optimal*, Journal of Approximation Theory, 65 (1991), pp. 261–272.

[129] B. FISCHER AND L. REICHEL, *A stable Richardson iteration method for complex linear systems*, Numerische Mathematik, 54 (1988), pp. 225–241.

[130] R. FLETCHER, *Conjugate gradient methods for indefinite systems*, in Proceedings of the Dundee Biennial Conference on Numerical Analysis, 1974, G. A. Watson, ed., Springer-Verlag, New York, 1975, pp. 73–89.

[131] I. T. FOSTER, *Designing and Building Parallel Programs: Concepts and Tools for Parallel Software Engineering*, Addison-Wesley, Reading, MA, 1995.

[132] R. W. FREUND, *Conjugate gradient-type methods for linear systems with complex symmetric coefficient matrices*, SIAM Journal on Scientific and Statistical Computing, 13 (1992), pp. 425–448.

[133] R. W. FREUND, *Quasi-kernel polynomials and convergence results for quasi-minimal residual iterations*, in Numerical Methods of Approximation Theory, Volume 9, D. Braess and L. L. Schumaker, eds., International series of numerical mathematics, Birkhäuser-Verlag, Basel, 1992, pp. 1–19.

[134] R. W. FREUND, *A transpose-free quasi-minimal residual algorithm for non-Hermitian linear systems*, SIAM Journal on Scientific Computing, 14 (1993), pp. 470–482.

[135] R. W. FREUND, M. H. GUTKNECHT, AND N. M. NACHTIGAL, *An implementation of the look-ahead Lanczos algorithm for non-Hermitian matrices*, SIAM Journal on Scientific and Statistical Computing, 14 (1993), pp. 137–158.

[136] R. W. FREUND AND N. M. NACHTIGAL, *QMR: a quasi-minimal residual method for non-Hermitian linear systems*, Numerische Mathematik, 60 (1991), pp. 315–339.

[137] K. GALLIVAN, A. SAMEH, AND Z. ZLATEV, *A parallel hybrid sparse linear system solver*, Computing Systems in Engineering, 1 (1990), pp. 183–195.

[138] E. GALLOPOULOS AND Y. SAAD, *Parallel block cyclic reduction algorithm for the fast solution of elliptic equations*, Parallel Computing, 10 (1989), pp. 143–160.

[139] F. R. GANTMACHER, *The Theory of Matrices*, Chelsea, New York, 1959.

[140] N. GASTINEL, *Analyse Numérique Linéaire*, Hermann, Paris, 1966.

[141] W. GAUTSCHI, *On generating orthogonal polynomials*, SIAM Journal on Scientific and Statistical Computing, 3 (1982), pp. 289–317.

[142] A. GEORGE, *Computer implementation of the finite element method*, Tech. Rep. STAN-CS-208, Department of Computer Science, Stanford University, Stanford, CA, 1971.

[143] A. GEORGE AND J. W.-H. LIU, *The evolution of the minimum degree ordering algorithm*, SIAM Review, 31 (March 1989), pp. 1–19.

[144] J. A. GEORGE AND J. W. LIU, *Computer Solution of Large Sparse Positive Definite Systems*, Prentice-Hall, Englewood Cliffs, NJ, 1981.

[145] J. R. GILBERT, G. L. MILLER, AND S.-H. TENG, *Geometric mesh partitioning: Implementation and experiments*, SIAM Journal on Scientific Computing, 19 (1998), pp. 2091–2110.

[146] J. R. GILBERT AND T. PEIERLS, *Sparse partial pivoting in time proportional to arithmetic operations*, SIAM Journal on Scientific Computing, 9 (1988), pp. 862–874.

[147] S. K. GODUNOV AND G. P. PROKOPOV, *A method of minimal iteration for evaluating the eigenvalues of an elliptic operator*, Zhurnal Vychislitel'noĭ Matematiki i Matematicheskoĭ Fiziki, 10 (1970), pp. 1180–1190.

[148] T. GOEHRING AND Y. SAAD, *Heuristic algorithms for automatic graph partitioning*, Tech. Rep. umsi-94-29, University of Minnesota Supercomputer Institute, Minneapolis, MN, 1994.

[149] G. H. GOLUB AND C. VAN LOAN, *Matrix Computations*, The John Hopkins University Press, Baltimore, 1996.

[150] G. H. GOLUB AND M. L. OVERTON, *The convergence of inexact Chebyshev and Richardson iterative methods for solving linear systems*, Numerische Mathematik, 53 (1988), pp. 571–593.

[151] G. H. GOLUB AND R. S. VARGA, *Chebyshev semi iterative methods, successive overrelaxation iterative methods and second order Richardson iterative methods*, Numerische Mathematik, 3 (1961), pp. 147–168.

[152] G. H. GOLUB AND A. J. WATHEN, *An iteration for indefinite systems and its application to the Navier–Stokes equations*, SIAM Journal on Scientific Computing, 19 (1998), pp. 530–539.

[153] G. H. GOLUB AND Q. YE, *Inexact preconditioned conjugate gradient method with inner-outer iterations*, SIAM Journal on Scientific Computing, 21 (1999), pp. 1305–1320.

[154] A. GREENBAUM, *Iterative Methods for Solving Linear Systems*, SIAM, Philadelpha, 1997.

[155] A. GREENBAUM, C. LI, AND H. Z. CHAO, *Parallelizing preconditioned conjugate gradient algorithms*, Computer Physics Communications, 53 (1989), pp. 295–309.

[156] W. GROPP, E. LUSK, AND A. SKJELLUM, *Using MPI: Portable Parallel Programming with the Message-Passing Interface*, MIT Press, Cambridge, MA, 1994.

[157] M. GROTE AND T. HUCKLE, *Effective parallel preconditioning with sparse approximate inverses*, in Proceedings of the Seventh SIAM Conference on Parallel Processing for Scientific Computing, D. H. Bailey et al., eds., SIAM, Philadelphia, 1995, pp. 466–471.

[158] M. GROTE AND H. D. SIMON, *Parallel preconditioning and approximate inverses on the connection machine*, in Parallel Processing for Scientific Computing Volume 2, R. F. Sincovec, D. E. Keyes, L. R. Petzold, and D. A. Reed, eds., SIAM, Philadelphia, 1993, pp. 519–523.

[159] M. J. GROTE AND T. HUCKLE, *Parallel preconditioning with sparse approximate inverses*, SIAM Journal on Scientific Computing, 18 (1997), pp. 838–853.

[160] M. H. GUTKNECHT, *A completed theory of the unsymmetric Lanczos process and related algorithms*. Part I, SIAM Journal on Matrix Analysis and Applications, 13 (1992), pp. 594–639.

[161] M. H. GUTKNECHT, *A completed theory of the unsymmetric Lanczos process and related algorithms*. Part II, SIAM Journal on Matrix Analysis and Applications, 15 (1994), pp. 15–58.

[162] W. HACKBUSCH, *Multi-Grid Methods and Applications*, Volume 4, Springer Series in Computational Mathematics, Springer-Verlag, Berlin, 1985.

[163] W. HACKBUSCH, *Iterative Solution of Large Linear Systems of Equations*, Springer-Verlag, New York, 1994.

[164] P. R. HALMOS, *Finite-Dimensional Vector Spaces*, Springer-Verlag, New York, 1958.

[165] S. HAMMOND AND R. SCHREIBER, *Efficient ICCG on a shared memory multiprocessor*, Tech. Rep. 89. 24, RIACS, NASA Ames Research Center, Moffett Field, CA, 1989.

[166] B. HENDRICKSON AND R. LELAND, *An improved spectral graph partitioning algorithm for mapping parallel computations*, Tech. Rep. SAND92-1460, UC-405, Sandia National Laboratories, Albuquerque, NM, 1992.

[167] M. R. HESTENES AND E. L. STIEFEL, *Methods of conjugate gradients for solving linear systems*, Journal of Research of the National Bureau of Standards, Section B, 49 (1952), pp. 409–436.

[168] C. HIRSCH, *Numerical Computation of Internal and External Flows*, John Wiley and Sons, New York, 1988.

[169] R. W. HOCKNEY, *A fast direct solution of Poisson's equation using Fourier analysis*, Journal of the Association for Computing Machinery, 12 (1965), pp. 95–113.

[170] R. W. HOCKNEY, *The potential calculation and some applications*, Methods of Computational Physics, 9 (1970), pp. 135–211.

[171] R. A. HORN AND C. R. JOHNSON, *Matrix Analysis*, Cambridge University Press, Cambridge, 1985.

[172] A. S. HOUSEHOLDER, *Theory of Matrices in Numerical Analysis*, Blaisdell Publishing Company, Johnson, CO, 1964.

[173] T. J. R. HUGHES, R. M. FERENCZ, AND J. O. HALLQUIST, *Large-scale vectorized implicit calculations in solid mechanics on a Cray X-MP/48 utilizing EBE precon-*

ditioned conjugate gradients, Computer Methods in Applied Mechanics and Engineering, 61 (1987), pp. 215–248.

[174] K. JBILOU, *Projection minimization methods for nonsymmetric linear systems*, Linear Algebra and Its Applications, 229 (1995), pp. 101–125.

[175] K. JBILOU, A. MESSAOUDI, AND H. SADOK, *Global FOM and GMRES algorithms for matrix equations*, Applied Numerical Mathematics, 31 (1999), pp. 49–63.

[176] K. JBILOU AND H. SADOK, *Analysis of some vector extrapolation methods for solving systems of linear equations*, Numerische Mathematik, (1995), pp. 73–89.

[177] K. C. JEA AND D. M. YOUNG, *Generalized conjugate gradient acceleration of nonsymmetrizable iterative methods*, Linear Algebra and Its Applications, 34 (1980), pp. 159–194.

[178] C. JOHNSON, *Numerical Solutions of Partial Differential Equations by the Finite Element Method*, Cambridge University Press, Cambridge, U.K., 1987.

[179] O. G. JOHNSON, C. A. MICCHELLI, AND G. PAUL, *Polynomial preconditioners for conjugate gradient calculations*, SIAM Journal on Numerical Analysis, 20 (1983), pp. 362–376.

[180] S. KACZMARZ, *Angenäherte auflösung von systemen linearer gleichungen*, Bulletin international de l'Académie polonaise des Sciences et Lettres, IIIA (1937), pp. 355–357.

[181] C. KAMATH AND A. SAMEH, *A projection method for solving nonsymmetric linear systems on multiprocessors*, Parallel Computing, 9 (1988), pp. 291–312.

[182] R. M. KARP, *Reducibility among combinatorial problems*, in Complexity of Computer Computations, R. E. Miller and J. W. Thatcher, eds., Plenum Press, New York, 1972, pp. 85–104.

[183] T. I. KARUSH, N. K. MADSEN, AND G. H. RODRIGUE, *Matrix multiplication by diagonals on vector/parallel processors*, Tech. Rep. UCUD, Lawrence Livermore National Lab., Livermore, CA, 1975.

[184] G. KARYPIS AND V. KUMAR, *A fast and high-quality multilevel scheme for partitioning irregular graphs*, SIAM Journal on Scientific Computing, 20 (1998), pp. 359–392.

[185] C. T. KELLY, *Iterative Methods for Linear and Nonlinear Equations*, SIAM, Philadelphia, 1995.

[186] D. S. KERSHAW, *The incomplete Choleski conjugate gradient method for the iterative solution of systems of linear equations*, Journal of Computational Physics, 26 (1978), pp. 43–65.

[187] R. KETTLER, *Analysis and comparison of relaxation schemes in robust multigrid and preconditioned conjugate gradient methods*, in Multigrid methods: Proceedings of

Koln-Porz, November 1981, W. Hackbusch and U. Trottenberg, eds., Lecture Notes in Mathematics 1228, Springer-Verlag, Berlin, 1982, pp. 502–534.

[188] D. E. KEYES AND W. D. GROPP, *A comparison of domain decomposition techniques for elliptic partial differential equations and their parallel implementation*, SIAM Journal on Scientific and Statistical Computing, 8 (1987), pp. s166–s202.v

[189] D. E. KEYES, Y. SAAD, AND D. G. TRUHLAR, eds., *Domain-Based Parallelism and Problem Decomposition Methods in Computational Science and Engineering*, SIAM, Philadelphia, 1995 (conference proceedings).

[190] L. Y. KOLOTILINA AND A. Y. YEREMIN, *On a family of two-level preconditionings of the incomplete block factorization type*, Soviet Journal of Numerical Analysis and Mathematical Modeling, 1 (1986), pp. 293–320.

[191] M. A. KRASNOSELSKII, G. M. VAINNIKO, P. P. ZABREIKO, YA. B. RUTITSKII, AND V. YA. STETSENKO, *Approximate Solutions of Operator Equations*, Wolters-Nordhoff, Gröningen, Netherlands, 1972.

[192] L. KRONSJÖ AND G. DAHLQUIST, *On the design of nested iterations for elliptic difference equations*, BIT, 11 (1971), pp. 63–71.

[193] V. KUMAR, A. GRAMA, A. GUPTA, AND G. KAPYRIS, *Parallel Computing*, 2nd edition, Benjamin Cummings, Redwood City, CA, 2003.

[194] C. LANCZOS, *An iteration method for the solution of the eigenvalue problem of linear differential and integral operators*, Journal of Research of the National Bureau of Standards, 45 (1950), pp. 255–282.

[195] C. LANCZOS, *Chebyshev polynomials in the solution of large-scale linear systems*, in Proceedings of the ACM, 1952, pp. 124–133.

[196] C. LANCZOS, *Solution of systems of linear equations by minimized iterations*, Journal of Research of the National Bureau of Standards, 49 (1952), pp. 33–53.

[197] P. LETALLEC, *Domain decomposition methods in computational mechanics*, Computational Mechanics Advances, 1 (1994), pp. 121–220.

[198] R. LEUZE, *Independent set orderings for parallel matrix factorizations by Gaussian elimination*, Parallel Computing, 10 (1989), pp. 177–191.

[199] J. G. LEWIS, B. W. PEYTON, AND A. POTHEN, *A fast algorithm for reordering sparse matrices for parallel factorizations*, SIAM Journal on Scientific and Statistical Computing, 6 (1989), pp. 1146–1173.

[200] J. G. LEWIS AND H. D. SIMON, *The impact of hardware gather-scatter on sparse Gaussian elimination*, SIAM Journal on Scientific and Statistical Computing, 9 (1988), pp. 304–311.

[201] N. LI, Y. SAAD, AND E. CHOW, *Crout versions of ILU for general sparse matrices*, Tech. Rep. umsi-2002-021, Minnesota Supercomputer Institute, University of Minnesota, Minneapolis, MN, 2002.

[202] Z. LI, Y. SAAD, AND M. SOSONKINA, *pARMS: a parallel version of the algebraic recursive multilevel solver*, Tech. Rep. umsi-2001-100, Minnesota Supercomputer Institute, University of Minnesota, Minneapolis, MN, 2001.

[203] J. W.-H. LIU, *Modification of the minimum degree algorithm by multiple elimination*, ACM Transactions on Mathematical Software, 11 (1985), pp. 141–153.

[204] S. MA, *Parallel block preconditioned Krylov subspace methods for partial differential equations*, PhD thesis, Department of Computer Science, University of Minneapolis, Minneapolis, MN, 1993.

[205] T. A. MANTEUFFEL, *An incomplete factorization technique for positive definite linear systems*, Mathematics of Computation, 34 (1980), pp. 473–497.

[206] S. F. MCCORMICK, ed., *Multigrid Methods*, SIAM Philadelphia, 1987.

[207] J. A. MEIJERINK AND H. A. VAN DER VORST, *An iterative solution method for linear systems of which the coefficient matrix is a symmetric M-matrix*, Mathematics of Computation, 31 (1977), pp. 148–162.

[208] G. MEURANT, *Computer Solution of Large Linear Systems*, Volume 28, Studies in Mathematics and Its Application, North-Holland, Amsterdam, 1999.

[209] C. D. MEYER, *Matrix Analysis and Applied Linear Algebra*, SIAM, Philadelphia, 2000.

[210] G. L. MILLER, S. H. TENG, W. THURSTON, AND S. A. VAVASIS, *Automatic mesh partitioning*, in Graph Theory and Sparse Matrix Computation, A. George, J. Gilbert, and J. Liu, eds., Springer, New York, 1993, pp. 57–84.

[211] N. MUNKSGAARD, *Solving sparse symmetric sets of linear equations by preconditioned conjugate gradient method*, ACM Transactions on Mathematical Software, 6 (1980), pp. 206–219.

[212] N. M. NACHTIGAL, *A look-ahead variant of the Lanczos algorithm and its application to the quasi-minimal residual method for non-Hermitian linear systems*, PhD thesis, Applied Mathematics, Cambridge, 1991.

[213] Y. NOTAY, *Flexible conjugate gradients*, SIAM Journal on Scientific Computing, 22 (2000), p. 1444–1460.

[214] D. O'LEARY, *The block conjugate gradient algorithm and related methods*, Linear Algebra and Its Applications, 29 (1980), pp. 243–322.

[215] C. W. OOSTERLEE AND T. WASHIO, *An evaluation of parallel multigrid as a solver and a preconditioner for singularly perturbed problems*, SIAM Journal on Scientific and Statistical Computing, 19 (1998), pp. 87–110.

[216] J. ORTEGA, *Efficient implementations of certain iterative methods*, SIAM Journal on Scientific and Statistical Computing, 9 (1988), pp. 882–891.

[217] J. ORTEGA, *Orderings for conjugate gradient preconditionings*, SIAM Journal on Optimization, 11 (1991), pp. 565–582.

[218] J. M. ORTEGA, *Introduction to Parallel and Vector Solution of Linear Systems*, Plenum Press, New York, 1988.

[219] J. M. ORTEGA AND R. G. VOIGT, *Solution of partial differential equations on vector and parallel computers*, SIAM Review, 27 (1985), pp. 149–240.

[220] O. OSTERBY AND Z. ZLATEV, *Direct Methods for Sparse Matrices*, Springer-Verlag, New York, 1983.

[221] C. C. PAIGE, *Computational variants of the Lanczos method for the eigenproblem*, Journal of the Institute of Mathematics and Its Applications, 10 (1972), pp. 373–381.

[222] C. C. PAIGE AND M. A. SAUNDERS, *Solution of sparse indefinite systems of linear equations*, SIAM Journal on Numerical Analysis, 12 (1975), pp. 617–629.

[223] B. N. PARLETT, *The Symmetric Eigenvalue Problem*, Prentice-Hall, Englewood Cliffs, NJ, 1980.

[224] B. N. PARLETT, D. R. TAYLOR, AND Z. S. LIU, *A look-ahead Lanczos algorithm for nonsymmetric matrices*, Mathematics of Computation, 44 (1985), pp. 105–124.

[225] D. PEACEMAN AND H. RACHFORD, *The numerical solution of elliptic and parabolic differential equations*, Journal of SIAM, 3 (1955), pp. 28–41.

[226] S. PISSANETZKY, *Sparse Matrix Technology*, Academic Press, New York, 1984.

[227] E. L. POOLE AND J. M. ORTEGA, *Multicolor ICCG methods for vector computers*, SIAM Journal on Numerical Analysis, 24 (1987), pp. 1394–1418.

[228] A. POTHEN, H. D. SIMON, AND K. P. LIOU, *Partitioning sparse matrices with eigenvectors of graphs*, SIAM Journal on Matrix Analysis and Applications, 11 (1990), pp. 430–452.

[229] J. S. PRZEMIENIECKI, *Matrix structural analysis of substructures*, American Institute of Aeronautics and Astronautics Journal, 1 (1963), pp. 138–147.

[230] J. K. REID, *On the method of conjugate gradients for the solution of large sparse systems of linear equations*, in Large Sparse Sets of Linear Equations, J. K. Reid, ed., Academic Press, 1971, pp. 231–254.

[231] T. J. RIVLIN, *The Chebyshev Polynomials: From Approximation Theory to Algebra and Number Theory*, J. Wiley and Sons, New York, 1990.

[232] D. J. ROSE AND R. E. TARJAN, *Algorithmic aspects of vertex elimination on directed graphs*, SIAM Journal on Applied Mathematics, 34 (1978), pp. 176–197.

[233] F. X. ROUX, *Acceleration of the outer conjugate gradient by reorthogonalization for a domain decomposition method with Lagrange multiplier*, in Proceedings of the Third International Symposium on Domain Decomposition Methods, Houston, March 1989, T. Chan et al., eds., SIAM, Philadelphia, 1990, pp. 314–321.

[234] A. RUGE AND K. STÜBEN, *Algebraic multigrid*, in Multigrid Methods, S. McCormick, ed., Frontiers in Applied Mathematics, Volume 3, SIAM, Philadelphia, 1987, Ch. 4.

[235] A. RUHE, *Implementation aspects of band Lanczos algorithms for computation of eigenvalues of large sparse symmetric matrices*, Mathematics of Computation, 33 (1979), pp. 680–687.

[236] H. RUTISHAUSER, *Theory of gradient methods*, in Refined Iterative Methods for Computation of the Solution and the Eigenvalues of Self-Adjoint Boundary Value Problems, Basel-Stuttgart, Institute of Applied Mathematics, Birkhäuser Verlag, Zurich, 1959, pp. 24–49.

[237] Y. SAAD, *Krylov subspace methods for solving large unsymmetric linear systems*, Mathematics of Computation, 37 (1981), pp. 105–126.

[238] Y. SAAD, *The Lanczos biorthogonalization algorithm and other oblique projection methods for solving large unsymmetric systems*, SIAM Journal on Numerical Analysis, 19 (1982), pp. 485–506.

[239] Y. SAAD, *Iterative solution of indefinite symmetric linear systems by methods using orthogonal polynomials over two disjoint intervals*, SIAM Journal on Numerical Analysis, 20 (1983), pp. 784–811.

[240] Y. SAAD, *Practical use of polynomial preconditionings for the conjugate gradient method*, SIAM Journal on Scientific and Statistical Computing, 6 (1985), pp. 865–881.

[241] Y. SAAD, *Least squares polynomials in the complex plane and their use for solving sparse nonsymmetric linear systems*, SIAM Journal on Numerical Analysis, 24 (1987), pp. 155–169.

[242] Y. SAAD, *On the Lanczos method for solving symmetric linear systems with several right-hand sides*, Mathematics of Computation, 48 (1987), pp. 651–662.

[243] Y. SAAD, *Krylov subspace methods on supercomputers*, SIAM Journal on Scientific and Statistical Computing, 10 (1989), pp. 1200–1232.

[244] Y. SAAD, *SPARSKIT: A basic tool kit for sparse matrix computations*, Tech. Rep. RIACS-90-20, Research Institute for Advanced Computer Science, NASA Ames Research Center, Moffet Field, CA, 1990.

[245] Y. SAAD, *Numerical Methods for Large Eigenvalue Problems*, Halsted Press, New York, 1992.

[246] Y. SAAD, *A flexible inner-outer preconditioned GMRES algorithm*, SIAM Journal on Scientific and Statistical Computing, 14 (1993), pp. 461–469.

[247] Y. SAAD, *Highly parallel preconditioners for general sparse matrices*, in Recent Advances in Iterative Methods, IMA Volumes in Mathematics and Its Applications, Volume 60, G. Golub, M. Luskin, and A. Greenbaum, eds., Springer-Verlag, New York, 1994, pp. 165–199.

[248] Y. SAAD, *ILUT: a dual threshold incomplete ILU factorization*, Numerical Linear Algebra with Applications, 1 (1994), pp. 387–402.

[249] Y. SAAD, *Analysis of augmented Krylov subspace methods*, SIAM Journal on Matrix Analysis and Applications, 18 (1997), pp. 435–449.

[250] Y. SAAD AND M. H. SCHULTZ, *GMRES: a generalized minimal residual algorithm for solving nonsymmetric linear systems*, SIAM Journal on Scientific and Statistical Computing, 7 (1986), pp. 856–869.

[251] Y. SAAD AND M. H. SCHULTZ, *Parallel implementations of preconditioned conjugate gradient methods*, in Mathematical and Computational Methods in Seismic Exploration and Reservoir Modeling, W. E. Fitzgibbon, ed., SIAM, Philadelphia, 1986, pp. 108–127.

[252] Y. SAAD AND B. SUCHOMEL, *ARMS: An algebraic recursive multilevel solver for general sparse linear systems*, Numerical Linear Algebra with Applications, 9 (2002), pp. 359–378.

[253] Y. SAAD AND H. A. VAN DER VORST, *Iterative solution of linear systems in the 20th century*, Journal of Computational and Applied Mathematics, 123 (2000), pp. 1–33.

[254] Y. SAAD AND K. WU, *DQGMRES: a direct quasi-minimal residual algorithm based on incomplete orthogonalization*, Numerical Linear Algebra with Applications, 3 (1996), pp. 329–343.

[255] H. SADOK, *Méthodes de projection pour les systèmes linéaires et non linéaires*, PhD thesis, University of Lille 1, Lille, France, 1981.

[256] J. SALTZ, R. MIRCHANDANEY, AND K. CROWLEY, *Run-time paralellization and scheduling of loops*, IEEE Transactions on Computers, 40 (1991), pp. 603–612.

[257] J. H. SALTZ, *Automated problem scheduling and reduction of synchronization delay effects*, Tech. Rep. 87-22, ICASE, Hampton, VA, 1987.

[258] W. SCHONAUER, *The efficient solution of large linear systems, resulting from the FDM for 3-D PDE's, on vector computers*, in Proceedings of the 1st International Colloquium on Vector and Parallel Computing in Scientific Applications, Paris, March, 1983.

[259] W. SCHÖNAUER, *Scientific Computing on Vector Computers*, North-Holland, New York, 1987.

[260] H. A. SCHWARZ, *Gesammelte Mathematische Abhandlungen*, Volume 2, Springer-Verlag, Berlin, 1890, pp. 133–143. First published in Vierteljahrsschrift der Naturforschenden Gesellschaft in Zürich, Volume 15, 1870, pp. 272–286.

[261] F. SHAKIB, *Finite element analysis of the compressible Euler and Navier Stokes equations*, PhD thesis, Department of Aeronautics, Stanford University, Stanford, CA, 1989.

[262] A. SIDI, *Extrapolation vs. projection methods for linear systems of equations*, Journal of Computational and Applied Mathematics, 22 (1988), pp. 71–88.

[263] D. SILVESTER AND A. WATHEN, *Fast iterative solution of stabilized Stokes problems, Part II: Using general block preconditioners*, SIAM Journal on Numerical Analysis, 30 (1994), pp. 1352–1367.

[264] H. D. SIMON, *Incomplete LU preconditioners for conjugate gradient type iterative methods*, in Proceedings of the SPE 1985 Reservoir Simulation Symposium, Dallas, TX, 1988, Society of Petroleum Engineers of AIME, pp. 302–306. Paper number 13533.

[265] V. SIMONCINI AND E. GALLOPOULOS, *Convergence properties of block GMRES and matrix polynomials*, Linear Algebra and Its Applications, 247 (1996), pp. 97–120.

[266] V. SIMONCINI AND E. GALLOPOULOS, *An iterative method for nonsymmetric systems with multiple right-hand sides*, SIAM Journal on Scientific Computing, 16 (1995), pp. 917–933.

[267] B. SMITH, P. BJØRSTAD, AND W. GROPP, *Domain Decomposition: Parallel Multilevel Methods for Elliptic Partial Differential Equations*, Cambridge University Press, New York, 1996.

[268] B. F. SMITH, *An optimal domain decomposition preconditioner for the finite element solution of linear elasticity problems*, SIAM Journal on Scientific and Statistical Computing, 13 (1992), pp. 364–378.

[269] D. A. SMITH, W. F. FORD, AND A. SIDI, *Extrapolation methods for vector sequences*, SIAM Review, 29 (1987), pp. 199–233.

[270] D. C. SMOLARSKI AND P. E. SAYLOR, *An optimum iterative method for solving any linear system with a square matrix*, BIT, 28 (1988), pp. 163–178.

[271] P. SONNEVELD, *CGS, A fast Lanczos-type solver for nonsymmetric linear systems*, SIAM Journal on Scientific and Statistical Computing, 10 (1989), pp. 36–52.

[272] G. W. STEWART, *Introduction to Matrix Computations*, Academic Press, New York, 1973.

[273] G. W. STEWART, *Matrix Algorithms*, Volumes 1 to 5, SIAM, Philadelphia, 2002.

[274] W. J. STEWART, *Introduction to the Numerical Solution of Markov Chains*, Princeton University Press, Princeton, NJ, 1994.

[275] E. L. STIEFEL, *Kernel polynomials in linear algebra and their applications*, U.S. National Bureau of Standards, Applied Mathematics Series, 49 (1958), pp. 1–24.

[276] G. STRANG AND G. J. FIX, *An Analysis of the Finite Element Method*, Prentice-Hall, Englewood Cliffs, NJ, 1973.

[277] K. STUBEN AND U. TROTTENBERG, *Multi-grid methods: Fundamental algorithms, model problem analysis and applications*, in Multigrid Methods, Lecture Notes in Mathematics, Volume 960, W. Hackbusch and U. Trottenberg, eds., Springer-Verlag, Berlin, 1982, pp. 1–176.

[278] P. N. SWARTZRAUBER, *A direct method for the discrete solution of separable elliptic equations*, SIAM Journal on Numerical Analysis, 11 (1974), pp. 1136–1150.

[279] P. N. SWARTZRAUBER, *The methods of cyclic reduction, Fourier analysis, and the FACR algorithm for the discrete solution of Poisson's equation on a rectangle*, SIAM Review, 19 (1977), pp. 490–501.

[280] R. SWEET, *A parallel and vector variant of the cyclic reduction algorithm*, Supercomputer, 22 (1987), pp. 18–25.

[281] D. B. SZYLD AND J. A. VOGEL, *FQMR: A flexible quasi-minimal residual method with inexact preconditioning*, SIAM Journal on Scientific Computing, 23 (2001), pp. 363–380.

[282] P. L. TALLEC, Y.-H. D. ROECK, AND M. VIDRASCU, *Domain-decomposition methods for large linearly elliptic three dimensional problems*, Journal of Computational and Applied Mathematics, 34 (1991).

[283] D. TAYLOR, *Analysis of the look-ahead Lanczos algorithm*, PhD thesis, Department of Computer Science, University of California at Berkeley, Berkeley, CA, 1983.

[284] L. N. TREFETHEN, *Approximation theory and numerical linear algebra*, Tech. Rep. Numerical Analysis Report 88-7, Massachussetts Institute of Technology, Cambridge, MA, 1988.

[285] U. TROTTENBERG, C. OOSTERLEE, AND A. SCHÜLLER, *Multigrid*, Academic Press, New York, 2001.

[286] R. UNDERWOOD, *An iterative block Lanczos method for the solution of large sparse symmetric eigenproblems*, Tech. Rep. Stan-CS74-469, Stanford University, Stanford, CA, 1975.

[287] H. A. VAN DER VORST, *The performance of FORTRAN implementations for preconditioned conjugate gradient methods on vector computers*, Parallel Computing, 3 (1986), pp. 49–58.

[288] H. A. VAN DER VORST, *Large tridiagonal and block tridiagonal linear systems on vector and parallel computers*, Parallel Computing, 5 (1987), pp. 303–311.

[289] H. A. VAN DER VORST, *Bi-CGSTAB: A fast and smoothly converging variant of Bi-CG for the solution of non-symmetric linear systems*, SIAM Journal on Scientific and Statistical Computing, 13 (1992), pp. 631–644.

[290] H. A. VAN DER VORST AND C. VUIK, *GMRESR: A family of nested GMRES methods*, Numerical Linear Algebra with Applications, 1 (1994), pp. 369–386.

[291] R. S. VARGA, *Factorizations and normalized iterative methods*, in Boundary Problems in Differential Equations, R. E. Langer, ed., University of Wisconsin Press, Madison, WI, 1960, pp. 121–142.

[292] R. S. VARGA, *Matrix Iterative Analysis*, Prentice Hall, Englewood Cliffs, NJ, 1962.

[293] V. VENKATAKRISHNAN, *Preconditioned conjugate gradient methods for the compressible Navier Stokes equations*, AIAA Journal, 29 (1991), pp. 1092–1100.

[294] V. VENKATAKRISHNAN, *Parallel implicit methods for aerodynamic applications on unstructured grids*, in Domain-Based Parallelism and Problem Decomposition

Methods in Computational Science and Engineering, D. E. Keyes, Y. Saad, and D. G. Truhlar, eds., SIAM, Philadelphia, 1995, pp. 57–74.

[295] V. VENKATAKRISHNAN AND D. J. MAVRIPLIS, *Implicit solvers for unstructured meshes*, Journal of Computational Physics, 105 (1993), pp. 83–91.

[296] V. VENKATAKRISHNAN, H. D. SIMON, AND T. J. BARTH, *An MIMD implementation of a parallel Euler solver for unstructured grids*, The Journal of Supercomputing, 6 (1992), pp. 117–137.

[297] P. K. W. VINSOME, *ORTHOMIN, an iterative method for solving sparse sets of simultaneous linear equations*, in Proceedings of the Fourth Symposium on Reservoir Simulation, Society of Petroleum Engineers of AIME, 1976, pp. 149–159.

[298] V. V. VOEVODIN, *The problem of a non-selfadjoint generalization of the conjugate gradient method has been closed*, U.S.S.R. Computational Mathematics and Mathematical Physics, 23 (1983), pp. 143–144.

[299] E. L. WACHSPRESS, *Iterative Solution of Elliptic Systems and Applications to the Neutron Equations of Reactor Physics*, Prentice-Hall, Englewood Cliffs, NJ, 1966.

[300] C. WAGNER, *Introduction to Algebraic Multigrid*—Course notes of an algebraic multigrid course at the University of Heidelberg in the winter semester 1998/99.

[301] C. WAGNER, W. KINZELBACH, AND G. WITTUM, *Schur-complement multigrid, a robust method for groundwater flow and transport problems*, Numerische Mathamatik, 75 (1997), pp. 523–545.

[302] H. F. WALKER, *Implementation of the GMRES method using Householder transformations*, SIAM Journal on Scientific Computing, 9 (1988), pp. 152–163.

[303] X. WANG, K. GALLIVAN, AND R. BRAMLEY, *CIMGS: An incomplete orthogonal factorization preconditioner*, SIAM Journal on Scientific Computing, 18 (1997), pp. 516–536.

[304] A. WATHEN AND D. SILVESTER, *Fast iterative solution of stabilized Stokes problems, Part I: Using simple diagonal preconditioners*, SIAM Journal on Numerical Analysis, 30 (1993), pp. 630–649.

[305] J. W. WATTS III, *A conjugate gradient truncated direct method for the iterative solution of the reservoir simulation pressure equation*, Society of Petroleum Engineers Journal, 21 (1981), pp. 345–353.

[306] R. WEISS, *Convergence behavior of generalized conjugate gradient methods*, PhD thesis, Karlsruhe, Germany, 1990.

[307] R. WEISS, *A theoretical overview of Krylov subspace methods*, Applied Numerical Mathematics, 19 (1995), pp. 33–56.

[308] P. WESSELING, *A robust and efficient multigrid method*, in Multigrid Methods: Proceedings of Koln-Porz, November 1981, W. Hackbusch and U. Trottenberg, eds., Lecture Notes in Mathematics 1228, Springer-Verlag, Berlin, 1982, pp. 614–630.

[309] P. WESSELING, *Theoretical and practical aspects of a multigrid method*, SIAM Journal of Scientific and Statistical Computing, 3 (1982), pp. 387–407.

[310] P. WESSELING, *An Introduction to Multigrid Methods*, J. Wiley and Sons, Chichester, U.K., 1992.

[311] P. WESSELING AND P. SONNEVELD, *Numerical experiments with a multiple-grid and a preconditioned Lanczos-type method*, in Approximation Methods for Navier-Stokes Problems, Lecture Notes in Mathematics, Volume 771, Springer-Verlag, Berlin, Heidelberg, New York, 1980, pp. 543–562.

[312] O. WIDLUND, *A Lanczos method for a class of non-symmetric systems of linear equations*, SIAM Journal on Numerical Analysis, 15 (1978), pp. 801–812.

[313] L. B. WIGTON, *Application of MACSYMA and sparse matrix technology to multi-element airfoil calculations*, in Proceedings of the AIAA-87 conference, Honolulu, Hawai, 1987, AIAA, New York, 1987, pp. 444–457. Paper number AIAA-87-1142-CP.

[314] P. WILDERS AND E. BRAKKEE, *Schwarz and Schur: an algebraic note on equivalence properties*, SIAM Journal on Scientific Computing, 20 (1999), pp. 2297–2303.

[315] B. WILKINSON AND C. M. ALLEN, *Parallel Programming: Techniques and Applications Using Networked Workstations and Parallel Computers*, Prentice-Hall, Englewood Cliffs, NJ, 1998.

[316] J. H. WILKINSON, *The Algebraic Eigenvalue Problem*, Clarendon Press, Oxford, U.K., 1965.

[317] O. WING AND J. W. HUANG, *A computation model of parallel solutions of linear equations*, IEEE Transactions on Computers, C-29 (1980), pp. 632–638.

[318] C. H. WU, *A multicolor SOR method for the finite-element method*, Journal of Computational and Applied Mathematics, 30 (1990), pp. 283–294.

[319] J. XU, *Iterative methods by space decomposition and subspace correction*, SIAM Review, 34 (1992), pp. 581–613.

[320] Q. YE, *A breakdown-free variation of the Lanczos algorithm*, Mathematics of Computation, 62 (1994), pp. 179–207.

[321] D. M. YOUNG, *Iterative Solution of Large Linear Systems*, Academic Press, New York, 1971.

[322] D. P. YOUNG, R. G. MELVIN, F. T. JOHNSON, J. E. BUSSOLETTI, L. B. WIGTON, AND S. S. SAMANT, *Application of sparse matrix solvers as effective preconditioners*, SIAM Journal on Scientific and Statistical Computing, 10 (1989), pp. 1186–1199.

[323] L. ZHOU AND H. F. WALKER, *Residual smoothing techniques for iterative methods*, SIAM Journal on Scientific Computing, 15 (1994), pp. 297–312.

[324] Z. ZLATEV, *Use of iterative refinement in the solution of sparse linear systems*, SIAM Journal on Numerical Analysis, 19 (1982), pp. 381–399.

Index

A-norm, 31, 133, 204
Adams, L. M., 101, 404
additive projection procedure, 143
ADI, 124
    Peaceman–Rachford algorithm, 124
adjacency graph, 75
    of PDE matrices, 76
adjoint of a matrix, 6
AINV, 331
algebraic multigrid, 437–445
    coarsening, 437
algebraic multiplicity, 14
Aliga, J. I., 243
Allen, M. C., 375
alternating direction implicit, *see* ADI
AMG, *see* algebraic multigrid
Anderson, E., 375
angle between a vector and a subspace, 137
anisotropic medium, 48
approximate inverse, *see* AINV
approximate inverse preconditioners, 320
    column-oriented, 323
    global iteration, 321
    for improving a preconditioner, 331
approximate inverse techniques, 399
Arnoldi, W. E., 153
Arnoldi's method, 153–164
    basic algorithm, 154
    breakdown of, 155
    with Householder orthogonalization, 156
    for linear systems, 159
    lucky breakdown, 156
    with modified Gram–Schmidt, 156
    practical implementation, 156
Arrow–Hurwicz's algorithm, 256
assembled matrix, 65
assembly process, 64
Axelsson, O., 72, 216, 280, 349, 350

Bakhvalov, N. S., 449
banded matrices, 5
bandwidth
    of a bus, 356
    of a matrix, 5
Bank, R., 449
Barker, V. A., 72, 350
Barth, T. J., 243
basis of a subspace, 9
BCG, 222–226
    algorithm, 223
    transpose-free variants, 228–240
BCR, *see* block cyclic reduction
Benantar, M., 101
Benzi, M., 331, 337, 350
BFS, *see* breadth first search
BICGSTAB, 231
biconjugate gradient, *see* BCG
bidiagonal matrices, 5
bilinear form, 61
binary search trees, 317
biorthogonal bases, 34
biorthogonal vectors, 34, 218
biorthogonalization, 217
bipartite graph, 87, 119, 335
bipartite matching, 335
bipartite transversal, 335
Birkhoff, G., 126
Björk, A., 259
Bjørstad, P., 492
block Arnoldi
    algorithm, 208
    Ruhe's variant, 209
block cyclic reduction, 57
    Buneman's variant, 58
block diagonal matrices, 5
block FOM, 210–211
block Gaussian elimination, 453–457
    algorithm, 456

block GMRES, 210–212
    multiple right-hand sides, 211
block Gram–Schmidt, 208
block independent sets, 443
block Jacobi, 110
    as a preconditioner, 378
block Krylov subspace methods, 152, 208–212
block preconditioners, 337
block relaxation, 106
block tridiagonal matrices, 5, 337
    preconditioning, 337
Bodewig, E., 349
Boley, D. L., 243
Bollhöfer, M., 350
bottleneck transversal, 336
boundary conditions, 46, 47
    Dirichlet, 47
    mixed, 47
    Neumann, 47
Brakhage, H., 449
Brakkee, W., 493
Bramble, J. H., 493
Bramley, R., 259
Brandt, A., 449
breadth first search, 81
Brezinski, C., 243, 244
Brinkkemper, S., 350
Brown, P. N., 178, 181, 216, 228
Buleev, N. I., 350
Buneman, O., 58
Buneman's algorithm, 58

cache memory, 357
canonical form, 14
    Jordan, 15
    Schur, 16
Cauchy–Schwarz inequality, 6, 7
Cayley–Hamilton theorem, 152
cell-centered scheme, 69
cell-vertex scheme, 69
centered difference approximation, 49
centered difference formula, 49
centerpoint, 485
Cesari, L., 349, 404
CG algorithm, *see* conjugate gradient algorithm
CG for normal equations, 252, 253

CGNE, 253
    algorithm, 254
    optimality, 254
CGNR, 252
    algorithm, 252
    optimality, 252
CGS, 229–231
    algorithm, 231
CGW algorithm, 278
Chan, T. F., 243, 467, 480, 493
characteristic polynomial, 3
Chebyshev
    acceleration, 382
Chebyshev polynomials, 199–204, 206, 381–388
    complex, 200, 216
    and ellipses, 200
    optimality, 201–203
    for preconditioning, 381
    real, 199
Cheney, C. C., 200
Chow, E., 316, 350
Ciarlet, P. G., 72
Cimmino, G., 259
Cimmino's method, 249
circuit switching, 357
coarse-grain, 378
coarse-grid correction, 425
coarsening, 437, 442
coefficient matrix, 103
coloring vertices, 87
column reordering, 78
complex GMRES, 184
compressed sparse column storage, *see* CSC
compressed sparse row storage, *see* CSR
Concus, Golub, and Widlund algorithm, *see* CGW algorithm
Concus, P., 278, 350
condition number, 38
    for normal equations systems, 246
condition numbers and CG, 192
conjugate gradient algorithm, 187–194
    algorithm, 190
    alternative formulations, 191
    convergence, 203, 204
    derivation, 187, 190
    eigenvalue estimates, 192
    for the normal equations, 251
    preconditioned, 262
conjugate gradient squared, *see* CGS

Index

conjugate residual algorithm, 194
consistent matrix norms, 8
consistent orderings, 119–123
control volume, 68
convection diffusion equation, 48
convergence
    factor, 113
        general, 113
        specific, 113
    of GMRES, 205
    of the minimal residual method, 141
    rate, 113
    of relaxation methods, 112
    of Schwarz procedures, 473
coordinate storage format, 89
Cormen, T. H., 317
Cosgrove, J. D. F., 350
Courant characterization, 25
Craig's method, 254
CSC storage format, 90
    matvecs in, 362
CSR storage format, 90, 291
    matvecs in, 362
Cullum, J., 179
Cuthill–McKee ordering, 81
    queue implementation, 82

Dahlquist, G., 449
data coherence, 357
data-parallel, 355
Datta, B. N., 43
Davis, T. A., 405
defective eigenvalue, 15
Delong, M. A., 405
Demmel, J., 43
derogatory, 15
determinant, 3
Deuflhard, P., 244
diagonal
    compensation, 305
    dominance, 116, 117
    form of matrices, 15
    matrices, 4
diagonal storage format, 91, 364
    matvecs in, 364
diagonalizable matrix, 15
diagonally dominant matrix, 117
diagonally structured matrices, 91
diameter of a graph, 299, 487
diameter of a triangle, 62

Diaz, J. C., 350
difference equation, 464
DIOM, 161–164, 188
    algorithm, 163
direct IOM, *see* DIOM
direct sum of subspaces, 9, 32
directed graph, 75
Dirichlet boundary conditions, 46, 47
distributed
    computing, 355
    ILU, 396
    memory, 357
    sparse matrices, 366, 396
divergence of a vector, 47
divergence operator, 47
domain decomposition
    convergence, 473
    and direct solution, 456
    full matrix methods, 481
    induced preconditioners, 478
    Schur complement approaches, 477
    Schwarz alternating procedure, 465
domain sweep, 467
double orthogonalization, 156
double-striping, 488
DQGMRES, 172–177, 181, 276
    algorithm, 174
Dryja, M., 477
Duff, I. S., 96, 101, 350, 405

EBE preconditioner, 399
EBE regularization, 401
edge cuts, 486
edge in a graph, 75
eigenspace, 10
eigenvalues, 2
    from CG iteration, 192
    definition, 3
    index, 16
    of an orthogonal projector, 36
eigenvector, 3
    left, 4
    right, 4
Eisenstat's implementation, 266, 280
Eisenstat's trick, *see* Eisenstat's implementation
Eisenstat, S. C., 280, 319
element-by-element preconditioner, *see* EBE preconditioner
Elfving, T., 259

elliptic operators, 46
Ellpack-Itpack storage format, 91, 364
    matvecs in, 364
Elman, H. C., 350, 405
energy norm, 31, 133, 252, 254
Erisman, A. M., 101
error projection methods, 135
Euclidean inner product, 6
Euclidean norm, 7

Faber, V., 198, 216
Faber–Manteuffel theorem, 196
FACR, 59
factored approximate inverse, 329
Fan, K., 288
fan-in
    in multigrid, 422
fan-out
    in multigrid, 422
Farhat, C., 493
fast Poisson solvers, 48, 55–59, 452
    block cyclic reduction, 57
    Buneman's algorithm, 58
    FACR, 59
    FFT based, 56, 57
Fedorenko, R. P., 449
FFT, 57, 465
FFT solvers, 55
FGMRES, 273–275
    algorithm, 273
fictitious domain methods, 455
Fiedler vector, 486
field of values, 22
fill factor, 334
fill-in elements, 293
fill path, 298
fine-grain algorithms, 378
finite difference scheme, 48
    for 1-D problems, 51
    for 2-D problems, 54
    for the Laplacian, 50
    upwind schemes, 52
finite element method, 45, 60
finite volume method, 68
Fischer, B., 203, 216
Fix, G. J., 72
Flaherty, J. E., 101
Fletcher, R., 222, 243
flexible GMRES, *see* FGMRES
flexible iteration, 273

flux vector, 68
FMG, 424, 429
FOM, 159
    algorithm, 159
    with restarting, 160
Foster, I. T., 375
FPS, *see* fast Poisson solvers
Freund, R. W., 176, 203, 216, 234, 243, 244
Frobenius norm, 8
frontal methods, 65, 400
full matrix methods, 481–483
full multigrid, 424, 429
full orthogonalization method, *see* FOM
full weighting, *see* FW
FW, 421, 422

Galerkin conditions, 131
Galerkin projection, 423
    in multigrid, 423
Gallopoulos, E., 216, 243
Gastinel, N., 149, 216
Gastinel's method, 146
gather operation, 362
Gauss–Seidel iteration, 103
    backward, 105
    for normal equations, 247
    in parallel, 402
    symmetric, 105
Gaussian elimination, 65, 188, 288–292, 297,
        303, 305–307, 392, 393, 452
    block, 453
    frontal methods, 400
    IKJ variant, 290
    in IOM and DIOM, 164
    in Lanczos process, 189
    parallel, 479
    parallelism in, 76
    reordering in, 80
    in skyline format, 314
    sparse, 75
Gautschi, W., 187
GCR, 194–196
generalized conjugate residual, *see* GCR
geometric multiplicity, 15
George, A., 84, 94, 101, 299, 487
Gershgorin discs, 117
Gershgorin's theorem, 117
Gilbert, J. R., 317
Gilbert–Peierls algorithm, 317
global iteration, 321–323, 328

global reduction operations, 361
GMRES, 164–177, 196, 197, 205–207
    algorithm, 165
    block algorithm, 211
    breakdown, 171
    complex version, 184
    convergence, 205
    flexible variant, 268, 273–275
    Householder version, 165
    lucky breakdown, 171
    with polynomial preconditioning, 387
    practical implementation, 167
    relation with FOM, 177, 181
    via residual smoothing, 183
    with restarting, 171
    stagnation, 172
    truncated, 172
Godunov, S. K., 216
Golub, G. H., 43, 278, 281, 350, 351
Goovaerts, D., 467, 493
grade of a vector, 152
Gram–Schmidt algorithm, 10–11, 342
    block, 208
    cancellations in, 156
    modified, 11
    standard, 10
graph, 75
    bipartite, 87, 335
    coloring, 87, 474
    diameter, 299
    directed, 75
    edges, 75
    Laplacian of a, 486
    partitioning, 451, 483
        geometric, 484
        graph theory techniques, 487
        spectral techniques, 486
        type, 453
    separator, 82
    undirected, 75
    vertices, 75
graph separator, 82
Greenbaum, A., 179, 375
grid transfer, 414
Griewank, A., 350
Gropp, W. D., 359, 493
Grote, M., 350
group independent sets, 443
Gutknecht, M. H., 243

Hackbusch, W., 449, 493
Halmos, P. R., 16
Hankel matrix, 221
harmonic functions, 47
Harwell–Boeing collection, 94, 96
Hausdorff's convex hull theorem, 22
Haws, J. C., 337
heap-sort, in ILUT, 311
heaps, 317
Hendrickson, B., 493
Hermitian inner product, 5
Hermitian matrices, 4, 23
Hermitian positive definite, 30
Hessenberg matrices, 5
Hesteness, M. R., 215
high frequency, 413
high frequency modes, 407
high modes, 413
Hindmarsh, A. C., 216
Hirsch, C., 72
Hockney, R. W., 72
Hölder norms, 7
Horn, R. A., 23
Householder, A. S., 149
Householder algorithm, 11
Householder orthogonalization
    in Arnoldi's method, 156
Householder reflectors, 11
HPD, *see* Hermitian positive definite
Huckle, T., 350
hypercube, 358

idempotent, 9, 32
if and only if, 3
iff, *see* if and only if
Il'in, V. P., 350
ILQ
    factorization, 342
    preconditioning, 342
ILU, 287–316
    Crout variant, 316
    distributed, 396
    factorization, 287
        instability in, 313, 321
    general algorithm, 289
    IKJ version, 291
    ILUC, 316
    ILUS, 314–316
        algorithm, 316
    modified, 305–306

with multi-elimination, *see* ILUM
preconditioner, 287
    for Schur complement, 479
reordering, 333
static pattern, 292
with threshold, *see* ILUT and ILUTP
zero pattern, 289
ILU(0), 284, 287, 293–295
algorithm, 294
distributed factorization, 398
for distributed sparse matrices, 397
for red-black ordering, 390
ILU(1), 297
ILUM, 394, 442
ILUT, 306–313
algorithm, 307
analysis, 308
implementation, 310
with pivoting, *see* ILUTP
ILUTP, 312
for normal equations, 339
incomplete factorization, 284, 287
Gram–Schmidt, 342
ILQ, 342
QR, 342
incomplete Gram–Schmidt, 343
incomplete LQ, *see* ILQ
incomplete LU, *see* ILU
incomplete orthogonalization
algorithm, 161
incomplete orthogonalization method, *see* IOM
indefinite inner product, 220
independent set orderings, 84, 442
independent sets, 85, 392
maximal, 85
index of an eigenvalue, 16
indirect addressing, 75
induced norm, 8
induced preconditioners, 478
inhomogeneous medium, 48
injection, 421
inner products
    B inner product, 31
    indefinite, 220
invariant subspace, 10, 136
inverse LU factors, 329
IOM, 161
algorithm, 162
direct version, 161
irreducibility, 89
irreducible, 26

ISO, *see* independent set orderings
isometry, 7
iteration matrix, 110, 111

Jacobi iteration, 103
    for the normal equations, 249
JAD storage format, 365
definition, 365
in level scheduling, 372
matvecs in, 366
jagged diagonal format, *see* JAD storage format
jagged diagonals, 365
Jbilou, K., 216
Jea, K. C., 196, 216
Johnson, C., 72
Johnson, C. R., 23
Johnson, O. G., 385
Jordan block, 16
Jordan box, 16
Jordan canonical form, 15
Jordan, H., 101
Jordan submatrix, 16
Joukowski mapping, 200

Kaczmarz, S., 259
Kamath, C., 259
Karp, R. M., 85
Karush, T. I., 375, 404
Karypis, G., 493
kernel, 9
Kershaw, D. S., 351
Keyes, D. E., 493
Kolotilina, L. Y., 350
Koster, J., 350
Kranoselskii, M. A., 149
Kronsjö, L., 449
Krylov subspace, 152
    dimension of a, 152
    invariant, 153
    methods, 151, 217
    versus multigrid, 445
Kumar, V., 375, 493

Lanczos algorithm, 185, 186
algorithm, 186, 217
biorthogonalization, 217
breakdown, 219
    incurable, 220
    lucky, 220
    serious, 220
for linear systems, 221

Index

look-ahead version, 220
loss of orthogonality, 186
modified Gram–Schmidt version, 186
nonsymmetric, 217
and orthogonal polynomials, 186
partial reorthogonalization, 186
practical implementations, 220
selective reorthogonalization, 186
symmetric case, 185
Lanczos, C., 215, 222, 243, 404
Laplacian, *see* Laplacian operator
Laplacian operator, 47, 60
of a graph, 486
least-squares polynomials, 383
least-squares problem, 245
left eigenvector, 4
left versus right preconditioning, 272
Leisersen, C. E., 317
Leland, R., 493
LeTallec, P., 493
Leuze, R., 405
level of fill-in, 297
level scheduling, 370–372
for five-point matrices, 370
for general matrices, 370
level-set orderings, 81, 487
Lewis, J. G., 375, 405
line relaxation, 107
linear mappings, 2
linear span, 9
linear system, 37, 103
existence of a solution, 37
right-hand side of a, 37
singular, 37
unknown of a, 37
linked lists, 93
Liou, K. P., 493
Liu, J. W-H., 94, 299, 487
Liu, Z. S., 221
local Schur complement, 462
look-ahead Lanczos algorithm, 220
low frequency modes, 407
lower triangular matrices, 4
LQ factorization, 342
algorithm, 343
lucky breakdowns, 156
Lusk, E., 359

M-matrix, 25, 288, 338
Manteuffel, T., 198, 216, 243, 351
Marín, J., 350
mask, 348

matching, 335
Mathew, T. P., 493
matrix, 1
addition, 2
adjoint of a, 6
banded, 5
bidiagonal, 5
canonical forms, 14
characteristic polynomial, 3
diagonal, 4
diagonal dominant, 116
diagonal form, 15
diagonalizable, 15
Hermitian, 4, 20, 23
Hessenberg, 5
irreducible, 89
Jordan canonical form, 15
M-, 25
multiplication, 2
nonnegative, 4, 25
nonsingular, 3
norm of a, 7
normal, 4, 20
orthogonal, 4
outer product, 5
positive definite, 29–31
powers of a, 18
reduction, 14
Schur form, 16
self-adjoint, 6, 473
singular, 3
skew-Hermitian, 4
skew-symmetric, 4
spectral radius, 3
spectrum, 3
square, 2
symmetric, 4
symmetric positive definite, 30, 119
trace, 3
transpose, 2
transpose conjugate, 2
triangular, 4
tridiagonal, 5
unitary, 4
matrix-by-vector product, 361
dense matrices, 361
in diagonal format, 364
for distributed matrices, 368
in Ellpack format, 365
in triad form, 364
matrix norm, 8
maximum transversal, 335

McCormick, S. F., 449
Meijerink, J. A., 280, 349
mesh generation, 66
mesh refinement, 66
mesh size, 62
message passing, 358
Meurant, G. A., 350
Meyer, C. D., 43
MG, see multigrid
Micchelli, C. A., 385
Miller, G. L., 484, 485
MILU, 305–306
minimal residual iteration, 140
    algorithm, 140
    convergence, 141
minimal residual smoothing, 181–184, 228
minimum degree ordering, 93, 333
min-max theorem, 23
mixed boundary conditions, 46, 47
modified Gram–Schmidt, 156
modified ILU, see MILU
modified sparse row storage, see MSR
molecule, 49
moment matrix, 221
    in Lanczos procedure, 221
moment of intertia, 484
MR iteration, see minimal residual iteration
MRS, see minimal residual smoothing
MSR storage format, 90
multi-elimination, 392, 393
multicolor orderings, 87
multicoloring, 389–392
    for general sparse matrices, 391
multifrontal methods, 405
multigrid methods, 407–446
    algebraic multigrid, see AMG
    AMG, 437
    FMG, 429
    full multigrid, see FMG
    Galerkin projection, 423
    nested iteration, 424, 429
    V-cycle, 427
    W-cycle, 427
multinode expansion algorithm, 488
multiple eigenvalue, 14
multiple vector pipelines, 355
multiplicative projection process, 145
multiplicative Schwarz preconditioning, 470
multiprocessing, 355
Munksgaard, N., 350

Nachtigal, N. M., 176, 216, 243
natural ordering, 54
near singularity, 38
nested dissection, 333
nested dissection ordering, 93
nested iteration, 424, 429
Neumann boundary conditions, 46, 47
Neumann polynomials, 380
nonnegative matrix, 4, 25
nonsingular matrix, 3
norm
    A-norm, energy norm, 31
    energy norm, A-norm, 133
    Euclidean, 7
    Hölder, 7
    induced, 8
    of matrices, 7
    p-norm, 8
    of vectors, 5
normal derivative, 60
normal equations, 245
normal matrix, 4, 20
Notay, Y., 281
null space, 9
    of a projector, 32
numerical radius, 22

O'Leary, D., 216
oblique projection methods, 129, 133
oblique projector, 34
operator
    elliptic, 46
    Laplacian, 47
optimality of projection methods, 133
order relation for matrices, 25
Ortega, J., 280
Ortega, J. M., 101, 404
ORTHODIR, 194–196
orthogonal
    complement, 10
    matrix, 4
    projector, 10, 34
    vectors, 10
orthogonal bases, 10
orthogonality, 10
    between vectors, 10
    of a vector to a subspace, 10
ORTHOMIN, 194–196
orthonormal, 10
orthonormal bases, 10

Index

oscillatory modes, 407
Osterby, O., 101, 350
outer product matrices, 5
overdetermined systems, 245
overlapping domains, 455
overrelaxation, 105
Overton, M. L., 281

p-norm, 8
packet switching, 357
Paige, C. C., 189, 216
parallel architectures, 355
parallel sparse techniques, 77
parallelism, 354
    forms of, 354
Parlett, B. N., 156, 186, 220, 221
partial differential equations, 45
partial Schur decomposition, 17
partition, 108
partition vector, 486
partitioning, 453
Pasciak, J. E., 493
Paul, G., 385
PDE, *see* partial differential equations
PE, *see* processing element
Peaceman, D., 124, 126, 128
Peaceman–Rachford algorithm, 124
Peierls, T., 317
peripheral node, 487
permutation matrices, 5, 78
permutations, 77
Perron–Frobenius theorem, 26
perturbation analysis, 38
Petrov–Galerkin conditions, 129–131
Peyton, B. W., 405
physical mesh versus graph, 77
pipelining, 354
polynomial approximation, 152
polynomial preconditioning, 377, 379–388
Poole, E. L., 101, 405
positive definite matrix, 5, 24, 29–31
positive matrix, 25
positive real matrix, *see* positive definite matrix
positive semidefinite, 24
Pothen, A., 405, 493
preconditioned
    CG, 262
        efficient implementations, 265
        left, 263

    for the normal equations, 276
    parallel implementation, 359
    split, 264
    symmetry in, 262
    fixed-point iteration, 111
    GMRES, 267
        comparison, 271
        flexible variant, 272, 273
        left preconditioning, 268
        right preconditioning, 269
        split preconditioning, 270
preconditioner, 111
preconditioning, 110, 172, 261
    approximate inverse, 320
    block, 337
    EBE, 399
    by ILQ, 342
    incomplete LU, 287
    induced, 478
    Jacobi, 284
    left, 261
    for normal equations, 339
    polynomial, 379–388
        with Chebyshev polynomials, 381
        with least-squares polynomials, 383
        with Neumann polynomials, 380
    and relaxation scheme, 111
    right, 261
    SOR, 284
    split, 261
    SSOR, 284
probing, 480
processing element, 355
profile, 84
projection
    operator, *see* projector
    orthogonal to, 32
    parallel to, 32
projection methods, 129
    additive, 143
    approximate problem, 130
    definitions, 129
    error bounds, 135
    general, 130
    matrix representation, 131
    multiplicative, 145
    oblique, 129, 217
    one-dimensional, 137
    optimality, 133
    orthogonal, 129, 131

prototype, 131
residual, 134
theory, 132
projector, 9, 32–36, 109
  existence, 33
  matrix representation, 34
  oblique, 34
  orthogonal, 34
    eigenvalues, 36
    properties, 36
prolongation operator, 109, 468
property A, 119
Propkopov, G. P., 216
Przemieniecki, J. S., 492
pseudo-peripheral node, 487

QMR, 222–226
  algorithm, 225
  approximation, 225
  peaks and plateaus of residuals, 228
  from residual smoothing, 228
QMRS, *see* quasi-minimal residual smoothing
QR decomposition, 11
quasi-GMRES, 172–173
  algorithm, 172
  direct version, 172
  DQGMRES, 172
quasi-minimal residual, *see* QMR
quasi-minimal residual smoothing, 184, 228
quasi-residual norm, 174
quasi-residual smoothing, 228
quasi-Schur form, 17
quick-split, in ILUT, 311
quotient graph, 77

Rachford, H., 124, 126, 128
range, 2, 9
  of a projector, 32
rank, 9
  full, 10
rapid elliptic solvers, *see* fast Poisson solvers
Rayleigh quotient, 22, 23
RCM, 333
real Schur form, 17
recursive graph bisection, 487
red-black ordering, 389
Redivo Zaglia, M., 243
reduced system, 346, 456
reducible, 26
reduction of matrices, 14

reduction operations, 361
refinement, 66
reflectors, 11
regular splitting, 115
regularization, 257
Reid, J. K., 101, 216
relaxation methods
  block, 106
  convergence, 112
reordering, 78, 333
  for ILU, 333
reordering rows, columns, 77
reorthogonalization, 10
residual norm steepest descent, 142
residual projection methods, 134
residual smoothing, 181, 182
restarted FOM, 160
restriction, 421
restriction operator, 109, 468
reverse Cuthill–McKee ordering, *see* RCM
Richardson's iteration, 114, 250, 412
right-hand side, 37, 103
  multiple, 211
right versus left preconditioning, 272
Rivest, R. L., 317
Rivlin, T. J., 201
Rose–Tarjan theorem, 298
Roux, J. X., 493
row projection methods, 247, 401
  parallel, 401
row reordering, 78
row sum, 305
Ruge, A., 449
Ruhe, A., 209
Rutishauser, H., 404

saddle-point problems, 254
Sadok, H., 216, 243
Saltz, J., 375
Sameh, A., 259
Saunders, M. A., 189
SAXPY, 137, 324, 360
  parallel, 360
  sparse, 324
scatter and gather operations, 362–363
Schönauer, W., 216, 243
Schultz, M. H., 319
Schur complement, 456
  approaches, 477
  and direct solution, 456

for finite element partitionings, 460
local, 460
methods, 477
for model problems, 463
properties, 457
for vertex partitionings, 458
Schur form, 16
example, 17
nonuniqueness, 18
partial, 17
quasi, 17
real, 17
Schwarz alternating procedure, 455, 465
additive, 472
algorithm, 465
multiplicative, 465
Schwarz, H. A., 492
search subspace, 129
section of an operator, 153
self-adjoint, 6, 473
self-preconditioning, 324
convergence behavior, 326
semisimple, 15
separators, 484
set decomposition, 108
shared memory computers, 356
Sherman, A. H., 319
Sidi, A., 244
similarity transformation, 14
Simon, H. D., 350, 375, 493
Simoncini, V., 216, 243
simple eigenvalue, 14
singular matrix, 3
singular values, 9
sites (in graph partitioning), 488
skew-Hermitian
matrices, 4, 20, 198
part, 30
skew-symmetric matrices, 4
Skjellum, A., 359
skyline solvers, 84
Smith, B., 493
smooth modes, 407
smoother, 423
smoothing operator, 423
smoothing property, 435
Sonneveld, P., 229, 243
SOR, 105
convergence, 119
iteration, 103

multicolor sweep, 392
for SPD matrices, 119
span of q-vectors, 9
sparse, 63
sparse Gaussian elimination, 75, 93
sparse matrices
adjacency graph, 75
basic operations, 92
direct methods, 93
graph representation, 75
matrix-by-vector operation, 92
permutation and reordering, 77
storage, 89–92
sparse matrix-by-vector product, 92
sparse skyline storage format, *see* SSK
sparse-sparse mode computations, 323
sparse triangular system solution, 93
sparsity, 73
SPARSKIT, 94–97
SPD, *see* symmetric positive definite
spectral bisection, 486
spectral radius, 3
spectrum of a matrix, 3
split preconditioning, 262
splitting, 105
square matrices, 2
SSK storage format, 314
SSOR, 105
steepest descent, 138
stencil, 49
stereographic projection, 485
Stewart, G. W., 43
Stewart, W. J., 449
Stiefel, E. L., 215, 380, 404
Stieltjes algorithm, 187
stiffness matrix, 64, 65
Stokes problem, 256
storage format
coordinate, 89
CSC, 90
CSR, 90, 291
Ellpack-Itpack, 91
MSR, 90
SSK, 314
storage of sparse matrices, 89–92
Strang, G., 72
structural nonsingularity, 336
structured sparse matrix, 73
Stüben, K., 449
subdomain, 397

subspace, 9
    of approximants, 129
    of constraints, 129
    direct sum, 9
    orthogonal, 10
    sum, 9
successive overrelaxation, *see* SOR
Swartzrauber, P. N., 59
Sweet, R., 72
symbolic factorization, 93
symmetric Gauss–Seidel, 105
symmetric matrices, 4
symmetric positive definite, 30, 119
symmetric SOR, *see* SSOR
symmetric squaring, 342
symmetry in preconditioned CG, 262
Szyld, D. B., 281, 350

Tang, W. P., 350
Taylor, D. R., 220, 221
Teng, S. H., 484, 485
tensor product, 411
tensor sum, 411
test problems, 94
TFQMR, 234
    algorithm, 239
topological sorting, 370
trace, 3
transpose-free QMR, *see* TFQMR
transversal, 335
triad operation, 364
triangular systems, 369
    distributed, 399
    level scheduling, 370
    sparse, 369
tridiagonal matrices, 5
Trottenberg, U., 449
Tůma, M. T., 331, 337, 350

unassembled matrix, 65
underdetermined, 246
undirected graph, 75
unitary matrices, 4
unstructured sparse matrix, 73
upper triangular matrices, 4
upwind schemes, 52
Uzawa's method, 254, 255

van der Vorst, H. A., 233, 243, 280, 349, 375
Varga, R. S., 43, 121, 126, 128, 350, 405
variable preconditioner, 273
Vassilevski, P., 280
Vavasis, S. A., 484, 485
vector
    computers, 355
    operations, 360
    orthogonality, 10
    processors, 355
    of unknowns, 103
    updates, 137, 360
        parallel, 360
vertex (in a graph), 75
Vinsome, P. K. W., 216
Voevodin, V. V., 216

Wachspress, E. L., 128
Wagner, C., 449
Walker, H. F., 156, 183, 216, 228, 244
Walter, A., 244
Wang, X., 345, 351
Wathens, A. J., 350
Watts, J. W., 350
wave-fronts, 370
weak formulation, 61
weakly diagonally dominant matrix, 116
Weiss, R., 182, 216, 243
Wesseling, P., 243, 449
Widlund, O. B., 278, 477
Wigton, L. B., 350
Wilders, P., 493
Wilkinson, B., 375
Wilkinson, J. H., 216, 220
Winget regularization, 401
Wu, C. H., 101

Xu, J., 477, 493

Ye, Q., 281
Yeremin, A. Y., 350
Young, D. M., 121, 123, 126, 128, 196, 405
Young, D. P., 350

Zarantonello's lemma, 201
Zhou, L., 183, 216, 228, 244
Zlatev, Z., 101, 350